An Introduction to Linear Statistical Models

VOLUME I

McGRAW-HILL SERIES IN PROBABILITY AND STATISTICS

DAVID BLACKWELL, *Consulting Editor*

Bharucha-Reid Elements of the Theory of Markov Processes and Their Applications
Drake Fundamentals of Applied Probability Theory
Dubins and Savage How to Gamble If You Must
Ehrenfeld and Littauer Introduction to Statistical Methods
Graybill Introduction to Linear Statistical Models, Volume I
Jeffrey The Logic of Decision
Li Introduction to Experimental Methods
Mood and Graybill Introduction to the Theory of Statistics
Pearce Biological Statistics: An Introduction
Pfeiffer Concepts of Probability Theory
Wadsworth and Bryan Introduction to Probability and Random Variables
Weiss Statistical Decision Theory
Wolf Elements of Probability and Statistics

AN INTRODUCTION TO LINEAR STATISTICAL MODELS

Volume I

FRANKLIN A. GRAYBILL

Professor of Mathematical Statistics
Colorado State University
Fort Collins, Colorado

McGRAW-HILL BOOK COMPANY, INC.

New York Toronto London

1961

AN INTRODUCTION TO LINEAR STATISTICAL MODELS
VOLUME I

 Library of Congress Catalog Card Number 60-9841

ISBN 07-024331-x

11 12 13 14 15 16-MAMM-7 6 5 4 3 2 1

TO JEANNE

Preface

This book was written with the intention of fulfilling three needs: (1) for a theory textbook in *experimental* statistics, for undergraduates or first-year graduate students; (2) for a reference book in the area of regression, correlation, least squares, experimental design, etc., for consulting statisticians with limited mathematical training; and (3) for a reference book for experimenters with limited mathematical training who use statistics in their research. This is not a book on mathematics, neither is it an advanced book on statistical theory. It is intended to be an introductory mathematical treatment of topics which are important to experimenters, statistical consultants, and those who are training to be statistical consultants. The only mathematics that is required is generally obtained before the senior year in college. The mathematical level of this book is about the same as "An Introduction to the Theory of Statistics" by A. M. Mood, except for the fact that a great deal of use is made of matrix theory. Chapter 1 is a review of the mathematical concepts, especially matrices, that will be used in the book. For students who have had a course in matrix algebra, Chapter 1 can be omitted, since in most cases the theorems are stated without proof and simple problems are given to illustrate them. Chapter 2 is a review of those statistical concepts which are usually obtained in a first course in mathematical statistics. In these two chapters the important concepts are presented in the form of theorems so that they can be referred to as they are needed in later chapters. In the remainder of the book some of the proofs are omitted, but most of them are presented in detail, since this book was written for those who will have limited mathematical training.

Among other things, the statistical consultant is involved in aiding in the interpretation of observations that are collected by experimenters. This interpretation can often be materially improved by the use of "statistical models." This use involves three parts. The first is the specification of the structure of the observations, or, in other words, the selection of a model to represent the real-world situation that the observations describe. The second part is the mathematical treatment of the model to obtain results about various components of it. This is generally in the nature of point estimation, interval estimation, hypothesis testing, etc. The third part is the use of the results of the mathematical treatment of the model to make decisions in the real world. This book is concerned only with the second part, namely, the mathematical treatment of statistical models, and no attempt is made to justify any model for a given real-world situation. Five important statistical models are discussed, which are the basis for many of the applications in statistics, and these five models are presented from the point of view of infinite-model theory. Throughout the book emphasis is placed on the power of tests and on the width of confidence intervals.

A set of notes from which this book was taken has served as a two-semester course for seniors and first-year graduate students at Oklahoma State University; the prerequisites are one year of calculus, one semester of matrix algebra, one year of introductory statistical theory from a book such as "Introduction to the Theory of Statistics," by A. M. Mood, or "Statistical Theory in Research," by R. L. Anderson and T. A. Bancroft, and one year of statistical methods. Many of the students who were not statistics majors had no formal course in matrix theory, but some had audited a matrix course, and some were self-taught. These students were required to work the pertinent problems in Chapter 1.

Volume II will contain such topics as sample size, multiple comparisons, multivariate analysis of variance, response surfaces, discriminant functions, partially balanced incomplete block designs, orthogonal latin squares, randomization theory, split-plot models, and some nonlinear models.

This book was written with the direct and indirect help of many people. I owe a great debt to my mother, Mrs. Lula Graybill, and to my two sisters and their husbands, Mr. and Mrs. Wayne Carr and Mr. and Mrs. Homer Good, for their help and encouragement. Had it not been for them, I probably would not have been able to attend college.

I want to thank Professor Carl E. Marshall for introducing me to statistics and for encouraging me to do further work in this field, and

Professor Oscar Kempthorne, first for his help during my graduate work at Iowa State College, and later for reading the entire manuscript. I also want to express my appreciation to Professor L. Wayne Johnson, the head of the Mathematics Department of Oklahoma State University, who made personnel available for typing and proofreading of the manuscript, and to David Weeks and Vanamamal Seshadri, graduate students in statistics at Oklahoma State University, who read the entire manuscript.

I wish to thank the typists who helped to put my handwritten notes into typescript; Miss Nancy Adams, Miss Carolyn Hundley, Mrs. Carlotta Fine, and Miss Biruta Stakle.

I am indebted to Mr. F. W. Brown, Director of the National Bureau of Standards at Boulder, Colorado, for permission to reproduce a portion of the material in NBS report 5069, Graphs and Tables of the Percentage Points $F(v_1,v_2,p)$ for the Fisher-Snedecor Variance Ratio, to be published by Lewis E. Vogler and Kenneth A. Norton.

I am also indebted to Professor E. S. Pearson, J. Neyman, P. C. Tang, and University College for permission to reprint the tables from The Power Function of the Analysis of Variance Tests with Tables and Illustrations for Their Use, from *Statistical Research Memoirs*, volume 1.

I am also indebted to the Biometrika Trustees for permission to reproduce a portion of "Tables of Percentage Points of the Inverted Beta (F) Distribution" by M. Merrington and C. M. Thompson.

Finally, I wish to express my thanks to my wife, Jeanne, who relieved me of most of the household chores during the many evenings that I worked on the manuscript.

Franklin A. Graybill

Contents

1

Mathematical Concepts

1.1 Matrices

The theory of linear statistical models that will be developed in this book will require some of the mathematical techniques of calculus, matrix algebra, etc. In this chapter we shall give some of the important mathematical theorems that are necessary to develop this theory. Most of the theorems in this chapter will be given without proof; for some, however, the proof will be supplied.

A matrix $\mathbf{A}$ will have elements denoted by a_{ij}, where i refers to the row and j to the column. If $\mathbf{A}$ denotes a matrix, then $\mathbf{A}'$ will denote the transpose of $\mathbf{A}$, and $\mathbf{A}^{-1}$ will denote the inverse of $\mathbf{A}$. The symbol $|\mathbf{A}|$ will be used to denote the determinant of $\mathbf{A}$. The identity matrix will be denoted by $\mathbf{I}$, and $\mathbf{0}$ will denote the null matrix, i.e., a matrix whose elements are all zeros. The dimension of a matrix is the number of its rows by the number of its columns. For example, a matrix $\mathbf{A}$ of dimension $n \times m$, or an $n \times m$ matrix $\mathbf{A}$, will be a matrix $\mathbf{A}$ with n rows and m columns. If $m = 1$, the matrix will be called an $n \times 1$ vector. The rank of the matrix $\mathbf{A}$ will sometimes be denoted by $\rho(\mathbf{A})$.

Given the matrices $\mathbf{A} = (a_{ij})$ and $\mathbf{B} = (b_{ij})$, the product $\mathbf{AB} = \mathbf{C} = (c_{ij})$ is defined as the matrix $\mathbf{C}$ with pqth element equal to $\sum_{s=1}^{n} a_{ps}b_{sq}$. For $\mathbf{AB}$ to be defined, the number of columns in $\mathbf{A}$ must equal the number of rows in $\mathbf{B}$. For $\mathbf{A} + \mathbf{B}$ to be defined, $\mathbf{A}$ and $\mathbf{B}$ must have the same dimension; $\mathbf{A} + \mathbf{B} = \mathbf{C}$ gives $c_{ij} = a_{ij} + b_{ij}$. If k is a scalar and $\mathbf{A}$ is a matrix, then $k\mathbf{A}$ means the matrix such that each element is the corresponding element of $\mathbf{A}$ multiplied by k.

A diagonal matrix $\mathbf{D}$ is defined as a square matrix whose off-diagonal elements are all zero; that is, if $\mathbf{D} = (d_{ij})$, then $d_{ij} = 0$ if $i \neq j$.

◆ **Theorem 1.1** The transpose of $\mathbf{A}'$ equals $\mathbf{A}$; that is, $(\mathbf{A}')' = \mathbf{A}$.

◆ **Theorem 1.2** The inverse of $\mathbf{A}^{-1}$ is $\mathbf{A}$; that is, $(\mathbf{A}^{-1})^{-1} = \mathbf{A}$.

◆ **Theorem 1.3** The transpose and inverse symbols may be permuted; that is, $(\mathbf{A}')^{-1} = (\mathbf{A}^{-1})'$.

◆ **Theorem 1.4** $(\mathbf{AB})' = \mathbf{B}'\mathbf{A}'$.

◆ **Theorem 1.5** $(\mathbf{AB})^{-1} = \mathbf{B}^{-1}\mathbf{A}^{-1}$ if $\mathbf{A}$ and $\mathbf{B}$ are each nonsingular.

◆ **Theorem 1.6** A scalar commutes with every matrix; that is, $k\mathbf{A} = \mathbf{A}k$.

◆ **Theorem 1.7** For any matrix $\mathbf{A}$ we have $\mathbf{IA} = \mathbf{AI} = \mathbf{A}$.

◆ **Theorem 1.8** All diagonal matrices of the same dimension are commutative.

◆ **Theorem 1.9** If $\mathbf{D}_1$ and $\mathbf{D}_2$ are diagonal matrices, then the product is diagonal; that is, $\mathbf{D}_1\mathbf{D}_2 = \mathbf{D}_2\mathbf{D}_1 = \mathbf{D}$, where $\mathbf{D}$ is diagonal. The ith diagonal element of $\mathbf{D}$ is the product of the ith diagonal element of $\mathbf{D}_1$ and the ith diagonal element of $\mathbf{D}_2$.

◆ **Theorem 1.10** If $\mathbf{X}$ and $\mathbf{Y}$ are vectors and if $\mathbf{A}$ is a nonsingular matrix and if the equation $\mathbf{Y} = \mathbf{AX}$ holds, then $\mathbf{X} = \mathbf{A}^{-1}\mathbf{Y}$.

◆ **Theorem 1.11** The rank of the product $\mathbf{AB}$ of the two matrices $\mathbf{A}$ and $\mathbf{B}$ is less than or equal to the rank of $\mathbf{A}$ and is less than or equal to the rank of $\mathbf{B}$.

◆ **Theorem 1.12** The rank of the sum of $\mathbf{A} + \mathbf{B}$ is less than or equal to the rank of $\mathbf{A}$ plus the rank of $\mathbf{B}$.

◆ **Theorem 1.13** If $\mathbf{A}$ is an $n \times n$ matrix and if $|\mathbf{A}| = 0$, then the rank of $\mathbf{A}$ is less than n. ($|\mathbf{A}|$ denotes the determinant of the matrix $\mathbf{A}$.)

◆ **Theorem 1.14** If the rank of $\mathbf{A}$ is less than n, then the rows of $\mathbf{A}$ are not independent; likewise, the columns of A are not independent. ($\mathbf{A}$ is $n \times n$.)

◆ **Theorem 1.15** If the rank of $\mathbf{A}$ is $m \leqslant n$, then the number of linearly independent rows is m; also, the number of linearly independent columns is m. ($\mathbf{A}$ is $n \times n$.)

◆ **Theorem 1.16** If $\mathbf{A}'\mathbf{A} = \mathbf{0}$, then $\mathbf{A} = \mathbf{0}$.

◆ **Theorem 1.17** The rank of a matrix is unaltered by multiplication by a nonsingular matrix; that is, if $\mathbf{A}$, $\mathbf{B}$, and $\mathbf{C}$ are matrices such that **AB** and **BC** exist and if **A** and **C** are nonsingular, then $\rho(\mathbf{AB}) = \rho(\mathbf{BC}) = \rho(\mathbf{B})$.

◆ **Theorem 1.18** If the product **AB** of two square matrices is **0**, then $\mathbf{A} = \mathbf{0}$ or $\mathbf{B} = \mathbf{0}$ or **A** and **B** are both singular.

◆ **Theorem 1.19** If **A** and **B** are $n \times n$ matrices of rank r and s, respectively, then the rank of **AB** is greater than or equal to $r + s - n$.

◆ **Theorem 1.20** The rank of $\mathbf{AA}'$ equals the rank of $\mathbf{A}'\mathbf{A}$ equals the rank of **A** equals the rank of $\mathbf{A}'$.

1.2 Quadratic Forms

If **Y** is an $n \times 1$ vector with ith element y_i and if **A** is an $n \times n$ matrix with ijth element equal to a_{ij}, then the *quadratic form* $\mathbf{Y}'\mathbf{AY}$ is defined as $\sum_{i=1}^{n} \sum_{j=1}^{n} y_i y_j a_{ij}$. The rank of the quadratic form $\mathbf{Y}'\mathbf{AY}$ is defined as the rank of the matrix **A**. The quadratic form $\mathbf{Y}'\mathbf{AY}$ is said to be *positive definite* if and only if $\mathbf{Y}'\mathbf{AY} > 0$ for all vectors **Y** where $\mathbf{Y} \neq \mathbf{0}$. A quadratic form $\mathbf{Y}'\mathbf{AY}$ is said to be *positive semidefinite* if and only if $\mathbf{Y}'\mathbf{AY} \geqslant 0$ for all **Y**, and $\mathbf{Y}'\mathbf{AY} = 0$ for some vector $\mathbf{Y} \neq \mathbf{0}$. The matrix **A** of a quadratic form $\mathbf{Y}'\mathbf{AY}$ is said to be positive definite (semidefinite) when the quadratic form is positive definite (semidefinite). If **C** is an $n \times n$ matrix such that $\mathbf{C}'\mathbf{C} = \mathbf{I}$, then **C** is said to be an *orthogonal* matrix, and $\mathbf{C}' = \mathbf{C}^{-1}$.

Consider the transformation from the vector **Z** to the vector **Y** by the matrix **P** such that $\mathbf{Y} = \mathbf{PZ}$. Then, $\mathbf{Y}'\mathbf{AY} = (\mathbf{PZ})'\mathbf{A}(\mathbf{PZ}) = \mathbf{Z}'\mathbf{P}'\mathbf{APZ}$. Thus, by the transformation $\mathbf{Y} = \mathbf{PZ}$, the quadratic form $\mathbf{Y}'\mathbf{AY}$ is transformed into the quadratic form $\mathbf{Z}'(\mathbf{P}'\mathbf{AP})\mathbf{Z}$.

◆ **Theorem 1.21** If **P** is a nonsingular matrix and if **A** is positive definite (semidefinite), then $\mathbf{P}'\mathbf{AP}$ is positive definite (semidefinite).

◆ **Theorem 1.22** A necessary and sufficient condition for the symmetric matrix **A** to be positive definite is that there exist a nonsingular matrix **P** such that $\mathbf{A} = \mathbf{PP}'$.

◆ **Theorem 1.23** A necessary and sufficient condition that the matrix **A** be positive definite, where

$$\mathbf{A} = \begin{pmatrix} a_{11} & a_{12} & \cdots & a_{1n} \\ a_{21} & a_{22} & \cdots & a_{2n} \\ \cdots & \cdots & \cdots & \cdots \\ a_{n1} & a_{n2} & \cdots & a_{nn} \end{pmatrix}$$

is that the following inequalities hold:

$$a_{11} > 0, \quad \begin{vmatrix} a_{11} & a_{12} \\ a_{21} & a_{22} \end{vmatrix} > 0, \quad \ldots, \quad \begin{vmatrix} a_{11} & a_{12} & \cdots & a_{1n} \\ a_{21} & a_{22} & \cdots & a_{2n} \\ \cdots & \cdots & \cdots & \cdots \\ a_{n1} & a_{n2} & \cdots & a_{nn} \end{vmatrix} > 0$$

◆ **Theorem 1.24** If $\mathbf{A}$ is an $n \times m$ matrix of rank $m < n$, then $\mathbf{A}'\mathbf{A}$ is positive definite and $\mathbf{A}\mathbf{A}'$ is positive semidefinite.

◆ **Theorem 1.25** If $\mathbf{A}$ is an $n \times m$ matrix of rank $k < m$ and $k < n$, then $\mathbf{A}'\mathbf{A}$ and $\mathbf{A}\mathbf{A}'$ are each positive semidefinite.

◆ **Theorem 1.26** If $\mathbf{C}$ is an orthogonal matrix, and if the transformation $\mathbf{Y} = \mathbf{CZ}$ is made on $\mathbf{Y}'\mathbf{Y}$, we get $\mathbf{Y}'\mathbf{Y} = \mathbf{Y}'\mathbf{IY} = \mathbf{Z}'\mathbf{C}'\mathbf{ICZ} = \mathbf{Z}'\mathbf{C}'\mathbf{CZ} = \mathbf{Z}'\mathbf{Z}$.

In order to develop the theory of quadratic forms, it is necessary to define a characteristic root of a matrix. A *characteristic root* of a $p \times p$ matrix $\mathbf{A}$ is a scalar λ such that $\mathbf{AX} = \lambda\mathbf{X}$ for some vector $\mathbf{X} \neq \mathbf{0}$. The vector $\mathbf{X}$ is called the *characteristic vector* of the matrix $\mathbf{A}$. It follows that, if λ is a characteristic root of $\mathbf{A}$, then $\mathbf{AX} - \lambda\mathbf{X} = \mathbf{0}$ and $(\mathbf{A} - \lambda\mathbf{I})\mathbf{X} = \mathbf{0}$. Thus, λ is a scalar such that the above homogeneous set of equations has a nontrivial solution, i.e., a solution other than $\mathbf{X} = \mathbf{0}$. It is known from elementary matrix theory that this implies $|\mathbf{A} - \lambda\mathbf{I}| = 0$. Thus the characteristic root of a matrix $\mathbf{A}$ could be defined as a scalar λ such that $|\mathbf{A} - \lambda\mathbf{I}| = 0$. It is easily seen that $|\mathbf{A} - \lambda\mathbf{I}|$ is a pth degree polynomial in λ. This polynomial is called the *characteristic polynomial*, and its roots are the characteristic roots of the matrix $\mathbf{A}$. We shall now give a few theorems concerning characteristic roots, characteristic vectors, and characteristic polynomials. In this book the elements of a matrix will be real.

◆ **Theorem 1.27** The number of nonzero characteristic roots of a matrix $\mathbf{A}$ is equal to the rank of $\mathbf{A}$.

◆ **Theorem 1.28** The characteristic roots of $\mathbf{A}$ are identical with the characteristic roots of $\mathbf{CAC}^{-1}$. If $\mathbf{C}$ is an orthogonal matrix, it follows that $\mathbf{A}$ and $\mathbf{CAC}'$ have identical characteristic roots.

◆ **Theorem 1.29** The characteristic roots of a symmetric matrix are real; i.e., if $\mathbf{A} = \mathbf{A}'$, the characteristic polynomial of $|\mathbf{A} - \lambda\mathbf{I}| = 0$ has all real roots.

◆ **Theorem 1.30** The characteristic roots of a positive definite matrix $\mathbf{A}$ are positive; the characteristic roots of a positive semidefinite matrix are nonnegative.

◆ **Theorem 1.31** For every symmetric matrix $\mathbf{A}$ there exists an orthogonal matrix $\mathbf{C}$ such that $\mathbf{C}'\mathbf{A}\mathbf{C} = \mathbf{D}$, where $\mathbf{D}$ is a diagonal matrix whose diagonal elements are the characteristic roots of $\mathbf{A}$.

◆ **Theorem 1.32** Let $\mathbf{A}_1, \mathbf{A}_2, \ldots, \mathbf{A}_t$ be a collection of symmetric $n \times n$ matrices. A necessary and sufficient condition that there exists an orthogonal transformation $\mathbf{C}$ such that $\mathbf{C}'\mathbf{A}_1\mathbf{C}, \mathbf{C}'\mathbf{A}_2\mathbf{C}, \ldots, \mathbf{C}'\mathbf{A}_t\mathbf{C}$ are all diagonal is that $\mathbf{A}_i\mathbf{A}_j$ be symmetric for all i and j. Since all the $\mathbf{A}_i$ are symmetric, it follows that $\mathbf{A}_i\mathbf{A}_j$ is symmetric if and only if $\mathbf{A}_i$ and $\mathbf{A}_j$ commute.

◆ **Theorem 1.33** Let an $n \times n$ matrix $\mathbf{C}$ be written

$$\mathbf{C} = \begin{pmatrix} \mathbf{c}_1 \\ \mathbf{c}_2 \\ \cdots \\ \mathbf{c}_n \end{pmatrix}$$

where $\mathbf{c}_i$ is the ith row of $\mathbf{C}$. Thus $\mathbf{c}_i$ is the transpose of an $n \times 1$ column vector. The following two conditions are necessary and sufficient for $\mathbf{C}$ to be orthogonal:

(1) $\quad \mathbf{c}_i\mathbf{c}_j' = 0 \qquad \text{for all } i \neq j$

(2) $\quad \mathbf{c}_i\mathbf{c}_i' = 1 \qquad \text{for all } i$

That is to say, any two rows of any orthogonal matrix are orthogonal (their inner product is zero), and the inner product of any row with itself is unity.

◆ **Theorem 1.34** Let $\underset{p \times n}{\mathbf{C}_1} = \begin{pmatrix} \mathbf{c}_1 \\ \mathbf{c}_2 \\ \cdots \\ \mathbf{c}_p \end{pmatrix}$ be p rows of an $n \times n$ orthogonal matrix. That is to say, let $\mathbf{c}_1, \mathbf{c}_2, \ldots, \mathbf{c}_p$ be the transposes of p vectors such that $\mathbf{c}_i\mathbf{c}_j' = 0$ $(i \neq j = 1, 2, \ldots, p)$ and $\mathbf{c}_i\mathbf{c}_i' = 1$ $(i = 1, 2, \ldots, p)$. Then there exist $n - p$ vectors $\mathbf{f}_j$ such that $\mathbf{c}_i\mathbf{f}_j = 0$ for all i and j and $\mathbf{f}_i'\mathbf{f}_i = 1$ $(i = 1, 2, \ldots, n - p)$, $\mathbf{f}_i'\mathbf{f}_j = 0$ if $i \neq j$. Thus the theorem states that, if we are given a matrix such as $\mathbf{C}_1$, there exists a matrix $\mathbf{C}_2$ of dimension $(n - p) \times n$ such that $\begin{pmatrix} \mathbf{C}_1 \\ \mathbf{C}_2 \end{pmatrix} = \mathbf{C}$ ($\mathbf{C}_1$ forms the first p rows of $\mathbf{C}$ and $\mathbf{C}_2$ the last $n - p$ rows of $\mathbf{C}$), where $\mathbf{C}$ is orthogonal.

1.3 Determinants

In this section a few of the important theorems on determinants will be given. It will be assumed that the student knows the definition of a determinant and knows how to evaluate small ones. In linear-hypothesis applications it is often necessary to solve systems involving a great many equations. It might at times be necessary to evaluate large determinants. There are many methods of doing these things that are adaptable to automatic and semiautomatic computing machines. These methods will be discussed in detail later. It will be assumed here that the student knows how to evaluate determinants by the method of minors or by some other simple method.

◆ **Theorem 1.35** The determinant of a diagonal matrix is equal to the product of the diagonal elements.

◆ **Theorem 1.36** If $\mathbf{A}$ and $\mathbf{B}$ are $n \times n$ matrices, then $|\mathbf{AB}| = |\mathbf{BA}| = |\mathbf{A}|\,|\mathbf{B}|$.

◆ **Theorem 1.37** If $\mathbf{A}$ is singular, $|\mathbf{A}| = 0$.

◆ **Theorem 1.38** If $\mathbf{C}$ is an orthogonal matrix, then $|\mathbf{C}| = +1$ or $|\mathbf{C}| = -1$.

◆ **Theorem 1.39** If $\mathbf{C}$ is an orthogonal matrix, then $|\mathbf{C}'\mathbf{AC}| = |\mathbf{A}|$.

◆ **Theorem 1.40** The determinant of a positive definite matrix is positive.

◆ **Theorem 1.41** The determinant of a triangular matrix is equal to the product of the diagonal elements.

◆ **Theorem 1.42** $|\mathbf{D}^{-1}| = 1/|\mathbf{D}|$, if $|\mathbf{D}| \neq 0$.

◆ **Theorem 1.43** If $\mathbf{A}$ is a square matrix such that

$$\mathbf{A} = \begin{pmatrix} \mathbf{A}_{11} & \mathbf{A}_{12} \\ \mathbf{A}_{21} & \mathbf{A}_{22} \end{pmatrix}$$

where $\mathbf{A}_{11}$ and $\mathbf{A}_{22}$ are square matrices, and if $\mathbf{A}_{12} = \mathbf{0}$ or $\mathbf{A}_{21} = \mathbf{0}$, then $|\mathbf{A}| = |\mathbf{A}_{11}|\,|\mathbf{A}_{22}|$.

◆ **Theorem 1.44** If $\mathbf{A}_1$ and $\mathbf{A}_2$ are symmetric and $\mathbf{A}_2$ is positive definite and if $\mathbf{A}_1 - \mathbf{A}_2$ is positive semidefinite (or positive definite), then $|\mathbf{A}_1| \geqslant |\mathbf{A}_2|$.

1.4 Miscellaneous Theorems on Matrices

In this section we shall discuss some miscellaneous theorems concerning matrices, which we shall use in later chapters.

The *trace* of a matrix **A**, which will be written tr(**A**), is equal to the sum of the diagonal elements of **A**; that is, $\text{tr}(\mathbf{A}) = \sum_{i=1}^{n} a_{ii}$.

◆ **Theorem 1.45** $\text{tr}(\mathbf{AB}) = \text{tr}(\mathbf{BA})$.

Proof: By definition, tr(**AB**) is equal to $\sum_{ij} a_{ij}b_{ji}$. By definition, tr(**BA**) is equal to $\sum_{ik} b_{ik}a_{ki}$. But it is clear that $\sum_{ij} a_{ij}b_{ji} = \sum_{ik} b_{ik}a_{ki}$; therefore, $\text{tr}(\mathbf{AB}) = \text{tr}(\mathbf{BA})$.

◆ **Theorem 1.46** $\text{tr}(\mathbf{ABC}) = \text{tr}(\mathbf{CAB}) = \text{tr}(\mathbf{BCA})$; that is, the trace of the product of matrices is invariant under any cyclic permutation of the matrices.

Proof: By Theorem 1.45, $\text{tr}[(\mathbf{AB})\mathbf{C}] = \text{tr}[\mathbf{C}(\mathbf{AB})]$.

◆ **Theorem 1.47** $\text{tr}(\mathbf{I}) = n$, where I is an $n \times n$ identity matrix.

◆ **Theorem 1.48** If **C** is an orthogonal matrix, $\text{tr}(\mathbf{C}'\mathbf{AC}) = \text{tr}(\mathbf{A})$.

Proof: By Theorem 1.46, $\text{tr}(\mathbf{C}'\mathbf{AC}) = \text{tr}(\mathbf{CC}'\mathbf{A}) = \text{tr}(\mathbf{IA}) = tr(\mathbf{A})$.

It is sometimes advantageous to break a matrix up into submatrices. This is called *partitioning* a matrix into submatrices, and a matrix can be partitioned in many ways. For example, **A** might be partitioned into submatrices as follows:

$$\mathbf{A} = \begin{pmatrix} \mathbf{A}_{11} & \mathbf{A}_{12} \\ \mathbf{A}_{21} & \mathbf{A}_{22} \end{pmatrix}$$

where **A** is $m \times n$, $\mathbf{A}_{11}$ is $m_1 \times n_1$, $\mathbf{A}_{12}$ is $m_1 \times n_2$, $\mathbf{A}_{21}$ is $m_2 \times n_1$, and $\mathbf{A}_{22}$ is $m_2 \times n_2$, and where $m_1 + m_2 = m$ and $n_1 + n_2 = n$.

The product **AB** of two matrices can be made symbolically even if **A** and **B** are broken into submatrices. The multiplication proceeds as if the submatrices were single elements of the matrix. However, the dimensions of the matrices and of the submatrices must be such that they will multiply. For example, if **B** is an $n \times p$ matrix such that

$$\mathbf{B} = \begin{pmatrix} \mathbf{B}_{11} & \mathbf{B}_{12} \\ \mathbf{B}_{21} & \mathbf{B}_{22} \end{pmatrix}$$

where $\mathbf{B}_{ij}$ is an $n_i \times p_j$ matrix, then the product **AB** exists; and the corresponding submatrices will multiply, since $\mathbf{A}_{ij}$ is of dimension $m_i \times n_j$ and $\mathbf{B}_{jk}$ is of dimension $n_j \times p_k$. The resulting matrix is as follows:

$$\mathbf{AB} = \begin{pmatrix} \mathbf{A}_{11} & \mathbf{A}_{12} \\ \mathbf{A}_{21} & \mathbf{A}_{22} \end{pmatrix}\begin{pmatrix} \mathbf{B}_{11} & \mathbf{B}_{12} \\ \mathbf{B}_{21} & \mathbf{B}_{22} \end{pmatrix} = \begin{pmatrix} \mathbf{A}_{11}\mathbf{B}_{11} + \mathbf{A}_{12}\mathbf{B}_{21} & \mathbf{A}_{11}\mathbf{B}_{12} + \mathbf{A}_{12}\mathbf{B}_{22} \\ \mathbf{A}_{21}\mathbf{B}_{11} + \mathbf{A}_{22}\mathbf{B}_{21} & \mathbf{A}_{21}\mathbf{B}_{12} + \mathbf{A}_{22}\mathbf{B}_{22} \end{pmatrix}$$

◆ **Theorem 1.49** If $\mathbf{A}$ is a positive definite symmetric matrix such that

$$\mathbf{A} = \begin{pmatrix} \mathbf{A}_{11} & \mathbf{A}_{12} \\ \mathbf{A}_{21} & \mathbf{A}_{22} \end{pmatrix}$$

and if $\mathbf{B}$ is the inverse of $\mathbf{A}$ such that

$$\mathbf{B} = \begin{pmatrix} \mathbf{B}_{11} & \mathbf{B}_{12} \\ \mathbf{B}_{21} & \mathbf{B}_{22} \end{pmatrix}$$

and if $\mathbf{B}_{ii}$ and $\mathbf{A}_{ii}$ are each of dimension $m_i \times m_i$, etc.,

$$\mathbf{A}_{11}^{-1} = \mathbf{B}_{11} - \mathbf{B}_{12}\mathbf{B}_{22}^{-1}\mathbf{B}_{21}$$

Proof: Since $\mathbf{A} = \mathbf{B}^{-1}$, then $\mathbf{AB} = \mathbf{I}$. Thus,

$$\begin{pmatrix} \mathbf{A}_{11} & \mathbf{A}_{12} \\ \mathbf{A}_{21} & \mathbf{A}_{22} \end{pmatrix}\begin{pmatrix} \mathbf{B}_{11} & \mathbf{B}_{12} \\ \mathbf{B}_{21} & \mathbf{B}_{22} \end{pmatrix} = \mathbf{I}$$

and we get the following two matrix equations:

$$\mathbf{A}_{11}\mathbf{B}_{11} + \mathbf{A}_{12}\mathbf{B}_{21} = \mathbf{I} \qquad \text{and} \qquad \mathbf{A}_{11}\mathbf{B}_{12} + \mathbf{A}_{12}\mathbf{B}_{22} = \mathbf{0}$$

Solving the second equation for $\mathbf{A}_{12}$, we get $\mathbf{A}_{12} = -\mathbf{A}_{11}\mathbf{B}_{12}\mathbf{B}_{22}^{-1}$. Substituting this value for $\mathbf{A}_{12}$ into the first equation, we get

$$\mathbf{A}_{11}\mathbf{B}_{11} - \mathbf{A}_{11}\mathbf{B}_{12}\mathbf{B}_{22}^{-1}\mathbf{B}_{21} = \mathbf{I}$$

Multiplying by $\mathbf{A}_{11}^{-1}$ gives the desired result.

It is known that $\mathbf{B}_{22}^{-1}$ and $\mathbf{A}_{11}^{-1}$ exist, since $\mathbf{A}$ and $\mathbf{B}$ are positive definite matrices and since $\mathbf{A}_{11}$ and $\mathbf{B}_{22}$ are principal minors of $\mathbf{A}$ and $\mathbf{B}$, respectively. By Theorem 1.23, the determinant of the principal minor of a positive definite matrix is positive. Similar equations can be derived for $\mathbf{A}_{22}^{-1}$, $\mathbf{B}_{11}^{-1}$, and $\mathbf{B}_{22}^{-1}$.

◆ **Theorem 1.50** Let the square matrix $\mathbf{A}$ be such that

$$\mathbf{A} = \begin{pmatrix} \mathbf{A}_{11} & \mathbf{A}_{12} \\ \mathbf{A}_{21} & \mathbf{A}_{22} \end{pmatrix}$$

If $\mathbf{A}_{22}$ is nonsingular, then $|\mathbf{A}| = |\mathbf{A}_{22}|\,|\mathbf{A}_{11} - \mathbf{A}_{12}\mathbf{A}_{22}^{-1}\mathbf{A}_{21}|$.

Proof: The determinant of $\mathbf{A}$ can be written as follows:

$$|\mathbf{A}| = |\mathbf{A}_{22}|\,|\mathbf{A}|\,|\mathbf{B}|$$

where

$$\mathbf{B} = \begin{pmatrix} \mathbf{I} & \mathbf{0} \\ -\mathbf{A}_{22}^{-1}\mathbf{A}_{21} & \mathbf{A}_{22}^{-1} \end{pmatrix}$$

This is clear since, by Theorem 1.43,

$$|\mathbf{B}| = |\mathbf{I}\mathbf{A}_{22}^{-1}| = |\mathbf{A}_{22}^{-1}|$$

so

$$|\mathbf{A}| = |\mathbf{A}_{22}|\,|\mathbf{A}|\,|\mathbf{A}_{22}^{-1}| = |\mathbf{A}|\,|\mathbf{A}_{22}^{-1}|\,|\mathbf{A}_{22}| = |\mathbf{A}|$$

Replacing **A** and **B** by their submatrices, we get

$$|\mathbf{A}| = |\mathbf{A}_{22}| \begin{vmatrix} \mathbf{A}_{11} & \mathbf{A}_{12} \\ \mathbf{A}_{21} & \mathbf{A}_{22} \end{vmatrix} \begin{vmatrix} \mathbf{I} & \mathbf{0} \\ -\mathbf{A}_{22}^{-1}\mathbf{A}_{21} & \mathbf{A}_{22}^{-1} \end{vmatrix}$$

$$= |\mathbf{A}_{22}| \left| \begin{pmatrix} \mathbf{A}_{11} & \mathbf{A}_{12} \\ \mathbf{A}_{21} & \mathbf{A}_{22} \end{pmatrix} \begin{pmatrix} \mathbf{I} & \mathbf{0} \\ -\mathbf{A}_{22}^{-1}\mathbf{A}_{21} & \mathbf{A}_{22}^{-1} \end{pmatrix} \right|$$

The corresponding submatrices are such that they multiply; so

$$|\mathbf{A}| = |\mathbf{A}_{22}| \begin{vmatrix} \mathbf{A}_{11} - \mathbf{A}_{12}\mathbf{A}_{22}^{-1}\mathbf{A}_{21} & \mathbf{A}_{12}\mathbf{A}_{22}^{-1} \\ \mathbf{0} & \mathbf{I} \end{vmatrix} = |\mathbf{A}_{22}|\,|\mathbf{A}_{11} - \mathbf{A}_{12}\mathbf{A}_{22}^{-1}\mathbf{A}_{21}|$$

as was to be shown.

Consider the system of equations $\mathbf{AX} = \mathbf{Y}$, where **A** is an $n \times m$ matrix, **X** is an $m \times 1$ vector, and **Y** is an $n \times 1$ vector. Writing this linear system in detail, we get

$$\begin{aligned} a_{11}x_1 + a_{12}x_2 + \cdots + a_{1m}x_m &= y_1 \\ a_{21}x_1 + a_{22}x_2 + \cdots + a_{2m}x_m &= y_2 \\ \cdots\cdots\cdots\cdots\cdots&\cdots\cdots \\ a_{n1}x_1 + a_{n2}x_2 + \cdots + a_{nm}x_m &= y_n \end{aligned}$$

For a given set of a_{ij} and y_j (that is to say, for a given matrix **A** and vector **Y**), does there exist a set of elements x_i (that is, a vector **X**) such that the equations $\mathbf{AX} = \mathbf{Y}$ are satisfied? Three cases must be considered:

1. The equations have no solution. In this case there exists no vector **X** such that the system of equations is satisfied, and the system is said to be *inconsistent*.

2. There is just one set of x_i that satisfies the system. In this case, there is said to exist a unique solution.

3. There is more than one vector **X** that satisfies the system. If more than one such vector **X** exists, then an infinite number of vectors exist that satisfy the system of equations.

We shall consider two matrices: the coefficient matrix **A** and the augmented matrix $\mathbf{B} = (\mathbf{A}, \mathbf{Y})$, which is the matrix $\mathbf{A}$ with the vector $\mathbf{Y}$ joined to it as the $(m + 1)$st column; that is to say,

$$\mathbf{B} = \begin{pmatrix} a_{11} & a_{12} & \cdots & a_{1m} & y_1 \\ a_{21} & a_{22} & \cdots & a_{2m} & y_2 \\ \cdot & \cdot & \cdot & \cdot & \cdot \\ a_{n1} & a_{n2} & \cdots & a_{nm} & y_n \end{pmatrix}$$

We shall now state some important theorems concerning solutions to the system of equations $\mathbf{AX} = \mathbf{Y}$.

◆ **Theorem 1.51** A necessary and sufficient condition that the system of equations $\mathbf{AX} = \mathbf{Y}$ be consistent (have at least one vector $\mathbf{X}$ satisfying it) is that the rank of the coefficient matrix $\mathbf{A}$ be equal to the rank of the augmented matrix $\mathbf{B} = (\mathbf{A}, \mathbf{Y})$.

◆ **Theorem 1.52** If $\rho(\mathbf{A}) = \rho(\mathbf{B}) = p$, then $m - p$ of the unknowns x_i can be assigned any desired value and the remaining p of the x_i will be uniquely determined. It is essential that the $m - p$ of the unknown x_i that are assigned given values be chosen such that the matrix of the coefficients of the remaining p unknowns have rank p.

◆ **Theorem 1.53** If $\rho(\mathbf{A}) = \rho(\mathbf{B}) = m < n$, there is a unique vector $\mathbf{X}$ that satisfies $\mathbf{AX} = \mathbf{Y}$.

As an example, consider the system of equations

$$x_1 - x_2 = 6$$

$$2x_1 - 2x_2 = 3$$

This can be put into matrix form as

$$\begin{pmatrix} 1 & -1 \\ 2 & -2 \end{pmatrix} \begin{pmatrix} x_1 \\ x_2 \end{pmatrix} = \begin{pmatrix} 6 \\ 3 \end{pmatrix}$$

It can easily be verified that the rank of the augmented matrix

$$\begin{pmatrix} 1 & -1 & 6 \\ 2 & -2 & 3 \end{pmatrix}$$

is 2. Therefore, the system of equations is not consistent, and there exist no values x_1 and x_2 that satisfy it. This fact is also easily seen if we multiply the first equation by 2 and subtract it from the second equation. We get $0 = -9$, an impossible result.

1.5 The Derivatives of Matrices and Vectors

We shall now discuss some theorems relating to the differentiation of quadratic forms and bilinear forms. It will sometimes be advantageous in taking derivatives of quadratic and bilinear forms to be able to take derivatives of matrices and vectors.

Let **X** be a $p \times 1$ vector with elements x_i, let **A** be a $p \times 1$ vector with elements a_i, and let $Z = \mathbf{X}'\mathbf{A} = \mathbf{A}'\mathbf{X}$ (Z is a scalar). The derivative of Z with respect to the vector **X**, which will be written $\partial Z/\partial \mathbf{X}$, will mean the vector

$$\begin{pmatrix} \dfrac{\partial Z}{\partial x_1} \\ \dfrac{\partial Z}{\partial x_2} \\ \cdots \\ \dfrac{\partial Z}{\partial x_p} \end{pmatrix}$$

◆ **Theorem 1.54** If **X**, **A**, and Z are as defined above, then $\partial Z/\partial \mathbf{X} = \mathbf{A}$.

Proof: To find the ith element of the vector $\partial Z/\partial \mathbf{X}$, we find

$$\frac{\partial Z}{\partial x_i} = \frac{\partial\left(\sum_{j=1}^{p} a_j x_j\right)}{\partial x_i}$$

which equals a_i. Thus the ith element of $\partial Z/\partial \mathbf{X}$ is a_i; so $\partial Z/\partial \mathbf{X} = \mathbf{A}$.

◆ **Theorem 1.55** Let **A** be a $p \times 1$ vector, **B** be a $q \times 1$ vector, and **X** be a $p \times q$ matrix whose ijth element equals x_{ij}. Let

$$Z = \mathbf{A}'\mathbf{X}\mathbf{B} = \sum_{m=1}^{q} \sum_{n=1}^{p} a_n x_{nm} b_m$$

Then $\partial Z/\partial \mathbf{X} = \mathbf{A}\mathbf{B}'$.

Proof: $\partial Z/\partial \mathbf{X}$ will be a $p \times q$ matrix whose ijth element is $\partial Z/\partial x_{ij}$. Assuming that **X** is not symmetric and that the elements of **X** are independent,

$$\frac{\partial Z}{\partial x_{ij}} = \frac{\partial\left(\sum_{m=1}^{q} \sum_{n=1}^{p} a_n x_{nm} b_m\right)}{\partial x_{ij}} = a_i b_j$$

Thus the ijth element of $\partial Z/\partial \mathbf{X}$ is $a_i b_j$. Therefore, it follows that

$$\frac{\partial Z}{\partial \mathbf{X}} = \mathbf{AB}'$$

◆ **Theorem 1.56** Let $\mathbf{X}$ be a $p \times 1$ vector, let $\mathbf{A}$ be a $p \times p$ symmetric matrix, and let $Z = \mathbf{X}'\mathbf{AX} = \sum_{i=1}^{p} \sum_{j=1}^{p} x_i x_j a_{ij}$; then $\partial Z/\partial \mathbf{A} = 2\mathbf{XX}' - D(\mathbf{XX}')$, where $D(\mathbf{XX}')$ is a diagonal matrix whose diagonal elements are the diagonal elements of $\mathbf{XX}'$.

Proof: By $\partial Z/\partial \mathbf{A}$ we shall mean a matrix whose ijth element is $\partial Z/\partial a_{ij}$. Thus,

$$\frac{\partial Z}{\partial a_{ij}} = \frac{\partial\left(\sum_{m=1}^{p} \sum_{n=1}^{p} x_m x_n a_{mn}\right)}{\partial a_{ij}}$$

If $i = j$, $\partial Z/\partial a_{ii} = x_i^2$. If $i \neq j$, then $\partial Z/\partial a_{ij} = 2x_i x_j$ (remembering that $a_{ij} = a_{ji}$). Thus $\partial Z/\partial \mathbf{A} = 2\mathbf{XX}' - D(\mathbf{XX}')$.

◆ **Theorem 1.57** Let $\mathbf{X}$ be a $p \times 1$ vector and let $\mathbf{A}$ be a $p \times p$ symmetric matrix such that $Z = \mathbf{X}'\mathbf{AX}$; then $\partial Z/\partial \mathbf{X} = 2\mathbf{AX}$.

Proof: The derivative of the scalar Z with respect to the vector $\mathbf{X}$ will mean a $p \times 1$ vector whose ith element is $\partial Z/\partial x_i$.

$$\frac{\partial Z}{\partial x_i} = \frac{\partial(\mathbf{X}'\mathbf{AX})}{\partial x_i} = \frac{\partial\left(\sum_{m=1}^{p} \sum_{n=1}^{p} x_m x_n a_{mn}\right)}{\partial x_i}$$

$$= \frac{\partial\left(\sum_{m=1}^{p} x_m^2 a_{mm} + \sum_{\substack{m=1 \\ m\neq n}}^{p} \sum_{n=1}^{p} x_m x_n a_{mn}\right)}{\partial x_i}$$

$$= 2x_i a_{ii} + 2 \sum_{\substack{n=1 \\ n\neq i}}^{p} x_n a_{in} = 2 \sum_{n=1}^{p} x_n a_{in}$$

but

$$2\mathbf{AX} = 2 \sum_{n=1}^{p} x_n a_{in}$$

1.6 Idempotent Matrices

We shall now prove some theorems concerning a special type of matrix, the *idempotent* matrix. Since many elementary textbooks on matrix algebra include few theorems on idempotent matrices and since these theorems will play so important a part in the theory to follow, we shall supply the proofs to the theorems. We shall make extensive use of idempotent matrices in our ensuing work. A square matrix $\mathbf{A}$ is a *symmetric idempotent matrix* if the following two conditions hold:

(1) $$\mathbf{A} = \mathbf{A}'$$

(2) $$\mathbf{A} = \mathbf{A}^2$$

For brevity we shall omit the word "symmetric." That is to say, when we say a matrix is idempotent, we shall mean symmetric idempotent. We shall make no use whatsoever of idempotent matrices that are not symmetric.

◆ **Theorem 1.58** The characteristic roots of an idempotent matrix are either zero or unity.

Proof: If $\mathbf{A}$ is idempotent and if λ is a characteristic root of $\mathbf{A}$, there exists a vector $\mathbf{X} \neq \mathbf{0}$ such that $\mathbf{AX} = \lambda\mathbf{X}$. If we multiply both sides by $\mathbf{A}$, we get

$$\mathbf{A}^2\mathbf{X} = \lambda\mathbf{AX} = \lambda^2\mathbf{X}$$

But $\mathbf{A}^2\mathbf{X} = \mathbf{AX} = \lambda\mathbf{X}$; so we have

$$\lambda\mathbf{X} = \lambda^2\mathbf{X}$$

$$(\lambda^2 - \lambda)\mathbf{X} = \mathbf{0}$$

But $\mathbf{X} \neq \mathbf{0}$; so $\lambda^2 - \lambda$ must be zero. Thus $\lambda = 0$ or $\lambda = 1$.

◆ **Theorem 1.59** If $\mathbf{A}$ is idempotent and nonsingular, then $\mathbf{A} = \mathbf{I}$.

Proof: $\mathbf{AA} = \mathbf{A}$. Multiply both sides by $\mathbf{A}^{-1}$.

◆ **Theorem 1.60** If $\mathbf{A}$ is idempotent of rank r, there exists an orthogonal matrix $\mathbf{P}$ such that $\mathbf{P}'\mathbf{AP} = \mathbf{E}_r$, where $\mathbf{E}_r$ is a diagonal matrix with r diagonal elements equal to unity and the remaining diagonal elements equal to zero.

Proof: This follows immediately from Theorem 1.31.

◆ **Theorem 1.61** All idempotent matrices not of full rank are positive semidefinite.

Proof: Since $\mathbf{A} = \mathbf{A}'\mathbf{A}$, the result follows from Theorem 1.25. This theorem permits us to state that no idempotent matrix can have negative elements on its diagonal.

◆ **Theorem 1.62** If $\mathbf{A}$ is idempotent with elements a_{ij} and if the ith diagonal element of $\mathbf{A}$ is zero, then the elements of the ith row and the ith column of $\mathbf{A}$ are all identically zero.

Proof: Since $\mathbf{A} = \mathbf{A}^2$, we get for the ith diagonal element of $\mathbf{A}$

$$a_{ii} = \sum_{j=1}^{n} a_{ij}a_{ji}$$

But $a_{ij} = a_{ji}$; so

$$a_{ii} = \sum_{j=1}^{n} a_{ij}^2$$

But if $a_{ii} = 0$, then $a_{ij} = 0$ (for $j = 1, 2, \ldots, n$); that is, the elements of the ith row are all zero. But $\mathbf{A} = \mathbf{A}'$; so the elements of the ith column are also all zero.

◆ **Theorem 1.63** If $\mathbf{A}$ is idempotent of rank r, then $\text{tr}(\mathbf{A}) = r$.

Proof: By Theorem 1.31, there exists an orthogonal matrix $\mathbf{P}$ such that $\mathbf{P}'\mathbf{A}\mathbf{P} = \mathbf{E}_r$. But $\text{tr}(\mathbf{P}'\mathbf{A}\mathbf{P}) = \text{tr}(\mathbf{A})$; thus $\text{tr}(\mathbf{A}) = \text{tr}(\mathbf{P}'\mathbf{A}\mathbf{P}) = \text{tr}(\mathbf{E}_r) = r$.

◆ **Theorem 1.64** If $\mathbf{A}$ is an idempotent matrix and $\mathbf{B}$ is an idempotent matrix, then $\mathbf{AB}$ is idempotent if $\mathbf{AB} = \mathbf{BA}$.

Proof: If $\mathbf{AB} = \mathbf{BA}$, then $(\mathbf{AB})(\mathbf{AB}) = (\mathbf{AA})(\mathbf{BB}) = \mathbf{AB}$.

◆ **Theorem 1.65** If $\mathbf{A}$ is idempotent and $\mathbf{P}$ is orthogonal, $\mathbf{P}'\mathbf{A}\mathbf{P}$ is idempotent.

Proof: $(\mathbf{P}'\mathbf{A}\mathbf{P})(\mathbf{P}'\mathbf{A}\mathbf{P}) = (\mathbf{P}'\mathbf{A})(\mathbf{A}\mathbf{P}) = \mathbf{P}'\mathbf{A}\mathbf{P}$.

◆ **Theorem 1.66** If $\mathbf{A}$ is idempotent and $\mathbf{A} + \mathbf{B} = \mathbf{I}$, $\mathbf{B}$ is idempotent and $\mathbf{AB} = \mathbf{BA} = \mathbf{0}$.

Proof: We shall first show that $\mathbf{B}$ is idempotent. Now $\mathbf{B} = \mathbf{I} - \mathbf{A}$, and we must show that $\mathbf{B}^2 = \mathbf{B}$. We get $(\mathbf{I} - \mathbf{A})^2 = (\mathbf{I} - \mathbf{A})(\mathbf{I} - \mathbf{A}) = \mathbf{I} - \mathbf{IA} - \mathbf{AI} + \mathbf{A}^2 = \mathbf{I} - \mathbf{A}$. Thus $\mathbf{B}^2 = \mathbf{B}$.

We must now show that $\mathbf{AB} = \mathbf{BA} = \mathbf{0}$. We have $\mathbf{A} + \mathbf{B} = \mathbf{I}$. Multiply on the right by $\mathbf{B}$ and obtain $\mathbf{AB} + \mathbf{B}^2 = \mathbf{B}$ or $\mathbf{AB} = \mathbf{0}$. If we multiply the quantity $\mathbf{A} + \mathbf{B} = \mathbf{I}$ on the right by $\mathbf{A}$, the result $\mathbf{BA} = \mathbf{0}$ follows.

◆ **Theorem 1.67** If $\mathbf{A}_1, \mathbf{A}_2, \ldots, \mathbf{A}_n$ are $p \times p$ idempotent matrices, a necessary and sufficient condition that there exist an orthogonal matrix $\mathbf{P}$ such that $\mathbf{P}'\mathbf{A}_1\mathbf{P}, \mathbf{P}'\mathbf{A}_2\mathbf{P}, \ldots, \mathbf{P}'\mathbf{A}_n\mathbf{P}$ are each diagonal is that $\mathbf{A}_i\mathbf{A}_j = \mathbf{A}_j\mathbf{A}_i$ for all i and j.

Proof: This theorem is a very special case of Theorem 1.32. Because of its importance, we have stated it as a separate theorem.

◆ **Theorem 1.68** If $\mathbf{A}_1, \mathbf{A}_2, \ldots, \mathbf{A}_m$ are symmetric $p \times p$ matrices, any two of the following conditions imply the third:

(1) $\mathbf{A}_1, \mathbf{A}_2, \ldots, \mathbf{A}_m$ are each idempotent.

(2) The sum $\mathbf{B} = \sum_{i=1}^{m} \mathbf{A}_i$ is idempotent.

(3) $\mathbf{A}_i\mathbf{A}_j = \mathbf{0}$ for all $i \neq j$.

Proof: Suppose (1) and (2) are given. Then $\mathbf{B} = \sum_{i=1}^{m} \mathbf{A}_i$ is idempotent, and there exists an orthogonal matrix $\mathbf{P}$ such that

$$\mathbf{P}'\mathbf{B}\mathbf{P} = \begin{pmatrix} \mathbf{I}_r & \mathbf{0} \\ \mathbf{0} & \mathbf{0} \end{pmatrix}$$

where we suppose that $\mathbf{B}$ is of rank r and $\mathbf{I}_r$ is the $r \times r$ identity matrix. Thus we have

$$\mathbf{P}'\mathbf{B}\mathbf{P} = \begin{pmatrix} \mathbf{I}_r & \mathbf{0} \\ \mathbf{0} & \mathbf{0} \end{pmatrix} = \sum_{i=1}^{m} \mathbf{P}'\mathbf{A}_i\mathbf{P}$$

But $\mathbf{P}'\mathbf{A}_i\mathbf{P}$ are each idempotent, by Theorem 1.65. By Theorem 1.61, the last $p - r$ diagonal elements of each $\mathbf{P}'\mathbf{A}_i\mathbf{P}$ must be zero. This follows since the diagonal elements of an idempotent matrix are nonnegative; so, if their sum is zero, they must all be identically zero. Also by Theorem 1.62, the last $p - r$ rows and $p - r$ columns of each $\mathbf{P}'\mathbf{A}_i\mathbf{P}$ must be zero. Thus we may write

$$\mathbf{P}'\mathbf{A}_i\mathbf{P} = \begin{pmatrix} \mathbf{B}_i & \mathbf{0} \\ \mathbf{0} & \mathbf{0} \end{pmatrix}$$

So, using only the first r rows and first r columns of

$$\mathbf{P}'\mathbf{B}\mathbf{P} = \sum_{i=1}^{m} \mathbf{P}'\mathbf{A}_i\mathbf{P}$$

we have

$$\mathbf{I} = \sum_{i=1}^{m} \mathbf{B}_i$$

where the $\mathbf{B}_i$ are idempotent. Let us assume that the rank of $\mathbf{B}_i$ is r_i. Then there exists an $r \times r$ orthogonal matrix $\mathbf{C}$ such that

$$\mathbf{C}'\mathbf{B}_t\mathbf{C} = \begin{pmatrix} \mathbf{I}_t & \mathbf{0} \\ \mathbf{0} & \mathbf{0} \end{pmatrix}$$

Then

$$\mathbf{C}'\mathbf{I}\mathbf{C} = \mathbf{I} = \sum_{\substack{i=1 \\ i \neq t}}^{m} \mathbf{C}'\mathbf{B}_i\mathbf{C} + \begin{pmatrix} \mathbf{I}_t & \mathbf{0} \\ \mathbf{0} & \mathbf{0} \end{pmatrix}$$

Since $\mathbf{C}'\mathbf{B}_i\mathbf{C}$ is idempotent, by Theorems 1.61 and 1.62 we have

$$\mathbf{C}'\mathbf{B}_i\mathbf{C} = \begin{pmatrix} \mathbf{0} & \mathbf{0} \\ \mathbf{0} & \mathbf{K}_i \end{pmatrix} \qquad i = 1, 2, \ldots, m;\ i \neq t$$

where $\mathbf{K}_i$ is an $(r - t) \times (r - t)$ matrix. Thus we see that

$$\mathbf{C}'\mathbf{B}_i\mathbf{C}\mathbf{C}'\mathbf{B}_t\mathbf{C} = \mathbf{0}$$

which implies

$$\mathbf{B}_i\mathbf{B}_t = \mathbf{0} \qquad \text{and} \qquad \mathbf{A}_i\mathbf{A}_t = \mathbf{0} \qquad i = 1, 2, \ldots, m; i \neq t$$

Since t was arbitrary, the proof of condition (3) is complete.

Now suppose (1) and (3) are given. Then we have

$$\mathbf{B}^2 = \left(\sum_{i=1}^{m} \mathbf{A}_i\right)^2 = \sum_{i=1}^{m} \mathbf{A}_i^2 + \sum_{i \neq j} \mathbf{A}_i\mathbf{A}_j = \sum_{i=1}^{m} \mathbf{A}_i = \mathbf{B}$$

We have shown that the sum is idempotent, and condition (2) is satisfied.

Finally, suppose (2) and (3) are given. By Theorem 1.67, there exists an orthogonal matrix $\mathbf{P}$ such that $\mathbf{P}'\mathbf{A}_1\mathbf{P}$, $\mathbf{P}'\mathbf{A}_2\mathbf{P}$, $\ldots$, $\mathbf{P}'\mathbf{A}_m\mathbf{P}$ are each diagonal (since $\mathbf{A}_i\mathbf{A}_j = \mathbf{A}_j\mathbf{A}_i = \mathbf{0}$). Since the sum of diagonal matrices is a diagonal matrix, it also follows that $\mathbf{P}'\mathbf{B}\mathbf{P}$ is diagonal. By condition 3, we know that $\mathbf{P}'\mathbf{A}_i\mathbf{P}\mathbf{P}'\mathbf{A}_j\mathbf{P} = \mathbf{0}$ for all $i \neq j$. But the product of two diagonal matrices $\mathbf{P}'\mathbf{A}_i\mathbf{P}$ and $\mathbf{P}'\mathbf{A}_j\mathbf{P}$ is zero if and only if the corresponding nonzero diagonal elements of $\mathbf{P}'\mathbf{A}_i\mathbf{P}$ are zero in $\mathbf{P}'\mathbf{A}_j\mathbf{P}$. Thus, if the tth diagonal element of $\mathbf{P}'\mathbf{A}_i\mathbf{P}$ is nonzero, the tth diagonal element of $\mathbf{P}'\mathbf{A}_j\mathbf{P}$ (for all $j \neq i$) must be zero. Since $\mathbf{P}'\mathbf{B}\mathbf{P} = \mathbf{E}_r$, the tth diagonal element must be 0 or 1 for each $\mathbf{P}'\mathbf{A}_i\mathbf{P}$. For, if the tth element of $\mathbf{P}'\mathbf{A}_j\mathbf{P}$ is equal to $k \neq 0$, then the tth diagonal element of the remaining $\mathbf{P}'\mathbf{A}_i\mathbf{P}$ $(i \neq j)$ is zero. But the tth diagonal element of $\mathbf{B}$ is 0 or 1 and is the sum of the tth diagonal elements of $\mathbf{P}'\mathbf{A}_i\mathbf{P}$ $(i = 1, 2, \ldots, m)$. Thus $k = 0$ or $k = 1$. Since $\mathbf{P}'\mathbf{A}_i\mathbf{P}$ is diagonal, the characteristic roots of $\mathbf{A}_i$ are displayed down the diagonal, and, since these roots are either 0 or 1, $\mathbf{A}_i$ is idempotent, and the proof is complete.

It is of special interest to note that, if

$$\sum_{i=1}^{m} \mathbf{A}_i = \mathbf{I}$$

then condition (2) of Theorem 1.68 is satisfied. In this situation, condition (1) implies condition (3) and vice versa.

◆ **Theorem 1.69** If any two of the three conditions of Theorem 1.68 hold, the rank of $\sum_{i=1}^{m} \mathbf{A}_i$ equals the sum of the ranks of the $\mathbf{A}_i$.

Proof: If any two conditions of Theorem 1.68 hold, this implies that there exists an orthogonal matrix **P** such that the following are true:

(*a*) $$\mathbf{P'BP} = \begin{pmatrix} \mathbf{I}_r & \mathbf{0} \\ \mathbf{0} & \mathbf{0} \end{pmatrix} \quad \text{where the rank of } \mathbf{B} \text{ is } r$$

(*b*) $$\mathbf{P'A}_1\mathbf{P} = \begin{pmatrix} \mathbf{I}_{r_1} & \mathbf{0} \\ \mathbf{0} & \mathbf{0} \end{pmatrix}; \ \mathbf{P'A}_2\mathbf{P} = \begin{pmatrix} \mathbf{0} & \mathbf{0} & \mathbf{0} \\ \mathbf{0} & \mathbf{I}_{r_2} & \mathbf{0} \\ \mathbf{0} & \mathbf{0} & \mathbf{0} \end{pmatrix}; \ldots;$$

$$\mathbf{P'A}_m\mathbf{P} = \begin{pmatrix} \mathbf{0} & \mathbf{0} & \mathbf{0} \\ \mathbf{0} & \mathbf{I}_{r_m} & \mathbf{0} \\ \mathbf{0} & \mathbf{0} & \mathbf{0} \end{pmatrix}$$

where the rank of $\mathbf{A}_i = r_i$. Thus the result follows.

1.7 Maxima, Minima, and Jacobians

We shall now state some theorems concerning the maxima and minima of functions.

◆ **Theorem 1.70** If $y = f(x_1, x_2, \ldots, x_n)$ is a function of n variables and if all partial derivatives $\partial y / \partial x_i$ are continuous, then y attains its maxima and minima only at the points where

$$\frac{\partial y}{\partial x_1} = \frac{\partial y}{\partial x_2} = \cdots = \frac{\partial y}{\partial x_n} = 0$$

In particular, if $f(x_1, x_2, \ldots, x_n)$ is a quadratic form, then $f(x_1, x_2, \ldots, x_n)$ is continuous and has continuous derivatives of all orders; so the theorem applies.

◆ **Theorem 1.71** If $f(x_1, x_2, \ldots, x_n)$ is such that all the first and second partial derivatives are continuous, then at the point where

$$\frac{\partial f}{\partial x_1} = \frac{\partial f}{\partial x_2} = \cdots = \frac{\partial f}{\partial x_n} = 0$$

the function has

(1) a minimum, if the matrix **K**, where the ijth element of **K** is $\partial^2 f / \partial x_i \partial x_j$, is positive definite.

(2) a maximum, if the matrix $-\mathbf{K}$ is positive definite.

In the above two theorems on maxima and minima, it must be remembered that the x_i are independent variables. Many times it is desired to maximize or minimize a function $f(x_1,x_2, \ldots ,x_n)$ where the x_i are not independent but are subject to constraints. For example, suppose it is necessary to minimize the function $f(x_1,x_2, \ldots ,x_n)$ subject to the condition $h(x_1,x_2, \ldots ,x_n) = 0$. Since the x_i are not independent, Theorems 1.70 and 1.71 will not necessarily give the desired result. If the equation $h(x_1,x_2, \ldots ,x_n) = 0$ could be solved for x_t such that

$$x_t = h_t(x_1,x_2, \ldots ,x_n)$$

then this value of x_t could be substituted into $f(x_1,x_2, \ldots ,x_n)$ and Theorems 1.70 and 1.71 could be applied.

As an example, suppose we want to find the minimum of $f = x_1^2 - 2x_1 + x_2^2 - 6x_2 + 16$. Using Theorem 1.70, we get

$$\frac{\partial f}{\partial x_1} = 2x_1 - 2 = 0$$

$$\frac{\partial f}{\partial x_2} = 2x_2 - 6 = 0$$

The solutions yield $x_1 = 1$, $x_2 = 3$.

The matrix $\mathbf{K}$ is given by

$$\mathbf{K} = \begin{pmatrix} \dfrac{\partial^2 f}{\partial x_1^2} & \dfrac{\partial^2 f}{\partial x_1\, \partial x_2} \\ \dfrac{\partial^2 f}{\partial x_2\, \partial x_1} & \dfrac{\partial^2 f}{\partial x_2^2} \end{pmatrix} = \begin{pmatrix} 2 & 0 \\ 0 & 2 \end{pmatrix}$$

$\mathbf{K}$ is positive definite; so f has a minimum at the point $x_1 = 1$, $x_2 = 3$.

If we want to find the minimum of f subject to the condition $x_1 + x_2 = 1$, we proceed as follows. Substitute $x_1 = 1 - x_2$ into the f function, and proceed as before. This gives

$$f = (1 - x_2)^2 - 2(1 - x_2) + x_2^2 - 6x_2 + 16$$

$$\frac{\partial f}{\partial x_2} = -2(1 - x_2) + 2 + 2x_2 - 6 = 0$$

The solution gives $x_2 = \frac{3}{2}$. In this case the matrix $\mathbf{K}$ consists of a single term $\partial^2 f/\partial x_2^2$. So $\mathbf{K} = 4$ and is positive definite. Thus f, subject to the constraint $x_1 + x_2 = 1$, attains a minimum at the point $x_1 = -\frac{1}{2}$, $x_2 = \frac{3}{2}$.

If the constraint is a complicated function or if there are many constraining equations, this method may become cumbersome. An alternative is the *method of Lagrange multipliers*. For example, if we want to minimize $f(x_1,x_2, \ldots ,x_n)$ subject to the constraint $h(x_1,x_2, \ldots ,x_n) = 0$, we form the equation

$$F = f(x_1,x_2, \ldots ,x_n) - \lambda h(x_1,x_2, \ldots ,x_n)$$

The x_1, $x_2, \ldots ,x_n, \lambda$ can now be considered $n + 1$ independent variables. We now state the theorem:

◆ **Theorem 1.72** If $f(x_1,x_2, \ldots ,x_n)$ and the constraint $h(x_1,x_2, \ldots , x_n) = 0$ are such that all first partial derivatives are continuous, then the maximum or minimum of $f(x_1,x_2, \ldots ,x_n)$ subject to the constraint $h(x_1,x_2, \ldots ,x_n) = 0$ can occur only at a point where the derivatives of $F = f(x_1,x_2, \ldots ,x_n) - \lambda h(x_1,x_2, \ldots ,x_n)$ vanish; i.e., where

$$\frac{\partial F}{\partial x_1} = \frac{\partial F}{\partial x_2} = \cdots = \frac{\partial F}{\partial x_n} = \frac{\partial F}{\partial \lambda} = 0$$

if $\partial h/\partial x_i \neq 0$ for all i at the point.

Thus we now have $n + 1$ equations and $n + 1$ unknowns, and we need not worry about which variables are independent, for we treat all $n + 1$ as if they were independent variables.

This will be generalized in the following.

◆ **Theorem 1.73** To find the maximum or minimum of the function $f(x_1,x_2, \ldots ,x_n)$ subject to the k constraints $h_i(x_1,x_2, \ldots ,x_n) = 0$ $(i = 1, 2, \ldots , k)$, form the function $F = f(x_1,x_2, \ldots ,x_n) - \sum_{i=1}^{k} \lambda_i h_i(x_1,x_2, \ldots ,x_n)$. If $\partial F/\partial x_i$ $(i = 1, 2, \ldots , n)$ are continuous, then f, subject to the constraints, can have its maxima and minima only at the points where the following equations are satisfied (if a Jacobian $|\partial h_i/\partial x_j| \neq 0$ at the point):

$$\frac{\partial F}{\partial x_1} = \frac{\partial F}{\partial x_2} = \cdots = \frac{\partial F}{\partial x_n} = \frac{\partial F}{\partial \lambda_1} = \frac{\partial F}{\partial \lambda_2} = \cdots = \frac{\partial F}{\partial \lambda_k} = 0$$

Let $g(x_1,x_2, \ldots ,x_n)$, where $-\infty < x_i < \infty$ $(i = 1, 2, \ldots , n)$, represent a frequency-density function. This is equivalent to the following two conditions on the function $g(x_1,x_2, \ldots ,x_n)$:

(1) $$g(x_1,x_2, \ldots ,x_n) \geqslant 0$$

(2) $$\int_{-\infty}^{\infty}\int_{-\infty}^{\infty} \cdots \int_{-\infty}^{\infty} g(x_1,x_2, \ldots ,x_n)\, dx_1\, dx_2 \cdots dx_n = 1$$

If we make the transformation $x_i = h_i(y_1,y_2, \ldots ,y_n)$, where $i = 1, 2, \ldots, n$, the frequency-density function in terms of the new variables $y_1, y_2, \ldots, y_n$ is given by

$$k(y_1,y_2, \ldots ,y_n) = g(h_1,h_2, \ldots ,h_n)\ |J|$$

(we shall assume certain regularity conditions on the transformation equations). The symbol $|J|$ denotes the absolute value of the Jacobian of the transformation. The Jacobian is the determinant of a matrix **K** whose ijth element is $\partial x_i/\partial y_j$.

For example, if

$$f(x_1,x_2) = \frac{1}{\pi} e^{-x_1^2-x_2^2} \qquad -\infty < x_1 < \infty; -\infty < x_2 < \infty$$

is a frequency-density function and if we want to find the corresponding frequency-density function $k(y_1,y_2)$, where

$$x_1 = 4y_1 + y_2$$
$$x_2 = 2y_1 - y_2$$

we have

$$\mathbf{K} = \begin{pmatrix} \dfrac{\partial x_1}{\partial y_1} & \dfrac{\partial x_1}{\partial y_2} \\[2ex] \dfrac{\partial x_2}{\partial y_1} & \dfrac{\partial x_2}{\partial y_2} \end{pmatrix} = \begin{pmatrix} 4 & 1 \\ 2 & -1 \end{pmatrix}$$

$J = -6$ and $|J| = 6$. So we have

$$k(y_1,y_2) = \frac{6}{\pi} e^{-[(4y_1+y_2)^2+(2y_1-y_2)^2]}$$

Thus it is quite clear that the Jacobian J will play an important part in the theory of probability distributions.

◆ **Theorem 1.74.** If a set of transformations is given by $x_i = h_i(y_1,y_2, \ldots ,y_n)$, where $i = 1, 2. \ldots, n$, and the Jacobian J is the determinant of the matrix **K** whose ijth element is $\partial x_i/\partial y_j$, and if the equations satisfy mild regularity conditions, and the solutions for the y_i yield

$$y_i = d_i(x_1,x_2, \ldots ,x_n) \qquad i = 1, 2, \ldots, n$$

then the Jacobian J (if $J \neq 0$) can also be given by $1/|\mathbf{K}^*|$, where the ijth element of $\mathbf{K}^*$ is $\partial y_i/\partial x_j$.

This theorem might be useful if the equations were such that the $\partial x_i/\partial y_j$ were difficult to obtain but the $\partial y_i/\partial x_j$ were relatively easy to obtain.

In the example above, if we solve the equations for y_1 and y_2, we get

$$y_1 = \frac{x_1 + x_2}{6} \qquad y_2 = \frac{x_1 - 2x_2}{3}$$

and

$$\mathbf{K}^* = \begin{pmatrix} \frac{1}{6} & \frac{1}{6} \\ \frac{1}{3} & -\frac{2}{3} \end{pmatrix}$$

So

$$J = \frac{1}{|\mathbf{K}^*|} = \frac{1}{-\frac{2}{18} - \frac{1}{18}} = -6$$

Thus we see that $|J| = 6$, the same value as that we obtained before.

◆ **Theorem 1.75** If the transformation from the vector $\mathbf{X}$ to the vector $\mathbf{Y}$ is given by $\mathbf{Y} = \mathbf{AX}$, where $\mathbf{A}$ is an $n \times n$ nonsingular matrix, then $|\mathbf{A}^{-1}| = J$.

Proof: The Jacobian is the determinant of the inverse of the matrix $\mathbf{A}$ whose ijth element is $\partial y_i / \partial x_j = a_{ij}$; the result follows. If $\mathbf{A}$ is an orthogonal matrix, then $J = |\mathbf{A}'| = \pm 1$ and $|J| = 1$.

Problems

1.1 Find the rank of the two matrices

$$\begin{pmatrix} 2 & 6 & 3 & 9 \\ 1 & 4 & 8 & 6 \\ 3 & 10 & 11 & 15 \\ 1 & 9 & 8 & 4 \\ 1 & 6 & 2 & 1 \end{pmatrix} \qquad \begin{pmatrix} 1 & 6 & 3 \\ 2 & 1 & 4 \\ 1 & 8 & 9 \\ 4 & 13 & 10 \end{pmatrix}$$

1.2 Find the inverse of the matrix

$$\mathbf{A} = \begin{pmatrix} 1 & 2 & 1 \\ 3 & 4 & 8 \\ 6 & 2 & 5 \end{pmatrix}$$

1.3 Given

$$\mathbf{A} = \begin{pmatrix} 1 & 2 & 3 \\ 6 & 1 & 4 \\ 2 & -7 & -8 \end{pmatrix}$$

verify that rank $\mathbf{A}$ = rank $\mathbf{A}'$ = rank $\mathbf{AA}'$ = rank $\mathbf{A}'\mathbf{A}$.

1.4 Prove Theorem 1.16.

1.5 Let

$$\mathbf{A} = \begin{pmatrix} 1 & 2 & 4 \\ 6 & -3 & -1 \\ 8 & 1 & 7 \end{pmatrix} \qquad \mathbf{B} = \begin{pmatrix} 2 & 1 & 4 \\ 2 & 2 & 1 \\ 4 & 3 & 5 \end{pmatrix}$$

Find **AB**, **BA**, $\rho(\mathbf{A})$, $\rho(\mathbf{B})$, $\rho(\mathbf{AB})$, and $\rho(\mathbf{A}+\mathbf{B})$, and verify Theorems 1.11, 1.12, and 1.19. Find a set of linearly independent rows of **A**.

1.6 If

$$\mathbf{A} = \begin{pmatrix} 6 & 2 \\ 1 & 3 \end{pmatrix} \quad \text{and} \quad \mathbf{B} = \begin{pmatrix} 2 & 1 \\ -1 & 4 \end{pmatrix}$$

show that $(\mathbf{AB})' = \mathbf{B}'\mathbf{A}'$. Also show that $(\mathbf{AB})^{-1} = \mathbf{B}^{-1}\mathbf{A}^{-1}$.

1.7 Given

$$\mathbf{A} = \begin{pmatrix} \frac{1}{2} & \frac{1}{2} \\ \frac{1}{2} & \frac{1}{2} \end{pmatrix} \quad \mathbf{B} = \begin{pmatrix} \frac{1}{2} & -\frac{1}{2} \\ -\frac{1}{2} & \frac{1}{2} \end{pmatrix}$$

show that $\mathbf{AB} = \mathbf{0}$, and thus verify Theorem 1.18.

1.8 If

$$\mathbf{D}_1 = \begin{pmatrix} 6 & 0 & 0 \\ 0 & 4 & 0 \\ 0 & 0 & -2 \end{pmatrix} \quad \text{and} \quad \mathbf{D}_2 = \begin{pmatrix} 3 & 0 & 0 \\ 0 & -1 & 0 \\ 0 & 0 & 8 \end{pmatrix}$$

show that $\mathbf{D}_1\mathbf{D}_2 = \mathbf{D}_2\mathbf{D}_1$ is diagonal, and thus verify Theorems 1.8 and 1.9.

1.9 If **A** is any $n \times m$ matrix, prove that $\mathbf{AA}'$ and $\mathbf{A}'\mathbf{A}$ are symmetric.

1.10 Given the quadratic form $2X_1^2 + 3X_2^2 + 6X_1X_2$, find its matrix **A**.

1.11 If **C** is any matrix and **A** is a symmetric matrix, prove that $\mathbf{C}'\mathbf{AC}$ is symmetric.

1.12 Given the matrix

$$\mathbf{C} = \begin{pmatrix} \frac{1}{\sqrt{4}} & \frac{1}{\sqrt{4}} & \frac{1}{\sqrt{4}} & \frac{1}{\sqrt{4}} \\ \frac{1}{\sqrt{2}} & -\frac{1}{\sqrt{2}} & 0 & 0 \\ \frac{1}{\sqrt{6}} & \frac{1}{\sqrt{6}} & -\frac{2}{\sqrt{6}} & 0 \\ \frac{1}{\sqrt{12}} & \frac{1}{\sqrt{12}} & \frac{1}{\sqrt{12}} & -\frac{3}{\sqrt{12}} \end{pmatrix}$$

prove that **C** is orthogonal.

1.13 Given the quadratic form $\frac{1}{2}X_1^2 + 3X_1X_2 + 4X_2^2 + 6X_3^2$, find its matrix. Test to see whether the matrix is positive definite.

1.14 Given the matrix

$$\mathbf{P} = \begin{pmatrix} 6 & 2 & 1 \\ 3 & 1 & 4 \end{pmatrix}$$

verify that $\mathbf{PP}'$ is positive definite and that $\mathbf{P}'\mathbf{P}$ is positive semidefinite.

1.15 Given the matrix

$$\mathbf{A} = \begin{pmatrix} 4 & 0 & 2 \\ 0 & 3 & 0 \\ 2 & 0 & 4 \end{pmatrix}$$

find the characteristic polynomial and the characteristic roots.

Show that the orthogonal matrix

$$\mathbf{C}' = \begin{pmatrix} \frac{1}{\sqrt{2}} & 0 & \frac{1}{\sqrt{2}} \\ -\frac{1}{\sqrt{2}} & 0 & \frac{1}{\sqrt{2}} \\ 0 & 1 & 0 \end{pmatrix}$$

is such that **C′AC** is a diagonal matrix with characteristic roots on the diagonal.

1.16 Prove the following: If **C** is an orthogonal matrix, then **C′** is an orthogonal matrix and $\mathbf{C}^{-1}$ is an orthogonal matrix.

1.17 If **A** and **B** are orthogonal matrices, prove that **AB** is an orthogonal matrix.

1.18 Let

$$\mathbf{A} = \begin{pmatrix} 6 & 2 \\ 1 & 4 \end{pmatrix} \qquad \text{and} \qquad \mathbf{C} = \begin{pmatrix} \frac{1}{\sqrt{2}} & -\frac{1}{\sqrt{2}} \\ \frac{1}{\sqrt{2}} & \frac{1}{\sqrt{2}} \end{pmatrix}$$

Show that **C** is orthogonal. Find the characteristic roots of **A** and of **C′AC**, and show they are equal.

1.19 Let

$$\mathbf{C}_1 = \begin{pmatrix} \frac{1}{3} & \frac{1}{3} & \frac{1}{3} & \frac{\sqrt{2}}{\sqrt{3}} \\ \frac{1}{\sqrt{6}} & \frac{1}{\sqrt{6}} & -\frac{2}{\sqrt{6}} & 0 \\ \frac{1}{\sqrt{2}} & -\frac{1}{\sqrt{2}} & 0 & 0 \end{pmatrix}$$

Find a vector $\mathbf{C}_2$ such that $\mathbf{C} = \begin{pmatrix} \mathbf{C}_1 \\ \mathbf{C}_2 \end{pmatrix}$ is orthogonal, thus illustrating Theorem 1.34.

1.20 Prove that, if a matrix **A** is positive definite, then **A′** and $\mathbf{A}^{-1}$ are also positive definite.

1.21 Prove that, if **D** is a diagonal matrix, where d_i is the ith diagonal element, and if $d_i \neq 0$ for any i, then the inverse of **D** exists, and the ith diagonal element of $\mathbf{D}^{-1}$ is $1/d_i$.

1.22 Given the following systems of equations, find which are consistent and which are inconsistent. Find solutions where possible.

(a)
$$\begin{aligned} X_1 + 2X_2 + 3X_3 &= 6 \\ X_1 - X_2 \qquad\quad &= 2 \\ X_1 \qquad\quad - X_3 &= -1 \end{aligned}$$

(b)
$$\begin{aligned} X_1 - X_2 + 2X_3 &= 2 \\ X_1 - X_2 - X_3 &= -1 \\ 2X_1 - 2X_2 + X_3 &= 2 \end{aligned}$$

(c)
$$\begin{aligned} X_1 + X_2 + X_3 + X_4 &= 8 \\ X_1 - X_2 - X_3 - X_4 &= 6 \\ 3X_1 + X_2 + X_3 + X_4 &= 22 \end{aligned}$$

1.23 Given the quadratic form

$$Q = 6X_1^2 + 3X_2^2 + 4X_1X_2 + X_3^2 + 2X_1X_3 + X_4^2 + X_1X_4 + 2X_2X_4$$

find $\partial Q/\partial \mathbf{X}$.

1.24 Given the matrix

$$\mathbf{A} = \begin{pmatrix} 1 & 1 & 2 \\ 2 & 3 & 1 \end{pmatrix}$$

show that the system of equations $\mathbf{AA'X} = \mathbf{Y}$ has a vector $\mathbf{X}$ satisfying the system when the vector $\mathbf{Y}$ is equal to any column of the matrix $\mathbf{A}$. That is to say, show that the rank of the coefficient matrix $\mathbf{AA'}$ equals the rank of the augmented matrix $(\mathbf{AA'}, \mathbf{A}_i)$, where $\mathbf{A}_i$ is any column of $\mathbf{A}$.

1.25 If

$$\mathbf{A} = \begin{pmatrix} 1 & 2 & -1 & 2 \\ 2 & 1 & 2 & -2 \\ -1 & 2 & 1 & -2 \\ 2 & -2 & -2 & 1 \end{pmatrix}, \quad \mathbf{B} = \begin{pmatrix} 1 & -3 & 0 & 0 \\ 2 & 4 & 0 & 0 \\ 2 & 1 & 3 & 5 \\ 2 & 1 & -4 & 2 \end{pmatrix}, \quad \mathbf{A}_{11} = \begin{pmatrix} 1 & 2 \\ 2 & 1 \end{pmatrix}, \text{ etc.}$$

and

$$\mathbf{C} = \begin{pmatrix} \frac{1}{2} & \frac{1}{2} & \frac{1}{2} & \frac{1}{2} \\ \frac{1}{2} & \frac{1}{2} & -\frac{1}{2} & -\frac{1}{2} \\ \frac{1}{2} & -\frac{1}{2} & \frac{1}{2} & -\frac{1}{2} \\ +\frac{1}{2} & -\frac{1}{2} & -\frac{1}{2} & \frac{1}{2} \end{pmatrix}$$

show:

(*a*) $|\mathbf{AB}| = |\mathbf{BA}| = |\mathbf{A}|\,|\mathbf{B}|$, thus illustrating Theorem 1.36.
(*b*) $\mathbf{C}$ is orthogonal.
(*c*) $\text{tr}(\mathbf{C'AC}) = \text{tr}(\mathbf{A})$, thus illustrating Theorem 1.48.
(*d*) $|\mathbf{B}| = |\mathbf{B}_{11}|\,|\mathbf{B}_{22}|$, thus illustrating Theorem 1.43.
(*e*) $|\mathbf{C}| = -1$, thus illustrating Theorem 1.38.
(*f*) $|\mathbf{C'AC}| = |\mathbf{A}|$, thus illustrating Theorem 1.39.
(*g*) $|\mathbf{B}^{-1}| = |\mathbf{B}|^{-1}$, thus illustrating Theorem 1.42.
(*h*) $\text{tr}(\mathbf{AB}) = \text{tr}(\mathbf{BA})$, thus illustrating Theorem 1.45.
(*i*) $\mathbf{A}_{11}^{-1} = \mathbf{D}_{11} - \mathbf{D}_{12}\mathbf{D}_{22}^{-1}\mathbf{D}_{21}$, thus illustrating Theorem 1.49 (where $\mathbf{D} = \mathbf{A}^{-1}$).
(*j*) $|\mathbf{A}| = |\mathbf{A}_{22}|\,|\mathbf{A}_{11} - \mathbf{A}_{12}\mathbf{A}_{22}^{-1}\mathbf{A}_{21}|$, thus illustrating Theorem 1.50.

1.26 If

$$\mathbf{A} = \begin{pmatrix} \frac{1}{3} & \frac{1}{3} & \frac{1}{3} \\ \frac{1}{3} & \frac{1}{3} & \frac{1}{3} \\ \frac{1}{3} & \frac{1}{3} & \frac{1}{3} \end{pmatrix}$$

show that $\mathbf{A}$ is idempotent, and find the characteristic roots of $\mathbf{A}$, the trace of $\mathbf{A}$, and the rank of $\mathbf{A}$.

1.27 If $\mathbf{X}$ is a $p \times 1$ vector with elements x_i, find the matrix $\mathbf{A}$ of the quadratic form $\mathbf{X}'\mathbf{A}\mathbf{X} = \left(\frac{1}{p}\sum_{i=1}^{p} x_i\right)^2$. *Solution:*

$$\mathbf{X}'\mathbf{A}\mathbf{X} = \frac{1}{p^2}\sum_{i=1}^{p} x_i^2 + \frac{1}{p^2}\sum_{i=1}^{p}\sum_{\substack{j=1\\ i\neq j}}^{p} x_i x_j$$

But
$$\mathbf{X}'\mathbf{A}\mathbf{X} = \sum_{i=1}^{p}\sum_{j=1}^{p} x_i x_j a_{ij} = \sum_{i=1}^{p} x_i^2 a_{ii} + \sum_{i=1}^{p}\sum_{\substack{j=1\\ i\neq j}}^{p} x_i x_j a_{ij}$$

The coefficients of the x_i^2 term are the diagonal elements of $\mathbf{A}$, which are equal to p^{-2} in this case. The coefficient of $x_i x_j$ $(i \neq j)$ is the ijth element of the matrix $\mathbf{A}$, which in this case equals p^{-2}. Thus, in the example, the matrix $\mathbf{A}$ has every element equal to p^{-2}.

1.28 Use Prob. 1.27 to find the matrix of the quadratic form $p\bar{x}^2 \left(\bar{x} = \frac{1}{p}\sum_{i=1}^{p} x_i\right)$, and show that the matrix of this quadratic form is idempotent.

1.29 Prove $\sum_{i=1}^{p} x_i^2 = \sum_{i=1}^{p} (x_i - \bar{x})^2 + p\bar{x}^2$.

1.30 In Prob. 1.29, find the matrix of the quadratic form for each of the three quantities $\sum_{i=1}^{p} x_i^2$, $\sum_{i=1}^{p} (x_i - \bar{x})^2$, $p\bar{x}^2$.

1.31 In Prob. 1.30, show that each matrix is idempotent.

1.32 Let $\mathbf{B}_1$ represent the matrix for $p\bar{x}^2$. Let $\mathbf{B}_2$ represent the matrix of the quadratic form $\sum_{i=1}^{p} (x_i - \bar{x})^2$. Show that $\mathbf{B}_1 + \mathbf{B}_2 = \mathbf{I}$ and $\mathbf{B}_1\mathbf{B}_2 = \mathbf{B}_2\mathbf{B}_1 = \mathbf{0}$ for $p = 5$.

1.33 Without using Theorem 1.68, prove the following: If $\mathbf{A}_1 + \mathbf{A}_2 = \mathbf{I}$ and if $\mathbf{A}_1\mathbf{A}_2 = \mathbf{0}$, then $\mathbf{A}_1$ and $\mathbf{A}_2$ are each idempotent matrices.

1.34 If $\mathbf{A}$ is a matrix such that $\mathbf{A}\mathbf{A} = k^2\mathbf{A}$, where k is a scalar, find a scalar m such that $m\mathbf{A}$ is idempotent.

1.35 If

$$\mathbf{X} = \begin{pmatrix} 1 & 0 \\ 1 & 0 \\ 0 & 1 \\ 0 & 1 \end{pmatrix}$$

find $\mathbf{X}\mathbf{X}'$. Find the constant k such that $k\mathbf{X}\mathbf{X}'$ is idempotent.

1.36 Given the matrices

$$\mathbf{A}_1 = \begin{pmatrix} \frac{2}{3} & 0 & \frac{\sqrt{2}}{3} \\ 0 & 1 & 0 \\ \frac{\sqrt{2}}{3} & 0 & \frac{1}{3} \end{pmatrix} \qquad \mathbf{A}_2 = \begin{pmatrix} \frac{1}{3} & 0 & -\frac{\sqrt{2}}{3} \\ 0 & 0 & 0 \\ -\frac{\sqrt{2}}{3} & 0 & \frac{2}{3} \end{pmatrix}$$

show that $\mathbf{A}_1 + \mathbf{A}_2 = \mathbf{I}$. Show that $\mathbf{A}_1$ and $\mathbf{A}_2$ are each idempotent. Show that $\mathbf{A}_1\mathbf{A}_2 = \mathbf{A}_2\mathbf{A}_1 = \mathbf{0}$, and find the rank of $\mathbf{A}_1$ and $\mathbf{A}_2$.

1.37 For the matrices in Prob. 1.36, it can be seen that the conditions of Theorem 1.67 are satisfied. So there must exist an orthogonal transformation $\mathbf{C}$ such that $\mathbf{C}'\mathbf{A}_1\mathbf{C} = \mathbf{E}_2$ and $\mathbf{C}'\mathbf{A}_2\mathbf{C} = \mathbf{E}_1$, where $\mathbf{E}_2$ and $\mathbf{E}_1$ have ranks of 2 and 1, respectively. Verify the fact that the orthogonal matrix

$$\mathbf{C}' = \begin{pmatrix} \frac{1}{\sqrt{3}} & \frac{1}{\sqrt{2}} & \frac{1}{\sqrt{6}} \\ \frac{1}{\sqrt{3}} & -\frac{1}{\sqrt{2}} & \frac{1}{\sqrt{6}} \\ \frac{1}{\sqrt{3}} & 0 & -\frac{2}{\sqrt{6}} \end{pmatrix}$$

will transform $\mathbf{A}_1$ and $\mathbf{A}_2$ into $\mathbf{E}_1$ and $\mathbf{E}_2$, respectively.

1.38 Find the minimum value of the function $2X_1^2 + X_2^2 + 2X_1X_2 + 3X_3^2$ subject to the constraint $X_1 + X_2 = 1$.

1.39 Given the transformation equations

$$X_1 = Y_1 \cos Y_2$$
$$X_2 = Y_1 \sin Y_2$$

evaluate the Jacobian $J = |\partial x_i/\partial y_j|$. Also evaluate $K^* = |\partial y_i/\partial x_j|$ by solving the transformation equations for Y_1 and Y_2. Show that $|K^*| = |J|^{-1}$.

Further Reading

1 P. S. Dwyer and M. S. Macphail: Symbolic Matrix Derivatives, *Ann. Math. Statist.*, vol. 19, pp. 517–534, 1948.

2 G. Birkhoff and S. MacLane: "A Survey of Modern Algebra," The Macmillan Company, New York, 1953.

3 M. J. Weiss: "Higher Algebra for the Undergraduate," John Wiley & Sons, Inc., New York, 1949.

4 W. L. Ferrar: "Algebra," Oxford University Press, New York, 1946.

5 S. Perlis: "Theory of Matrices," Addison-Wesley Publishing Company, Cambridge, Mass., 1952.

6 W. Kaplan: "Advanced Calculus," Addison-Wesley Publishing Company, Cambridge, Mass., 1953.

7 R. Courant: "Differential and Integral Calculus," vols., I, II, Interscience Publishers, Inc., New York, 1947.

8 D. V. Widder: "Advanced Calculus," Prentice-Hall, Inc., Englewood Cliffs, N.J., 1947.

2

Statistical Concepts

In this chapter we shall review some of the elementary notions of mathematical statistics, including random variables, frequency functions, estimation of parameters, and testing of statistical hypotheses.

2.1 Sample Space, Random Variable, Frequency Function

2.1.1 Sample Space. One way to obtain knowledge is to observe the outcome of experiments. Since these observations are generally subject to unpredictable factors, it may be desirable to build these factors into some kind of a theoretical model. The unpredictable or uncontrollable element in an experiment can often be subjected to some objectivity by the use of probability. A very useful concept in probability methods is sample space. Suppose an experiment is contemplated in which every outcome is known. The set consisting of all possible outcomes of the experiment will be called the *sample space*, and the outcome of one particular trial of the experiment will be called a *sample point* in this sample space. In this book, the sample space will always be the Euclidean line, the two-dimensional Euclidean plane, or the n-dimensional Euclidean hyperplane.

For example, suppose an experiment consists in throwing a die once and observing the number that appears. The sample space consists of six points, since the outcome will be a 1, 2, 3, 4, 5, or 6. Or suppose an experiment consists in measuring the weight in pounds of an individual from the city of Stillwater, Oklahoma. The sample space could be the Euclidean half-line consisting of all positive real numbers, since in that half-line there would be a number available to represent each person's weight. Of course, *every* number on this half-line could not represent a person's weight, since it is inconceivable that anyone should weigh, say, 1 million pounds. It does not, however, matter if there are points in the sample space for which there is no outcome, but for every conceivable outcome of the experiment there must be a point in the sample space.

For another example, suppose an experiment is run to test two fertilizer treatments on pecan trees. Six trees will be selected from an orchard, the fertilizers will be applied, and the weights of the pecans from these six trees will be observed. The outcome of this experiment will consist of a set of six numbers $(x_1,x_2,x_3,x_4,x_5,x_6)$. The sample space could be the six-dimensional Euclidean plane, since every conceivable outcome is a point in this six-dimensional space.

2.1.2 Random Variables. Another concept important in probability methods is that of a random variable. A *random variable* is any function defined on a sample space. For example, in the experiment on throwing a die in Art. 2.1.1, if x represents the outcome, then x can equal 1, 2, 3, 4, 5, or 6. The function $u(x) = x^2$ is a random variable; in fact, the outcome itself is also a random variable, as is $\log x$, $e^x - 1$, etc.

A *statistic* is defined as any function of observable random variables that does not involve unknown parameters.

2.1.3 Density Function. A *density function* on a sample space is a nonnegative function whose integral over the sample space is unity. For example, if the sample space is the Euclidean line; i.e., if the outcome of an experiment is the number x_0, where x_0 is any number x such that $-\infty < x < \infty$, then any function $f(X)$ is a density function on the sample space if

(1) $$f(X) \geqslant 0 \qquad \text{for all } X \text{ such that } -\infty < X < \infty$$

(2) $$\int_{-\infty}^{\infty} f(X)\, dX = 1$$

Given a random variable y, it is said to have a density function $g(Y)$ if

$$\int_{-\infty}^{A} g(Y)\, dY = P(y \leqslant A)$$

where $P(y \leqslant A)$ is the probability that the random variable y is less than or equal to A. In this case we shall say that the random variable *y is distributed as $g(Y)$*.

These ideas can be extended to functions of two random variables x_1 and x_2. When we make the statement that x_1 and x_2 *are jointly distributed as $f(X_1,X_2)$*, we mean that, for any a, b, c, and d, the probability that $a \leqslant x_1 \leqslant b$ and $c \leqslant x_2 \leqslant d$, denoted by $P(a \leqslant x_1 \leqslant b, c \leqslant x_2 \leqslant d)$, is given by $\int_c^d \int_a^b f(X_1,X_2)\, dX_1\, dX_2$.

These ideas can readily be extended from two variables X_1 and X_2 to k variables $X_1, X_2, \ldots, X_k$. For a function of k variables to be a joint frequency function (sometimes called a *multivariate frequency function*), the following conditions must hold:

(1) $$f(X_1,X_2,\ldots,X_k) \geqslant 0 \qquad -\infty < X_i < \infty;\ i = 1, 2, \ldots, k$$

(2) $$\int_{-\infty}^{\infty}\int_{-\infty}^{\infty}\cdots\int_{-\infty}^{\infty} f(X_1,X_2,\ldots,X_k)\,dX_1\,dX_2\cdots dX_k = 1$$

In multivariate distributions we must also consider marginal and conditional distributions. If the chance (random) variables $x_1, x_2, \ldots, x_k$ have the joint frequency $f(X_1,X_2,\ldots,X_k)$, then the marginal distribution of a subset of $x_1, x_2, \ldots, x_k$, say, $x_1, x_2, \ldots, x_p$ $(p < k)$, is given by

$$g(X_1,X_2,\ldots,X_p) = \int_{-\infty}^{\infty}\int_{-\infty}^{\infty}\cdots\int_{-\infty}^{\infty} f(X_1,X_2,\ldots,X_k)\,dX_{p+1}\,dX_{p+2}\cdots dX_k$$

That is to say, the marginal distribution of a subset is obtained by integrating the joint frequency function over the random variables that are not in the subset. For example, suppose the random variables x_1, x_2, x_3 have the joint frequency function

$$f(X_1,X_2,X_3) = e^{-(X_1+X_2+X_3)} \qquad \text{for } 0 < X_i < \infty;\ i = 1, 2, 3$$

then the marginal distribution of x_1 is obtained by integrating $f(X_1,X_2,X_3)$ over the variables X_2, X_3; that is, the marginal of x_1 is

$$g_1(X_1) = e^{-X_1} \qquad 0 < X_1 < \infty$$

A marginal distribution is also a frequency function. A conditional distribution of a certain subset of the chance variables $x_1, x_2, \ldots, x_k$ is the joint distribution of this subset under the condition that the remaining variables are given certain values. The conditional distribution of $x_1, x_2, \ldots, x_p$, given $X_{p+1}, X_{p+2}, \ldots, X_k$, is

$$f(X_1,X_2,\ldots,X_p \mid X_{p+1},X_{p+2},\ldots,X_k) = \frac{f_1(X_1,X_2,\ldots,X_k)}{g_1(X_{p+1},X_{p+2},\ldots,X_k)}$$

if $g_1(X_{p+1},\ldots,X_k) \neq 0$.

For example, in the above illustration,

$$f(X_1 \mid X_2,X_3) = \frac{f_1(X_1,X_2,X_3)}{g_1(X_2,X_3)} = \frac{e^{-(X_1+X_2+X_3)}}{\int_0^{\infty} e^{-(X_1+X_2+X_3)}\,dX_1}$$

$$= \frac{e^{-(X_1+X_2+X_3)}}{e^{-(X_2+X_3)}} = e^{-X_1} \qquad 0 < X_1 < \infty$$

◆ **Definition 2.1** Two random variables x_1 and x_2 are said to be statistically independent if and only if the conditional distribution of x_1, given x_2, equals the marginal distribution of x_1, i.e., if

$$f_1(X_1 \mid X_2) = g_1(X_1)$$

This can be extended to

◆ **Definition 2.2** A set of random variables $x_1, x_2, \ldots, x_k$ is said to be jointly statistically independent if and only if the joint frequency function equals the product of the marginal frequency functions; i.e., if

$$f(X_1, X_2, \ldots, X_k) = f_1(X_1) f_2(X_2) \cdots f_k(X_k)$$

where $f_i(X_i)$ is the marginal distribution of x_i.

We note that, if two random variables x_1 and x_2 are independent, then the covariance of x_1 and x_2 ,written $\text{cov}(x_1, x_2)$, is zero.

Suppose a random variable has a frequency function $f(X)$. Suppose we desire to sample, say, k items from this frequency function. The value of the k items will be denoted by $X_1, X_2, \ldots, X_k$, where the subscript refers to the sample numbers. In future work we shall use the same symbols for random variables and for values of random variables.

◆ **Definition 2.3** If $X_1, X_2, \ldots, X_k$ is a random sample of size k from a population with density function $f(X)$, the joint density of the X_i is $f(X_1) f(X_2) \cdots f(X_k)$.

In many applied problems where statistics is used, a random sample is selected from a population in which the form of the frequency function is assumed to be known, and from the joint distribution of the random variables a distribution of a function of the random variables is derived. For example, if a sample of size k is drawn from a frequency function $f(X)$, it might be desirable to obtain the distribution of the mean $\bar{X}$ of the k observations. The theory of derived distributions is extremely important. There are many methods of deriving the distributions of functions of random variables, such as the geometric method, the use of transformations, the method of the moment-generating function, and many others. We shall assume that the student is acquainted with the more elementary methods and with some of the important derived distributions. Some of the important distributions will be given below.

Normal Distribution. One of the most important and most frequently used distributions in statistics is the *normal distribution.* A chance variable y is said to be distributed as the univariate normal distribution with mean μ and variance σ^2 if the frequency function of y is given as

$$f(y) = \frac{1}{\sqrt{2\pi\sigma^2}} e^{-(y-\mu)^2/2\sigma^2} \qquad -\infty < y < \infty$$

From this basic distribution many important distributions are derived.

The Chi-square Distribution. If $y_1, y_2, \ldots, y_n$ are normally and independently distributed with mean 0 and variance 1, then

$$u = \sum_{i=1}^{n} y_i^2$$

is distributed as chi-square with n degrees of freedom, designated by $\chi^2(n)$. The functional form of chi-square is

$$f(u) = \frac{u^{\frac{1}{2}(n-2)}}{2^{n/2}\Gamma(n/2)} e^{-u/2} \qquad 0 < u < \infty$$

If X is distributed as $\chi^2(n)$ and Z is distributed as $\chi^2(m)$ and if X and Z are independent, then $X + Z$ is distributed as $\chi^2(m + n)$. This is called the reproductive property of chi-square and can be extended to any finite number of independent chi-square variables.

Snedecor's F Distribution. If X is distributed as $\chi^2(n)$ and Y is distributed as $\chi^2(m)$ and if X and Y are independent, then $u = (n/m)(Y/X)$ is distributed as Snedecor's F and m and n degrees of freedom, denoted by $F(m, n)$. The frequency function of the F distribution is given by

$$f(u) = \frac{\Gamma\left(\frac{m+n}{2}\right)\left(\frac{m}{n}\right)^{m/2} u^{\frac{1}{2}(m-2)}}{\Gamma\left(\frac{m}{2}\right)\Gamma\left(\frac{n}{2}\right)\left(1 + \frac{m}{n}u\right)^{\frac{1}{2}(m+n)}} \qquad 0 < u < \infty$$

Student's t Distribution. If X is distributed normally with mean 0 and variance σ^2 and if Y^2/σ^2 is distributed as chi-square with n degrees of freedom and if X and Y are independent, then $u = X\sqrt{n}/Y$ is distributed as Student's t with n degrees of freedom, denoted by $t(n)$. The frequency function is given by

$$f(u) = \frac{\Gamma\left(\frac{n+1}{2}\right)}{\sqrt{n\pi}\,\Gamma\left(\frac{n}{2}\right)\left(1 + \frac{u^2}{n}\right)^{\frac{1}{2}(n+1)}} \qquad -\infty < u < \infty$$

If $y_1, y_2, \ldots, y_n$ are independently distributed normal variables with means μ and variances σ^2, then

1. $\bar{y} = \frac{1}{n} \sum_{i=1}^{n} y_i$ is distributed normally with mean μ and variance σ^2/n.

2. $\frac{(n-1)s^2}{\sigma^2} = \sum_{i=1}^{n} \frac{(y_i - \bar{y})^2}{\sigma^2}$ is distributed as $\chi^2(n-1)$.

3. $\bar{y}$ and s^2 are independent.

The distributions described above are all *derived distributions* that evolve from the basic normal distribution. They can also be looked upon as basic distributions in their own right, but in the theory of linear models we shall regard them as derived from the normal distribution.

We shall have occasion to derive some distributions later and shall make use of the method of the moment-generating function, for which the following theorem will be useful:

◆ **Theorem 2.1** If the two frequency functions $f_1(X)$ and $f_2(X)$ have the same moment-generating function $m(t)$ and if there exists a positive number d such that $m(t)$ exists at each point in the interval $-d < t < d$, then $f_1(X)$ and $f_2(X)$ are equal.

For example, if a random variable y is distributed normally with mean μ and variance σ^2, then the moment-generating function of y is

$$m_y(t) = e^{\mu t + \frac{1}{2}\sigma^2 t^2}$$

Thus, if we have a frequency function $f(y)$ whose functional form is unknown but if we know that the moment-generating function of y is e^{-2t+6t^2}, then, since the conditions of the theorem hold, $f(y)$ will be a normal distribution with mean -2 and variance 12.

2.2 Statistical Inference

In conducting scientific investigations the investigator often knows or is willing to assume that the population from which he takes observations is of a certain functional form $f(x;\theta)$, where the parameter θ is not known but is the quantity that his investigation is attempting to describe. For example, it might be known from theoretical considerations or from previous investigations that the frequency of weights of the 100,000 inhabitants of a given city as an approximation has the functional form

$$f(x;\mu) = \frac{1}{\sqrt{2\pi}} e^{-\frac{1}{2}(x-\mu)^2} \qquad -\infty < x < \infty$$

where the average weight μ is unknown. In general, if we say that x is distributed as $f(x;\mu)$, we shall mean that μ is the unknown parameter in the frequency function $f(x)$.

A scientist selects observations from the population under study, and on the basis of these observations he tries (1) to ascertain the value of the unknown parameter θ, or (2) to decide whether θ or some function of θ is equal to some preconceived value, say, θ_0. The first procedure is known as *estimating the parameter* θ and the second as *testing a hypothesis* about the parameter θ.

In speaking about estimating a parameter θ, we consider two types of estimation: *point estimation* and *interval estimation.*

2.3 Point Estimation

In point estimation we use certain prescribed methods to arrive at a number $\hat{\theta}_0$, which we accept as the point estimate of θ. The procedure will be to select a random sample of n observations $x_1, x_2, \ldots, x_n$ from $f(x;\theta)$ and take some function of the sample values, say $\hat{\theta} = h(x_1, x_2, \ldots, x_n)$, as the "best" value (estimate) of θ that we can get from the information available. Since $\hat{\theta}$ is a function of the n random variables, it is itself a random variable, and the "best" or "closest" estimate of θ must be expressed in terms of probability. That is to say, a random sample of n observations will be drawn from the frequency function $f(x;\theta)$, and out of all the functions of n variables available we shall select one function $h(x_1,x_2, \ldots ,x_n)$ such that, if we let this function be the estimate of θ, then we shall have a "better" estimate than if we had selected any other function $g(x_1,x_2, \ldots ,x_n)$ to represent the estimate of θ.

It would be very desirable to be able to select an estimator $\hat{\theta}$ for θ such that, for every pair of numbers λ_1 and λ_2 ($0 < \lambda_1 < \infty$, $0 < \lambda_2 < \infty$),

$$P(\theta - \lambda_1 < \hat{\theta} < \theta + \lambda_2) \geqslant P(\theta - \lambda_1 < \theta^* < \theta + \lambda_2) \tag{2.1}$$

where θ^* is any other estimator for θ. That is to say, since each random sample from $f(x;\theta)$ will give rise to a different estimate, it seems that a "good" estimating function $\hat{\theta} = h(x_1,x_2, \ldots ,x_n)$ would be a function such that a large percentage of the samples produce estimates that fall very close to θ. In other words, it seems desirable to have an estimator $\hat{\theta}$ such that the probability of the estimate falling in the interval $\theta - \lambda_1$ to $\theta + \lambda_2$ is larger than the probability of any other estimate of θ falling in that interval. Although we should like to have an estimator that satisfied (2.1), for practical situations methods for finding such functions do not exist. Therefore, we shall enumerate various other properties that we think are desirable in an estimator. On the basis of the nature of each specific scientific investigation, we must select those properties which seem the most reasonable substitutes for condition (2.1) from among those properties which are attainable in the situation at hand. In many cases there is controversy about which criteria should be preferred.

Important desirable properties for estimators include: (1) sufficiency, (2) unbiasedness, (3) consistency, (4) efficiency, (5) minimum variance, (6) completeness, and (7) invariance under transformation. These will be discussed briefly.

Sufficiency. Let $f(x_1, \ldots, x_n;\ \theta_1, \ldots, \theta_k)$ be a joint frequency function involving k parameters. The statistics $\hat{\theta}_1 = h_1(x_1, \ldots, x_n)$, $\hat{\theta}_2 = h_2(x_1, \ldots, x_n), \ldots, \hat{\theta}_m = h_m(x_1, x_2, \ldots, x_n)$ are a *set of sufficient statistics* if $g(x_1, x_2, \ldots, x_n \mid \hat{\theta}_1, \ldots, \hat{\theta}_k)$ is independent of the parameters θ_i, where $g(x_1, \ldots, x_n \mid \hat{\theta}_1, \ldots, \hat{\theta}_m)$ is the conditional density function of the observations, given the statistics.

We see that in any problem there may be many sets of sufficient statistics. The original observations $x_1, x_2, \ldots, x_n$ always form a set of sufficient statistics, but we generally want a smaller set since this will be easier to work with. We shall then be interested in a "smallest" set, which is referred to as a *minimal sufficient set.* A set of sufficient statistics $\hat{\theta}_1, \hat{\theta}_2, \ldots, \hat{\theta}_m$ is a minimal sufficient set if m is the smallest number of statistics that satisfy the definition of sufficiency.

Next we shall give a criterion to aid us in finding a set of sufficient statistics.

◆ **Theorem 2.2** Assume that the range of the x_i's in the frequency function is independent of the parameters (it will always be independent of the parameters for the problems in this book). If the frequency function $f(x_1, \ldots, x_n;\ \theta_1, \ldots, \theta_k)$ factors into the product of two functions such that

$$f(x_1, \ldots, x_n; \theta_1, \ldots, \theta_k) = f_1(x_1, \ldots, x_n) f_2(\hat{\theta}_1, \ldots, \hat{\theta}_m;\ \theta_1, \ldots, \theta_k)$$

where $f_1(x_1, \ldots, x_n)$ does not contain the parameters, then $\hat{\theta}_1, \hat{\theta}_2, \ldots, \hat{\theta}_m$ is a set of sufficient statistics.

Unbiasedness. An estimator $\hat{\theta}$ is said to be an *unbiased estimator of* θ if $E(\hat{\theta}) = \theta$, where E denotes mathematical expectation.

Thus, if we have a great many unbiased estimates of θ, then the average of these estimates equals the parameter θ. This criterion is quite important in our present system of scientific investigation and reporting, where many observers report estimates of the same parameter. These can then be combined into one "good" estimate if they are unbiased. In many cases, however, the concept of unbiasedness leads to very poor estimates. For example, a sample might be selected from the population $f(x;\theta)$ to estimate θ where it is known that

$\theta \geqslant 0$. If one hunts for an unbiased estimate $\hat{\theta}$, it may be that some values of $\hat{\theta}$ will be negative. If one does arrive at a negative estimate of θ under these conditions it would obviously be better to take $\hat{\theta}_0 = 0$. However, if a great many estimates are to be obtained, it is also wise to report the unbiased estimate of θ even though it is an impossible result in itself.

Consistency. An estimator $\hat{\theta}_n$ (more accurately, a sequence of estimators $\{\hat{\theta}_n\}$) is said to be *consistent for* θ if the limit of the probability that $|\hat{\theta}_n - \theta| < \epsilon$ is 1 as n approaches infinity. $\hat{\theta}_n$ is the estimate of θ based on a sample of size n.

When large-size samples are used to estimate θ, consistency seems to be an important property for an estimator to possess. However, if a relatively small number of observations is used, the concept of consistency is not too important unless the above limit gets close to unity while n is relatively small.

Efficiency. An estimator $\hat{\theta}_n$ (more accurately, a sequence of estimators $\{\hat{\theta}_n\}$) is said to be *efficient* when the following two conditions are satisfied:

1. $\sqrt{n}(\hat{\theta}_n - \theta)$ is asymptotically normally distributed with mean 0 and variance, say, σ^2, where n is the sample size.

2. The variance σ^2 is less than the variance of any other estimator θ_n^* that satisfies condition 1.

We see that the concept of efficiency is a large-sample concept.

Minimum Variance. An estimator $\hat{\theta}$ is said to be a *minimum-variance estimator of* θ if $E[\hat{\theta} - E(\hat{\theta})]^2 \leqslant E[\theta^* - E(\theta^*)]^2$, where θ^* is any other estimator for θ.

If an estimator is efficient, then it is consistent and unbiased in the limit but need not be unbiased for finite sample sizes. An unbiased estimator is not necessarily consistent. The concept of minimum variance is not a large-sample concept.

Completeness. An estimator $\hat{\theta}$ is *complete* if there exists no unbiased estimator of zero in the frequency function $f(\hat{\theta};\theta)$ (except zero itself).

Invariance. An estimator $\hat{\theta}$ of θ is said to be an *invariant estimator for a certain class of transformations* g if the estimator is $g(\hat{\theta})$ when the transformation changes the parameter to $g(\theta)$.

This principle of an invariance estimator seems very reasonable. However, because of the advanced nature of the mathematics involved, we shall not do much with this principle of estimation in this book.

The following theorem will aid in finding minimum-variance unbiased estimators.

◆ **Theorem 2.3** If $\hat{\theta}$ is a sufficient statistic for θ, if $\hat{\theta}$ is complete (no function of the sufficient statistic for $\hat{\theta}$ is an unbiased estimate of zero except zero itself), and if $E[g(\hat{\theta})] = h(\theta)$, then $g(\hat{\theta})$ is the unique minimum-variance unbiased estimate of $h(\theta)$ for every size of sample.

For a full discussion of these criteria the reader is referred to the references in the bibliography.

Even if one has decided upon "good" criteria for estimators to have, one must still devise methods for arriving at estimating functions $h(x_1,x_2, \ldots ,x_n)$ that meet these criteria. There are many important methods, two of which we shall discuss, since they will be used in later chapters: (1) the method of maximum likelihood, and (2) the method of least squares.

Maximum Likelihood. The *likelihood function* for a random sample of n observations from the frequency function $f(x;\mu)$ is defined as the joint frequency function of the n variables, which is equal to

$$L(x_1,x_2, \ldots ,x_n; \mu) = f(x_1;\mu)f(x_2;\mu) \cdots f(x_n;\mu)$$

since the x_i are independent. The *maximum-likelihood estimator of* μ is defined as the value μ^* such that

$$L(x_1,x_2, \ldots ,x_n; \mu^*) \geqslant L(x_1,x_2, \ldots ,x_n; \mu)$$

That is, the maximum-likelihood estimate of μ is the value, say, μ^*, that maximizes the likelihood function when it is considered as a function of μ.

Intuitively, this method of estimation seems to be a good method. Under certain quite general conditions the maximum-likelihood method gives rise to estimators that are consistent, efficient, and sufficient. It is important to note, however, that, if the method of maximum likelihood is to be used, the form of the frequency function must be known.

An important property of the maximum-likelihood method of estimation that will sometimes be helpful is given in the next theorem.

◆ **Theorem 2.4** Let $\hat{\theta}_1, \hat{\theta}_2, \ldots, \hat{\theta}_k$ be the respective maximum-likelihood estimates of $\theta_1, \theta_2, \ldots, \theta_k$. Let $\alpha_1 = \alpha_1(\theta_1, \ldots ,\theta_k)$, $\ldots, \alpha_k = \alpha_k(\theta_1, \ldots ,\theta_k)$ be a set of transformations that are one-to-one. Then the maximum-likelihood estimates of $\alpha_1, \ldots,$ α_k are $\hat{\alpha}_1 = \alpha_1(\hat{\theta}_1,\hat{\theta}_2, \ldots ,\hat{\theta}_k), \ldots, \hat{\alpha}_k = \alpha_k(\hat{\theta}_1,\hat{\theta}_2, \ldots ,\hat{\theta}_k)$.

This property will be referred to as the *invariant property* of maximum-likelihood estimators.

Least Squares. In the method of least squares, the observations $x_i (i = 1, 2, \ldots, n)$ are assumed to be of the form

$$x_i = f_i(\theta_1,\theta_2, \ldots ,\theta_k) + e_i$$

where $f_i(\theta_1,\theta_2, \ldots ,\theta_k)$ is some known function of the unknown parameters and where the e_i are random variables. The sum of squares

$$\sum_{i=1}^{n} e_i^2 = \sum_{i=1}^{n} [x_i - f_i(\theta_1,\theta_2, \ldots ,\theta_k)]^2$$

is formed, and the value of the θ_i that minimizes this sum of squares is the least-squares estimator for the θ_i.

In the method of least squares, the form of the frequency function need not be known, but the method can be used if it is known. In many important cases, even if the form of the frequency function is unknown, the method of least squares gives rise to estimators that are unbiased and consistent and under certain conditions minimum-variance unbiased. Also, in at least one very important case, the method of least squares and the method of maximum likelihood give rise to the same estimators. These cases will be discussed in greater detail in a later chapter.

2.4 Interval Estimation

In point estimation the procedure is to select a function of the sample random variables that will "best" represent the parameter being estimated. In scientific studies it is generally not essential to obtain the *exact* value of a parameter under investigation. For example, in ascertaining the tensile strength of wire it may well be that a knowledge of the exact average tensile strength is not necessary but that a value within, say, $\frac{1}{2}$ lb of the true average tensile strength will be adequate. It is desirable, however, to have some confidence that the value obtained is within the desired limits. A point estimate will not do this. This suggests using an interval estimate.

A *confidence interval* is a random interval whose end points f_1 and f_2 $(f_1 < f_2)$ are functions of the observed random variables such that the probability that the inequalities $f_1 < \theta < f_2$ are satisfied is a predetermined number $1 - \alpha$. In the above formulation, θ is the parameter whose value is desired, and $1 - \alpha$ is generally taken as .80, .90, .95, etc. The statement is written

$$P(f_1 < \theta < f_2) = 1 - \alpha \tag{2.2}$$

The frequency interpretation is: If the same functions but different sets of observed random variables are used (each set gives a different interval), then, on the average, $1 - \alpha$ of all the intervals will cover the parameter θ. Even though we observe only one interval, we have a confidence level of $1 - \alpha$ that this interval covers the unknown parameters θ. There may be many functions f_1 and f_2 that will give rise to a confidence interval on θ with coefficient $1 - \alpha$. The question is: Of all pairs of functions that satisfy (2.2), which pair of functions gives the "best" interval? It seems that in a certain sense the "best" interval is the *shortest* interval. Unfortunately, there almost never exist functions that give uniformly shortest confidence intervals for every value of the unknown parameter θ.

Therefore, it may be reasonable to minimize not d, but some function of the length d of the interval. For example, since the length $d = f_2 - f_1$ is a random variable, we may like to use functions in the confidence interval such that the average length $E(d)$ is a minimum for all θ. However, certain functions may give rise to an average length that is minimal for some θ but not for other θ; in other words, there may not exist functions that give rise to confidence intervals uniformly smallest in expected length.

We shall illustrate the above ideas with an example. Suppose we have N sample values $X_1, X_2, \ldots, X_N$ from a normal population with unknown mean μ and variance 1. Suppose we wish to set a confidence interval on μ with confidence coefficient of .95. We shall use $\bar{X}$ to obtain this confidence interval. Now $\bar{X}$ is distributed normally with mean μ and variance $1/N$. Hence, if α_1 and α_2 are any two numbers such that

$$\int_{\alpha_1}^{\alpha_2} \frac{1}{\sqrt{2\pi}} e^{-y^2/2}\, dy = .95$$

then

$$\bar{X} - \frac{\alpha_2}{\sqrt{N}} \leqslant \mu \leqslant \bar{X} - \frac{\alpha_1}{\sqrt{N}}$$

is a confidence interval on μ with confidence coefficient of .95. We see that

$$f_1 = \bar{X} - \frac{\alpha_2}{\sqrt{N}}$$

$$f_2 = \bar{X} - \frac{\alpha_1}{\sqrt{N}}$$

and the length of the interval is

$$d = f_2 - f_1 = \frac{\alpha_2 - \alpha_1}{\sqrt{N}}$$

We want to choose constants α_1 and α_2 that satisfy the above integral and make d a minimum. Clearly, they must be such that $-\alpha_1 = +\alpha_2 = 1.96$. In this example the expected length $E(d)$ is also $(\alpha_2 - \alpha_1)/\sqrt{N}$, but in general this is not true.

Since uniformly smallest-width confidence intervals almost never exist and since confidence intervals with uniformly smallest expected width almost never exist, we shall list other desirable properties for confidence intervals to have. The experimenter must choose those properties which he feels will be most useful in a given situation. It is desirable for interval estimators to be: (1) minimum-length, (2) minimum-expected-length, (3) shortest (in the Neyman-Pearson sense), (4) unbiased.

The reader should keep in mind that the above concepts have meaning within the framework of a given sample. That is to say, the experimenter chooses one criterion over another in order to get the most information from the sample at hand. If he has some control over his observations he will, as far as possible, pick those which will give a confidence interval with small width. The desirable thing to do is to decide how long he wants his confidence interval to be in order to make a decision about the parameter under observation. He then selects his samples in such a way as to attain the given length if possible. A confidence interval that is too long does not give him the necessary precision; an interval shorter than necessary wastes information and resources. We shall say more about this when we discuss the design of experiments. However, it is important for the researcher to realize that there are two very important questions to remember in interval estimation: (1) Does the interval cover the parameter? (2) Is the interval narrow enough? In an experiment we want the probability of a yes answer to each of these questions to be as high as possible.

Just as in point estimation, so in interval estimation sets of sufficient statistics play an important part. The principle involved is given in the next theorem.

◆ **Theorem 2.5** Let $x_1, \ldots, x_n$ be a random sample from the density $f(x;\theta)$. Let g_1, g_2 be two functions of the x_i such that $P(g_1 < \theta < g_2) = 1 - \alpha$. Let $t = t(x_1, \ldots, x_n)$ be a sufficient statistic. There exist two functions f_1 and f_2 of t such that $P(f_1 < \theta < f_2) = 1 - \alpha$, and such that the two random variables (lengths) $l_1 = g_2 - g_1$ and $l_2 = f_2 - f_1$ have the same distribution.

This says that, so far as length of confidence intervals is concerned, one might as well use sufficient statistics.

2.5 Testing Hypotheses

One aspect of the scientific method is the testing of scientific hypotheses. At this point in the scientific method, statistics, especially the method of analysis of variance, plays an important role. For example, an investigator sets up a hypothesis, which we shall call H_0, concerning a phenomenon in nature. This hypothesis may be arrived at in many ways, including theoretical considerations and empirical observations. The hypothesis H_0 is set up; then the researcher conducts an experiment and takes observations to see how closely they conform to the hypothesis, and on the basis of these observations he decides to accept H_0 as true or to reject H_0 as false. It should be pointed out that a *general* hypothesis can never be proved but can be disproved. For example, one might try to prove the hypothesis that one can always draw an ace on the first draw from an ordinary deck of cards. No matter how often an investigator performs this feat, he can never prove the hypothesis. One failure, however, will disprove it. On the other hand, if a subject does pull an ace a great many times without failure, we might be willing to say that, for all practical purposes, our subject can always select an ace on the first draw—but the hypothesis as it was set up has not been proved.

As another example, suppose one has a new process of manufacture that one hopes will give added length of life to electron tubes. An investigator may form the hypothesis that tubes manufactured by this process will have a longer life than tubes manufactured by a standard process. He proceeds to test this hypothesis by observing the length of life of a certain number of tubes of each manufacturing method. Clearly he can never prove his hypothesis, but he can disprove it. However, if a great number of tubes manufactured by the new process each have a longer life than tubes manufactured by the standard process, he may be willing to accept the new process, but he cannot prove the hypothesis as it was formed.

One may form rules about when a hypothesis will be accepted and when it will be rejected. In the example above, one might use the following formula: Accept the hypothesis if 1,000 electron tubes are examined from each manufacturing process and if 950 or more of those manufactured by the new process have longer life than the average of the 1,000 manufactured by the standard method. Here the accepting or rejecting of a hypothesis is identified with the experimenter's willingness to accept or reject the new manufacturing process and not with proof or disproof of the hypothesis.

Testing scientific hypotheses is much more complicated than has been illustrated here. It is not our intention to go into great detail concerning the philosophical aspects of the problem, but our intention is to point out that in testing hypotheses one uses inductive reasoning—reasoning from the particular to the general. That is to say, from a small set of observations one draws conclusions about what would happen if the totality of observations were taken, where the number of observations is probably unlimited and many of them are in the future. Thus the scientific hypothesis is concerned with taking a relatively few observations and predicting what will happen when future observations are taken. This type of reasoning—from the particular to the general—cannot be made with certainty, and probability measures its uncertainty.

When the hypothesis H_0 is formed and the experimenter takes observations and uses them as a basis for either accepting or rejecting H_0, he is liable to two kinds of error, as follows:

1. The hypothesis H_0 might be true; yet on the basis of his observations the experimenter might reject H_0 and say that it is false. This is generally referred to as *error of the first kind* or *type I error*; briefly, it is the error of rejecting a true hypothesis.

2. The hypothesis H_0 might be false; yet on the basis of his observations the experimenter might accept H_0 as true. This is generally referred to as *error of the second kind* or *type II error*; briefly, it is the error of accepting a false hypothesis.

We should like to minimize the likelihood or probability of making either of these two errors. However, in general, for a fixed number of observations, if we decrease the probability of making an error of one type, we increase the chances of making the other. $P(\text{I})$ and $P(\text{II})$ will represent the probability of making an error of type I and type II, respectively. We shall generally use the power of the test instead of the probability of type II error where

$$\beta = \text{power of the test} = 1 - P(\text{II})$$

Since it is desirable for $P(\text{II})$ to be as small as possible, we shall want β to be as large as possible. The power of a test $\beta(\theta^*)$, where θ is the parameter under test, is the probability of rejecting the hypothesis H_0: $\theta = \theta_0$ when the parameter θ actually equals θ^*.

In most scientific investigations it seems desirable for the probability of rejecting the hypothesis $\theta = \theta_0$ to increase as the distance between θ and θ_0 increases. That is to say, if the true value θ of the parameter is close to θ_0, it may not be too disastrous if the test procedure fails to reject H_0. However, if the true value of the parameter is not close but quite distant from θ_0, we would want the probability of rejecting H_0 to be relatively large. The power of the test is a function of the parameter as denoted by $\beta(\theta)$, and it is clear that $0 \leqslant \beta(\theta) \leqslant 1$. It might be pointed out that, if the hypothesis H_0 is $\theta = \theta_0$, then $\beta(\theta_0) = P(\text{I})$.

The procedure for testing a hypothesis H_0: $\theta = \theta_0$ about a parameter θ in the frequency function $f(x;\theta)$ is as follows:

1. A random sample of values is taken from $f(x;\theta)$, and some function of these values is formed, say, $\hat{\theta} = g(x_1,x_2, \ldots ,x_n)$.

2. The distribution of $\hat{\theta}$, say, $h(\hat{\theta};\theta)$, is found. This distribution will not be independent of θ.

3. The $\hat{\theta}$ axis is divided into two regions R_r and R_N, where R_r is called the region of rejection, and is such that $P(\hat{\theta} \text{ is in } R_r \mid \theta = \theta_0) = P(\text{I})$. That is to say, the probability that $\hat{\theta}$ is in R_r when H_0 is true is equal to the probability of the type I error. This is expressed by

$$P(\text{I}) = \int_{R_r} h(\hat{\theta};\theta_0)\, d\hat{\theta}$$

4. The power of the test $\beta(\theta)$ is expressed as

$$\beta(\theta^*) = \int_{R_r} h(\hat{\theta};\theta^*)\, d\hat{\theta}$$

and is the probability that $\hat{\theta}$ falls in the region R_r when $\theta = \theta^*$.

Of course, we should like to minimize $P(\text{I})$ and maximize $\beta(\theta)$. In general it can be shown for a fixed size of sample that, when $\beta(\theta)$ increases, $P(\text{I})$ increases, and vice versa. Therefore, the general procedure is to choose $P(\text{I})$ in advance and then, from all functions $\hat{\theta}$ and all regions R_r such that $P(\text{I})$ is fixed, choose a particular function and region that maximize $\beta(\theta)$.

In examining the "goodness" properties of a test function we need to examine only the power function. If the power function of a given test is $\beta(\theta)$ and if $\beta(\theta) \geqslant \beta_1(\theta)$, where $\beta_1(\theta)$ is the power function of any other test function, then we say the test function corresponding to $\beta(\theta)$ is *uniformly most powerful*. Unfortunately, a uniformly most powerful test function exists only in very few cases; $\beta(\theta)$ may be greater than any other power function for some values of θ but not for others.

To examine these ideas further we shall assume that we have a sample of size n from the distribution $f(x;\ \theta_1,\theta_2)$. The ideas will hold for a function of s parameters $\theta_1, \theta_2, \ldots, \theta_s$, but we shall discuss the function for $s = 2$. We consider a parameter space designated by Ω, in this case two-dimensional, that is, θ_1 and θ_2. This space contains all the values that it is possible for θ_1 and θ_2 to attain. If we wish to test a certain hypothesis H_0, this means that we are going to make a decision on the basis of observations of whether the parameters occur within a certain region ω of the parameter space Ω; ω is called the *parameter space under the null hypothesis* H_0. If ω is a single point, the hypothesis is called *simple*; otherwise it is called *composite*. The test function is a function of the observations, call it $\phi(x)$, which is a rule for stating whether the parameters are in ω or in $\Omega - \omega$.

Of course we should like to obtain a test that is uniformly most powerful, but, since such a test rarely exists, we may restrict our class of tests and find a uniformly most powerful test in this smaller class.

The experimenter should become well acquainted with the idea of a power function in testing hypotheses.

We need a method for constructing tests that has some optimum properties. The method we shall discuss is based on the likelihood-ratio principle. Suppose we have a frequency function $f(x;\ \theta_1,\theta_2)$. Suppose a sample of n values $x_1, x_2, \ldots, x_n$ is taken and that the likelihood function $L(x_1,x_2, \ldots ,x_n;\ \theta_1,\theta_2)$ is formed. Suppose we desire to test the hypothesis H_0: θ_1 and θ_2 are in ω and that the alternative hypothesis is H_1: θ_1 and θ_2 are in $\Omega - \omega$. We form the ratio

$$L = \frac{L(\hat{\omega})}{L(\hat{\Omega})}$$

where $L(\hat{\Omega})$ is the maximum of L with respect to θ_1 and θ_2 subject to the condition that θ_1 and θ_2 are in Ω. $L(\hat{\omega})$ is the maximum of L with respect to θ_1 and θ_2 subject to the condition that θ_1 and θ_2 are in ω. If $L(\hat{\omega})$ is close to $L(\hat{\Omega})$, we expect H_0 to be true and we have no reason to reject H_0; if, however, $L(\hat{\omega})$ is quite distant from $L(\hat{\Omega})$, we expect H_0 to be false and we reject H_0. In other words we reject H_0 if L is small and not if L is large. Clearly, L is such that $0 \leqslant L \leqslant 1$. So we need to find a constant A and reject H_0 if our observed L is less than or equal to A, where A is so chosen that the probability of type I error is α. To do this we need to find the frequency distribution of L when H_0 is true. Let this be $g(L;\mathrm{H}_0)$, with $0 \leqslant L \leqslant 1$. Then let A be such that

$$P(\mathrm{I}) = \int_0^A g(L;\mathrm{H}_0)\, dL = \alpha$$

The power of the test is

$$\beta(\mathrm{H}_1) = \int_0^A g(L;\mathrm{H}_1)\, dL$$

Let $y = h(L)$, where $h(L)$ is a monotonic function of L. Then, if we change variables from L to y, we get [let $L = f(y)$]

$$\alpha = \int_0^A g(L;\mathrm{H}_0)\, dL = \int_{h(0)}^{h(A)} g[f(y)]\, |f'(y)|\, dy$$

This states that we can use any monotonic function of the likelihood-ratio test as an equivalent test function; i.e., we shall reject or not reject using y when we reject or do not reject using L.

The likelihood-ratio test in many cases has desirable properties as follows:

1. If a uniformly most powerful test exists, it is often given by the likelihood-ratio test.

2. $-2 \log L$ is approximately distributed as a chi-square variable if the sample size is large.

3. For most of the models we shall discuss, the likelihood ratio will give a test that has reasonably "good" properties.

4. The likelihood test is always a function of sufficient statistics. The importance of property 4 will be given in Theorem 2.6.

We shall illustrate with an example. Suppose a random sample of size n is selected from a normal population with mean μ and variance σ^2. Let the sample value be $x_1, x_2, \ldots, x_n$. Suppose we desire to test the hypothesis H_0: $\mu = 0$ and that the alternative is $\mu \neq 0$. Now the points in the Ω space are all points μ, σ^2 such that $-\infty < \mu < \infty$, $0 \leqslant \sigma^2 < \infty$. The points in the ω space are all points μ, σ^2 such that $\mu = 0$, $0 \leqslant \sigma^2 < \infty$. Since the ω space is not just a single point, the hypothesis is composite. The likelihood is

$$\frac{1}{(\sigma^2 2\pi)^{n/2}} \exp\left[-\frac{1}{2\sigma^2}\Sigma(x_i - \mu)^2\right]$$

Clearly, $$L(\hat{\Omega}) = \frac{1}{(2\pi\hat{\sigma}^2)^{n/2}} \exp\left[-\frac{1}{2\hat{\sigma}^2}\Sigma(x_i - \bar{x})^2\right] = \frac{e^{-n/2}}{(2\pi\hat{\sigma}^2)^{n/2}}$$

where $$\hat{\sigma}^2 = \frac{1}{n}\Sigma(x_i - \bar{x})^2$$

In the ω space, $\mu = 0$; so the only quantity we need to maximize with respect to is σ^2. Clearly,

$$L(\hat{\omega}) = \frac{n^{n/2}}{(2\pi\Sigma x_i^2)^{n/2}} e^{-n/2}$$

So $$L = \left[\frac{\Sigma(x_i - \bar{x})^2}{\Sigma x_i^2}\right]^{n/2}$$

$$= \left[\frac{\Sigma(x_i - \bar{x})^2}{\Sigma(x_i - \bar{x})^2 + n\bar{x}^2}\right]^{n/2}$$

$$= \left[\frac{1}{1 + n\dfrac{\bar{x}^2}{\Sigma(x_i - \bar{x})^2}}\right]^{n/2}$$

$$= \left[\frac{1}{1 + u/(n-1)}\right]^{n/2}$$

where $$u = \frac{n(n-1)\bar{x}^2}{\Sigma(x_i - \bar{x})^2}$$

Now L is a monotonic decreasing function of u; hence we can use u as our test function. Now the critical region of L is $0 \leqslant L \leqslant A$. This gives $c \leqslant u < \infty$ as the critical region for u, where

$$c = \left(\frac{1}{A^{2/n}} - 1\right)(n-1)$$

Also, we see that u is distributed as $F(1, n-1)$ if H_0 is true.

We see that the test function in the above example is based on the sufficient statistics $\bar{x}$, $\Sigma(x_i - \bar{x})^2$. We shall now state an important theorem concerning the power of a test and a set of sufficient statistics.

◆ **Theorem 2.6** Suppose we desire to test a hypothesis H_0 about the parameters in the frequency function $f(x;\ \theta_1, \ldots, \theta_s)$. For any test function $\phi(x)$, there exists a test function based on the set of sufficient statistics that has a power function equivalent to the power function of the original test function $\phi(x)$.

The discussion above has been based on the assumption that the sample size is fixed. That is to say, for a given set of observations, an experimenter should use the "best" test function available to test the hypothesis H_0.

However, if the experimenter has some control over his observations, he should take them in such a manner as to satisfy the following two conditions:

1. The probability of a type I error should be equal to a preassigned number α.

2. The power of the test should be equal to a preassigned number β, when the true value of the parameter is larger than specified.

For example, if a researcher is testing the hypothesis H_0 that two varieties of wheat yield the same, he may not want to say that they yield differently unless the difference is at least 5 bushels/acre. On the other hand, if the difference in yield is 5 bushels/acre or more, he may then want the probability of rejecting H_0 to be quite high; i.e., he may want β to be close to 1.

We shall illustrate some of these ideas with an example. Suppose a random variable X is distributed normally with mean θ and variance 1. We wish to test H_0: $\theta = 0$. A random sample of size n is selected, and the sample mean $\bar{X}$ is used as the test function. Therefore, the distribution of $\bar{X}$ when $\theta = 0$ must be found. Under these conditions $\bar{X}$ is distributed normally with mean 0 and variance $1/n$. We shall take $P(\mathrm{I}) = .05$ and R_r as the two intervals $-\infty$ to $-\alpha_0$ and α_0 to $+\infty$. We have

$$.05 = P(\mathrm{I}) = \int_{-\infty}^{-\alpha_0} \frac{\sqrt{n}}{\sqrt{2\pi}} e^{-\frac{1}{2}n\bar{X}^2}\, d\bar{X} + \int_{\alpha_0}^{\infty} \frac{\sqrt{n}}{\sqrt{2\pi}} e^{-\frac{1}{2}n\bar{X}^2}\, d\bar{X}$$

which can be written more concisely

$$P(\mathrm{I}) = 1 - \int_{-1.96/\sqrt{n}}^{1.96/\sqrt{n}} \frac{\sqrt{n}}{\sqrt{2\pi}} e^{-\frac{1}{2}n\bar{X}^2}\, d\bar{X}$$

Also,

$$\beta(\theta) = 1 - \int_{-1.96/\sqrt{n}}^{1.96/\sqrt{n}} \frac{\sqrt{n}}{\sqrt{2\pi}} e^{-\frac{1}{2}n(\bar{X}-\theta)^2}\, d\bar{X}$$

If we let $Y = (\bar{X} - \theta)\sqrt{n}$, we get

$$\beta(\theta) = 1 - \int_{-1.96-\theta\sqrt{n}}^{1.96-\theta\sqrt{n}} \frac{1}{\sqrt{2\pi}} e^{-Y^2/2}\, dY$$

Thus it is clear that, as θ increases, $\beta(\theta)$ also increases, and the same situation holds for increasing n.

2.6 Conclusion

The ideas presented in this chapter have induced a great deal of excellent mathematics in the past thirty years. The literature on these subjects is now vast and increasing rapidly. Many of the topics are the subject of much discussion, and, when an experimenter selects criteria to aid him in making decisions, he must settle for some degree of arbitrariness. Many important topics have been omitted here, and those presented have been reviewed very briefly.

Problems

2.1 If y_1 and y_2 are independent normal variables with means 0 and variance 1, find:

(a) The joint distribution of y_1 and y_2.

(b) The conditional distribution of y_1, given $\bar{y}$, where $\bar{y} = (y_1 + y_2)/2$.

2.2 If the joint distribution of y_1 and y_2 is

$$f(y_1, y_2) = Cy_1y_2 \qquad 0 \leqslant y_1 \leqslant y_2 \leqslant 1$$

where C is a constant,

(a) Find C.

(b) Find the marginal distribution of y_1.

(c) Find the conditional distribution of y_1, given y_2.

2.3 If x_i ($i = 1, 2, \ldots, n$) are normally and independently distributed with mean μ and variance σ^2 and if Z is distributed normally with mean 0 and variance 1 and if x_i and Z are jointly independent, find the distribution of

(a) $\dfrac{1}{\sigma^2}\sum_{i=1}^{n}(x_i - \mu)^2$

(b) $\dfrac{1}{\sigma^2}\sum_{i=1}^{n}(x_i - \mu)^2 + Z^2$

(c) $\sum_{i=1}^{n}\dfrac{(x_i - \mu)^2}{nZ^2\sigma^2}$

2.4 Find the moment-generating function of $\sum_{i=1}^{n}\dfrac{(x_i - \mu)^2}{\sigma^2}$.

2.5 If a random sample of size n is taken from the distribution e^{-X}, where $X \geqslant 0$, find the moment-generating function of the mean of the n observations. Using Theorem 2.1, find the distribution of the mean of the n observations.

2.6 If $y_1, \ldots, y_n$ are independent and each is distributed $N(\mu,\sigma^2)$, find the maximum-likelihood estimates of μ and σ^2.

2.7 In Prob. 2.6, show that the estimates are sufficient statistics.

2.8 In Prob. 2.6, find the likelihood-ratio test of H_0: $\mu = 0$ with the alternative hypothesis $\mu \neq 0$.

2.9 In Prob. 2.8, show that the likelihood-ratio test is equivalent to the conventional t test.

2.10 In Prob. 2.6, it can be shown that $\bar{x}$ and s^2 are sufficient complete statistics. Find the minimum-variance unbiased estimate of $\mu + 3\sigma^2$.

2.11 In Prob. 2.6, set a 95 per cent confidence interval on μ.

2.12 In Prob. 2.11, find the width and the expected width of the interval.

2.13 In Prob. 2.6, find the likelihood-ratio test for testing H_0: $\sigma^2 = 5$ with the alternative H_A: $\sigma^2 \neq 5$.

Further Reading

1 A. M. Mood: "Introduction to the Theory of Statistics," McGraw-Hill Book Company, Inc., New York, 1950.

2 P. G. Hoel: "Introduction to Mathematical Statistics," John Wiley & Sons, Inc., New York, 1956.

3 M. G. Kendall: "The Advanced Theory of Statistics," Charles Griffin & Co., Ltd., London, vols. I, II, 1946.

4 H. Cramér: "Mathematical Methods of Statistics," Princeton University Press, Princeton, N.J., 1945.

5 R. L. Anderson and T. A. Bancroft: "Statistical Theory in Research," McGraw-Hill Book Company, Inc., New York, 1952.

6 T. W. Anderson: "An Introduction to Multivariate Statistical Analysis," John Wiley & Sons, Inc., New York, 1958.

7 S. S. Wilks: "Mathematical Statistics," Princeton University Press, Princeton, N.J., 1943.

8 C. R. Rao: "Advanced Statistical Methods in Biometric Research," John Wiley & Sons, Inc., New York, 1952.

9 S. Kolodziejczyk: On an Important Class of Statistical Hypotheses, *Biometrika*, vol. 27, 1935.

10 D. Blackwell: Conditional Expectation and Unbiased Sequential Estimation, *Ann. Math. Statist.*, vol. 18, pp. 105–110, 1947.

11 E. L. Lehmann and H. Scheffé: Completeness, Similar Regions and Unbiased Estimation, *Sankhyā*, vol. 10, pp. 305–340, 1950.

12 R. A. Fisher: "Statistical Methods for Research Workers," 10th ed., Table IV, Oliver & Boyd, Ltd., London, 1946.

13 R. A. Fisher: Applications of "Student's" Distribution, *Metron*, vol. 5, 1925.

14 G. W. Snedecor: "Statistical Methods," 5th ed., Iowa State College Press, Ames, Iowa, 1957.

15 "Student": Probable Error of the Mean, *Biometrika*, vol. 6, 1908.

16 R. A. Fisher: On the Mathematical Foundations of Theoretical Statistics, *Phil. Trans. Roy. Soc., London*, ser. A, vol. 222, 1922.

17 R. A. Fisher: Theory of Statistical Estimation, *Proc. Cambridge Phil. Soc.*, vol. 22, 1925.

3

The Multivariate Normal Distribution

3.1 Definition

Just as the normal distribution is one of the most important frequency functions of one variable used in applied and theoretical statistics, so the multivariate normal distribution is one of the most important frequency functions of more than one variable. In this section we shall investigate the multivariate normal distribution and discuss some of the important properties that will be needed in succeeding work. In this chapter we shall not discuss the estimating of parameters or the testing of hypotheses about parameters in the multivariate normal. We shall reserve this discussion for a later chapter.

We shall now define the multivariate normal distribution.

◆ **Definition 3.1** If $y_1, y_2, \ldots, y_p$ are p random variables and if $\mathbf{Y}$ is the $p \times 1$ vector of these variables, we shall define the function

$$f(y_1,y_2,\ldots,y_p) = Ke^{-\frac{1}{2}(\mathbf{Y}-\boldsymbol{\mu})'\mathbf{R}(\mathbf{Y}-\boldsymbol{\mu})} \qquad -\infty < y_i < \infty; \quad i = 1, 2, \ldots, p \tag{3.1}$$

to be a multivariate (p-variate) normal frequency function if the following conditions hold:

(*a*) $\mathbf{R}$ is a positive definite matrix whose elements r_{ij} are constants.

(*b*) K is a positive constant. (3.2)

(*c*) μ_i, which is the ith element of the vector $\boldsymbol{\mu}$, is a constant.

We shall now show that the function (3.1) satisfies conditions that qualify it for a frequency function, i.e.,

$$(a) \qquad f(y_1,y_2, \ldots ,y_p) \geqslant 0 \qquad \text{for all values of } y_i$$

$$(b) \qquad \int_{-\infty}^{\infty} \int_{-\infty}^{\infty} \cdots \int_{-\infty}^{\infty} Ke^{-\frac{1}{2}(\mathbf{Y}-\boldsymbol{\mu})'\mathbf{R}(\mathbf{Y}-\boldsymbol{\mu})} \, dy_1 \, dy_2 \cdots dy_p = 1 \tag{3.3}$$

Since the exponential function cannot assume negative values and since $K > 0$, condition (3.3*a*) is satisfied. We shall show that there exists a positive constant K such that condition (3.3*b*) is satisfied. First let us make the transformation from the $\mathbf{Y} - \boldsymbol{\mu}$ vector to a vector $\mathbf{Z}$ by the relation $\mathbf{Y} - \boldsymbol{\mu} = \mathbf{Z}$, which can be written

$$y_i - \mu_i = z_i \qquad i = 1, 2, \ldots , p$$

Since $dy_i/dz_j = 0$ if $i \neq j$ and $dy_i/dz_i = 1$, we see that the Jacobian of the transformation is unity. Thus we get

$$\int_{-\infty}^{\infty} \int_{-\infty}^{\infty} \cdots \int_{-\infty}^{\infty} Ke^{-\frac{1}{2}(\mathbf{Y}-\boldsymbol{\mu})'\mathbf{R}(\mathbf{Y}-\boldsymbol{\mu})} \, dy_1 \, dy_2 \cdots dy_p$$
$$= \int_{-\infty}^{\infty} \int_{-\infty}^{\infty} \cdots \int_{-\infty}^{\infty} Ke^{-\frac{1}{2}\mathbf{Z}'\mathbf{R}\mathbf{Z}} \, dz_1 \, dz_2 \cdots dz_p$$

The transformation is one to one, and the entire $\mathbf{Y}$ space transforms into the entire $\mathbf{Z}$ space, as indicated by the limits on the integrals on the right side of the equation. Since $\mathbf{R}$ is positive definite, we know by Theorem 1.31 that there exists an orthogonal matrix $\mathbf{P}$ such that

$$\mathbf{P}'\mathbf{R}\mathbf{P} = \mathbf{D}$$

where $\mathbf{D}$ is diagonal with the characteristic roots d_i of $\mathbf{R}$ (which are all positive) displayed on the diagonal. Thus we shall make the transformation from the vector $\mathbf{Z}$ to the vector $\mathbf{T} = (t_i)$ by the equation $\mathbf{Z} = \mathbf{PT}$. By Theorem 1.75, the Jacobian is unity. The range of integration on each t_i is from minus infinity to plus infinity, since an orthogonal transformation is simply a rotation. Thus,

$$\int_{-\infty}^{\infty} \int_{-\infty}^{\infty} \cdots \int_{-\infty}^{\infty} Ke^{-\frac{1}{2}\mathbf{Z}'\mathbf{R}\mathbf{Z}} \, dz_1 \, dz_2 \cdots dz_p$$
$$= K \int_{-\infty}^{\infty} \int_{-\infty}^{\infty} \cdots \int_{-\infty}^{\infty} e^{-\frac{1}{2}\mathbf{T}'(\mathbf{P}'\mathbf{R}\mathbf{P})\mathbf{T}} dt_1 \, dt_2 \cdots dt_p$$
$$= K \int_{-\infty}^{\infty} \int_{-\infty}^{\infty} \cdots \int_{-\infty}^{\infty} e^{-\frac{1}{2}\mathbf{T}'\mathbf{D}\mathbf{T}} \, dt_1 \, dt_2 \cdots dt_p$$
$$= K \int_{-\infty}^{\infty} \int_{-\infty}^{\infty} \cdots \int_{-\infty}^{\infty} \exp\left(-\tfrac{1}{2} \Sigma d_i t_i^2\right) dt_1 \, dt_2 \cdots dt_p$$

The last integral can be written as

$$\int_{-\infty}^{\infty}\int_{-\infty}^{\infty}\cdots\int_{-\infty}^{\infty} K \exp\left(-\tfrac{1}{2}\,\Sigma d_i t_i^2\right) dt_1\,dt_2\cdots dt_p$$
$$= K\int_{-\infty}^{\infty} e^{-\frac{1}{2}d_1t_1^2}\,dt_1 \int_{-\infty}^{\infty} e^{-\frac{1}{2}d_2t_2^2}\,dt_2 \cdots \int_{-\infty}^{\infty} e^{-\frac{1}{2}d_pt_p^2}\,dt_p$$

By using the frequency function of a univariate normal distribution, we see that

$$\int_{-\infty}^{\infty} e^{-\frac{1}{2}d_it_i^2}\,dt_i = \sqrt{\frac{2\pi}{d_i}}$$

Thus we get

$$K\int_{-\infty}^{\infty} e^{-\frac{1}{2}d_1t_1^2}\,dt_1 \int_{-\infty}^{\infty} e^{-\frac{1}{2}d_2t_2^2}\,dt_2 \cdots \int_{-\infty}^{\infty} e^{-\frac{1}{2}d_pt_p^2}\,dt_p$$
$$= K\sqrt{\frac{2\pi}{d_1}}\sqrt{\frac{2\pi}{d_2}}\cdots\sqrt{\frac{2\pi}{d_p}} = \frac{K(2\pi)^{p/2}}{\prod_{i=1}^{p} d_i^{\frac{1}{2}}} \tag{3.4}$$

For condition (3.3*b*) to be satisfied, we must set (3.4) equal to unity. We get

$$K = \frac{\prod_{i=1}^{p} d_i^{\frac{1}{2}}}{(2\pi)^{p/2}} \tag{3.5}$$

Thus the function given by (3.1) is a frequency function, for the conditions (3.3) are satisfied. We point out that K is not a function of $\boldsymbol{\mu}$; (3.1) is a frequency function regardless of the value of the elements of the vector $\boldsymbol{\mu}$ or of the positive definite matrix $\mathbf{R}$.

The joint distribution of $y_1, y_2, \ldots, y_p$ will sometimes be referred to as the distribution of the vector $\mathbf{Y}$, and we shall sometimes write $f(\mathbf{Y})$ in place of $f(y_1,y_2, \ldots ,y_p)$.

◆ **Theorem 3.1** In the p-variate normal as given in Definition 3.1,

$$K = \frac{|\mathbf{R}|^{\frac{1}{2}}}{(2\pi)^{p/2}}$$

Proof: By Theorem 1.41,

$$|\mathbf{D}| = \prod_{i=1}^{p} d_i$$

Thus

$$|\mathbf{D}|^{\frac{1}{2}} = \prod_{i=1}^{p} d_i^{\frac{1}{2}}$$

Since $\mathbf{P}$ is orthogonal, we use Theorem 1.39 and get $|\mathbf{R}| = |\mathbf{P'RP}| = |\mathbf{D}|$; the result follows.

◆ **Corollary 3.1.1** The multiple integral

$$\int_{-\infty}^{\infty}\int_{-\infty}^{\infty}\cdots\int_{-\infty}^{\infty} e^{-\frac{1}{2}(\mathbf{Y}-\boldsymbol{\mu})'\mathbf{R}(\mathbf{Y}-\boldsymbol{\mu})}\,dy_1\,dy_2\cdots dy_p = \frac{1}{K}$$

and does not depend on the value of the elements of the vector $\boldsymbol{\mu}$.

◆ **Definition 3.2** In the multivariate normal distribution

$$Ke^{-\frac{1}{2}(\mathbf{Y}-\boldsymbol{\mu})'\mathbf{R}(\mathbf{Y}-\boldsymbol{\mu})}$$

we shall define $(\mathbf{Y} - \boldsymbol{\mu})'\mathbf{R}(\mathbf{Y} - \boldsymbol{\mu})$ as the quadratic form of the multivariate normal.

3.2 Marginal Distributions

We shall sometimes be interested in the marginal distribution of one of the random variables in the vector **Y**. We shall also have occasion to need the marginal distribution of a subset of the random variables in the vector **Y**. This will be the text of the next few theorems.

◆ **Theorem 3.2** The marginal distribution of any random variable in the p-variate normal is the simple normal (univariate normal) distribution.

The proof will consist in showing that the marginal distribution of y_1 is normal (the proof is general if we use y_1 instead of y_i).

Proof: By definition,

$$g(y_1) = \int_{-\infty}^{\infty}\int_{-\infty}^{\infty}\cdots\int_{-\infty}^{\infty} Ke^{-\frac{1}{2}(\mathbf{Y}-\boldsymbol{\mu})'\mathbf{R}(\mathbf{Y}-\boldsymbol{\mu})}\,dy_2\,dy_3\cdots dy_p \qquad (3.6)$$

Partitioning the vectors and the matrix of the exponent gives

$$(\mathbf{Y} - \boldsymbol{\mu})'\mathbf{R}(\mathbf{Y} - \boldsymbol{\mu}) = [(y_1 - \mu_1),(\mathbf{Y}_2 - \mathbf{U}_2)']\begin{pmatrix}\mathbf{R}_{11} & \mathbf{R}_{12}\\ \mathbf{R}_{21} & \mathbf{R}_{22}\end{pmatrix}\begin{pmatrix}y_1 - \mu_1\\ \mathbf{Y}_2 - \mathbf{U}_2\end{pmatrix}$$

where $\mathbf{Y}_2$ and $\mathbf{U}_2$ are the vectors containing the last $p - 1$ elements of **Y** and $\boldsymbol{\mu}$, respectively. $\mathbf{R}_{11}$ is the element in the first row and column of **R** (that is, $\mathbf{R}_{11} = r_{11}$). The dimensions of the remaining matrices $\mathbf{R}_{12}$, $\mathbf{R}_{21}$, and $\mathbf{R}_{22}$ are, therefore, determined. Multiplication gives

$$\begin{aligned}(\mathbf{Y} - \boldsymbol{\mu})'\mathbf{R}(\mathbf{Y} - \boldsymbol{\mu}) &= (y_1 - \mu_1)r_{11}(y_1 - \mu_1) + (y_1 - \mu_1)\mathbf{R}_{12}(\mathbf{Y}_2 - \mathbf{U}_2)\\ &\quad + (\mathbf{Y}_2 - \mathbf{U}_2)'\mathbf{R}_{21}(y_1 - \mu_1) + (\mathbf{Y}_2 - \mathbf{U}_2)'\mathbf{R}_{22}(\mathbf{Y}_2 - \mathbf{U}_2) \qquad (3.7)\end{aligned}$$

Since $\mathbf{R}$ is symmetric and positive definite, we know that $\mathbf{R}_{12} = \mathbf{R}'_{21}$, that $\mathbf{R}_{22}^{-1}$ exists and is symmetric, and that $\mathbf{R}_{11} = r_{11} > 0$. Equation (3.7) can be written

$$(\mathbf{Y} - \boldsymbol{\mu})'\mathbf{R}(\mathbf{Y} - \boldsymbol{\mu}) = (y_1 - \mu_1)(\mathbf{R}_{11} - \mathbf{R}_{12}\mathbf{R}_{22}^{-1}\mathbf{R}_{21})(y_1 - \mu_1)$$
$$+ [(\mathbf{Y}_2 - \mathbf{U}_2) + \mathbf{R}_{22}^{-1}\mathbf{R}_{21}(y_1 - \mu_1)]'\mathbf{R}_{22}[(\mathbf{Y}_2 - \mathbf{U}_2) + \mathbf{R}_{22}^{-1}\mathbf{R}_{21}(y_1 - \mu_1)] \tag{3.8}$$

Note that the first term on the right-hand side of (3.8) does not involve any random variable except y_1 and, hence, that (3.6) can be written

$$g(y_1) = KFe^{-\frac{1}{2}(y_1-\mu_1)(\mathbf{R}_{11}-\mathbf{R}_{12}\mathbf{R}_{22}^{-1}\mathbf{R}_{21})(y_1-\mu_1)} \tag{3.9}$$

where F is the multiple integral

$$\int_{-\infty}^{\infty}\int_{-\infty}^{\infty}\cdots\int_{-\infty}^{\infty} e^{-\frac{1}{2}\{\mathbf{Y}_2-[\mathbf{U}_2-\mathbf{R}_{22}^{-1}\mathbf{R}_{21}(y_1-\mu_1)]\}'\mathbf{R}_{22}\{\mathbf{Y}_2-[\mathbf{U}_2-\mathbf{R}_{22}^{-1}\mathbf{R}_{21}(y_1-\mu_1)]\}}\, dY_2$$

where $$d\mathbf{Y}_2 = dy_2\, dy_3 \cdots dy_p$$

y_1 is not a variable of integration; therefore, if y_1 occurs in the integral, it will be considered a constant with respect to the integration. If we let the vector $\mathbf{U}_2 - \mathbf{R}_{22}^{-1}\mathbf{R}_{21}(y_1 - \mu_1) = \mathbf{h}$, we can write (3.9)

$$g(y_1) = KFe^{-\frac{1}{2}(y_1-\mu_1)(\mathbf{R}_{11}-\mathbf{R}_{12}\mathbf{R}_{22}^{-1}\mathbf{R}_{21})(y_1-\mu_1)}$$

where $$F = \int_{-\infty}^{\infty}\int_{-\infty}^{\infty}\cdots\int_{-\infty}^{\infty} e^{-\frac{1}{2}(\mathbf{Y}_2-\mathbf{h})'\mathbf{R}_{22}(\mathbf{Y}_2-\mathbf{h})}\, dy_2\, dy_3 \cdots dy_p$$

Using Corollary 3.1.1 to evaluate the multiple integral, we get

$$g(y_1) = \frac{K}{K_1} e^{-\frac{1}{2}(y_1-\mu_1)'(\mathbf{R}_{11}-\mathbf{R}_{12}\mathbf{R}_{22}^{-1}\mathbf{R}_{21})(y_1-\mu_1)} \qquad -\infty < y_1 < \infty$$

where $$\frac{1}{K_1} = \frac{(2\pi)^{\frac{1}{2}(p-1)}}{|\mathbf{R}_{22}|^{\frac{1}{2}}}$$

But $$\frac{K}{K_1} = \frac{|\mathbf{R}|^{\frac{1}{2}}}{(2\pi)^{p/2}} \frac{(2\pi)^{\frac{1}{2}(p-1)}}{|\mathbf{R}_{22}|^{\frac{1}{2}}} = \frac{|\mathbf{R}_{11} - \mathbf{R}_{12}\mathbf{R}_{22}^{-1}\mathbf{R}_{21}|^{\frac{1}{2}}}{\sqrt{2\pi}}$$

since, by Theorem 1.50, $|\mathbf{R}| = |\mathbf{R}_{22}|\,|\mathbf{R}_{11} - \mathbf{R}_{12}\mathbf{R}_{22}^{-1}\mathbf{R}_{21}|$. We also see that $y_1 - \mu_1$ and $\mathbf{R}_{11} - \mathbf{R}_{12}\mathbf{R}_{22}^{-1}\mathbf{R}_{21}$ are both scalar quantities and that $\mathbf{R}_{11} - \mathbf{R}_{12}\mathbf{R}_{22}^{-1}\mathbf{R}_{21} = 1/\sigma^2$ (say) is positive definite, by Theorem 1.49; so $\sigma^2 > 0$. Thus

$$g(y_1) = \frac{1}{\sigma\sqrt{2\pi}} e^{-(y_1-\mu_1)^2/2\sigma^2} \qquad \infty < y_1 < \infty$$

which is the frequency function of the univariate normal, and the proof is complete.

The result of this theorem gives us a logical reason to call the frequency function (3.1) a p-variate normal. We shall now generalize this theorem.

◆ **Theorem 3.3** If the random variables $y_1, y_2, \ldots, y_p$ are jointly normal, the joint marginal distribution of any subset of $s < p$ of the random variables is the s-variate normal.

The theorem in vector terminology would be stated: If the vector $\mathbf{Y}$ is normally distributed and is such that

$$\mathbf{Y} = \begin{pmatrix} \mathbf{Y}_1 \\ \mathbf{Y}_2 \end{pmatrix}$$

where $\mathbf{Y}_1$ is a vector containing s of the elements of $\mathbf{Y}$, then the marginal distribution of the vector $\mathbf{Y}_1$ is the s-variate normal.

Proof: The proof follows the method of Theorem 3.2 almost exactly, the only difference being in the partitioning of the matrix and vectors in the exponent. That is to say, $\mathbf{R}_{11}$ is no longer of dimension 1×1, but is now $s \times s$. Also, the scalar $y_1 - \mu_1$ is replaced by the $s \times 1$ vector $\mathbf{Y}_1 - \mathbf{U}_1$. The resulting marginal distribution of the vector $\mathbf{Y}_1$ is

$$h(\mathbf{Y}_1) = \frac{e^{-\frac{1}{2}(\mathbf{Y}_1 - \mathbf{U}_1)'(\mathbf{R}_{11} - \mathbf{R}_{12}\mathbf{R}_{22}^{-1}\mathbf{R}_{21})(\mathbf{Y}_1 - \mathbf{U}_1)}}{(2\pi)^{s/2}|\mathbf{R}_{11} - \mathbf{R}_{12}\mathbf{R}_{22}^{-1}\mathbf{R}_{21}|^{-\frac{1}{2}}}$$

$$\infty < y_i < \infty; \; i = 1, 2, \ldots, s$$

and it is clear that $h(\mathbf{Y}_1)$ is an s-variate normal frequency function, for it satisfies Definition 3.1.

3.3 Moments of the Multivariate Normal

The moments of any distribution are often useful, since the parameters are generally a function of the moments. In the multivariate normal distribution, all the parameters are functions of only the first two moments. We shall prove some pertinent theorems relating moments and parameters of the multivariate normal distribution.

◆ **Definition 3.3** The expected value of a matrix or vector $\mathbf{A}$, which we shall write $E(\mathbf{A})$, will be defined as the expected value of each element of $\mathbf{A}$. Thus, if $\mathbf{A}$ has elements a_{ij}, $E(\mathbf{A}) = (E(a_{ij}))$.

◆ **Theorem 3.4** If the $p \times 1$ random vector $\mathbf{Y}$ is normally distributed with frequency function given by Definition 3.1, then $E(\mathbf{Y}) = \boldsymbol{\mu}$.

Proof: By virtue of Definition 3.3 we must show that $E(y_i) = \mu_i$ for all $i = 1, 2, \ldots, p$. The definition of the expected value of the random variable y_1 (again we shall use y_1 instead of y_i, to facilitate notation—the proof is clearly general) gives

$$E(y_1) = K\int_{-\infty}^{\infty}\int_{-\infty}^{\infty}\cdots\int_{-\infty}^{\infty} y_1 e^{-\frac{1}{2}(\mathbf{Y}-\boldsymbol{\mu})'\mathbf{R}(\mathbf{Y}-\boldsymbol{\mu})}\,dy_1\,dy_2\cdots dy_p$$

We have

$$E(y_1) = \int_{-\infty}^{\infty} y_1\left[\int_{-\infty}^{\infty}\int_{-\infty}^{\infty}\cdots\int_{-\infty}^{\infty} Ke^{-\frac{1}{2}(\mathbf{Y}-\boldsymbol{\mu})'\mathbf{R}(\mathbf{Y}-\boldsymbol{\mu})}\,dy_2\,dy_3\cdots dy_p\right]dy_1 \tag{3.10}$$

By Theorem 3.2, the bracketed term is the marginal distribution of y_1, and (3.10) becomes

$$E(y_1) = \int_{-\infty}^{\infty}\frac{y_1}{\sigma\sqrt{2\pi}}\,e^{-(y_1-\mu_1)^2/2\sigma^2}dy_1 = \mu_1$$

and the proof is complete.

We shall now investigate the variance of the random variables y_i and the covariance of the random variables y_i and y_j in the multivariate normal distribution as given in Definition 3.1.

In a p-variate normal there will be p variances, one for each random variable y_i, and $\frac{1}{2}p(p-1)$ covariances, one for each combination of two of the p random variables, such as y_i and y_j $(i \neq j)$. We shall form a $p \times p$ matrix whose diagonal elements will be the variances of the y_i and whose ijth element $(i \neq j)$ will be the covariance of y_i and y_j. The matrix will be symmetric. If we let $\mathbf{V}$ represent the variance-covariance matrix of a vector $\mathbf{Y}$ and if the elements of $\mathbf{V}$ are v_{ij}, then, by definition of variance and covariance, we have for the frequency function given in Definition 3.1

$$E\{[\mathbf{Y} - E(\mathbf{Y})][\mathbf{Y} - E(\mathbf{Y})]'\} = E[(\mathbf{Y} - \boldsymbol{\mu})(\mathbf{Y} - \boldsymbol{\mu})'] = \mathbf{V}$$

◆ **Theorem 3.5** $\mathbf{V} = \mathbf{R}^{-1}$. That is to say, the matrix in the exponent of a p-variate normal distribution is the inverse of the variance-covariance matrix of the vector $\mathbf{Y}$.

Proof: We must show that the covariance of y_i and y_j $(i \neq j)$ is the ijth element of $\mathbf{R}^{-1}$ and that the variance of y_i is the ith diagonal element of $\mathbf{R}^{-1}$. The proof will consist in finding the joint moment-generating function of

$$(y_1 - \mu_1), (y_2 - \mu_2), \ldots, (y_p - \mu_p)$$

If $\mathbf{T}$ is a $p \times 1$ vector with elements t_i, then, by definition, the joint moment-generating function of $y_i - \mu_i$, where $i = 1, 2, \ldots, p$, is

$$M_{\mathbf{Y}-\boldsymbol{\mu}}(\mathbf{T}) = E \exp\left[\sum_{i=1}^{p} t_i(y_i - \mu_i)\right] = E[e^{(\mathbf{Y}-\boldsymbol{\mu})'\mathbf{T}}]$$
$$= \int_{-\infty}^{\infty}\int_{-\infty}^{\infty}\cdots\int_{-\infty}^{\infty} Ke^{(\mathbf{Y}-\boldsymbol{\mu})'\mathbf{T}-\frac{1}{2}(\mathbf{Y}-\boldsymbol{\mu})'\mathbf{R}(\mathbf{Y}-\boldsymbol{\mu})}\, dy_1\, dy_2 \cdots dy_p \tag{3.11}$$

Examining the exponent, we see that the following relationship holds:

$$(\mathbf{Y} - \boldsymbol{\mu})'\mathbf{T} - \tfrac{1}{2}(\mathbf{Y} - \boldsymbol{\mu})'\mathbf{R}(\mathbf{Y} - \boldsymbol{\mu})$$
$$= -\tfrac{1}{2}(\mathbf{Y} - \boldsymbol{\mu} - \mathbf{R}^{-1}\mathbf{T})'\mathbf{R}(\mathbf{Y} - \boldsymbol{\mu} - \mathbf{R}^{-1}\mathbf{T}) + \tfrac{1}{2}\mathbf{T}'\mathbf{R}^{-1}\mathbf{T}$$

If we let $\boldsymbol{\mu} + \mathbf{R}^{-1}\mathbf{T} = \mathbf{h}$, we get

$$M_{\mathbf{Y}-\boldsymbol{\mu}}(\mathbf{T}) = e^{\frac{1}{2}\mathbf{T}'\mathbf{R}^{-1}\mathbf{T}}\int_{-\infty}^{\infty}\cdots\int_{-\infty}^{\infty} Ke^{-\frac{1}{2}(\mathbf{Y}-\mathbf{h})'\mathbf{R}(\mathbf{Y}-\mathbf{h})}\, dy_1\, dy_2 \cdots dy_p$$

which, by Corollary 3.1.1 (if we let the ijth element of $\mathbf{R}^{-1}$ equal s_{ij}), gives

$$M_{\mathbf{Y}-\boldsymbol{\mu}}(\mathbf{T}) = e^{\frac{1}{2}\mathbf{T}'\mathbf{R}^{-1}\mathbf{T}} = \exp\left(\tfrac{1}{2}\sum_{i=1}^{p}\sum_{j=1}^{p} t_i t_j s_{ij}\right) \tag{3.12}$$

In Eq. (3.11), let us evaluate

$$\frac{\partial^2 M_{\mathbf{Y}-\boldsymbol{\mu}}(\mathbf{T})}{\partial t_m\, \partial t_n}$$

at the point $\mathbf{T} = \mathbf{0}$. We obtain the equation

$$\frac{\partial^2 M_{\mathbf{Y}-\boldsymbol{\mu}}(\mathbf{T})}{\partial t_m\, \partial t_n} = E[(y_m - \mu_m)(y_n - \mu_n)] \exp\left[\sum_{i=1}^{p} t_i(y_i - \mu_i)\right]$$

to be evaluated at the point $\mathbf{T} = \mathbf{0}$. The final result is

$$\left.\frac{\partial^2 M_{\mathbf{Y}-\boldsymbol{\mu}}(\mathbf{T})}{\partial t_m\, \partial t_n}\right|_{\mathbf{T}=\mathbf{0}} = E[(y_m - \mu_m)(y_n - \mu_n)] = \begin{cases}\operatorname{cov}(y_m, y_n) & \text{if } m \neq n \\ \operatorname{var}(y_m) & \text{if } m = n\end{cases}$$

The above is valid only if the function under the integral is such that the variances and covariances exist, such that differentiation can be performed under the integral, and such that integration and taking of limit of the vector $\mathbf{T}$ are interchangeable. The conditions hold for the p-variate normal. From (3.12) we see that

$$\left.\frac{\partial^2 M_{\mathbf{Y}-\boldsymbol{\mu}}(\mathbf{T})}{\partial t_m\, \partial t_n}\right|_{\mathbf{T}=\mathbf{0}} = s_{mn}$$

and the proof is complete.

The vector of means $\boldsymbol{\mu}$ and the variance-covariance matrix $\mathbf{V}$ completely specify the multivariate normal distribution. If we speak of a vector $\mathbf{Y}$ as being distributed normally with mean $\boldsymbol{\mu}$ and variance matrix $\mathbf{V}$, the density function can be written

$$f(Y) = \frac{1}{(2\pi)^{p/2}|\mathbf{V}|^{\frac{1}{2}}} e^{-\frac{1}{2}(\mathbf{Y}-\boldsymbol{\mu})'\mathbf{V}^{-1}(\mathbf{Y}-\boldsymbol{\mu})} \qquad -\boldsymbol{\infty}^* < \mathbf{Y} < \boldsymbol{\infty}^*$$

where $\boldsymbol{\infty}^*$ represents a vector with each entry equal to ∞. If we let $\mathbf{V} = \begin{pmatrix} \mathbf{V}_{11} & \mathbf{V}_{12} \\ \mathbf{V}_{21} & \mathbf{V}_{22} \end{pmatrix}$, where $\mathbf{V}_{11}$ has dimension $s \times s$, then, by Theorem 1.49, $\mathbf{V}_{11}^{-1} = \mathbf{R}_{11} - \mathbf{R}_{12}\mathbf{R}_{22}^{-1}\mathbf{R}_{21}$. If we are given that a random $p \times 1$ vector $\mathbf{Y} = \begin{pmatrix} \mathbf{Y}_1 \\ \mathbf{Y}_2 \end{pmatrix}$ is normal with mean $\boldsymbol{\mu} = \begin{pmatrix} \mathbf{U}_1 \\ \mathbf{U}_2 \end{pmatrix}$ and variance $\mathbf{V}$, then the marginal distribution of the $s \times 1$ vector $\mathbf{Y}_1$ has mean $\mathbf{U}_1$ and covariance $\mathbf{V}_{11}$, and the exponent matrix is $\mathbf{V}_{11}^{-1}$. So, if the mean and covariance matrices of a normal vector $\mathbf{Y}$ are known, it is a simple matter to get the mean and covariance matrices of the marginal distribution of any subset of $\mathbf{Y}$.

3.4 Linear Functions of Normal Variables

It is sometimes stated that, if $y_1, y_2, \ldots, y_p$ are each distributed as the univariate normal, then a linear combination of the y_i is distributed normally. This is not necessarily true. However, if the y_i have a multivariate normal distribution, then a linear combination of y_i is normal. This is the content of the following theorem.

◆ **Theorem 3.6** If the $p \times 1$ vector $\mathbf{Y}$ with elements y_i has the multivariate normal distribution with mean $\boldsymbol{\mu}$ and variance $\mathbf{V}$, then any linear combination of the y_i, say, $\sum_{i=1}^{p} a_i y_i$, where the a_i are constants, has the univariate normal distribution with mean $\sum_{i=1}^{p} a_i \mu_i$ and variance $\sum_{j=1}^{p} \sum_{i=1}^{p} a_i a_j v_{ij}$.

Proof: Let $Z = \sum_{i=1}^{p} a_i y_i = \mathbf{Y}'\mathbf{A}$ where $\mathbf{A} = (a_i)$. The proof will consist in finding the moment-generating function of Z and recognizing it as the moment-generating function of the normal distribution. By definition,

$$m_Z(t) = E(e^{Zt}) = E[e^{(\mathbf{Y}'\mathbf{A})t}] = E[e^{(\mathbf{Y}-\boldsymbol{\mu})'\mathbf{A}t+\boldsymbol{\mu}'\mathbf{A}t}] = e^{(\boldsymbol{\mu}'\mathbf{A})t}E[e^{(\mathbf{Y}-\boldsymbol{\mu})'\mathbf{A}t}]$$

From (3.11) and (3.12) we have

$$E[e^{(\mathbf{Y}-\boldsymbol{\mu})'\mathbf{A}t}] = e^{\frac{1}{2}t(\mathbf{A}'\mathbf{V}\mathbf{A})t} = e^{\frac{1}{2}t^2(\mathbf{A}'\mathbf{V}\mathbf{A})}$$

So

$$m_Z(t) = e^{(\mathbf{A}'\boldsymbol{\mu})t + \frac{1}{2}t^2(\mathbf{A}'\mathbf{V}\mathbf{A})} \tag{3.13}$$

But it is easily verified that, if a random variable Z is distributed as the univariate normal with mean μ and variance σ^2, then the moment-generating function of Z is

$$m_Z(t) = e^{\mu t + \frac{1}{2}t^2\sigma^2}$$

Using Theorem 2.1 completes the proof.

We emphasize the fact that, if the y_i are independently distributed as the univariate normal, then it follows that any linear combination of the y_i is normal, since, if the y_i are independent and normal, the joint distribution of the y_i is the multivariate normal and the conditions of Theorem 3.6 are satisfied.

3.5 Independence

In statistical theory and in applications it is often important to be able to tell whether random variables are independent. In fact, one of the activities of an experimenter is to determine which variables are independent of the variable under study. We shall state two theorems on independence of quantities in the multivariate normal distribution.

◆ **Theorem 3.7** If $\mathbf{Y}$ has the multivariate normal distribution with mean $\boldsymbol{\mu}$ and variance $\mathbf{V}$, then the components y_i are jointly independent if and only if the covariance of y_i and y_j for all $i \neq j$ is zero, that is, if and only if the covariance matrix $\mathbf{V}$ is diagonal.

Proof: First we shall show that, if $\mathbf{V}$ is diagonal, the y_i are jointly independent [it is clear that $\mathbf{V}$ being diagonal is equivalent to the covariance of y_i and y_j (for all $i \neq j$) being zero]. By Definition 2.2 it follows that we must show that, if $\mathbf{V}$ is diagonal, the joint frequency factors into the product of the marginal distributions, which, from Theorem 3.2, are each normal. That is to say, we must show that

$$f(y_1,y_2,\ldots,y_p) = f_1(y_1)f_2(y_2)\cdots f_p(y_p) \tag{3.14}$$

where

$$f(y_1,y_2,\ldots,y_p) = Ke^{-\frac{1}{2}(\mathbf{Y}-\boldsymbol{\mu})'\mathbf{V}^{-1}(\mathbf{Y}-\boldsymbol{\mu})}$$

and

$$f_i(y_i) = \frac{1}{\sqrt{2\pi v_{ii}}}\, e^{-(1/2v_{ii})(y_i-\mu_i)^2}$$

If $v_{ij} = 0$ for $i \neq j$, we have $(\mathbf{Y}-\boldsymbol{\mu})'\mathbf{V}^{-1}(\mathbf{Y}-\boldsymbol{\mu}) = \sum_{i=1}^{p}(y_i-\mu_i)^2 v_{ii}^{-1}$.

Thus we see that (3.14) is satisfied, and the joint independence is established.

Next we shall show that, if the y_i are jointly independent, $\mathbf{V}$ is diagonal. In other words, if (3.14) holds, $\mathbf{V}$ is diagonal. We have

$$v_{ij} = \text{cov}(y_i, y_j) = E[(y_i - \mu_i)(y_j - \mu_j)] \qquad i \neq j$$
$$= \int_{-\infty}^{\infty} \int_{-\infty}^{\infty} \cdots \int_{-\infty}^{\infty} (y_i - \mu_i)(y_j - \mu_j) f(y_1, y_2, \ldots, y_p)\, dy_1 \cdots dy_p$$

But, since y_i and y_j are independent, we can use (3.14), and the integral becomes

$$\int_{-\infty}^{\infty} f_1(y_1)\, dy_1 \int_{-\infty}^{\infty} f_2(y_2)\, dy_2 \cdots \int_{-\infty}^{\infty} (y_i - \mu_i) f_i(y_i)\, dy_i \cdots$$
$$\int_{-\infty}^{\infty} (y_j - \mu_j) f_j(y_j)\, dy_j \cdots \int_{-\infty}^{\infty} f_p(y_p)\, dy_p$$

But
$$\int_{-\infty}^{\infty} (y_i - \mu_i) f_i(y_i)\, dy_i = 0$$

so $v_{ij} = 0$ for all $i \neq j$, and the proof is complete.

This theorem gives us a relatively easy method of determining whether jointly normal variables are independent and in effect states that, if a set of variables are jointly normal, they are independent if and only if they are pairwise uncorrelated. The theorem is not true for distributions in general. And it should also be reemphasized that it is possible at the same time for y_1 and y_2 each to be normally distributed and for their joint distribution not to be the multivariate normal.

◆ **Definition 3.4** Let us suppose the $p \times 1$ vector $\mathbf{Y}$ with subvectors $\mathbf{Y}_1, \mathbf{Y}_2, \ldots, \mathbf{Y}_q$ $(q \leqslant p)$, where

$$\mathbf{Y} = \begin{pmatrix} \mathbf{Y}_1 \\ \mathbf{Y}_2 \\ \cdots \\ \mathbf{Y}_q \end{pmatrix}$$

has a multivariate distribution. We shall say that the vectors are jointly independent if the joint distribution of $\mathbf{Y}$ factors into the product of the joint marginal distribution of $\mathbf{Y}_1, \mathbf{Y}_2, \ldots, \mathbf{Y}_q$, that is, if

$$f(\mathbf{Y}) = f_1(\mathbf{Y}_1) f_2(\mathbf{Y}_2) \cdots f_q(\mathbf{Y}_q) \qquad (3.15)$$

where $f_1(\mathbf{Y}_1), f_2(\mathbf{Y}_2), \ldots, f_q(\mathbf{Y}_q)$, are the marginal distributions of $\mathbf{Y}_1, \mathbf{Y}_2, \ldots, \mathbf{Y}_q$, respectively.

It is clear that, if the two vectors $\mathbf{Y}_1$ and $\mathbf{Y}_2$ are independent, then every element in $\mathbf{Y}_1$ is independent of every element in $\mathbf{Y}_2$.

◆ **Theorem 3.8** If the $p \times 1$ vector $\mathbf{Y}$ is normally distributed with mean $\boldsymbol{\mu}$ and covariance matrix $\mathbf{V}$ and if $\mathbf{Y}_1, \mathbf{Y}_2, \ldots, \mathbf{Y}_q$ are subvectors of $\mathbf{Y}$ such that

$$\mathbf{Y} = \begin{pmatrix} \mathbf{Y}_1 \\ \mathbf{Y}_2 \\ \cdots \\ \mathbf{Y}_q \end{pmatrix}$$

then a necessary and sufficient condition that the subvectors be jointly independent is that all the submatrices $\mathbf{V}_{ij}$ $(i \neq j)$ be equal to the zero matrix.

Proof: By the definition of $\mathbf{V}$ it follows that

$$\mathbf{V}_{ij} = E[(\mathbf{Y}_i - \mathbf{U}_i)(\mathbf{Y}_j - \mathbf{U}_j)'] \qquad \text{where } E(\mathbf{Y}_i) = \mathbf{U}_i$$

To prove sufficiency, let us assume $\mathbf{V}_{ij} = \mathbf{0}$ for all $i \neq j$. If $\mathbf{V}_{ij} = \mathbf{0}$, for all $i \neq j$, then $\mathbf{R}_{ij} = \mathbf{0}$ $(i \neq j)$, where $\mathbf{R} = \mathbf{V}^{-1}$, and

$$f(\mathbf{Y}) = Ke^{-\frac{1}{2}(\mathbf{Y}-\boldsymbol{\mu})'\mathbf{V}^{-1}(\mathbf{Y}-\boldsymbol{\mu})} = K \exp\left[-\tfrac{1}{2}\sum_{i=1}^{q}(\mathbf{Y}_i - \mathbf{U}_i)'\mathbf{V}_{ii}^{-1}(\mathbf{Y}_i - \mathbf{U}_i)\right]$$

$$= \prod_{i=1}^{q} C_i e^{-\frac{1}{2}(\mathbf{Y}_i-\mathbf{U}_i)'\mathbf{V}_{ii}^{-1}(\mathbf{Y}_i-\mathbf{U}_i)} = \prod_{i=1}^{q} f_i(\mathbf{Y}_i)$$

where C_i is a constant, and (3.15) is satisfied.

To prove necessity, let us assume that (3.15) is true and find $\mathbf{V}_{ij}$ $(i \neq j)$. Let y_{im} denote the mth element of the subvector $\mathbf{Y}_i$. Then the mnth element of the matrix $\mathbf{V}_{ij}$ is $E[(y_{im} - \mu_{im})(y_{jn} - \mu_{jn})]$ where $E(y_{im}) = \mu_{im}$, and equals

$$\int_{-\infty}^{\infty}\int_{-\infty}^{\infty}\cdots\int_{-\infty}^{\infty}(y_{im} - \mu_{im})(y_{jn} - \mu_{jn})f_1(\mathbf{Y}_1)f_2(\mathbf{Y}_2)\cdots$$
$$f_q(\mathbf{Y}_q)\,dy_1\,dy_2\cdots dy_p$$

This will factor into the product of multiple integrals, and one factor of this product will be

$$\int_{-\infty}^{\infty}\int_{\infty}^{\infty}\cdots\int_{-\infty}^{\infty}(y_{im} - \mu_{im})f_i(\mathbf{Y}_i)\,d\mathbf{Y}_i$$

where $d\mathbf{Y}_i$ is the product of the differentials of the elements of $\mathbf{Y}_i$. This integral is clearly zero for all m and n; hence $\mathbf{V}_{ij} = 0$, and the proof is complete.

Before we illustrate the foregoing theory with examples, we shall state a theorem that will sometimes aid us in finding the vector mean of a multivariate normal distribution. Suppose $\mathbf{Y}$ is normal with mean $\boldsymbol{\mu}$ and variance $\mathbf{V}$, that the functional form of the distribution is given, and that we desire to find $E(\mathbf{Y})$. This can be done by finding each element $E(y_i)$, but that would involve a great deal of integrating. If the quadratic form Q is given in matrix-and-vector form, we can pick $E(\mathbf{Y})$ from Q. However, if Q is not given in matrix form but if the multiplication is performed and the terms collected, it may be difficult to factor so as to obtain $E(\mathbf{Y})$. For example, if Q is given as

$$Q = \begin{pmatrix} y_1 & -6 \\ y_2 & -2 \end{pmatrix}' \begin{pmatrix} 3 & -1 \\ -1 & 2 \end{pmatrix} \begin{pmatrix} y_1 & -6 \\ y_2 & -2 \end{pmatrix}$$

we see that $E(\mathbf{Y}) = \begin{pmatrix} 6 \\ 2 \end{pmatrix}$. But if the multiplication is effected so that Q is given as

$$Q = 3y_1^2 + 2y_2^2 - 2y_1y_2 - 32y_1 + 4y_2 + 92$$

it may be difficult to factor, and $E(\mathbf{Y})$ will not be easily obtained. A method for obtaining $E(\mathbf{Y})$ that may prove useful at times is as follows.

◆ **Theorem 3.9** If $y_1, y_2, \ldots, y_p$ are jointly normally distributed with quadratic form Q, the vector of means $E(\mathbf{Y}) = \boldsymbol{\mu}$ is the vector that is the solution to the system of equations $\partial Q/\partial \mathbf{Y} = \mathbf{0}$.

Proof: The joint frequency can be written

$$f(\mathbf{Y}) = Ke^{-Q/2}$$

and the value of $\mathbf{Y}$ that maximizes $f(\mathbf{Y})$ is clearly the value of $\mathbf{Y}$ for which $Q = 0$. Since $Q = (\mathbf{Y} - \boldsymbol{\mu})'\mathbf{R}(\mathbf{Y} - \boldsymbol{\mu})$ and since $\mathbf{R}$ is positive definite, we know that Q can be zero only at the point where $\mathbf{Y} - \boldsymbol{\mu} = \mathbf{0}$, in other words, at the point where $\mathbf{Y} = \boldsymbol{\mu}$. Therefore, $\boldsymbol{\mu}$ is the point that maximizes $f(\mathbf{Y})$. Since $f(\mathbf{Y})$ is continuous and has derivatives, it follows from Theorem 1.70 that the solution to the system of equations $\partial f(\mathbf{Y})/\partial \mathbf{Y} = \mathbf{0}$ gives the point that maximizes $f(\mathbf{Y})$ and that that point is $\boldsymbol{\mu}$ by the argument above. But, since $\partial f(\mathbf{Y})/\partial \mathbf{Y} = Ke^{-Q/2}(-1/2)(\partial Q/\partial \mathbf{Y}) = \mathbf{0}$, it follows that the vector satisfying

$$\frac{\partial f(\mathbf{Y})}{\partial \mathbf{Y}} = \mathbf{0}$$

is the same as the vector that satisfies $\partial Q/\partial \mathbf{Y} = \mathbf{0}$. That is to say, the vector $\boldsymbol{\mu}$ is the solution to the system $\partial Q/\partial \mathbf{Y} = \mathbf{0}$.

We shall now illustrate the theorems with two examples.

3.5.1 Example. Let y_1 and y_2 be jointly distributed as the bivariate normal with quadratic form

$$Q = y_1^2 + 2y_2^2 - y_1y_2 - 3y_1 - 2y_2 + 4$$

Suppose we desire to find the following quantities: $E(y_1)$, $E(y_2)$, cov (y_1,y_2), var (y_1), and var (y_2), where var and cov denote variance and covariance, respectively. These quantities will be found by getting the vector of means and the covariance matrix. Since the quadratic form Q can be written

$$Q = (\mathbf{Y} - \boldsymbol{\mu})'\mathbf{R}(\mathbf{Y} - \boldsymbol{\mu}) = \mathbf{Y}'\mathbf{R}\mathbf{Y} - \boldsymbol{\mu}'\mathbf{R}\mathbf{Y} - \mathbf{Y}'\mathbf{R}\boldsymbol{\mu} + \boldsymbol{\mu}'\mathbf{R}\boldsymbol{\mu}$$

we see that $\mathbf{Y}'\mathbf{R}\mathbf{Y}$ is the only term that involves only second-degree terms in y_1 and y_2. Thus, from Q we select only the second-degree terms and evaluate the matrix $\mathbf{R}$. We have

$$\mathbf{Y}'\mathbf{R}\mathbf{Y} = y_1^2 - y_1y_2 + 2y_2^2 = \begin{pmatrix} y_1 \\ y_2 \end{pmatrix}' \begin{pmatrix} 1 & -\frac{1}{2} \\ -\frac{1}{2} & 2 \end{pmatrix} \begin{pmatrix} y_1 \\ y_2 \end{pmatrix}$$

So $$\mathbf{R} = \begin{pmatrix} 1 & -\frac{1}{2} \\ -\frac{1}{2} & 2 \end{pmatrix} \quad \text{and} \quad \mathbf{R}^{-1} = \mathbf{V} = \begin{pmatrix} \frac{8}{7} & \frac{2}{7} \\ \frac{2}{7} & \frac{4}{7} \end{pmatrix}$$

From the matrix $\mathbf{V}$ we pick out the quantities $\text{var}(y_1) = \frac{8}{7}$, $\text{var}(y_2) = \frac{4}{7}$, $\text{cov}(y_1,y_2) = \frac{2}{7}$. The means can be found by solving the system of equations $\partial Q/\partial \mathbf{Y} = \mathbf{0}$.

We get $$\frac{\partial Q}{\partial y_1} = 2y_1 - y_2 - 3 = 0$$

$$\frac{\partial Q}{\partial y_2} = -y_1 + 4y_2 - 2 = 0$$

The solution yields $y_2 = 1$, $y_1 = 2$; so we have $E(y_1) = 2$, $E(y_2) = 1$. As a check we can write the quadratic form Q in factored form and get

$$Q = \begin{pmatrix} y_1 & -2 \\ y_2 & -1 \end{pmatrix}' \begin{pmatrix} 1 & -\frac{1}{2} \\ -\frac{1}{2} & 2 \end{pmatrix} \begin{pmatrix} y_1 & -2 \\ y_2 & -1 \end{pmatrix} = y_1^2 + 2y_2^2 - y_1y_2 - 3y_1 - 2y_2 + 4$$

If we want the marginal distribution of y_2, we see from Theorem 3.3 that y_2 is normal with mean 1 and variance $\frac{4}{7}$.

3.5.2 Example. Let y_1, y_2, and y_3 be jointly normal with means equal to 1, -1, and 2, respectively, and with covariance matrix

$$\mathbf{V} = \begin{pmatrix} 2 & -1 & 0 \\ -1 & 1 & 0 \\ 0 & 0 & 3 \end{pmatrix}$$

Suppose it is desired to write the joint marginal distribution of y_1 and y_3. From Theorem 3.3 it follows that y_1 and y_3 are jointly normal with means 1 and 2, respectively, and with covariance matrix

$$\mathbf{V}_{13}^* = \begin{pmatrix} 2 & 0 \\ 0 & 3 \end{pmatrix}$$

$\mathbf{V}_{13}^*$ is obtained by striking out the second row and second column from $\mathbf{V}$ (since the joint density of y_1, y_2, and y_3 is to be integrated over the y_2 space). Since $\mathbf{V}_{13}^*$ is diagonal, we see that y_1 and y_3 are independent. If it is desired to obtain the marginal distribution of y_1, we should strike out the rows and columns pertaining to y_2 and y_3 (the second and third) from $\mathbf{V}$ and obtain the variance of y_1, which is equal to 2. Using Theorem 3.8, we see that y_3 is independent of both y_1 and y_2 but that y_1 and y_2 are not independent of each other.

Suppose we wanted to find the distribution of the linear combination of $\mathbf{Y}$ given by $\mathbf{A}'\mathbf{Y}$, where $\mathbf{A}' = (1, -2, 1)$. By Theorem 3.6, we know that $\mathbf{A}'\mathbf{Y}$ has the univariate normal distribution with mean equal to $\mathbf{A}'\boldsymbol{\mu}$, which is 5, and variance $\mathbf{A}'\mathbf{V}\mathbf{A}$, which is 13. We can also find K by examining Q. From Theorem 3.1 we have

$$K = \frac{|\mathbf{R}|^{\frac{1}{2}}}{(2\pi)^{p/2}} = \frac{1}{|\mathbf{V}|^{\frac{1}{2}}(2\pi)^{p/2}}$$

Hence,
$$K = \frac{1}{(2\pi)^{3/2}\sqrt{3}}$$

3.6 The Conditional Distribution

Suppose we have two random variables y_1 and y_2 that have a bivariate normal distribution with means μ_1 and μ_2, respectively, and covariance matrix $\mathbf{V}$. Suppose we want to estimate the expected value of y_1 for a given value of y_2. This suggests the use of the conditional distribution of the multivariate normal.

In this section we shall prove some important theorems about conditional distributions.

◆ **Theorem 3.10** If the $p \times 1$ vector $\mathbf{Y}$ is normally distributed with mean $\boldsymbol{\mu}$ and covariance $\mathbf{V}$ and if the vector $\mathbf{Y}$ is partitioned into two subvectors such that $\mathbf{Y} = \begin{pmatrix} \mathbf{Y}_1 \\ \mathbf{Y}_2 \end{pmatrix}$ and if $\mathbf{Y}^* = \begin{pmatrix} \mathbf{Y}_1 \\ \mathbf{Y}_2^* \end{pmatrix}$

$$\boldsymbol{\mu} = \begin{pmatrix} \mathbf{U}_1 \\ \mathbf{U}_2 \end{pmatrix} \quad \text{and} \quad \mathbf{V} = \begin{pmatrix} \mathbf{V}_{11} & \mathbf{V}_{12} \\ \mathbf{V}_{21} & \mathbf{V}_{22} \end{pmatrix}$$

are the corresponding partitions of $\mathbf{Y}^*$, $\boldsymbol{\mu}$, and $\mathbf{V}$, then the conditional distribution of the $q \times 1$ vector $\mathbf{Y}_1$ given the vector $\mathbf{Y}_2 = \mathbf{Y}_2^*$ is the multivariate normal distribution with mean $\mathbf{U}_1 + \mathbf{V}_{12}\mathbf{V}_{22}^{-1}(\mathbf{Y}_2^* - \mathbf{U}_2)$ and covariance matrix $(\mathbf{V}_{11} - \mathbf{V}_{12}\mathbf{V}_{22}^{-1}\mathbf{V}_{21})$.

Proof: By the definition of a conditional distribution we get

$$g(\mathbf{Y}_1 \mid \mathbf{Y}_2^*) = \frac{f(\mathbf{Y}_1,\mathbf{Y}_2^*)}{f_2(\mathbf{Y}_2^*)} = \frac{[1/(2\pi)^{p/2}|\mathbf{V}|^{\frac{1}{2}}]e^{-\frac{1}{2}(\mathbf{Y}^*-\boldsymbol{\mu})'\mathbf{V}^{-1}(\mathbf{Y}^*-\boldsymbol{\mu})}}{[1/(2\pi)^{\frac{1}{2}(p-q)}|\mathbf{V}_{22}|^{\frac{1}{2}}]e^{-\frac{1}{2}(\mathbf{Y}_2^*-\mathbf{U}_2)'\mathbf{V}_{22}^{-1}(\mathbf{Y}_2^*-\mathbf{U}_2)}} \tag{3.16}$$

where $f(\mathbf{Y}_1,\mathbf{Y}_2^*)$ is the joint distribution of the vectors $\mathbf{Y}_1$ and $\mathbf{Y}_2$ with $\mathbf{Y}_2^*$ substituted for $\mathbf{Y}_2$, and $f_2(\mathbf{Y}_2^*)$ is the joint marginal distribution of the elements of $\mathbf{Y}_2$ evaluated at the point $\mathbf{Y}_2 = \mathbf{Y}_2^*$. If in (3.16) we replace $\mathbf{Y}^*$, $\boldsymbol{\mu}$, and $\mathbf{V}$ with the corresponding submatrices and simplify, we get (using Theorem 1.49)

$$g(\mathbf{Y}_1 \mid \mathbf{Y}_2^*) = \frac{1}{K^*} e^{-\frac{1}{2}[\mathbf{Y}_1-\mathbf{U}_1-\mathbf{V}_{12}\mathbf{V}_{22}^{-1}(\mathbf{Y}_2^*-\mathbf{U}_2)]'[\mathbf{V}_{11}-\mathbf{V}_{12}\mathbf{V}_{22}^{-1}\mathbf{V}_{21}]^{-1}[\mathbf{Y}_1-\mathbf{U}_1-\mathbf{V}_{12}\mathbf{V}_{22}^{-1}(\mathbf{Y}_2^*-\mathbf{U}_2)]} \tag{3.17}$$

where $K^* = (2\pi)^{q/2}|\mathbf{V}_{11} - \mathbf{V}_{12}\mathbf{V}_{22}^{-1}\mathbf{V}_{21}|^{1/2}$, and, since $\mathbf{V}_{11} - \mathbf{V}_{12}\mathbf{V}_{22}^{-1}\mathbf{V}_{21}$ is positive definite, the theorem is proved.

It should be pointed out that $g(\mathbf{Y}_1 \mid \mathbf{Y}_2^*)$ is a frequency function of $\mathbf{Y}_1$ for any value of $\mathbf{Y}_2^*$. If the matrix in the exponent of $f(\mathbf{Y})$ is $\mathbf{R} = \mathbf{V}^{-1}$ and if, corresponding to the subdivision of $\mathbf{V}$ in Theorem 3.10, we get

$$\mathbf{R} = \begin{pmatrix} \mathbf{R}_{11} & \mathbf{R}_{12} \\ \mathbf{R}_{21} & \mathbf{R}_{22} \end{pmatrix}$$

then
$$\mathbf{R}_{11} = [\mathbf{V}_{11} - \mathbf{V}_{12}\mathbf{V}_{22}^{-1}\mathbf{V}_{21}]^{-1}$$

So, if we know the exponent matrix of $f(\mathbf{Y})$ and want the exponent matrix of the conditional distribution of $\mathbf{Y}_1$ given $\mathbf{Y}_2 = \mathbf{Y}_2^*$, we just cross out the rows and columns of $\mathbf{R}$ pertaining to the variables in $\mathbf{Y}_2$. This leaves $\mathbf{R}_{11}$. The covariance matrix of the conditional distribution is $\mathbf{R}_{11}^{-1}$.

◆ **Theorem 3.11** $E(\mathbf{Y}_1 \mid \mathbf{Y}_2^*) = \mathbf{U}_1 + \mathbf{V}_{12}\mathbf{V}_{22}^{-1}(\mathbf{Y}_2^* - \mathbf{U}_2)$, where $E(\mathbf{Y}_1 \mid \mathbf{Y}_2^*)$ means the expected value of the random variables in the conditional distribution of $\mathbf{Y}_1$ given $\mathbf{Y}_2 = \mathbf{Y}_2^*$.

Proof: The proof is immediately obvious from inspection of Eq. (3.17).

◆ **Theorem 3.12** The covariance matrix of the conditional distribution of $\mathbf{Y}_1$ given $\mathbf{Y}_2 = \mathbf{Y}_2^*$ does not depend on $\mathbf{Y}_2^*$.

Proof: The proof of this theorem is obvious when we examine Eq. (3.17) and see that the covariance matrix of the conditional distribution is $\mathbf{R}_{11}^{-1} = \mathbf{V}_{11} - \mathbf{V}_{12}\mathbf{V}_{22}^{-1}\mathbf{V}_{21}$ and does not involve $\mathbf{Y}_2^*$.

We shall adopt the notation $v_{ij} = \sigma_{ij}$ to indicate the covariance of y_i and y_j (variance of y_i if $i = j$) in the joint distribution of $y_1, y_2, \ldots, y_p$. We have shown that the covariance of y_i and y_j equals σ_{ij} in the joint marginal distribution of any subset of $y_1, y_2, \ldots, y_p$ that contains y_i and y_j. However, as is stated in Theorem 3.10, the covariance of y_i and y_j in the conditional distribution of $\mathbf{Y}_1$ given $\mathbf{Y}_2 = \mathbf{Y}_2^*$ is the ijth element of $\mathbf{R}_{11}^{-1}$. The following notation will be adopted: $\sigma_{ij \cdot rst \cdots p}$ will mean the covariance of y_i and y_j in the conditional distribution of $\mathbf{Y}_1$ given $\mathbf{Y}_2 = \mathbf{Y}_2^*$, where the subscripts on the right-hand side of the dot all appear in $\mathbf{Y}_2$ and the subscripts i and j must appear in $\mathbf{Y}_1$. For example, $\sigma_{12 \cdot 456}$ means the covariance between y_1 and y_2 in the joint conditional distribution of y_1, y_2, and y_3 given y_4, y_5, and y_6.

◆ **Theorem 3.13** $\sigma_{11 \cdot 23 \cdots p} \leqslant \sigma_{11}$.

Proof: The proof is as follows: $\sigma_{11 \cdot 23 \cdots p} = R_{11}^{-1} = r_{11}^{-1} = v_{11} - \mathbf{V}_{12}\mathbf{V}_{22}^{-1}\mathbf{V}_{21} \leqslant v_{11} = \sigma_{11}$, since $\mathbf{V}_{12}\mathbf{V}_{22}^{-1}\mathbf{V}_{21}$ is positive semidefinite and, hence, cannot be negative.

In Eq. (3.17), if $\mathbf{Y}_1$ is 1×1 and we make the substitution

$$e = (y_1 - \mu_1) - \mathbf{V}_{12}\mathbf{V}_{22}^{-1}(\mathbf{Y}_2 - \mathbf{U}_2) \tag{3.18}$$

it follows that e is normally distributed with mean 0 and variance

$$v_{11} - \mathbf{V}_{12}\mathbf{V}_{22}^{-1}\mathbf{V}_{21} = \sigma^2 \text{ (say)}$$

In Eq. (3.17), if we let the $1 \times (p - 1)$ vector $\mathbf{V}_{12}\mathbf{V}_{22}^{-1} = \boldsymbol{\beta} = (\beta_i)$, we get (writing $\tilde{y}_1$ for the conditional random variable)

$$\tilde{y}_1 = \mu_1 + \sum_{i=2}^{p} \beta_i(y_i^* - \mu_i) + e \tag{3.19}$$

which is called the *multiple regression equation* of y_1 on $y_2, y_3, \ldots, y_p$. To indicate the conditional distribution of $\mathbf{Y}_1$ given $\mathbf{Y}_2 = \mathbf{Y}_2^*$, we shall simply write it as the conditional distribution of $\mathbf{Y}_1$ given $\mathbf{Y}_2$ and we shall not explicitly indicate the value of $\mathbf{Y}_2$. If we consider $y_2, y_3, \ldots, y_p$ as given, we get for the mean of $\tilde{y}_1$

$$E(\tilde{y}_1) = \mu_1 - \sum_{i=2}^{p} \beta_i \mu_i + \sum_{i=2}^{p} \beta_i y_i \tag{3.20}$$

In (3.20), we see that the rate of change of $E(\tilde{y}_1)$ per unit change in y_j is β_j. Thus β_j is called the *partial regression coefficient*; it indicates the change in $E(\tilde{y}_1)$ per unit change of y_j when the other y_i $(i \neq j)$ remain unchanged.

In the conditional distribution of $\mathbf{Y}_1$ given $\mathbf{Y}_2$, we shall define the *partial correlation coefficient* $\rho_{ij \cdot rst \cdots p}$ to be the correlation between y_i and y_j (which are in $\mathbf{Y}_1$) in the conditional distribution of $\mathbf{Y}_1$ given $\mathbf{Y}_2$, where $y_r, y_s, y_t, \ldots,$ and y_p are in $\mathbf{Y}_2$. For example, $\rho_{13 \cdot 567}$ will represent the correlation between y_1 and y_3 in the conditional distribution of y_1, y_2, y_3, and y_4 given y_5, y_6, and y_7.

From this it follows that

$$\rho_{ij \cdot rst \cdots p} = \frac{\sigma_{ij \cdot rst \cdots p}}{\sqrt{\sigma_{ii \cdot rst \cdots p}\sigma_{jj \cdot rst \cdots p}}}$$

The partial correlation coefficient $\rho_{ij \cdot rst \cdots p}$ indicates the correlation between y_i and y_j when the variables $y_r, y_s, y_t, \ldots$ and y_p are held constant.

The *multiple-correlation coefficient* ρ_1 will be defined as the correlation between the two random variables y_1 and z, where

$$z = \mathbf{V}_{12}\mathbf{V}_{22}^{-1}(\mathbf{Y}_2 - \mathbf{U}_2)$$

Thus,

$$\rho_1 = \frac{E\{(y_1 - \mu_1)[z - E(z)]\}}{\sqrt{E(y_1 - \mu_1)^2 E[z - E(z)]^2}}$$

Now $\quad E\{[y_1 - E(y_1)][z - E(z)]\} = E[(y_1 - \mu_1)\mathbf{V}_{12}\mathbf{V}_{22}^{-1}(\mathbf{Y}_2 - \mathbf{U}_2)]$

$$= E[(y_1 - \mu_1)(\mathbf{Y}_2 - \mathbf{U}_2)'\mathbf{V}_{22}^{-1}\mathbf{V}_{21}] = \mathbf{V}_{12}\mathbf{V}_{22}^{-1}\mathbf{V}_{21}$$

Also $\quad E(y_1 - \mu_1)^2 = v_{11} = \sigma_{11}$

and

$$\begin{aligned} E[z - E(z)]^2 &= E\{[z - E(z)][z - E(z)]'\} \\ &= E[\mathbf{V}_{12}\mathbf{V}_{22}^{-1}(\mathbf{Y}_2 - \mathbf{U}_2)(\mathbf{Y}_2 - \mathbf{U}_2)'\mathbf{V}_{22}^{-1}\mathbf{V}_{21}] \\ &= \mathbf{V}_{12}\mathbf{V}_{22}^{-1}\mathbf{V}_{21} \end{aligned}$$

So

$$\rho_1 = \frac{\sqrt{\mathbf{V}_{12}\mathbf{V}_{22}^{-1}\mathbf{V}_{21}}}{\sqrt{v_{11}}} = \sqrt{\frac{\sigma_{11} - \sigma_{11 \cdot 23 \cdots p}}{\sigma_{11}}} = \sqrt{1 - \frac{\sigma_{11 \cdot 23 \cdots p}}{\sigma_{11}}} \tag{3.21}$$

◆ **Theorem 3.14** The multiple-correlation coefficient satisfies the inequality $0 \leqslant \rho_1^2 \leqslant 1$.

Proof: $\rho_1^2 = 1 - (\sigma_{11\cdot 23\cdots p}/\sigma_{11})$. By Theorem 3.13, $0 \leqslant (\sigma_{11\cdot 23\cdots p}/\sigma_{11}) \leqslant 1$ and the result follows.

From Eq. (3.21) we obtain

$$\frac{\sigma_{11} - \sigma_{11\cdot 23\cdots p}}{\sigma_{11}} = \rho_1^2$$

σ_{11} is the variance of y_1 in the marginal distribution of y_1; $\sigma_{11.23\cdots p}$ is the variance of $\tilde{y}_1$ in the conditional distribution of y_1 given $y_2, y_3, \ldots, y_p$. The quantity $\sigma_{11} - \sigma_{11.23\cdots p}$ is the amount that the variance of y_1 (when $y_2, \ldots, y_p$ are ignored) can be reduced by sampling from the distribution of $\tilde{y}_1$ when $y_2, y_3, \ldots, y_p$ are not ignored but are known.

For example, let us assume that the heights and weights of individuals in a large city follow the bivariate normal distribution. If we ignore the weights of people and compute the variance of the heights of all the population of the city, we get a number which we shall call σ_{11}. If, now, we compute the variance of the heights of only those individuals who weigh 150 lb, it is quite clear that we shall get a number, say, $\sigma_{11\cdot 2}$, such that $\sigma_{11} \geqslant \sigma_{11.2}$. That is, it is evident that the variance of the heights of all people will be equal to or larger than the variance of the heights of people in the class restricted to persons who weigh 150 lb. If $\sigma_{11} = \sigma_{11\cdot 2}$, then $\sigma_{11} - \sigma_{11\cdot 2} = 0$, and the variance of heights is not reduced by sampling from the restricted class. If $\sigma_{11\cdot 2} = 0$, then $\sigma_{11} - \sigma_{11\cdot 2} = \sigma_{11}$, and the variance is reduced by an amount σ_{11} by sampling from the restricted class. In many investigations the important thing is not the actual amount that the variance is reduced but the fractional reduction. Therefore, the square of the multiple-correlation coefficient is an important quantity; it gives us a measure of the fraction of reduction in the variance of a random variable when we use related variables.

If we examine the multiple-regression equation (3.19) we can also see the utility of the multiple-correlation coefficient ρ_1. In (3.19), if $y_2, y_3, \ldots, y_p$ are given (fixed), then the variance of $\tilde{y}_1$ equals the variance of e, which is $v_{11} - \mathbf{V}_{12}\mathbf{V}_{22}^{-1}\mathbf{V}_{21}$. On the other hand, if the y_i are not fixed but all that is known is that $y_1, y_2, \ldots, y_p$ are jointly normal, then the variance of y_1 is v_{11}. As before, the fractional reduction in the variance of y_1 that is due to $y_2, y_3, \ldots, y_p$ is ρ_1^2.

Sometimes multiple regression is considered apart from the multivariate normal. That is to say, let us assume that the random variable e is normally distributed with mean 0 and variance σ^2 and that y_i ($i = 2, \ldots, p$) are known. Also assume that the relationship of the random variable $\tilde{y}_1$ to these quantities is given by (3.19). This could be stated as follows: Assume that $\tilde{y}_1$ is normally distributed with mean $\mu_1 + \sum_{j=2}^{p} \beta_j(y_j - \mu_j)$ and variance σ^2. It is then sometimes desired to find the increase in the variance of $\tilde{y}_1$ if regression is not used; that is to say, it is desired to find the marginal distribution of y_1. It is impossible to obtain this from the conditional distribution of y_1 given $\mathbf{Y}_2$ unless either (1) the joint distribution of $y_2, y_3, \ldots, y_p$ is known, or (2) $y_2, y_3, \ldots, y_p$ are jointly independent of y_1. This can be seen from the fact that by the variance of y_1 when regression is not used we mean the variance of the marginal distribution of y_1. It is clear that the joint distribution of $y_1, y_2, \ldots, y_p$ (from which the marginal distribution of y_1 must be obtained) is equal to the product of the conditional distribution of y_1 given $y_2, y_3, \ldots, y_p$ and the joint marginal distribution of $y_2, y_3, \ldots, y_p$. Of course, if y_1 is independent of $y_2, y_3, \ldots, y_p$, then the conditional of y_1 equals the marginal of y_1.

Let us consider the case in which $p = 2$, and let $\tilde{y}_1 = y$ and $y_2 = x$. The regression equation is

$$y = \mu_y + \beta(x - \mu_x) + e \tag{3.22}$$

where x is known and where e is distributed normally with mean 0 and variance σ^2. This indicates that we know that, when $x = x_0$, $x = x_1$, or $x = x_2$, then y is normally distributed with mean $\mu_y + \beta(x_0 - \mu_x)$, $\mu_y + \beta(x_1 - \mu_x)$, or $\mu_y + \beta(x_2 - \mu_x)$, respectively, and with variance σ^2. The probability that y is between a and b (this requires the marginal distribution of y) is not known unless we know the relative probabilities that we are sampling from the distributions corresponding to x_0, x_1, and x_2. In other words, we need to know the joint distribution of x_0, x_1, and x_2.

Sometimes we are not primarily interested in how regression reduces the variance of y_1 in Eq. (3.19). We postulate that $\tilde{y}_1$ is distributed according to Eq. (3.19) for certain known y_j values, and we want to estimate μ_1, μ_i, the β_i, and σ^2 and to test certain hypotheses about these parameters. This will be discussed in detail later.

3.7 Additional Theorems

Some additional theorems will be stated in this section; some proofs will be asked for in the problems ($y_1, y_2, \ldots, y_p$ are assumed to be jointly normal).

◆ **Theorem 3.15** In the multivariate normal distribution, $y_2, y_3, \ldots, y_p$ are jointly independent of $y_1 - \mu_1 - \mathbf{V}_{12}\mathbf{V}_{22}^{-1}(\mathbf{Y}_2 - \mathbf{U}_2)$.

◆ **Theorem 3.16** If $\sigma_{12} = \sigma_{13} = \cdots = \sigma_{1p} = 0$, the partial regression coefficients β_j in (3.19) are all zero.

◆ **Theorem 3.17** If $\sigma_{12} = \sigma_{13} = \cdots = \sigma_{1p} = 0$, then $\sigma_{11} = \sigma_{11\cdot 23\cdots p}$.

◆ **Theorem 3.18** If $\sigma_{12} = \sigma_{13} = \cdots = \sigma_{1p} = 0$, the multiple correlation coefficient ρ_1^2 equals zero.

◆ **Theorem 3.19** If $\sigma_{13} = \sigma_{14} = \cdots = \sigma_{1p} = \sigma_{23} = \sigma_{24} = \cdots = \sigma_{2p} = 0$, then the partial correlation coefficient of y_1 and y_2 given $y_3, y_4, \ldots, y_p$ equals the simple correlation of y_1 and y_2; that is, $\rho_{12\cdot 34\cdots p} = \rho_{12}$.

◆ **Theorem 3.20** If $y_1, y_2, \ldots, y_p$ are jointly normal with covariance matrix $\mathbf{V}$, the covariance matrix of the conditional distribution of $\mathbf{Y}_1$ given $\mathbf{Y}_2$ can be obtained by striking out the rows and columns from $\mathbf{R} = \mathbf{V}^{-1}$ that correspond to the elements of $\mathbf{Y}_2$ and taking the inverse of the remaining matrix.

◆ **Theorem 3.21** The vector mean of the conditional distribution of $\mathbf{Y}_1$ given $\mathbf{Y}_2$, i.e., the regression function $E(\mathbf{Y}_1 \mid \mathbf{Y}_2)$, is the vector that is the solution of the equations $\partial Q/\partial \mathbf{Y}_1 = \mathbf{0}$, where Q is the quadratic form $(\mathbf{Y} - \boldsymbol{\mu})'\mathbf{R}(\mathbf{Y} - \boldsymbol{\mu})$ of the joint distribution of $y_1, y_2, \ldots, y_p$.

It was shown in Theorem 3.6 that, if $\mathbf{Y}$ is distributed as the multivariate normal, then any linear combination of $\mathbf{Y}$ is distributed normally. For example, if $\mathbf{A} = (\alpha_i)$ is a $p \times 1$ vector of constants, then the scalar quantity $\mathbf{A}'\mathbf{Y} = \sum_{i=1}^{p} \alpha_i y_i$ is normally distributed. This will be generalized to the case where $\mathbf{A}$ is not a vector but a $q \times p$ matrix.

◆ **Theorem 3.22** If the $p \times 1$ vector $\mathbf{Y}$ is distributed normally with mean $\boldsymbol{\mu}$ and covariance $\mathbf{V}$ and if $\mathbf{B}$ is a $q \times p$ matrix ($q \leqslant p$) of rank q, the vector $\mathbf{Z} = \mathbf{BY}$ is distributed as the q-variate normal distribution with mean $\mathbf{B}\boldsymbol{\mu}$ and covariance $\mathbf{BVB}'$.

The proof of this theorem will be left for the reader.

This theorem in effect says that the distribution of q linearly independent linear functions of variables that have a multivariate normal distribution is a q-variate normal.

Since a multivariate normal distribution is completely specified by its mean and covariance, we shall use the notation "$\mathbf{Y}$ is distributed $N(\boldsymbol{\mu},\mathbf{V})$" to indicate that the $p \times 1$ vector $\mathbf{Y}$ is distributed normally with mean $\boldsymbol{\mu}$ and covariance matrix $\mathbf{V}$.

If the covariance matrix of a normal distribution is diagonal, then, as shown in Theorem 3.7, the variables are mutually independent. In particular, if the covariance matrix of a multivariate normal is $\sigma^2\mathbf{I}$, where σ^2 is a scalar, it follows that the variables are independent and that each has a variance σ^2. If a vector $\mathbf{Y}$ is distributed $N(\boldsymbol{\mu},\sigma^2\mathbf{I})$, where $\mu_1 = \mu_2 = \cdots = \mu_p = \alpha$, say, then the frequency function of $\mathbf{Y}$ is equivalent to the joint distribution of p independent observations from the univariate normal with mean α and variance σ^2. The particular case in which $\alpha = 0$, that is, where $\mathbf{Y}$ is distributed $N(\mathbf{0},\sigma^2\mathbf{I})$ plays an important part in regression and analysis-of-variance theory.

◆ **Theorem 3.23** If a vector $\mathbf{Y}$ is distributed $N(\mathbf{0},\sigma^2\mathbf{I})$ and if $\mathbf{Z} = \mathbf{PY}$, where $\mathbf{P}$ is an orthogonal matrix, then $\mathbf{Z}$ is distributed $N(\mathbf{0},\sigma^2\mathbf{I})$, also. That is to say, if $\mathbf{Y}$ is distributed $N(\mathbf{0},\sigma^2\mathbf{I})$, any orthogonal transformation leaves the distribution unchanged.

Proof: By Theorem 3.22 we know that $\mathbf{Z} = \mathbf{PY}$ is normal; so we need only find the mean and covariance matrix of $\mathbf{Z}$, and the proof is complete.

3.7.1 Example. Suppose that the 4×1 vector $\mathbf{Y}$ is distributed $N(\boldsymbol{\mu},\mathbf{V})$ and that the quadratic form $(\mathbf{Y} - \boldsymbol{\mu})'\mathbf{R}(\mathbf{Y} - \boldsymbol{\mu})$, is given by

$$Q = 3y_1^2 + 2y_2^2 + 2y_3^2 + y_4^2 + 2y_1y_2 + 2y_3y_4 - 6y_1 - 2y_2 - 6y_3 - 2y_4 + 8$$

We shall find the following: (1) $f(y_1 \mid y_2, y_3, y_4)$, (2) $f_1(y_1)$, (3) ρ_{12}, (4) $\rho_{13\cdot2}$, and (5) the multiple-correlation coefficient ρ_1 of y_1 on y_2, y_3, and y_4.

First we shall find $\mathbf{R}$ and $\mathbf{V}$. These can easily be found by examining only the second-degree terms of Q, that is, $3y_1^2 + 2y_2^2 + 2y_3^2 + y_4^2 + 2y_1y_2 + 2y_3y_4$. This gives

$$\mathbf{R} = \begin{pmatrix} 3 & 1 & 0 & 0 \\ 1 & 2 & 0 & 0 \\ 0 & 0 & 2 & 1 \\ 0 & 0 & 1 & 1 \end{pmatrix} \qquad \mathbf{V} = \begin{pmatrix} \frac{2}{5} & -\frac{1}{5} & 0 & 0 \\ -\frac{1}{5} & \frac{3}{5} & 0 & 0 \\ 0 & 0 & 1 & -1 \\ 0 & 0 & -1 & 2 \end{pmatrix}$$

Next we shall find $\boldsymbol{\mu}$. This can be found by solving $\partial Q/\partial\mathbf{Y} = \mathbf{0}$. We get

$$\frac{\partial Q}{\partial y_1} = 6y_1 + 2y_2 - 6 = 0$$

$$\frac{\partial Q}{\partial y_2} = 2y_1 + 4y_2 - 2 = 0$$

$$\frac{\partial Q}{\partial y_3} = 4y_3 + 2y_4 - 6 = 0$$

$$\frac{\partial Q}{\partial y_4} = 2y_3 + 2y_4 - 2 = 0$$

The solution is $y_1 = 1$, $y_2 = 0$, $y_3 = 2$, $y_4 = -1$; so

$$\boldsymbol{\mu} = \begin{pmatrix} 1 \\ 0 \\ 2 \\ -1 \end{pmatrix}$$

1. We must exhibit the mean and variance of the distribution of $f(y_1 \mid y_2, y_3, y_4)$. We shall do this two ways.

a. We shall use Theorem 3.10. We get

$$\mathbf{V}_{11} = \tfrac{2}{5} \qquad \mathbf{V}_{12} = (-\tfrac{1}{5} \quad 0 \quad 0) \qquad \mathbf{U}_1 = 1 \qquad \mathbf{U}_2' = (0 \quad 2 \quad -1)$$

$$\mathbf{V}_{22} = \begin{pmatrix} \frac{3}{5} & 0 & 0 \\ 0 & 1 & -1 \\ 0 & -1 & 2 \end{pmatrix} \qquad \mathbf{V}_{22}^{-1} = \begin{pmatrix} \frac{5}{3} & 0 & 0 \\ 0 & 2 & 1 \\ 0 & 1 & 1 \end{pmatrix}$$

$$\mathbf{U}_1 + \mathbf{V}_{12}\mathbf{V}_{22}^{-1}(\mathbf{Y}_2 - \mathbf{U}_2) = 1 - \tfrac{1}{3}y_2$$

$$\mathbf{V}_{11} - \mathbf{V}_{12}\mathbf{V}_{22}^{-1}\mathbf{V}_{21} = \tfrac{2}{5} - \tfrac{1}{15} = \tfrac{1}{3}$$

Therefore, $f(y_1 \mid y_2, y_3, y_4)$ is a normal distribution with mean $1 - \frac{1}{3}y_2$ and variance $\frac{1}{3}$.

b. We shall use Theorems 3.20 and 3.21. In that case we get $\mathbf{R}_{11} = 3$; $\mathbf{R}_{11}^{-1} = \frac{1}{3}$ = variance. For the mean we get

$$\frac{\partial Q}{\partial y_1} = 6y_1 + 2y_2 - 6 = 0$$

Solving for y_1 gives $y_1 = -\frac{1}{3}y_2 + 1$, which is the mean of the distribution $f(y_1 \mid y_2, y_3, y_4)$.

2. To find the mean and variance of $f_1(y_1)$, we strike out the rows and columns of $\boldsymbol{\mu}$ and $\mathbf{V}$ that correspond to y_2, y_3, y_4. This shows us that y_1 is distributed $N(1, \frac{2}{5})$.

3. To find ρ_{12} we must first find the covariance of the joint distribution of y_1, y_2. We get (let $\mathbf{V}^*$ be the covariance matrix of y_1, y_2)

$$\mathbf{V}^* = \begin{pmatrix} \frac{2}{5} & -\frac{1}{5} \\ -\frac{1}{5} & \frac{3}{5} \end{pmatrix} \qquad \rho_{12} = \frac{-\frac{1}{5}}{\sqrt{(\frac{2}{5})(\frac{3}{5})}} = -\frac{1}{\sqrt{6}} = -\frac{\sqrt{6}}{6}$$

4. To find $\rho_{13\cdot 2}$ we must find the covariance of the conditional distribution of y_1, y_3, and y_4 given y_2. The covariance matrix is

obtained by striking out the row and column of $\mathbf{R}$ corresponding to y_2 (i.e., second row and second column). This gives

$$\mathbf{R}_{134} = \begin{pmatrix} 3 & 0 & 0 \\ 0 & 2 & 1 \\ 0 & 1 & 1 \end{pmatrix} \qquad \mathbf{R}_{134}^{-1} = \begin{pmatrix} \frac{1}{3} & 0 & 0 \\ 0 & 1 & -1 \\ 0 & -1 & 2 \end{pmatrix}$$

So
$$\rho_{13\cdot 2} = \frac{0}{\sqrt{(\frac{1}{3})(1)}} = 0$$

5. The multiple correlation coefficient of y_1 on y_2, y_3, y_4 is

$$\rho_1 = \frac{\sqrt{\mathbf{V}_{12}\mathbf{V}_{22}^{-1}\mathbf{V}_{21}}}{\sqrt{\mathbf{V}_{11}}}$$

where $\mathbf{V}_{ij}$ is defined in part 1 of this example. Thus

$$\rho_1 = \frac{\sqrt{\frac{1}{15}}}{\sqrt{\frac{2}{5}}} = \sqrt{\frac{1}{6}}$$

Another way to find ρ_1 is to find $\sigma_{11\cdot 234}$ first. This we do by striking out rows and columns 2, 3, 4 from $\mathbf{R}$ and inverting the result. We get $\sigma_{11\cdot 234} = \frac{1}{3}$. Also, $\mathbf{V}_{11} = \sigma_{11} = \frac{2}{5}$; so

$$\rho_1 = \sqrt{\frac{\sigma_{11} - \sigma_{11\cdot 234}}{\sigma_{11}}} = \sqrt{\frac{\frac{2}{5} - \frac{1}{3}}{\frac{2}{5}}} = \sqrt{\frac{1}{6}}$$

Problems

3.1 If y_1, y_2, and y_3 are jointly normal with quadratic form

$$Q = 2y_1^2 + 3y_2^2 + 4y_3^2 + 2y_1y_2 - 2y_1y_3 - 4y_2y_3 - 6y_1 - 6y_2 + 10y_3 + 8$$

(*a*) Find $\mathbf{R}$.
(*b*) Find $\boldsymbol{\mu}$.
(*c*) Find $f(y_1 \mid y_2, y_3)$.
(When a normal distribution is required it will be sufficient to state the vector mean and covariance matrix of the distribution.)

3.2 If a vector $\mathbf{Y}$ has a multivariate normal distribution with mean $\boldsymbol{\mu} = \mathbf{0}$ and covariance matrix

$$\mathbf{V} = \begin{pmatrix} 4 & 1 & 0 \\ 1 & 2 & 1 \\ 0 & 1 & 3 \end{pmatrix}$$

(*a*) Find $\mathbf{R}$.
(*b*) Find the marginal distribution of y_1.
(*c*) Find the joint marginal distribution of y_1 and y_2.
(*d*) Find the conditional distribution of y_1, given y_2 and y_3.
(*e*) Find ρ_{12}, ρ_{13}, and ρ_{23}.

(*f*) Find $\rho_{12 \cdot 3}$ and ρ_1^2.
(*g*) In (*d*), find β_3 and β_2.
(*h*) Find $\sigma_{11 \cdot 2}$, $\sigma_{11 \cdot 3}$, $\sigma_{11 \cdot 23}$, and $\sigma_{12 \cdot 3}$.
(*i*) Find the mean and variance of Z, where $Z = 4y_1 - 6y_2 + y_3$.

3.3 If a vector **Y** has a multivariate normal distribution with mean

$$\boldsymbol{\mu} = \begin{pmatrix} -1 \\ 0 \\ 1 \end{pmatrix}$$

and covariance matrix

$$\mathbf{V} = \begin{pmatrix} 2 & 0 & -1 \\ 0 & 4 & 0 \\ -1 & 0 & 3 \end{pmatrix}$$

find (*a*), (*b*), (*c*), (*d*), (*e*), (*f*), (*g*), and (*h*) of Prob. 3.2.

3.4 If a vector **Y** has a multivariate normal distribution with mean **0** and covariance matrix

$$\mathbf{V} = \begin{pmatrix} 1 & 0 & 0 \\ 0 & 2 & -3 \\ 0 & -3 & 5 \end{pmatrix}$$

(*a*) Find ρ_1^2, illustrating Theorem 3.18.
(*b*) Show that $\sigma_{11} = \sigma_{11 \cdot 23}$, illustrating Theorem 3.17.

3.5 If a vector **Y** has a multivariate normal distribution with mean **0** and covariance matrix

$$\mathbf{V} = \begin{pmatrix} 1 & 0 & 0 & 0 \\ 0 & 2 & 0 & 0 \\ 0 & 0 & 3 & -4 \\ 0 & 0 & -4 & 6 \end{pmatrix}$$

(*a*) Find ρ_{12}, ρ_{13}, ρ_{23}, ρ_{14}, ρ_{24}, and ρ_{34}.
(*b*) Find $\rho_{12 \cdot 34}$, illustrating Theorem 3.19.

3.6 If y_1, y_2, and y_3 are jointly normal and if
(*a*) $\rho_{12} \neq 0$ and $\rho_{13} \neq 0$, can $\rho_{23} = 0$?
(*b*) $\rho_{12} = 0$ and $\rho_{13} = 0$, can $\rho_{23} \neq 0$?

3.7 Prove Theorems 3.16, 3.17, 3.18, and 3.19, and state the hypothesis of each theorem as a condition on the covariance matrix **V**.

3.8 Prove Theorem 3.21.

3.9 The bivariate normal can be written

$$f(x,y) = \frac{1}{\sigma_x \sigma_y \sqrt{1 - \rho^2}(2\pi)} \exp\left\{ -\frac{1}{2(1-\rho^2)} \left[\left(\frac{x - \mu_x}{\sigma_x} \right)^2 - 2\rho \left(\frac{x - \mu_x}{\sigma_x} \right) \left(\frac{y - \mu_y}{\sigma_y} \right) + \left(\frac{y - \mu_y}{\sigma_y} \right)^2 \right] \right\}$$

Find **R** and **V**, and show that $\text{var}(x) = \sigma_x^2$, $\text{var}(y) = \sigma_y^2$, and $\text{cov}(x,y) = \sigma_x \sigma_y \rho$.

3.10 In the trivariate normal, show that

$$\rho_{12\cdot3} = \frac{\rho_{12} - \rho_{13}\rho_{23}}{\sqrt{1 - \rho_{13}^2}\sqrt{1 - \rho_{23}^2}}$$

3.11 Prove Theorem 3.22.

Further Reading

1 C. R. Rao: "Advanced Statistical Methods in Biometric Research," John Wiley & Sons, Inc., New York, 1952.

2 S. S. Wilks: "Mathematical Statistics," Princeton University Press, Princeton, N.J., 1943.

3 T. W. Anderson: "An Introduction to Multivariate Statistical Analysis," John Wiley & Sons, Inc., New York, 1958.

4 A. M. Mood: "Introduction to the Theory of Statistics," McGraw-Hill Book Company, Inc., New York, 1950.

5 M. G. Kendall: "The Advanced Theory of Statistics," vols. I, II, Charles Griffin & Co., Ltd., London, 1946.

6 H. Cramér: "Mathematical Methods of Statistics," Princeton University Press, Princeton, N.J., 1945.

7 S. N. Roy: "Some Aspects of Multivariate Analysis," John Wiley & Sons, Inc., New York, 1957.

4

Distribution of Quadratic Forms

4.1 Introduction

If the reader is acquainted with the arithmetical techniques of the analysis of variance, he knows that they involve the summing and squaring of observations and so lead to quadratic forms in the observations. The process known as the *analysis of variance* can be described as the partitioning of a sum of squares into a set of quadratic forms. It is, therefore, clear that quadratic forms play an important part in many segments of statistical applications. In this chapter we shall examine the distribution of quadratic forms when the variables are normal.

4.2 Noncentral Chi-square

Since $\mathbf{Y}'\mathbf{Y}$ is a sum of squares, we know that, if $\mathbf{Y}$ is distributed $N(\mathbf{0},\mathbf{I})$, then $\mathbf{Y}'\mathbf{Y} = \sum_{i=1}^{p} y_i^2$ is distributed as $\chi^2(p)$, that is, as a chi-square variate with p degrees of freedom. We shall also investigate the distribution of $\mathbf{Y}'\mathbf{Y}$ when the variables have the distribution $N(\boldsymbol{\mu},\mathbf{I})$.

◆ **Theorem 4.1** If $y_1, y_2, \ldots, y_p$ are independent normal variables with means $\mu_1, \mu_2, \ldots, \mu_p$, respectively, and with variance equal to unity, then $\sum_{i=1}^{p} y_i^2 = W$ is distributed as the noncentral chi-square with p degrees of freedom and parameter $\lambda = \frac{1}{2}\Sigma\mu_i^2$; and the frequency function is given by

$$f(W) = e^{-\lambda} \sum_{i=0}^{\infty} \frac{\lambda^i W^{\frac{1}{2}(p+2i)-1} e^{-W/2}}{i!\, 2^{\frac{1}{2}(p+2i)} \Gamma\left(\dfrac{p+2i}{2}\right)} \qquad 0 \leqslant W < \infty \qquad (4.1)$$

In vector notation this theorem could be restated: If $\mathbf{Y}$ is distributed $N(\boldsymbol{\mu},\mathbf{I})$, then $\mathbf{Y}'\mathbf{Y}$ is distributed as a noncentral chi-square with p degrees of freedom and parameter $\lambda = \frac{1}{2}\boldsymbol{\mu}'\boldsymbol{\mu}$.

Proof: The proof will consist in finding the moment-generating function of $\sum_{i=1}^{p} y_i^2$, finding the moment-generating function of the distribution given in (4.1), and then using Theorem 2.1. The moment-generating function of $W = \sum_{i=1}^{p} y_i^2$ is, by definition,

$$m_W(t) = E(e^{Wt}) = \int_{-\infty}^{\infty}\int_{-\infty}^{\infty}\cdots\int_{-\infty}^{\infty} \frac{1}{(2\pi)^{p/2}} \times \exp\left[-\tfrac{1}{2}\Sigma(y_i - \mu_i)^2 + t\Sigma y_i^2\right] dy_1\, dy_2 \cdots dy_p$$

We shall work with the exponent, to get it into a more suitable form. We have

$$\begin{aligned} -\tfrac{1}{2}\Sigma(y_i - \mu_i)^2 + t\Sigma y_i^2 &= -\tfrac{1}{2}(\Sigma y_i^2 - 2t\Sigma y_i^2 - 2\Sigma y_i\mu_i + \Sigma\mu_i^2) \\ &= -\frac{1-2t}{2}\left[\Sigma\left(y_i - \frac{\mu_i}{1-2t}\right)^2 - \frac{\Sigma\mu_i^2}{(1-2t)^2} + \frac{\Sigma\mu_i^2}{1-2t}\right] \\ &= -\frac{\Sigma\mu_i^2}{2}\left(1 - \frac{1}{1-2t}\right) - \frac{1-2t}{2}\Sigma\left(y_i - \frac{\mu_i}{1-2t}\right)^2 \end{aligned}$$

Thus we get

$$\begin{aligned} m_W(t) &= \exp\left[-\tfrac{1}{2}\Sigma\mu_i^2\left(1 - \frac{1}{1-2t}\right)\right]\frac{1}{(2\pi)^{p/2}}\int_{-\infty}^{\infty}\int_{-\infty}^{\infty}\cdots\int_{-\infty}^{\infty} \\ &\qquad \times \exp\left[-\frac{1-2t}{2}\Sigma\left(y_i - \frac{\mu_i}{1-2t}\right)^2\right] dy_1 \cdots dy_p \\ &= \frac{1}{(1-2t)^{p/2}}\exp\left[-\tfrac{1}{2}\Sigma\mu_i^2\left(1 - \frac{1}{1-2t}\right)\right] = \frac{\exp(-\frac{1}{2}\Sigma\mu_i^2)}{(1-2t)^{p/2}} \\ &\qquad \times \exp\left[\frac{\Sigma\mu_i^2}{2(1-2t)}\right] \end{aligned}$$

Let $\lambda = \frac{1}{2}\Sigma\mu_i^2$. If we expand $e^{\lambda/(1-2t)}$ into a Taylor series about zero, we get

$$m_W(t) = e^{-\lambda}(1-2t)^{-p/2}\sum_{i=0}^{\infty}\frac{\lambda^i}{i!(1-2t)^i} = e^{-\lambda}\sum_{i=0}^{\infty}\frac{\lambda^i}{i!}(1-2t)^{-(p/2+i)} \tag{4.2}$$

We shall now find the moment-generating function of the noncentral chi-square frequency function given in Eq. (4.1). We notice that the noncentral chi-square can be written

$$f(W) = \sum_{i=0}^{\infty} e^{-\lambda} \frac{\lambda^i}{i!} g_{p+2i}(W)$$

where $g_{p+2i}(W)$ is the central chi-square frequency with $p + 2i$ degrees of freedom. Since $e^{-\lambda}\lambda^i/i!$ is the ith term of a Poisson distribution, we see that the noncentral chi-square is the sum of an infinite number of chi-square frequencies each weighted by a term from the Poisson distribution. Using the relationship

$$\int_0^\infty g_{p+2i}(W)\, dW = 1$$

it is easily verified that

$$\int_0^\infty f(W)\, dW = \sum_{i=0}^{\infty} \frac{e^{-\lambda}\lambda^i}{i!} = 1$$

since we can interchange the operations of integration and infinite summation. The moment-generating function of the noncentral chi-square is

$$m_w(t) = E(e^{wt}) = \int_0^\infty e^{wt} \sum_{i=0}^{\infty} \frac{e^{-\lambda}\lambda^i (w)^{\frac{1}{2}(p+2i)-1} e^{-w/2}}{i!\, 2^{\frac{1}{2}(p+2i)} \Gamma\left(\dfrac{p+2i}{2}\right)}\, dw$$

$$= \sum_{i=0}^{\infty} \frac{e^{-\lambda}\lambda^i}{i!} \int_0^\infty e^{wt} \frac{(w)^{\frac{1}{2}(p+2i)-1} e^{-w/2}}{2^{\frac{1}{2}(p+2i)} \Gamma\left(\dfrac{p+2i}{2}\right)}\, dw$$

The integral is, by definition, the moment-generating function of the central chi-square distribution, and equals $(1 - 2t)^{-\frac{1}{2}(p+2i)}$. Thus

$$m_w(t) = e^{-\lambda} \sum_{i=0}^{\infty} \frac{\lambda^i}{i!} (1 - 2t)^{-\frac{1}{2}(p+2i)} \tag{4.3}$$

Comparing Eqs. (4.2) and (4.3), we find that

$$m_w(t) = m_W(t)$$

and, since they exist in some neighborhood of zero, we use Theorem 2.1. This completes the proof.

If $\lambda = 0$, that is, if $\mu_1 = \mu_2 = \cdots = \mu_p = 0$, the noncentral chi-square distribution degenerates into the central chi-square distribution.

Since the noncentral chi-square is completely specified by two parameters p and λ, we shall adopt the notation: "A random variable W is distributed as $\chi'^2(p, \lambda)$" to indicate that W has a frequency function given by (4.1).

The noncentral chi-square distribution, like the central chi-square, possesses the reproductive property. This is stated in the following theorem.

◆ **Theorem 4.2** If $W_1, W_2, \ldots, W_k$ are jointly independent and if each is distributed as the noncentral chi-square, so that W_i has p_i degrees of freedom and noncentrality parameter λ_i, then $W = \sum_{i=1}^{k} W_i$ has the noncentral chi-square distribution with degrees of freedom $p = \sum_{i=1}^{k} p_i$ and noncentrality parameter $\lambda = \sum_{i=1}^{k} \lambda_i$.

The proof is left to the reader.

◆ **Theorem 4.3** If a vector $\mathbf{Y}$ is distributed $N(\boldsymbol{\mu},\sigma^2\mathbf{I})$, then $\mathbf{Y}'\mathbf{Y}/\sigma^2$ has the noncentral chi-square distribution with $\lambda = \boldsymbol{\mu}'\boldsymbol{\mu}/2\sigma^2$.

This proof also is left to the reader.

◆ **Theorem 4.4** If a vector $\mathbf{Y}$ is distributed $N(\boldsymbol{\mu},\mathbf{D})$ where $\mathbf{D}$ is diagonal, then $\mathbf{Y}'\mathbf{D}^{-1}\mathbf{Y}$ has the noncentral chi-square distribution with p degrees of freedom and parameter $\lambda = \frac{1}{2}\boldsymbol{\mu}'\mathbf{D}^{-1}\boldsymbol{\mu}$.

Proof: Since $\mathbf{D}$ is positive definite, we know that there exists a nonsingular $p \times p$ matrix $\mathbf{B}$ (not orthogonal) such that $\mathbf{B}'\mathbf{D}\mathbf{B} = \mathbf{I}$. Let $\mathbf{Z} = \mathbf{B}'\mathbf{Y}$; by Theorem 3.22, $\mathbf{Z}$ is distributed $N(\mathbf{B}'\boldsymbol{\mu},\mathbf{I})$. So $\mathbf{Z}'\mathbf{Z}$ is distributed as the noncentral chi-square with p degrees of freedom and parameter $\lambda = \frac{1}{2}\boldsymbol{\mu}'\mathbf{B}\mathbf{B}'\boldsymbol{\mu}$. But from $\mathbf{B}'\mathbf{D}\mathbf{B} = \mathbf{I}$, we get $\mathbf{B}\mathbf{B}' = \mathbf{D}^{-1}$; so $\mathbf{Z}'\mathbf{Z} = \mathbf{Y}'\mathbf{B}\mathbf{B}'\mathbf{Y} = \mathbf{Y}'\mathbf{D}^{-1}\mathbf{Y}$ and $\lambda = \frac{1}{2}\boldsymbol{\mu}'\mathbf{B}\mathbf{B}'\boldsymbol{\mu} = \frac{1}{2}\boldsymbol{\mu}'\mathbf{D}^{-1}\boldsymbol{\mu}$, and the proof is complete.

4.3 Noncentral *F*

Since the noncentral chi-square has properties similar to those of the central chi-square, we consider the possibility that the ratio of two noncentral chi-squares might have some properties similar to Snedecor's F distribution. In fact, the ratio of a noncentral chi-square to a central chi-square will play an extremely important role in the theory of the power of tests in regression and analysis of variance. Therefore, we shall derive what is termed the *noncentral F distribution.*

◆ **Theorem 4.5** If a random variable w is distributed as $\chi'^2(p, \lambda)$, that is, as the noncentral chi-square with p degrees of freedom and parameter λ, and if another random variable z is distributed as $\chi'^2(q, 0)$, that is, as the central chi-square with q degrees of freedom, and if w and z are independent, then the quantity

$$u = \frac{w}{p}\frac{q}{z}$$

is distributed as the noncentral F distribution with p,q degrees of freedom and with noncentrality parameter λ. The frequency function of u is

$$f(u) = \sum_{i=0}^{\infty} \frac{\Gamma\left(\frac{2i+p+q}{2}\right)\left(\frac{p}{q}\right)^{\frac{1}{2}(2i+p)} \lambda^i e^{-\lambda}}{\Gamma\left(\frac{q}{2}\right)\Gamma\left(\frac{2i+p}{2}\right) i!} \frac{u^{\frac{1}{2}(2i+p-2)}}{\left(1+\frac{pu}{q}\right)^{\frac{1}{2}(2i+p+q)}} \qquad 0 \leqslant u < \infty \tag{4.4}$$

The noncentral F is generally denoted by F'. We shall adopt the following notation: "A random variable u is distributed as $F'(p, q, \lambda)$," this will indicate that u has the frequency function (4.4) with p degrees of freedom in the numerator, q degrees of freedom in the denominator, and noncentrality λ.

Proof: The joint distribution of w and z is

$$f(w, z) = f_1(w)f_2(z) = \frac{z^{\frac{1}{2}(q-2)}e^{-z/2}}{2^{q/2}\Gamma\left(\frac{q}{2}\right)} \sum_{i=0}^{\infty} \frac{e^{-\lambda}\lambda^i}{i!} \frac{w^{\frac{1}{2}(2i+p-2)}e^{-w/2}}{2^{\frac{1}{2}(2i+p)}\Gamma\left(\frac{2i+p}{2}\right)}$$

$$0 \leqslant w < \infty$$
$$0 \leqslant z < \infty$$

Let
$$\frac{w}{z} = x \qquad w = v$$

The Jacobian of the transformation is v/x^2. Substituting, we get

$$g(x, v) = \sum_{i=0}^{\infty} C_i \left(\frac{v}{x}\right)^{\frac{1}{2}(q-2)} e^{-v/2x} v^{\frac{1}{2}(2i+p-2)} e^{-v/2} \frac{v}{x^2} \qquad \begin{array}{l} 0 \leqslant v < \infty \\ 0 \leqslant x < \infty \end{array}$$

where
$$C_i = \frac{e^{-\lambda}\lambda^i}{i!\, 2^{q/2}\Gamma\left(\frac{q}{2}\right) 2^{\frac{1}{2}(2i+p)}\Gamma\left(\frac{2i+p}{2}\right)}$$

The marginal distribution of x is

$$h(x) = \int_0^{\infty} g(x, v)\, dv = \sum_{i=0}^{\infty} C_i x^{-\frac{1}{2}(q+2)} \int_0^{\infty} v^{\frac{1}{2}(2i+p+q-2)} e^{-\frac{1}{2}v(1+1/x)}\, dv$$

If we let $(v/2)(1 + 1/x) = r$, then $dv = 2x/(1 + x)\,dr$, and

$$h(x) = \sum_{i=0}^{\infty} C_i x^{-\frac{1}{2}(q+2)} \int_0^{\infty} \left(\frac{2x}{1+x}\right)^{\frac{1}{2}(2i+p+q)} r^{\frac{1}{2}(2i+p+q-2)} e^{-r}\, dr$$

$$= \sum_{i=0}^{\infty} \frac{C_i x^{\frac{1}{2}(2i+p-2)} \Gamma\left(\frac{2i+p+q}{2}\right) 2^{\frac{1}{2}(2i+p+q)}}{(1+x)^{\frac{1}{2}(2i+p+q)}}$$

If we make another substitution, $u = (q/p)x$, we get

$$f(u) = \sum_{i=0}^{\infty} \frac{\Gamma\left(\frac{2i+p+q}{2}\right)\left(\frac{p}{q}\right)^{\frac{1}{2}(2i+p)} \lambda^i e^{-\lambda}}{\Gamma\left(\frac{q}{2}\right)\Gamma\left(\frac{2i+p}{2}\right) i!} \frac{u^{\frac{1}{2}(2i+p-2)}}{\left(1+\frac{pu}{q}\right)^{\frac{1}{2}(2i+p+q)}} \qquad 0 \leqslant u < \infty$$

which is the desired result.

We note that, if $\lambda = 0$, $f(u)$ reduces to the frequency function of the central (Snedecor's) F. In the F' distribution there are three parameters: the degrees of freedom for the chi-square in the numerator of F', the degrees of freedom for the chi-square in the denominator of F', and the noncentrality λ of the chi-square in the numerator.

The frequency function of F' has not been extensively tabulated, but P. C. Tang has compiled tables that can be used to evaluate $\int_0^{F_\alpha} f(F')\,dF'$ for certain values of F_α. These tables are not given explicitly in terms of F', but in terms of E^2, where $E^2 = pF'/(q + pF')$. It is easily seen that the frequency function of E^2 is

$$g(E^2; p,q,\lambda) = \sum_{i=0}^{\infty} \frac{\Gamma\left(\frac{2i+p+q}{2}\right)}{\Gamma\left(\frac{q}{2}\right)\Gamma\left(\frac{2i+p}{2}\right)} \frac{\lambda^i}{i!} e^{-\lambda}(E^2)^{\frac{1}{2}(2i+p-2)}(1 - E^2)^{\frac{1}{2}(q-2)} \qquad 0 \leqslant E^2 \leqslant 1 \quad (4.5)$$

If $\lambda = 0$, $g(E^2;\ p,q,\lambda = 0)$ is the *beta distribution*. Similarly, when $\lambda \neq 0$, we term $g(E^2; p,q,\lambda)$ the *noncentral beta distribution*.

We are interested in evaluating the integral

$$\int_0^{E_\alpha^2} g(E^2; p,q,\lambda)\, dE^2 \qquad (4.6)$$

where the quantity E_α^2 is obtained from the integral

$$\int_{E_\alpha^2}^{1} g(E^2; p, q, \lambda = 0)\, dE^2 = \alpha \qquad (4.7)$$

where α is given. Before explaining how the tables prepared by P. C. Tang are used to evaluate (4.6) and (4.7), we shall discuss the implication of these integrals in the F' distribution. If we transform from E^2 to F' by the relationship

$$E^2 = \frac{pF'}{q + pF'}$$

we get $$\alpha = \int_{E_\alpha^2}^1 g(E^2; p, q, \lambda = 0)\, dE^2 = \int_{F_\alpha}^\infty f(F'; p, q, \lambda = 0)\, dF'$$

where $F_\alpha = qE_\alpha^2/(p - pE_\alpha^2)$. If $\alpha = .05$ (say), we can read F_α from Snedecor's tables for the F distribution with p and q degrees of freedom.

For example, suppose we wish to test the hypothesis $\lambda = 0$ in the F' distribution and use a type I error size α, where the interval $F_\alpha \leqslant F' < \infty$ is the critical region. In other words, if the observed F' falls in this interval, we reject the hypothesis $\lambda = 0$. The power of the test $\beta(\lambda)$ is the probability that the observed F' falls in the critical region $F_\alpha \leqslant F' < \infty$ when $\lambda \neq 0$, and is equal to the integral

$$\beta(\lambda) = \int_{F_\alpha}^\infty f(F'; p,q,\lambda)\, dF' \tag{4.8}$$

If in (4.8) we transform from F' to E^2, we get

$$\beta(\lambda) = \int_{E_\alpha^2}^1 g(E^2; p,q,\lambda)\, dE^2$$

So the integral in Eq. (4.6) equals $1 - \beta(\lambda)$, or unity minus the power of the test, which is the probability of a type II error.

Since most of the test functions in regression analysis and in the analysis of variance are based on the F distribution, it is apparent that the integral in (4.6) will be very important. The tables calculated by P. C. Tang allow us to obtain $\beta(\lambda)$ for various values of p, q, λ and for $\alpha = .05$ and .01.

Tang evaluated the integral

$$P(\text{II}) = 1 - \beta(\phi) = \int_0^{E_\alpha^2} g(E^2; f_1, f_2, \phi)\, dE^2 \tag{4.9}$$

for various values of f_1, f_2, ϕ, and E_α^2, where E_α^2 is obtained from Eq. (4.7) for $\alpha = .05$ and .01. In Tang's notation, the numerator degrees of freedom p is denoted by f_1, and the denominator degrees of freedom q is denoted by f_2. Instead of using the noncentrality parameter λ, Tang used the parameter ϕ, where

$$\phi = \sqrt{\frac{2\lambda}{f_1 + 1}} \tag{4.10}$$

where f_1 is the degrees of freedom in the numerator of the F statistic. The procedure for computing the power $\beta(\lambda)$ for a given λ is as follows (f_1 and f_2 will be given):

1. Choose the probability of type I error α; that is, set $\alpha = .01$ or $\alpha = .05$.
2. Find E^2_α from the Tang tables.
3. Compute $\phi = \sqrt{2\lambda/(f_1 + 1)}$.
4. Find $P(\text{II})$ for the appropriate values of f_1, f_2, and ϕ.
5. Then $\beta(\lambda) = 1 - P(\text{II})$.

Some examples will be given in the use of the tables.

4.3.1 Example. If $\alpha = .05$, $p = 6$, and $q = 10$, we see that $E^2_{.05} = .659$. Continuing the example, we see that, if $\lambda = 14$ (say), then $\phi = 2$, and $\beta(\lambda = 14) = 1 - .142 = .858$. That is, when $\lambda = 14$, the probability of rejecting the hypothesis that $\lambda = 0$ is equal to .858 when a type I error probability of .05 is used.

4.3.2 Example. Suppose $p = 2, q = 4$; find $\beta(\lambda = \frac{7.5}{2})$ when $\alpha = .01$. Now $E^2_{.01} = .900$, $\phi = \sqrt{\frac{7.5}{3}} = 5$; so $1 - \beta(\lambda = \frac{7.5}{2}) = .095$ and $\beta(\lambda = \frac{7.5}{2}) = .905$. Thus, if we are using F' as a test criterion, we compute E^2 and reject H_0: $\lambda = 0$ if $E^2 > .900$. The probability of rejecting H_0 when $\lambda = \frac{7.5}{2}$ is equal to .905.

4.3.3 Example. Suppose that a random variable y is distributed $N(\mu,1)$ and that we want to test the hypothesis H_0: $\mu = 0$. Suppose further that we decide to take a random sample of size n (denoted by $y_1, y_2, \ldots, y_n$) from the distribution and use

$$u - \frac{\bar{y}^2 n(n-1)}{\Sigma(y_i - \bar{y})^2}$$

as the test function. In order to find the critical region for a fixed type I error, it is necessary to determine the distribution of the random variable u when H_0 is true, i.e., when $\mu = 0$.

Since the basic variables $y_1, y_2, \ldots, y_n$ are independent and normal with mean 0 (assuming H_0 is true) and variance 1, it follows that

1. $\bar{y}$ is distributed normally with mean 0 and variance $1/n$. Therefore, $\bar{y}\sqrt{n}$ is distributed normally with mean 0 and variance 1, and $n\bar{y}^2$ has a chi-square distribution with 1 degree of freedom.
2. $\sum_{i=1}^{n} (y_i - \bar{y})^2$ is distributed as chi-square with $n - 1$ degrees of freedom.
3. $\sqrt{n}\bar{y}$ and $\sum_{i=1}^{n} (y_i - \bar{y})^2$ are independent.

Therefore,

$$u = \frac{\bar{y}^2 n(n-1)}{\Sigma(y_i - \bar{y})^2}$$

is distributed as Snedecor's F distribution with 1 and $n - 1$ degrees of freedom.

We shall choose for the critical region the right-hand tail of the F distribution such that the type I error is, say, 5 per cent. That is, if the observed u is greater than F_α, the tabulated F, then H_0 will be rejected. Of course, if μ does not equal zero, we *want* to reject H_0. We shall examine the probability of rejecting H_0 (getting a value of u in the critical region) when μ is not equal to zero. To do this we must find the distribution of the test function u when $\mu \neq 0$. From Theorem 4.1 it follows that $n\bar{y}^2$ has the noncentral chi-square distribution with 1 degree of freedom and parameter $\lambda = \frac{1}{2}n\mu^2$. Also, it is known that $\Sigma(y_i - \bar{y})^2$ has the central chi-square distribution even if $\mu \neq 0$, and the two are independent. Thus, when $\mu \neq 0$, the test function u is distributed as $F'(1, n - 1, n\mu^2/2)$. If $n = 7$, we reject H_0 if the calculated u is greater than 5.99. The power of the test can then be evaluated for different values of λ. For example, if $\lambda = 8$ (i.e., if $\mu^2 = \frac{16}{7}$), then $\phi = 2.8$ and β $(\lambda = 8) = .899$. If $n = 21$ instead of 7, then $E^2_{.05} = .179$ and β $(\lambda = 8)$ is about .955. This gives us some idea of the value of increasing the number of observations that are taken.

4.4 Distribution of Quadratic Forms

Since the distribution of quadratic forms will be very important to us in succeeding chapters, we shall prove some theorems that will enable us to ascertain the distribution by examining only the matrix of the quadratic form and the covariance matrix of the random variables.

◆ **Theorem 4.6** If the random vector $\mathbf{Y}$ is distributed $N(\mathbf{0},\mathbf{I})$, a necessary and sufficient condition that the quadratic form $\mathbf{Y}'\mathbf{AY}$ be distributed as $\chi^2(k)$ is that $\mathbf{A}$ be an idempotent matrix of rank k.

Proof: We shall prove sufficiency, but shall only sketch the proof of necessity. If $\mathbf{A}$ is idempotent of rank k, there exists an orthogonal matrix $\mathbf{P}$ such that $\mathbf{P}'\mathbf{AP} = \begin{pmatrix} \mathbf{I} & \mathbf{0} \\ \mathbf{0} & \mathbf{0} \end{pmatrix}$, where $\mathbf{I}$ is the $k \times k$ identity matrix. Define a $p \times 1$ vector $\mathbf{Z}$ by the relationship $\mathbf{Z} = \mathbf{P}'\mathbf{Y}$. Then $\mathbf{Y}'\mathbf{AY} = \mathbf{Z}'\mathbf{P}'\mathbf{APZ} = \mathbf{Z}_1'\mathbf{Z}_1$, where $\mathbf{Z} = \begin{pmatrix} \mathbf{Z}_1 \\ \mathbf{Z}_2 \end{pmatrix}$. By Theorem 3.22, $\mathbf{Z}$ is distributed $N(\mathbf{0},\mathbf{I})$, and it follows that $\mathbf{Z}_1'\mathbf{Z}_1 = \sum_{i=1}^{k} z_i^2$, which is a sum of squares of independent normal variables with means 0 and variances 1, is distributed as $\chi^2(k)$.

To prove necessity, we shall assume that $\mathbf{Y}'\mathbf{A}\mathbf{Y}$ is distributed as $\chi^2(k)$ and show that $\mathbf{A}$ is idempotent of rank k. Since $\mathbf{A}$ is symmetric, there exists an orthogonal matrix $\mathbf{C}$ such that $\mathbf{C}'\mathbf{A}\mathbf{C} = \mathbf{D} = (d_{ij})$, where $\mathbf{D}$ is a diagonal matrix. If we let $(w_i) = \mathbf{W} = \mathbf{C}'\mathbf{Y}$, we get $\mathbf{Y}'\mathbf{A}\mathbf{Y} = \mathbf{W}'\mathbf{C}'\mathbf{A}\mathbf{C}\mathbf{W} = \mathbf{W}'\mathbf{D}\mathbf{W} = \sum_{i=1}^{p} w_i^2 d_{ii}$. Since $\mathbf{C}$ is orthogonal, the w_i are independent normal variables with mean 0 and variance 1. Let $v = \sum_{i=1}^{p} w_i^2 d_{ii}$. If we equate the moment-generating function of v to the moment-generating function of $\mathbf{Y}'\mathbf{A}\mathbf{Y}$, we get

$$\prod_{i=1}^{p} (1 - 2td_{ii})^{-\frac{1}{2}} = (1 - 2t)^{-k/2}$$

It is clear that there exists a neighborhood of zero such that, for all t in that neighborhood, the quantities on the left and right sides of the identity exist. It can be shown that, for this to be true, k of the d_{ii} must be equal to unity and the remaining d_{ii} must be equal to zero. But, since d_{ii} are the characteristic roots of $\mathbf{A}$, this implies that $\mathbf{A}$ is idempotent of rank k.

We shall extend this theorem to the case in which the mean is not zero. The proof follows along the lines of the proof of Theorem 4.6.

◆ **Theorem 4.7** If $\mathbf{Y}$ is distributed $N(\boldsymbol{\mu},\mathbf{I})$, then $\mathbf{Y}'\mathbf{A}\mathbf{Y}$ is distributed as $\chi'^2(k,\lambda)$, where $\lambda = \frac{1}{2}\boldsymbol{\mu}'\mathbf{A}\boldsymbol{\mu}$, if and only if $\mathbf{A}$ is an idempotent matrix of rank k.

◆ **Corollary 4.7.1** If $\mathbf{Y}$ is distributed $N(\boldsymbol{\mu},\sigma^2\mathbf{I})$, then $\mathbf{Y}'\mathbf{A}\mathbf{Y}/\sigma^2$ is distributed as $\chi'^2(k,\lambda)$, where $\lambda = \boldsymbol{\mu}'\mathbf{A}\boldsymbol{\mu}/2\sigma^2$, if and only if $\mathbf{A}$ is an idempotent matrix of rank k.

We shall now prove some theorems that will help us to determine the distribution of a quadratic form $\mathbf{Y}'\mathbf{A}\mathbf{Y}$ when the vector $\mathbf{Y}$ has as its covariance the matrix $\mathbf{V}$ instead of the identity matrix $\mathbf{I}$.

◆ **Theorem 4.8** If a vector $\mathbf{Y}$ is distributed $N(\mathbf{0},\mathbf{V})$, then $\mathbf{Y}'\mathbf{B}\mathbf{Y}$ is distributed as $\chi^2(k)$ if and only if $\mathbf{B}\mathbf{V}$ is idempotent of rank k.

Proof: By Theorem 1.22, there exists a matrix $\mathbf{C}$ (not necessarily orthogonal) such that $\mathbf{C}'\mathbf{V}\mathbf{C} = \mathbf{I}$. If we let $\mathbf{Z} = \mathbf{C}'\mathbf{Y}$, $\mathbf{Z}$ is distributed $N(\mathbf{0},\mathbf{I})$. Also, $\mathbf{Y}'\mathbf{B}\mathbf{Y} = \mathbf{Z}'\mathbf{C}^{-1}\mathbf{B}\mathbf{C}'^{-1}\mathbf{Z}$, and, by Theorem 4.6, it follows that $\mathbf{Z}'(\mathbf{C}^{-1}\mathbf{B}\mathbf{C}'^{-1})\mathbf{Z}$ is distributed as $\chi^2(k)$ if and only if $\mathbf{C}^{-1}\mathbf{B}\mathbf{C}'^{-1}$ is idempotent. That is, we must show that $\mathbf{C}^{-1}\mathbf{B}\mathbf{C}'^{-1}$ is idempotent if and only if $\mathbf{B}\mathbf{V}$ is idempotent. If $\mathbf{B}\mathbf{V}$ is idempotent, then $\mathbf{BV} = \mathbf{BVBV}$; and $\mathbf{B} = \mathbf{BVB} = \mathbf{BC}'^{-1}\mathbf{C}^{-1}\mathbf{B}$, and $\mathbf{C}^{-1}\mathbf{BC}'^{-1} = \mathbf{C}^{-1}\mathbf{BC}'^{-1}\mathbf{C}^{-1}\mathbf{BC}'^{-1}$. Also, if $\mathbf{C}^{-1}\mathbf{BC}'^{-1}$ is idempotent, then $\mathbf{C}^{-1}\mathbf{BC}'^{-1} = \mathbf{C}^{-1}\mathbf{BC}'^{-1}\mathbf{C}^{-1}\mathbf{BC}'^{-1} = \mathbf{C}^{-1}\mathbf{BVBC}'^{-1}$ and $\mathbf{BV} = \mathbf{BVBV}$. This completes the proof.

◆ **Theorem 4.9** If $\mathbf{Y}$ is distributed $N(\boldsymbol{\mu},\mathbf{V})$, then $\mathbf{Y}'\mathbf{BY}$ is distributed as $\chi'^2(k,\lambda)$, where $\lambda = \frac{1}{2}\boldsymbol{\mu}'\mathbf{B}\boldsymbol{\mu}$ and k is the rank of $\mathbf{B}$, if and only if $\mathbf{BV}$ is idempotent.

Proof: As in Theorem 4.8, if we let $\mathbf{Z} = \mathbf{C}'\mathbf{Y}$, then $\mathbf{Z}$ is distributed $N(\mathbf{C}'\boldsymbol{\mu},\mathbf{I})$, and, by Theorem 4.7, $\mathbf{Y}'\mathbf{BY} = \mathbf{Z}'(\mathbf{C}^{-1}\mathbf{BC}'^{-1})\mathbf{Z}$ is distributed as $\chi'^2(k,\lambda)$ if and only if $\mathbf{C}^{-1}\mathbf{BC}'^{-1}$ is idempotent (where $\lambda = \frac{1}{2}\boldsymbol{\mu}'\mathbf{CC}^{-1}\mathbf{BC}'^{-1}\mathbf{C}'\boldsymbol{\mu} = \frac{1}{2}\boldsymbol{\mu}'\mathbf{B}\boldsymbol{\mu}$). It is necessary, then, to show that $\mathbf{C}^{-1}\mathbf{BC}'^{-1}$ is idempotent if and only if $\mathbf{BV}$ is idempotent; but this has been proved in Theorem 4.8.

4.5 Independence of Quadratic Forms

The preceding four theorems give us methods of determining when quadratic forms are distributed as the central or noncentral chi-square. It is also very important to have methods available to aid us in determining when quadratic forms are independent. This will be the text of the following seven theorems.

◆ **Theorem 4.10** If $\mathbf{Y}$ is distributed $N(\boldsymbol{\mu},\mathbf{I})$, the two positive semidefinite quadratic forms $\mathbf{Y}'\mathbf{AY}$ and $\mathbf{Y}'\mathbf{BY}$ are independent if and only if $\mathbf{AB} = \mathbf{0}$.

Proof: To prove sufficiency, assume $\mathbf{AB} = \mathbf{0}$. Taking the transpose of each side gives $\mathbf{B}'\mathbf{A}' = \mathbf{0}$, but, since $\mathbf{A}$ and $\mathbf{B}$ are symmetric, we get $\mathbf{AB} = \mathbf{BA} = \mathbf{0}$; that is to say, $\mathbf{A}$ and $\mathbf{B}$ commute. Hence, by Theorem 1.32, there exists an orthogonal transformation $\mathbf{P}$ such that $\mathbf{P}'\mathbf{AP} = \mathbf{D}_1^*$ and $\mathbf{P}'\mathbf{BP} = \mathbf{D}_2^*$, where $\mathbf{D}_1^*$ and $\mathbf{D}_2^*$ are each diagonal. If $\mathbf{AB} = \mathbf{0}$, it follows that $\mathbf{P}'\mathbf{APP}'\mathbf{BP} = \mathbf{0}$, which gives $\mathbf{D}_1^*\mathbf{D}_2^* = \mathbf{0}$. Thus, if the ith diagonal element of $\mathbf{D}_1^*$ is nonzero, the ith diagonal element of $\mathbf{D}_2^*$ must be equal to zero. We can then write, without loss of generality,

$$\mathbf{D}_1^* = \begin{pmatrix} \mathbf{D}_1 & \mathbf{0} & \mathbf{0} \\ \mathbf{0} & \mathbf{0} & \mathbf{0} \\ \mathbf{0} & \mathbf{0} & \mathbf{0}^* \end{pmatrix} \qquad \mathbf{D}_2^* = \begin{pmatrix} \mathbf{0} & \mathbf{0} & \mathbf{0} \\ \mathbf{0} & \mathbf{D}_2 & \mathbf{0} \\ \mathbf{0} & \mathbf{0} & \mathbf{0}^* \end{pmatrix}$$

where $\mathbf{D}_i$ is a $p_i \times p_i$ diagonal matrix with rank p_i and $\mathbf{0}^*$ is the $p_3 \times p_3$ null matrix where $p_1 + p_2 + p_3 = p$. If we let $\mathbf{Z} = \mathbf{P}'\mathbf{Y}$, then $\mathbf{Z}$ is distributed $N(\mathbf{P}'\boldsymbol{\mu},\mathbf{I})$. Also, $\mathbf{Y}'\mathbf{AY} = \mathbf{Z}'\mathbf{P}'\mathbf{APZ}$

$= \mathbf{Z}'\mathbf{D}_1^*\mathbf{Z} = \sum_{j=1}^{p_1} d_{1j}z_j^2$, where d_{1j} is the jth diagonal element of $\mathbf{D}_1$; if we apply the same transformation to $\mathbf{Y}'\mathbf{B}\mathbf{Y}$, we get $\mathbf{Y}'\mathbf{B}\mathbf{Y} = \mathbf{Z}'\mathbf{P}'\mathbf{B}\mathbf{P}\mathbf{Z} = \mathbf{Z}'\mathbf{D}_2^*\mathbf{Z} = \sum_{j=p_1+1}^{p_1+p_2} d_{2j}z_j^2$, where d_{2j} is the jth diagonal element of $\mathbf{D}_2$. Since $\mathbf{Y}'\mathbf{A}\mathbf{Y}$ depends only on the first p_1 element of $\mathbf{Z}$ and since $\mathbf{Y}'\mathbf{B}\mathbf{Y}$ depends only on the elements of $\mathbf{Z}$ from $p_1 + 1$ to p_2 and since all the elements of $\mathbf{Z}$ are independent, $\mathbf{Y}'\mathbf{A}\mathbf{Y}$ and $\mathbf{Y}'\mathbf{B}\mathbf{Y}$ are also independent. This completes the sufficiency part of the proof. The proof of necessity will be left for the reader.

◆ **Theorem 4.11** If $\mathbf{Y}$ is distributed $N(\boldsymbol{\mu},\mathbf{I})$, the set of positive semidefinite quadratic forms $\mathbf{Y}'\mathbf{B}_1\mathbf{Y}, \mathbf{Y}'\mathbf{B}_2\mathbf{Y}, \ldots, \mathbf{Y}'\mathbf{B}_k\mathbf{Y}$ are jointly independent if and only if $\mathbf{B}_i\mathbf{B}_j = \mathbf{0}$ for all $i \neq j$.

Proof: The proof is an extension of Theorem 4.10.

◆ **Theorem 4.12** If $\mathbf{Y}$ is distributed $N(\boldsymbol{\mu},\mathbf{I})$, the set of quadratic forms $\mathbf{Y}'\mathbf{A}_1\mathbf{Y}, \mathbf{Y}'\mathbf{A}_2\mathbf{Y}, \ldots, \mathbf{Y}'\mathbf{A}_k\mathbf{Y}$, where the rank of $\mathbf{A}_i$ is n_i, are jointly independent and $\mathbf{Y}'\mathbf{A}_i\mathbf{Y}$ is distributed as $\chi'^2(n_i,\lambda_i)$, where $\lambda_i = \frac{1}{2}\boldsymbol{\mu}'\mathbf{A}_i\boldsymbol{\mu}$, if any two of the following three conditions are satisfied:

(1) Each $\mathbf{A}_i$ is idempotent.

(2) $\sum_{i=1}^{k} \mathbf{A}_i$ is idempotent.

(3) $\mathbf{A}_i\mathbf{A}_j = \mathbf{0}$ for all $i \neq j$.

Proof: By Theorem 1.68, any two of the conditions imply the third. By Theorem 4.7, condition (1) implies that $\mathbf{Y}'\mathbf{A}_i\mathbf{Y}$ is distributed as $\chi'^2(n_i,\lambda_i)$ where n_i is the rank of $\mathbf{A}_i$ and $\lambda_i = \frac{1}{2}\boldsymbol{\mu}'\mathbf{A}_i\boldsymbol{\mu}$. By Theorem 4.11, condition (3) implies that the set is jointly independent. Thus the theorem is established.

An important case of Theorem 4.12 that occurs very frequently in the analysis of variance and in the theory of regression is when $\Sigma\mathbf{A}_i = \mathbf{I}$. In this case, condition (1) is necessary and sufficient for condition (3) (and, of course, vice versa). So we shall state the following very useful theorem.

◆ **Theorem 4.13** If $\mathbf{Y}$ is distributed $N(\boldsymbol{\mu},\mathbf{I})$ and if $\mathbf{Y}'\mathbf{Y} = \sum_{i=1}^{k} \mathbf{Y}'\mathbf{A}_i\mathbf{Y}$, either of the following two conditions is necessary and sufficient for $\mathbf{Y}'\mathbf{A}_j\mathbf{Y}$ to be distributed as $\chi'^2(n_j,\lambda_j)$, where n_j is the rank of $\mathbf{A}_j$ and $\lambda_j = \frac{1}{2}\boldsymbol{\mu}'\mathbf{A}_j\boldsymbol{\mu}$, and for the set $\mathbf{Y}'\mathbf{A}_1\mathbf{Y}, \mathbf{Y}'\mathbf{A}_2\mathbf{Y}, \ldots, \mathbf{Y}'\mathbf{A}_k\mathbf{Y}$ to be jointly independent:

(1) $\mathbf{A}_1, \mathbf{A}_2, \ldots, \mathbf{A}_k$ are each idempotent matrices.

(2) $\mathbf{A}_i\mathbf{A}_j = \mathbf{0}$ for all $i \neq j$.

The proof is immediate from Theorem 4.12. Another important form of this theorem is

◆ **Theorem 4.14** If $\mathbf{Y}$ is distributed $N(\boldsymbol{\mu},\mathbf{I})$ and if $\mathbf{Y}'\mathbf{Y} = \sum_{i=1}^{k} \mathbf{Y}'\mathbf{A}_i\mathbf{Y}$, a necessary and sufficient condition that $\mathbf{Y}'\mathbf{A}_i\mathbf{Y}$ be distributed as $\chi'^2(n_i,\lambda_i)$, where n_i is the rank of $\mathbf{A}_i$ and $\lambda_i = \frac{1}{2}\boldsymbol{\mu}'\mathbf{A}_i\boldsymbol{\mu}$, and for the $\mathbf{Y}'\mathbf{A}_1\mathbf{Y}, \mathbf{Y}'\mathbf{A}_2\mathbf{Y}, \ldots, \mathbf{Y}'\mathbf{A}_k\mathbf{Y}$ to be jointly independent, is that the rank of the sum of the $\mathbf{A}_i$ be equal to the sum of the ranks of the separate $\mathbf{A}_i$; that is to say, that $\Sigma\rho(\mathbf{A}_i) = \rho(\Sigma\mathbf{A}_i) = \rho(\mathbf{I})$.

Proof: It was shown in Theorem 1.69 that for the sum of the ranks of the $\mathbf{A}_i$ to be equal to the rank of the sum of the $\mathbf{A}_i$ is a necessary and sufficient condition for conditions 1 and 2 of Theorem 4.13. Hence the result.

The famous Cochran-Fisher theorem is a special case of Theorem 4.14, namely, for the special case when $\boldsymbol{\mu} = \mathbf{0}$. The proof of Theorem 4.14 was first given by Madow.

◆ **Theorem 4.15** If $\mathbf{Y}$ is distributed $N(\boldsymbol{\mu},\sigma^2\mathbf{I})$, the positive semidefinite quadratic forms $\mathbf{Y}'\mathbf{A}\mathbf{Y}$ and $\mathbf{Y}'\mathbf{B}\mathbf{Y}$ are independent if $\text{tr}(\mathbf{AB}) = 0$, or, in other words, if the covariance of $\mathbf{Y}'\mathbf{A}\mathbf{Y}$ and $\mathbf{Y}'\mathbf{B}\mathbf{Y}$ equals zero.

◆ **Theorem 4.16** Let $\mathbf{Y}$ be distributed $N(\boldsymbol{\mu},\sigma^2\mathbf{I})$, and let $\sum_{i=1}^{k} \mathbf{Y}'\mathbf{A}_i\mathbf{Y} = \mathbf{Y}'\mathbf{Y}$, where the rank of $\mathbf{A}_i$ is n_i. Any one of the three conditions listed below is a necessary and sufficient condition that the following two statements be true:

(1) $\mathbf{Y}'\mathbf{A}_i\mathbf{Y}/\sigma^2$ is distributed as $\chi'^2(n_i,\lambda_i)$, where $\lambda_i = \boldsymbol{\mu}'\mathbf{A}_i\boldsymbol{\mu}/2\sigma^2$.

(2) $\mathbf{Y}'\mathbf{A}_i\mathbf{Y}$ and $\mathbf{Y}'\mathbf{A}_j\mathbf{Y}$ are independent if $i \neq j$.

The conditions are:

(1) $\mathbf{A}_i$ is idempotent for all $i = 1, 2, \ldots, k$.

(2) $\mathbf{A}_i\mathbf{A}_j = \mathbf{0}$ for all $i \neq j$.

(3) $\sum_{i=1}^{k} n_i = n$; that is, the rank of the sum of the $\mathbf{A}_i$ is equal to the sum of the ranks of the $\mathbf{A}_i$.

The proofs of Theorems 4.15 and 4.16 will be left to the reader.

4.6 Independence of Linear and Quadratic Forms

By Theorem 3.22 we know that, if $\mathbf{Y}$ is distributed $N(\boldsymbol{\mu},\mathbf{V})$ and if $\mathbf{B}$ is a known matrix of constants, $\mathbf{BY}$ is distributed $N(\mathbf{B}\boldsymbol{\mu},\mathbf{BVB}')$. It will be important to be able to determine when the linear forms $\mathbf{BY}$ and a quadratic form $\mathbf{Y}'\mathbf{A}\mathbf{Y}$ are statistically independent. This is the substance of the next theorem.

◆ **Theorem 4.17** If $\mathbf{B}$ is a $q \times n$ matrix, $\mathbf{A}$ is an $n \times n$ matrix, and $\mathbf{Y}$ is distributed $N(\boldsymbol{\mu},\sigma^2\mathbf{I})$, then the linear forms $\mathbf{BY}$ are independent of the quadratic form $\mathbf{Y'AY}$ if $\mathbf{BA} = \mathbf{0}$.

Proof: Suppose $\mathbf{P}$ is an orthogonal matrix such that $\mathbf{P'AP} = \mathbf{D}$, where $\mathbf{D}$ is a diagonal matrix. Let $\mathbf{P'Y} = \mathbf{Z}$; then $\mathbf{Z}$ is distributed $N(\mathbf{P'}\boldsymbol{\mu},\sigma^2\mathbf{I})$. Let $\mathbf{C} = \mathbf{BP}$ and $\mathbf{D} = \begin{pmatrix} \mathbf{D}_1 & \mathbf{0} \\ \mathbf{0} & \mathbf{0} \end{pmatrix}$, where $\mathbf{D}_1$ is a diagonal matrix with nonzero elements on the diagonal. Now

$$\mathbf{0} = \mathbf{BA} = \mathbf{BAP} = \mathbf{BPP'AP} = \mathbf{CD} = \begin{pmatrix} \mathbf{C}_{11} & \mathbf{C}_{12} \\ \mathbf{C}_{21} & \mathbf{C}_{22} \end{pmatrix}\begin{pmatrix} \mathbf{D}_1 & \mathbf{0} \\ \mathbf{0} & \mathbf{0} \end{pmatrix} = \begin{pmatrix} \mathbf{0} \\ \mathbf{0} \end{pmatrix}$$

This implies that $\mathbf{C}_{11}\mathbf{D}_1 = \mathbf{0}$ and $\mathbf{C}_{21}\mathbf{D}_1 = \mathbf{0}$, which imply $\mathbf{C}_{11} = \mathbf{0}$ and $\mathbf{C}_{21} = \mathbf{0}$. So $\mathbf{C}$ can be written $\mathbf{C} = (\mathbf{0},\mathbf{C}_2)$, where $\mathbf{C}_2 = \begin{pmatrix} \mathbf{C}_{12} \\ \mathbf{C}_{22} \end{pmatrix}$. Now

$$\mathbf{BY} = \mathbf{BPP'Y} = \mathbf{CZ} = (\mathbf{0}, \mathbf{C}_2)\begin{pmatrix} \mathbf{Z}_1 \\ \mathbf{Z}_2 \end{pmatrix} = \mathbf{C}_2\mathbf{Z}_2$$

and

$$\mathbf{Y'AY} = \mathbf{Y'PP'APP'Y} = \mathbf{Z'DZ} = (\mathbf{Z}_1', \mathbf{Z}_2')\begin{pmatrix} \mathbf{D}_1 & \mathbf{0} \\ \mathbf{0} & \mathbf{0} \end{pmatrix}\begin{pmatrix} \mathbf{Z}_1 \\ \mathbf{Z}_2 \end{pmatrix} = \mathbf{Z}_1'\mathbf{D}_1\mathbf{Z}_1$$

So $\mathbf{BY}$ depends on the elements of $\mathbf{Z}$ in $\mathbf{Z}_2$, and $\mathbf{Y'AY}$ depends on the elements of $\mathbf{Z}$ in $\mathbf{Z}_1$, but all the elements in $\mathbf{Z}$ are independent; hence, the elements of $\mathbf{Z}_1$ are independent of the elements in $\mathbf{Z}_2$. Therefore, the result follows.

4.7 Expected Value of Quadratic Forms

We shall now state two theorems concerning the expected value of a quadratic form.

◆ **Theorem 4.18** If $\mathbf{Y}$ is distributed with mean $\mathbf{0}$ and covariance matrix $\sigma^2\mathbf{I}$, the expected value of the quadratic form $\mathbf{Y'AY}$ is equal to $\sigma^2 \operatorname{tr}(\mathbf{A})$.

Proof: $E(\mathbf{Y'AY}) = E(\sum_{ij} y_i y_j a_{ij}) = E(\sum_i a_{ii} y_i^2) + E(\sum_{i}\sum_{\substack{j \\ i \neq j}} y_i y_j a_{ij})$.

But, if $i \neq j$, then $E(y_i y_j) = E(y_i)E(y_j) = 0$; hence, $E(\mathbf{Y'AY}) = \sum_i a_{ii}E(y_i^2) = \sigma^2 \sum_i a_{ii} = \sigma^2 \operatorname{tr}(\mathbf{A})$.

◆ **Theorem 4.19** In Theorem 4.18, if $\mathbf{A}$ is an idempotent matrix of rank k, the expected value of $\mathbf{Y'AY}$ is equal to $k\sigma^2$.

The technique of analysis of variance can be viewed as the process of partitioning a sum of squares $\mathbf{Y'Y}$ into component parts such as $\mathbf{Y'Y} = \sum_{i=1}^{s} \mathbf{Y'A}_i\mathbf{Y}$. If $\mathbf{Y}$ is distributed $N(\boldsymbol{\mu},\sigma^2\mathbf{I})$ and if (say) $\boldsymbol{\mu}'\mathbf{A}_k = \mathbf{0}$ and if we desire to test the hypothesis $\boldsymbol{\mu}'\mathbf{A}_t = \mathbf{0}$ (say), then we can immediately see the utility of being able to determine the distribution and independence of the $\mathbf{Y'A}_i\mathbf{Y}$. For, if we choose as a test function $u = \mathbf{Y'A}_t\mathbf{Y}/\mathbf{Y'A}_k\mathbf{Y}$, we can determine the distribution of u; if, for example, $\mathbf{A}_t$ and $\mathbf{A}_k$ are idempotent and $\mathbf{A}_t\mathbf{A}_k = \mathbf{0}$, then $(n_k/n_t)u$ is distributed as the noncentral F with noncentrality $\lambda = \boldsymbol{\mu}'\mathbf{A}_t\boldsymbol{\mu}/2\sigma^2$. Also, we see that $\lambda = 0$ if and only if $\boldsymbol{\mu}'\mathbf{A}_t = \mathbf{0}$, and in that case $(n_k/n_t)u$ is distributed as Snedecor's F. Thus we can choose a suitable region of rejection and examine the power of the test by using Tang's tables. These ideas will be expanded in the ensuing chapters.

4.8 Additional Theorems

◆ **Theorem 4.20** Suppose that the $n \times 1$ vector $\mathbf{Y}$ is distributed $N(\boldsymbol{\mu},\mathbf{D})$ where $\mathbf{D}$ is diagonal. Then $\mathbf{Y'AY}$ is distributed as $\chi'^2(n - 1, \lambda)$, where $\lambda = \frac{1}{2}\boldsymbol{\mu}'\mathbf{A}\boldsymbol{\mu}$, if $\mathbf{A} = \mathbf{D}^{-1} - (\mathbf{D}^{-1}\mathbf{J}\mathbf{D}^{-1}/\mathbf{1}'\mathbf{D}^{-1}\mathbf{1})$. Also, $\lambda = 0$ if $\boldsymbol{\mu} = \mathbf{1}\mu$. ($\mathbf{1}$ is a vector with each element equal to 1, and $\mathbf{J} = \mathbf{1}\mathbf{1}'$.)

◆ **Theorem 4.21** Suppose that the $n \times 1$ vector $\mathbf{Y}$ is distributed $N(\boldsymbol{\mu},\mathbf{V})$; then $\mathbf{Y'AY}$ and $\mathbf{Y'BY}$ are independent if and only if $\mathbf{AVB} = \mathbf{0}$.

◆ **Theorem 4.22** Suppose that the $n \times 1$ vector $\mathbf{Y}$ is distributed $N(\boldsymbol{\mu},\mathbf{V})$.

Let $\mathbf{Y'AY} = \sum_{i=1}^{k+1} \mathbf{Y'A}_i\mathbf{Y}$, and suppose that $\mathbf{Y'AY}, \mathbf{Y'A}_1\mathbf{Y}, \ldots, \mathbf{Y'A}_k\mathbf{Y}$ are each noncentral chi-squares with parameters n,λ; n_1,λ_1; . . . ; n_k,λ_k; respectively. If $\mathbf{Y'A}_{k+1}\mathbf{Y}$ is nonnegative (i.e., semidefinite), it is distributed as $\chi'^2(n_{k+1}, \lambda_{k+1})$, where $n_{k+1} = n - \sum_{i=1}^{k} n_i$ and $\lambda_{k+1} = \lambda - \sum_{i=1}^{k} \lambda_i$, and $\mathbf{Y'A}_1\mathbf{Y}, \ldots, \mathbf{Y'A}_{k+1}\mathbf{Y}$ are independent.

The extent of Tang's tables is limited, and it might be desirable to find the value of the integral in Eq. (4.8) for values of p, q, and λ not included in the tables. Patnaik gave an approximation that will allow the noncentral F to be used to evaluate Eq. (4.8). This is given in the following theorem.

◆ **Theorem 4.23** Suppose u is distributed as $F'(p,q,\lambda)$; then the random variable $v = u/k$ is distributed approximately as $F(r,q)$, where $k = (p + 2\lambda)/p$ and $r = (p + 2\lambda)^2/(p + 4\lambda)$.

Using this theorem, we obtain for Eq. (4.8)

$$\beta(\lambda) = \int_{F_\alpha}^{\infty} f(u; p,q,\lambda)\, du$$

which is approximately equal to

$$\int_{(1/k)F_\alpha}^{\infty} g(v; r,q)\, dv$$

where F_α is such that

$$\int_{F_\alpha}^{\infty} f(u; p, q, \lambda = 0)\, du = \alpha$$

$f(u; p,q,\lambda)$ is the noncentral F distribution, and $g(v; r,q)$ and $f(u; p, q, \lambda = 0)$ are central F distributions.

4.8.1 Example. We shall illustrate Theorem 4.23 by referring to the example of Art. 4.3.1. We have $\alpha = .05$, $p = 6$, $q = 10$, and we want to find β $(\lambda = 14)$. The quantities in Theorem 4.23 are

$$k = \frac{p + 2\lambda}{p} = \frac{6 + 28}{6} = 5.67$$

$$r = \frac{(p + 2\lambda)^2}{p + 4\lambda} = \frac{(6 + 28)^2}{6 + 56} = 18.06$$

Also, $E^2_{.05} = .659$, and from the F table we get $F_{.05}(6,10) = 3.22$. So

$$\frac{1}{k} F_\alpha = \frac{3.22}{5.67} = .568$$

To find β $(\lambda = 14)$ we must find the probability of exceeding an F value of .568 with 18 and 10 degrees of freedom; i.e., we must evaluate

$$\int_{.568}^{\infty} g(v; 18,10)\, dv$$

where $g(v; 18,10)$ is the central F distribution with 18 and 10 degrees of freedom. The integral is approximately equal to .86, which agrees well with the value .858 obtained by using Tang's tables.

Problems

4.1 Find the mean and variance of a random variable u distributed as $\chi'^2(p,\lambda)$.
4.2 Find the mean of a random variable u distributed as $F'(p,q,\lambda)$.

4.3 Find the mean of a random variable u distributed as $E^2(p,q,\lambda)$.

4.4 Show, by using the transformation

$$E^2 = \frac{pF}{q + pF}$$

that $E^2_{.05}$ in Tang's tables corresponds to the $F_{.05}$ value of the central F for the following values of p and q:

p	6	7	2	8
q	3	10	20	60

4.5 Evaluate the following integral (and find $E^2_{.05}$):

$$\beta(\lambda) = \int_{E^2_{.05}}^{1} g(E^2;\, p,q,\lambda)\, dE^2$$

for the following values of the parameters (use Tang's tables):

p	2	4	5	6	3	7	7
q	6	2	18	30	5	2	4
λ	6	10	18.75	15	4.5	36	64

4.6 Prove Theorem 4.2 by the use of moment-generating functions.

4.7 Prove Theorem 4.3.

4.8 If $\mathbf{Y}$ is distributed $N(\boldsymbol{\mu},\mathbf{V})$, prove that the quadratic form of the p-variate normal

$$Q = (\mathbf{Y} - \boldsymbol{\mu})'\mathbf{V}^{-1}(\mathbf{Y} - \boldsymbol{\mu})$$

is distributed as $\chi^2(p)$.

4.9 If $\mathbf{Y}$ is distributed $N(\boldsymbol{\mu},\mathbf{V})$, prove that $\mathbf{Y}'\mathbf{V}^{-1}\mathbf{Y}$ is distributed as $\chi'^2(p,\lambda)$, where $\lambda = \frac{1}{2}\boldsymbol{\mu}'\mathbf{V}^{-1}\boldsymbol{\mu}$.

4.10 If $\mathbf{X}$ is an $n \times p$ matrix $(n > p)$ of rank p, show that $\mathbf{A} = \mathbf{X}(\mathbf{X}'\mathbf{X})^{-1}\mathbf{X}'$ is idempotent.

4.11 If the $n \times 1$ vector $\mathbf{Y}$ is distributed $N(\mathbf{0},\mathbf{I})$, find

$$E\{\mathbf{Y}'[\mathbf{I} - \mathbf{X}(\mathbf{X}'\mathbf{X})^{-1}\mathbf{X}']\mathbf{Y}\}$$

4.12 If the $n \times 1$ vector $\mathbf{Y}$ is distributed $N(\mathbf{0},\mathbf{I})$, find the distribution of

(*a*) $\mathbf{Y}'\mathbf{X}(\mathbf{X}'\mathbf{X})^{-1}\mathbf{X}'\mathbf{Y}$.

(*b*) $\mathbf{Y}'[\mathbf{I} - \mathbf{X}(\mathbf{X}'\mathbf{X})^{-1}\mathbf{X}']\mathbf{Y}$.

4.13 If the $n \times 1$ vector $\mathbf{Y}$ is distributed $N(\mathbf{X}\boldsymbol{\beta},\mathbf{I})$, where $\boldsymbol{\beta}$ is a $p \times 1$ unknown vector and $\mathbf{X}$ is as in Prob. 4.10, find the distribution of

(*a*) $\mathbf{Y}'[\mathbf{X}(\mathbf{X}'\mathbf{X})^{-1}\mathbf{X}']\mathbf{Y} = Q_1$

(*b*) $\mathbf{Y}'[\mathbf{I} - \mathbf{X}(\mathbf{X}'\mathbf{X})^{-1}\mathbf{X}']\mathbf{Y} = Q_2$

(*c*) and show that the quantities Q_1 and Q_2 are independent.

4.14 In Prob. 4.13, find the distribution of

$$u = \frac{n - p}{p}\,\frac{Q_1}{Q_2}$$

4.15 In Prob. 4.14, find $E(u)$.

4.16 If the 3×1 vector $\mathbf{Y}$ is distributed $N(\mathbf{0},\mathbf{I})$, find the following:

(*a*) $E(\mathbf{Y}'\mathbf{A}\mathbf{Y})$.

(*b*) $E(\mathbf{Y}'\mathbf{B}\mathbf{Y})$.

(*c*) The joint distribution of $\mathbf{Y}'\mathbf{A}\mathbf{Y}$ and $\mathbf{Y}'\mathbf{B}\mathbf{Y}$.

(*d*) The distribution of $u = \mathbf{Y}'\mathbf{A}\mathbf{Y}/\mathbf{Y}'\mathbf{B}\mathbf{Y}$, if

$$\mathbf{A} = \begin{pmatrix} \frac{2}{3} & -\frac{1}{3} & -\frac{1}{3} \\ -\frac{1}{3} & \frac{2}{3} & -\frac{1}{3} \\ -\frac{1}{3} & -\frac{1}{3} & \frac{2}{3} \end{pmatrix} \qquad \mathbf{B} = \begin{pmatrix} \frac{1}{3} & \frac{1}{3} & \frac{1}{3} \\ \frac{1}{3} & \frac{1}{3} & \frac{1}{3} \\ \frac{1}{3} & \frac{1}{3} & \frac{1}{3} \end{pmatrix}$$

4.17 If the $p \times 1$ vector $\mathbf{Y}$ is distributed $N(\mathbf{0},\mathbf{I})$,

(*a*) Find the matrix of $Q_1 = p\bar{y}^2$.

(*b*) Find the matrix of $Q_2 = \sum_{i=1}^{p} (y_i - \bar{y})^2$.

(*c*) Find the distribution of Q_1.

(*d*) Find the distribution of Q_2.

(*e*) Show that Q_1 and Q_2 are independent.

(*f*) Find the distribution of Q_1/Q_2.

(*g*) Show that $\mathbf{1}'\mathbf{Y}$ and $\sum_{i=1}^{p} (y_i - \bar{y})^2$ are independent.

(*h*) Find the expected value of Q_1.

(*i*) Find the expected value of Q_2.

4.18 Prove Theorem 4.19.

4.19 If the $n \times 1$ vector $\mathbf{Y}$ is distributed $N(\boldsymbol{\mu},\sigma^2\mathbf{I})$, find the expected value of $\mathbf{Y}'\mathbf{A}\mathbf{Y}$, where $\mathbf{A}$ is idempotent of rank p.

4.20 Using Prob. 4.1, find the noncentrality parameter λ of the distribution $\mathbf{Y}'\mathbf{A}\mathbf{Y}/\sigma^2$ if $\mathbf{Y}$ is distributed $N(\boldsymbol{\mu},\sigma^2\mathbf{I})$ and if $\mathbf{A}$ is idempotent.

4.21 Prove Theorem 4.20. *Hint:* Show that $\mathbf{AD}$ is idempotent.

4.22 Prove Theorem 4.21.

4.23 If the $n \times 1$ vector $\mathbf{Y}$ is distributed $N(\mathbf{0},\mathbf{A})$, where $\mathbf{A}$ is positive semidefinite of rank k, use Theorem 4.9 to prove that $\mathbf{Y}'\mathbf{Y}$ is distributed as $\chi'^2(k,0)$ if $\mathbf{A}$ is idempotent.

4.24 Suppose that y_i ($i = 1, 2, \ldots, n$) are independent and y_i is distributed $N(\mu_i,\sigma_i^2)$. Show that

$$\sum_{i=1}^{n} \frac{1}{\sigma_i^2} (y_i - y^*)^2$$

is distributed as $\chi'^2(n - 1, \lambda)$, where $\lambda = 0$, if $\mu_1 = \mu_2 = \cdots = \mu_n$. The quantity y^* is defined by

$$y^* = \frac{\Sigma\,(1/\sigma_i^2)y_i}{\Sigma 1/\sigma_i^2}$$

Hint: use Theorem 4.20.

4.25 Let $\mathbf{Y}' = (y,x)$ be distributed as a bivariate normal with mean $\boldsymbol{\mu}' = (\mu_1, \mu_2)$, variances 1 and correlation ρ. Let $\mathbf{Y}_1,\mathbf{Y}_2, \ldots, \mathbf{Y}_n$ be independent samples from this distribution. Show that for $\rho^2 \neq 1$

(*a*) $$q = \frac{\Sigma\,(x_i^2 - 2\rho x_i y_i + y_i^2)}{1 - \rho^2}$$

is distributed as $\chi'^2(2n,\lambda)$, where

$$\lambda = \frac{n(\mu_1^2 - 2\rho\mu_1\mu_2 + \mu_2^2)}{2(1 - \rho^2)}$$

(*b*)

$$q_1 = \frac{n\bar{x}^2 - 2n\rho\bar{x}\bar{y} + n\bar{y}^2}{1 - \rho^2}$$

is distributed as $\chi'^2(2,\lambda_1)$, where

$$\lambda_1 = \frac{n(\mu_1^2 - 2\rho\mu_1\mu_2 + \mu_2^2)}{2(1 - \rho^2)}$$

(*c*) Use the results of (*a*) and (*b*) and Theorem 4.22 to find the distribution of

$$q_2 = \frac{\Sigma\,[(x_i - \bar{x})^2 - 2\rho(x_i - \bar{x})(y_i - \bar{y}) + (y_i - \bar{y})^2]}{1 - \rho^2}$$

Notice that $q = q_1 + q_2$.

4.26 Use Theorem 4.23 to evaluate the integral in Eq. (4.8) for the values of p, q, and λ in Prob. 4.5.

Further Reading

1 A. C. Aitken: On the Statistical Independence of Quadratic Forms in Normal Variates, *Biometrika*, vol. 37, pp. 93–96, 1950.

2 A. T. Craig: Note on the Independence of Certain Quadratic Forms, *Ann. Math. Statist.*, vol. 14, pp. 195–197, 1943.

3 P. C. Tang: The Power Function of the Analysis of Variance Tests with Tables and Illustrations for Their Use, *Statist. Research Mem.*, vol. 2, pp. 126–146, 1938.

4 H. Hotelling: On A Matrix Theorem of A. T. Craig, *Ann. Math. Statist.*, vol. 15, pp. 427–429, 1944.

5 W. G. Cochran: The Distribution of Quadratic Forms in a Normal System, *Proc. Cambridge Phil. Soc.*, vol. 30, p. 178, 1934.

6 W. Madow: The Distribution of Quadratic Forms in Non-central Normal Random Variables, *Ann. Math. Statist.*, vol. 11, pp. 100–101, 1940.

7 F. A. Graybill and G. Marsaglia: Idempotent Matrices and Quadratic Forms in the General Linear Hypothesis, *Ann. Math. Statist.*, vol. 28, pp. 678–686, 1957.

8 R. V. Hogg and A. T. Craig: On the Decomposition of Certain χ^2 Variables, *Ann. Math. Statist.*, vol. 29, pp. 608–610, 1959.

9 O. Carpenter: Note on the Extension of Craig's Theorem to Non-central Variates, *Ann. Math. Statist.*, vol. 21, p. 455, 1950.

10 R. A. Fisher: "Statistical Methods for Research Workers," Oliver and Boyd, Ltd., London, 1946.

5

Linear Models

5.1 Introduction

One of the aims of science is to describe and predict events in the world in which we live. One way this is accomplished is by finding a formula or equation that relates quantities in the real world. We may be interested, for example, in the relation between temperature and pressure in a chemical process or in the relation between the number of apples on various trees in an orchard and the amount of fertilizer that each tree receives, etc.

Suppose[1] two variables U^* and V^* are functionally related by $f(U^*, V^*) = 0$. If U^* and V^* can take on every value in a given interval, then they *cannot be measured exactly*, and, instead of observing U^* and V^*, we actually observe u and v, where $u = U^* + d$ and $v = V^* + e$, where e and d are errors of measurement. However, if the error of measurement is extremely small, we may be willing to neglect it. Some may argue that $f(U^*, V^*) = 0$ is simply a mathematical abstraction and that no such functional relationship *can* exist in the *real* world. Be that as it may, it is nevertheless true that the *concept* of a functional relationship between sets of events in the real world is important. And, although the relationship may not be exact, it may be so close that the approximation is invaluable in prediction. Therefore, when we say that a functional relationship exists among

[1] Throughout this chapter we denote mathematical variables by starred capital letters U^*, V_1^*, V_2^*, . . . if they are not observable and by U, V_1, V_2, . . . if they can be observed. We denote random variables by lower-case starred letters y_1^*, u_1^*, v_2^*, . . . if they are not observable and by u, x_1, x_2, . . . if they can be observed. The letters a, b, d, e will be unobservable random variables. Greek letters will denote unknown parameters.

a set of variables, we shall mean that in the real world the function describes the relationship among the variables to a very close approximation.

For example, consider the relation between time and the distance that a particle falls under the influence of gravity; that is, $S = \beta T^2$. For many practical purposes we consider this a functional relationship, and we can predict the distance quite accurately in a real-world experience. Suppose S and T^2 cannot be observed. We should then write the equation as $S^* = \beta(T^*)^2$. If we can observe s and t^2, where $s = S^* + e$ and $t^2 = (T^*)^2 + d$, we can substitute and obtain $s = \beta t^2 - \beta d + e$ or $s = \beta t^2 + b$, where b is a random error. The problem here may be to estimate β by observing values of s and t^2.

There may be events in the real world that are related not *functionally* but in a more obscure manner. For example, consider the weight w and the height h of individuals. There seems to be no formula that will enable us to predict the height of an individual from his weight. Although there does not seem to be any functional relation between height and weight, there does seem to be some kind of relationship. For example, let us assume that w and h have a bivariate normal distribution. In the conditional distribution of h given w, there is a linear functional relationship between w and the expected value of h. In other words, if we divide the individuals into weight classes and evaluate the average height of the individuals in each weight class, we find that there is a functional (linear) relationship between weight and the average height of all individuals who have that weight. We see, therefore, that, even though we may not be able to predict w exactly by knowing only h, the quantity h may still be valuable for predicting w.

In scientific investigations the concept of cause and effect is often quite obscure, and we shall not enter into a discussion of it here. We shall not use cause-and-effect terminology to describe an equation such as $Y = \alpha + \beta X$; we shall say that X *can be used to predict* Y rather than that X *causes* Y.

In many fields of scientific endeavor, for example, in physics, relationships can often be expressed as functional equations $Y = F(X_1, X_2, \ldots, X_n)$. For instance, Ohm's law states that the electromotive force Y is equal to the resistance X_1 of the conductor times the current X_2; that is $Y = X_1 X_2$. The law of gravitation states that the force of attraction Y between two unit-mass points is inversely proportional to the square of the distance X_1 between the points; that is, $Y = \beta/X_1^2$. There are many other such relationships: Boyle's gas law, Kirchhoff's law in electricity, Newton's laws of force and acceleration, Newton's law of cooling, and so forth.

If a functional relationship can be found that relates observable variables, then a knowledge of $X_1, X_2, \ldots, X_n$ and of the function f can be used to predict Y accurately. For instance, the law of a body falling in a vacuum under the influence of gravity is $Y = \beta X^2$, where Y denotes the distance the body falls and X denotes the time the body falls. If an experiment is run where a body is dropped and the time and distance observed, then β can be evaluated. However, if another reading of time and distance is taken and β is evaluated again, the second β will be quite likely not to agree with the first. The reason for this disagreement may be either (1) that the relationship is not given by $Y = \beta X^2$; or (2) that, even though the relationship is given by $Y = \beta X^2$, the time and distance cannot be measured accurately, and so an error is introduced; this second type of error will be called *measurement error*. If reason 1 is operative, then the *real* functional relationship may be $Y = f(X,Z)$, where Z is the initial distance the body is from the center of the earth, or the distance may actually depend on many other factors, such as the mass of the body, the position of the moon, etc. If the distance depends on many factors $X_1, X_2, \ldots, X_n$, besides the time X, the relation could be written $Y = f(X,X_1,X_2, \ldots ,X_n)$ or, more explicitly,

$$Y = \beta X^2 + g(X,X_1, \ldots ,X_n)$$

If $Y = \beta X^2 + g(X,X_1, \ldots ,X_n)$ is a relationship that can be used to predict Y, then observations will not agree exactly when the equation $Y = \beta X^2$ is used instead. This disagreement of observation with theoretical relationship will be called *equation error*, or, in other words, error due to the use of the wrong equation. It may be realistic to assume that, for repeated observations of distance Y and time X, the variables $X_1, X_2, \ldots, X_n$ assume values such that $g(X,X_1, \ldots ,X_n)$ acts more or less as a random variable, and its distribution may be inferred. If this is the case, the relationship can be written $Y = \beta X^2 + e$, and Y cannot be exactly predicted from a knowledge of X only.

Although functional relationships are assumed to hold in many fields of science, such as physics, there are many scientific areas, such as biology, economics, meteorology, etc., where relationships are much more obscure.

For example, the yield of wheat in a given plot of ground cannot be predicted accurately. Many of the factors affecting this yield are known, but the equation relating the quantities is obscure. It is known that temperature X_1, rainfall X_2, amount of sunshine, X_3, fertility X_4, and *many* other factors influence the yield Y. Although all the factors that affect this yield are not known and although the relationship is not known, it is useful nevertheless to assume that there *exist* a finite number of factors $X_1, \ldots, X_n$ and a function g such that the yield Y can be exactly determined by

$$Y = g(X_1, X_2, \ldots, X_n)$$

Thus, to find relationships that aid in predicting events in the real world, we shall assume that the following proposition holds:

◆ **Proposition** Suppose Y is a variable quantity in the real world, which we want to predict. Then there exist a finite number of variables $X_1, X_2, \ldots, X_n$ and a function g such that $Y = g(X_1, X_2, \ldots, X_n)$.

We are not saying that $X_1, X_2, \ldots, X_n$ are capable of being observed, but that, if they could be observed, we could predict Y exactly; that is, $Y, X_1, X_2, \ldots, X_n$ are said to be functionally related.

For example, consider again the problem of predicting an individual's height. We know that we cannot predict his height exactly from a knowledge of his weight only. However, we may be able to predict his height exactly if we use a number of other quantities, such as his father's weight X_1, his father's height X_2, his mother's weight X_3, etc. We assume that the equation

$$h = g(w, X_1, X_2, X_3, \ldots, X_n)$$

holds. We might also want to assume that we can write this

$$h = \alpha + \beta w + f(X_1, X_2, \ldots, X_n)$$

and assume that, for a given value of w, the quantities $X_1, X_2, \ldots, X_n$ change and $f(X_1, X_2, \ldots, X_n)$ acts as a random error. The function g in the proposition above may be known and the elements $X_1, X_2, \ldots, X_n$ may be known, but it still may be impossible to predict Y exactly owing to our inability to measure some or all of the X_i exactly.

To sum up, we shall be interested in two types of error:

1. Equation error: the error of not knowing the function g or of not knowing all the variables $X_1, X_2, \ldots, X_n$.

2. Measurement error: the error of not being able to measure all the X_i exactly.

5.2 Linear Models

5.2.1 Definitions and Classification. We shall give the definition of a model that will be used in this book.

◆ **Definition 5.1** By a model we shall mean a mathematical equation involving random variables, mathematical variables, and parameters.

If the distribution of the random variables is given, this will be considered part of the model, and there may be unknown parameters in the distribution. For instance, an equation $S^* = \beta(T^*)^2$ for the distance that a body falls in a vacuum is a model that does not contain random variables. If $s = S^* + e$, where e is a random variable, substitution in the first equation gives $s = \beta(T^*)^2 + e$ as a model. If it is further assumed that e is a normal variable with mean 0 and variance σ^2, then this distributional property will be considered a part of the model.

In general there is no reason to restrict the equation in our definition to any particular type, such as quadratic, exponential, etc. However, there may be certain instances when we want to do so. The major part of our discussion will concern equations that are restricted to be linear in a certain fashion. This does not imply that other equations are not important but only that linear equations have received most attention from mathematical statisticians and that methods of treatment are available for them.

Therefore, in succeeding chapters we shall be mainly concerned with *linear* models, according to the following definition.

◆ **Definition 5.2** By a linear model we shall mean an equation that involves random variables, mathematical variables, and parameters and that is linear in the parameters and in the random variables.

For example, if $\gamma_0, \gamma_1, \gamma_2$ are unknown parameters, then $\gamma_0 + \gamma_1 X + \gamma_2 Y = 0$ is a linear model, and so is $\gamma_0 + \gamma_1 e^X + \gamma_1 Y \cos X = 0$. However, models such as $\gamma_0^2 + X \sin \gamma_1 + \gamma_2 = 0$ are not linear in the parameters $\gamma_0, \gamma_1, \gamma_2$.

In what follows we shall often use only two variables; however, the results will hold for any finite number of variables.

Let us consider the general linear model $\gamma_0 + \gamma_1 X + \gamma_2 Y = 0$. We could also have the model $\gamma_0 + \gamma_1 X^* + \gamma_2 Y^* = 0$ or $\gamma_0 + \gamma_1 x + \gamma_2 y = 0$ or even $\gamma_0 + \gamma_1 X + \gamma_2 Y^* = 0$, depending on whether the variables were random or fixed, observable or unobservable. However, we see that it is impossible to have models such as $\gamma_0 + \gamma_1 x + \gamma_2 Y = 0$, since this would imply that a random variable x was a function of parameters and of a mathematical variable Y.

We can write the linear model

$$\gamma_0 + \gamma_1 X + \gamma_2 Y = 0 \tag{5.1}$$

as

$$Y = -\frac{\gamma_0}{\gamma_2} - \frac{\gamma_1}{\gamma_2} X \qquad \text{if } \gamma_2 \neq 0 \tag{5.2}$$

or as
$$X = -\frac{\gamma_0}{\gamma_1} - \frac{\gamma_2}{\gamma_1} Y \qquad \text{if } \gamma_1 \neq 0 \tag{5.3}$$

In (5.1), X and Y enter the equation in symmetrical fashion. In (5.2), Y is called the dependent variable and X the independent variable; vice versa in (5.3). The words "independent" and "dependent" are used here in a mathematical sense and do not refer to the similarly named concepts in random variables. To a mathematician and generally to an engineer or physicist, these three equations are equivalent. If X and Y can be measured without error and if an experimenter wants to predict Y from a knowledge of X, he can use (5.2). If he wants to predict X from a knowledge of Y, (5.3) is appropriate. Since the γ_i are not known, three pairs of values X,Y can be observed, and the resulting three equations can be solved for the three unknown γ_i. If (5.2) or (5.3) is used, only two pairs of X,Y values need be observed to solve for the parameters γ_i/γ_j.

Suppose we want to predict Y^* in the model

$$Y^* = -\frac{\gamma_0}{\gamma_2} - \frac{\gamma_1}{\gamma_2} X$$

where γ_0/γ_2 and γ_1/γ_2 are unknown and Y^* is not observable. If we can observe y, where $y = Y^* + e$, we can write the model as

$$y = -\frac{\gamma_0}{\gamma_2} - \frac{\gamma_1}{\gamma_2} X + e$$

We can no longer proceed as we did before, i.e., observe two values of y and X and solve for γ_0/γ_2 and γ_1/γ_2. In fact, since there is no functional relationship between y and X, we cannot evaluate γ_i/γ_2 exactly, but we shall have to employ statistical procedures that will give us "estimates" of γ_i/γ_2.

There are many types of linear models that the scientist may want to use in getting information that will help him predict quantities in the real world. Five types, adequate for many experimental purposes, will be described. These are as follows: (1) functionally related models, (2) mean-related models, (3) regression models, (4) experimental design models, and (5) component-of-variance models. Each of the five types can be broken into various subdivisions, which we shall discuss in later chapters.

These divisions are made mainly for the purpose of aiding the experimenter. On the basis of his knowledge of the system that he wants to describe, the experimenter can choose from the catalogue of models the type that fits his situation. Having chosen a model and obtained data, he can use statistical techniques to analyze the data by means of the model. The statistician looks at a model in a somewhat different light; if an error term e is present in the model, he does not care whether it represents equation error or measurement error. To the experimenter, on the other hand, this question may be of great importance. Two models may be quite distinct so far as the experimenter is concerned but equivalent so far as the statistician is concerned.

We shall first discuss the classification above, and later we shall look at the models as a statistician would view them. If we solve the equation $\gamma_0 + \gamma_1 X + \gamma_2 Y = 0$ for Y, we shall write it $Y = \alpha_0 + \alpha_1 X$. If we solve it for X, we shall write it $X = \beta_0 + \beta_1 Y$.

5.2.2 Functionally Related Models. These models are characterized by a functional relationship among mathematical variables, some of which cannot be observed owing to errors of measurement. Suppose two quantities Y^* and X satisfy the equation $Y^* = \alpha_0 + \alpha_1 X$. This implies that we can observe X but not Y^*. Suppose there is a measurement error and, instead of observing Y^*, we observe y, where $y = Y^* + e$. Suppose e is a random error with mean 0; then $E(y) = Y^*$. We can write the model as $y - e = \alpha_0 + \alpha_1 X$ or as $y = \alpha_0 + \alpha_1 X + e$. It is important to notice the notation; y and e are random variables, y and X are observable.

For example, in electricity Ohm's law states that the voltage V^* in a circuit is equal to the product of the resistance ρ of the wire and the current I. The equation is $V^* = \rho I$. Suppose an experimenter wants to find the resistance ρ of a circuit. Suppose he cannot measure V^* but can observe v, where $v = V^* + e$. The equation becomes $v = \rho I + e$. Here v and e are random variables, I is a mathematical variable, and ρ is an unknown parameter to be determined.

5.2.3 Mean-related Models. These models are characterized by equation errors. Suppose that a functional relationship is given by

$$f(Y, X_1, X_2, \ldots, X_n) = 0$$

Suppose further that $f(Y, X_1, X_2, \ldots, X_n)$ can be written

$$Y = \alpha_0 + \alpha_1 X_1 + \alpha_2 X_2 + g(X_3, \ldots, X_n)$$

If the term $g(X_3, \ldots, X_n)$ is dropped, it is clear that Y is not exactly determined for fixed values of X_1, X_2, since the remaining variables may take on various values even though the first two variables are fixed.

Suppose, however, we are willing to assume that the variables X_1, X_2 do a fairly good job of predicting Y, realizing that the other variables $X_3, X_4, \ldots, X_n$ will be needed if we are to predict Y exactly. We may not know what these variables are, but it may be a fair approximation to assume that, for fixed values of the first two variables, the term $g(X_3, \ldots, X_n)$ acts as a random variable as the quantities $X_3, \ldots, X_n$ change. If this is true, we can then write

$$y = \alpha_0 + \alpha_1 X_1 + \alpha_2 X_2 + e$$

It should be pointed out that the X_i can be functions of other variables. For example, if $X_1 = \log t$, $X_2 = 3^t$, then

$$y = \alpha_0 + \alpha_1 \log t + \alpha_2 3^t + e$$

Or we could have $X_1 = t$, $X_2 = t^2$; then we should have the curvilinear model

$$y = \alpha_0 + \alpha_1 t + \alpha_2 t^2 + e$$

We stress the fact that, in this model, the dependent variable y is a random variable whereas the independent variables X_1 and X_2 are *not* random variables but are predetermined mathematical variables. Also, the expected value of the dependent variable y *does not* give a functional relationship with X_1 and X_2; that is, $E(y) \neq Y$. That is to say, the functional relationship is

$$Y = \alpha_0 + \alpha_1 X_1 + \alpha_2 X_2 + g(X_3 + \cdots + X_n)$$

whereas $E(y) = \alpha_0 + \alpha_1 X_1 + \alpha_2 X_2$, which is equal to the average value of y.

This model is important in problems where many factors influence the factor Y and where all the factors may not be known but a knowledge of some of them may be used to estimate an average value of y "close" to the true Y value. For example, suppose we want to predict the weight a pint of ice cream will lose when stored at very low temperatures. The researcher knows that many factors contribute to the loss of weight in the pint of ice cream, such as:

1. Storage time X_1
2. Temperature X_2
3. Humidity X_3
4. Butterfat content of the ice cream X_4, and many other factors, out to X_n

We assume that weight loss Y can be given exactly by $Y = f(X_1, X_2, \ldots, X_n)$ or by $Y = \alpha_0 + \alpha_1 X_1 + \alpha_2 X_2 + g(X_3, \ldots, X_n)$. However, we may feel that temperature and storage time are the important factors; so we form the prediction equation

$$y = \alpha_0 + \alpha_1 X_1 + \alpha_2 X_2 + e$$

where X_1 is storage time in weeks, X_2 is temperature in degrees Fahrenheit, and y is the weight loss in grams. Now y, X_1, and X_2

can be measured so accurately that, for all practical purposes, we assume that they are measured without error. The error term e is added to take into account all the factors besides temperature and storage time that affect the weight loss. Thus, the function $\alpha_0 + \alpha_1 X_1 + \alpha_2 X_2$ will predict the average value of y and not Y itself. However, if the variance of e is small, this function may be accurate enough to be valuable.

If the variance of e is zero, then this model has no equation error, and it is functionally related. If the variance of e is so large that the model is not useful, it may be possible to reduce the variance of e by including another variable X_3 in the model. This may have a stabilizing effect on the term $g(X_3, \ldots, X_n)$. Or, if the model is quite satisfactory but one of the variables is difficult or expensive to obtain, it may be desirable to omit that variable from the model and then test to see whether the new model is satisfactory.

We remark that, if there is an error in measuring y, the equation becomes $y^* = \alpha_0 + \alpha_1 X_1 + \alpha_2 X_2 + e$. If we let $y = y^* + d$, then we have $y = \alpha_0 + \alpha_1 X_1 + \alpha_2 X_2 + a$, where the a is a random error such that $a = e + d$.

The important thing to remember about this model is that we are replacing the model $Y = \alpha_0 + \alpha_1 X_1 + \alpha_2 X_2 + g(X_3, \ldots, X_n)$ by the model $y = \alpha_0 + \alpha_1 X_1 + \alpha_2 X_2 + e$. We are in effect not trying to predict Y directly but hoping instead that $E(y)$ is close enough to Y to be useful. It is important to remember also that we are assuming that y is a random variable and that the mean of y is $E(y) = \alpha_0 + \alpha_1 X_1 + \alpha_2 X_2$.

5.2.4 Regression Models. These models are characterized by the fact that the variables entering into them are random variables. Let $f(y,x)$ be the joint distribution of y and x such that $E(y \mid x = X) = \alpha_0 + \alpha_1 X$. We shall write this $y_x = \alpha_0 + \alpha_1 X + e$.

It may be that two unobservable random variables y^* and x^* have the joint distribution given above. If we observe $y = y^* + e$ and $x = x^* + d$, we then have a regression model with measurement errors.

For example, suppose it is desired to predict the temperature y in a certain locality by using only humidity x. We shall assume that y and x form a bivariate frequency function. We can use a regression model and predict the average temperature for a given value of humidity. If there are errors in measuring either temperature or humidity or both, then we can use the alternative model with measurement error.

5.2.5 Experimental Design Models. The models considered up to now have been such that they permitted us to say something about y or $E(y)$ or the parameters in the model $y = \alpha_0 + \alpha_1 X_1 + \alpha_2 X_2 + e$ when X_i varied in some interval. The experimental design model is somewhat different. *In this model X_i takes only the values 0 and 1.* For example, suppose we want to examine the durability of two different kinds of paint. The paints are each subjected to a friction machine, and the time it takes for this machine to wear the paint is recorded. We should like to have a formula for predicting the time it takes for each paint to wear away. Paint 1 and paint 2 bear no numerical relationship to each other. The prediction equation can, however, be written $y = \alpha_1 X_1 + \alpha_2 X_2$, where X_1 and X_2 take the values 0 and 1 and where α_i is the time it takes paint i to wear away. Thus, $y = \alpha_1$ when $X_1 = 1$, $X_2 = 0$; $y = \alpha_2$ when $X_2 = 1$, $X_1 = 0$. Therefore, if we let y_i be the time it takes for paint i to wear away, we can write the model $y = \alpha_1 X_1 + \alpha_2 X_2$ as $y_i = \alpha_i$. Actually, y is a function of X_i, but, since $X_i = 0$ or 1, X_i may not seem to appear in the formula. In addition, there may be an error in measuring y, or there may be other factors besides the effect of the paint that cause y to take on a particular value. We may, therefore, wish to write the equation $y_i = \alpha_i + e$, where e is a random error.

For another example, suppose a researcher has developed a new variety of wheat, which he wants to compare with a standard variety. He wants to examine the yield of the two varieties; so he plants both under uniform conditions. If we let α_1 be the average yield of the new variety and α_2 the average yield of the standard variety, we can write the model for the observed yield as

$$y = \alpha_1 X_1 + \alpha_2 X_2 + e$$

where y is the observed yield, e is an error, and X_1 and X_2 take on the values 0 and 1. Although the experimenter may attempt to plant the two varieties under very similar conditions, there are a great many factors, such as soil fertility, humidity, moisture, and so forth, that make the observed yield differ from the true average yield of the variety under observation. These effects constitute the random error term e. For example, if $X_1 = 1$ and $X_2 = 0$, then the corresponding y value is $y_1 = \alpha_1 + e$. We see that the observed value of y is not the true average effect α_1 but is equal to α_1 plus the error due to the uncontrolled factors. We may want to estimate α_1, which is $E(y_1)$. A similar situation holds for y when $X_1 = 0$ and $X_2 = 1$.

5.2.6 Components-of-variance Model. This model is similar to the experimental design model in that the X variables again take the values 0 and 1 only. The model can be written $y = a_1 X_1 + a_2 X_2 + e$. However, a_1 and a_2 are not parameters, but are unobservable random variables from distributions with variances σ_1^2 and σ_2^2, respectively. The object in this model is to observe values of y and estimate σ_1^2, σ_2^2, and σ^2 (σ^2 is the variance of e).

For example, in measuring the nitrogen content of the foliage on a certain tree, there are two major sources of variation: the variation of the leaves on the tree, and the variation due to the measurement or laboratory error. Suppose we take n leaves from the tree, where the actual nitrogen content of the ith leaf is a_i. In this case the a_i are random variables from a distribution with variance σ_1^2. Suppose m measurements are made on each leaf to determine the nitrogen content of that particular leaf. If we let y_{ij} denote the jth measurement of the ith leaf, the e_{ij} are random variables from a distribution with variance σ^2. The model can be written

$$y_{ij} = \mu + a_i + e_{ij} \qquad j = 1, 2, \ldots, m; \qquad i = 1, 2, \ldots, n$$

where y_{ij} is the observed nitrogen content of the jth measurement on the ith leaf. The quantity μ is a constant which is the average value of y_{ij}. One object of this model is to estimate σ_1^2 and σ^2 from the observed values of y_{ij}. We note that the variance of y_{ij} equals $\sigma_1^2 + \sigma^2$; that is, it is a linear function of the variances.

5.3 Model Classification

In the previous section we discussed the models from the point of view of a researcher. In this section we shall define models as a statistician looks at them. In subsequent chapters we shall discuss in detail the five models defined below.

5.3.1 Model 1. Suppose $X_1, X_2, \ldots, X_k$ are known mathematical variables, y is an observable random variable, $\beta_0, \beta_1, \beta_2, \ldots, \beta_k$ are unknown parameters, and e is an unobservable random variable with mean 0. Under these conditions the model

$$y = \beta_0 + \beta_1 X_1 + \beta_2 X_2 + \cdots + \beta_k X_k + e$$

will be defined as *model 1.*

Many of the models discussed in the last section will fit into the framework of this model. For example, suppose there is a functional relationship between Y^* and X_1, given by $Y^* = \beta_0 + \beta_1 X_1$. We cannot observe Y^*, but we observe y, where $y = Y^* + e$, where e is a measurement error. We then have $y = \beta_0 + \beta_1 X_1 + e$, which is *model 1.*

Suppose the functional relationship $Y = g(X_1, X_2, \ldots, X_k)$ can be written $Y = \beta_0 + \beta_1 X_1 + h(X_2, \ldots, X_k)$, which we approximate by

$$y = \beta_0 + \beta_1 X_1 + e$$

This is a mean-related model with equation error, but it fits the definition of model 1. If y is not observable, we write it

$$y^* = \beta_0 + \beta_1 X_1 + e$$

If $y = y^* + d$, we get $y = \beta_0 + \beta_1 X_1 + a$, where $a = e + d$. This also satisfies the requirements for model 1.

5.3.2 Model 2: Functionally Related Models with Variables Subject to Measurement Error. Suppose the mathematical variables $Y^*, X_1^*, \ldots, X_k^*$ are not observable but that $y, x_1, \ldots, x_k$ can be observed, where $y = Y^* + e$ and $x_i = X_i^* + e_i$ $(i = 1, 2, \ldots, k)$. Suppose further that there is a functional relationship among the mathematical variables, given by

$$Y^* = \beta_0 + \beta_1 X_1^* + \cdots + \beta_k X_k^*$$

We can substitute and write

$$y = \beta_0 + \beta_1 x_1 + \cdots + \beta_k x_k + e - \beta_1 e_1 - \beta_2 e_2 - \cdots - \beta_k e_k$$

When the above specifications are met, we shall define the model as *model 2*.

5.3.3 Model 3: Regression Models. Suppose $y, x_1, \ldots, x_k$ are a set of jointly distributed random variables such that the expected value in the conditional distribution of y given $x_i = X_i$ $(i = 1, 2, \ldots, k)$ is given by $\beta_0 + \beta_1 X_1 + \cdots + \beta_k X_k$. We can then write

$$y_x = \beta_0 + \beta_1 X_1 + \cdots + \beta_k X_k + e$$

When the above conditions are satisfied, we shall define the model as *model 3*.

The only difference between model 1 and model 3 is that the X_i in model 1 are mathematical variables, whereas in model 3 they are particular values of random values.

5.3.4 Model 4: Experimental Design Models. Let y be a random variable, and let $X_1, X_2, \ldots, X_k$ each be equal to either 0 or 1. If $\beta_1, \beta_2, \ldots, \beta_k$ are unknown parameters, then

$$y = \beta_1 X_1 + \beta_2 X_2 + \cdots + \beta_k X_k + e$$

will be called *model 4*. This model is a special case of model 1, but, because of its importance, it will be discussed separately.

This model will generally be written

$$y_{ij\cdots k} = \mu + \beta_i + \alpha_j + \cdots + e_{ij\cdots k}$$

where β_i, $\alpha_j, \ldots$ are unknown parameters and $e_{ij\cdots k}$ is a random variable.

5.3.5 Model 5: Components-of-variance Models. Let $a_1, a_2, \ldots, a_p$ be unobservable random observations from a distribution with mean 0 and variance σ_a^2; let $b_{11}, b_{12}, \ldots, b_{pq}$ be unobservable random observations from a distribution with mean 0 and variance σ_b^2. Let y_{ij} be an observable random variable such that

$$y_{ij} = \mu + a_i + b_{ij}$$

where μ is an unknown parameter. Models such as these will be called *model 5*. One object in this type of model is to estimate σ_a^2 and σ_b^2.

Further Reading

1 M. G. Kendall: Regression, Structure and Functional Relationship, *Biometrika*, parts I, II, vol. 39, pp. 96–108, 1952.
2 H. Scheffé: Alternative Models for Analysis of Variance, *Ann. Math. Statist.*, vol. 27, pp. 251–271, 1956.
3 C. P. Winsor: Which Regression? *Biometrics*, vol. 2, pp. 101–109, 1946.
4 H. F. Smith: "Variance Components, Finite Populations and Statistical Inference," *N. Carolina Inst. Statist. Mimeo. Ser.* 135, 1955.
5 M. B. Wilk and O. Kempthorne: Fixed, Mixed and Random Models, *J. Am. Statist. Assoc.*, vol. 50, pp. 1144–1178, 1955.

6

Model 1: The General Linear Hypothesis of Full Rank

6.1 Introduction

In this chapter we shall derive the distribution of pertinent statistics needed for estimation of certain parameters in model 1 and for testing hypotheses about them.

6.1.1 Definitions and Notation. Consider the frequency function $f(y; x_1,x_2, \ldots ,x_p; \beta_1, \ldots ,\beta_p)$ of a random variable y, which depends on p known quantities $x_1, \ldots , x_p$ and on p unknown parameters $\beta_1, \beta_2, \ldots , \beta_p$. In this chapter[1] we shall investigate various properties of the distribution of the random variable y. This frequency function will be denoted by $f(y;x;\beta)$, $f(y;\beta)$, or $f(y)$. We shall assume throughout the discussion that $E(y) = \sum_{i=1}^{p} \beta_i x_i$, where the β_i are unknown parameters, and that the variance of y equals σ^2, where σ^2 does not depend on the β_i or on the x_i. The vector $\mathbf{P}_j' = (y_j,x_{j1},x_{j2}, \ldots ,x_{jp})$ will represent an observation from this distribution. That is to say, when an observation y is taken from this distribution, the corresponding x_i values must be specified. Throughout this chapter, the x_i will be considered known constants, not random variables.

If in $f(y;x;\beta)$ we make the transformation $e = y - \Sigma\beta_i x_i$, then e is a random variable such that $E(e) = 0$ and $E(e^2) = \sigma^2$. This can be written

$$y = \sum_{i=1}^{p} \beta_i x_i + e \tag{6.1}$$

[1] We shall not necessarily adhere to the notation that lower-case letters represent random variables, etc., which we introduced in Chap. 5.

Equation (6.1) is sometimes referred to as a *prediction equation.* Suppose we wish to predict, for example, what the value of y will be for a given set of the x_i. Since e, and consequently y, is a random variable, we cannot in general predict the exact value of y for a given set of the x_i, but we may be able to predict an interval and establish the probability that the interval will contain y. On the other hand, we may not be particularly interested in predicting the value of y, but we may desire instead to estimate $E(y)$, the expected value of y, pertaining to a given set of the x_i.

Suppose a physicist, in studying the motion of a certain type of particle, concludes that there is a functional relationship between the distance d that the particle moves in a certain time interval t. That is to say, he is willing to assume that the functional equation $d = vt$ relates the distance and the time a particle moves. Suppose further that the velocity v is not known, so the experimenter decides to measure corresponding values of d and t and thus ascertain v. However, suppose he cannot measure d accurately. That is to say, instead of measuring d he can only measure y, where $y = d + e$, where e is a measurement error that is normally and independently distributed about a mean of 0 for each selected value of t. Therefore, the model fits the definition of $f(y;x;\beta)$.

If we can obtain an estimate of v (say, $\hat{v}$), then the estimated distance $\hat{d}$ that the particle travels during time t is given by $\hat{d} = \hat{v}t$. In this situation an experimenter may be interested in setting confidence limits on the distance the particle travels in time t; this would necessitate setting confidence limits on $E(y) = d$.

For another example, let us consider a large industrial firm that has many thousands of employees. Let us assume that an individual is contemplating accepting employment with this company and that he wants to get some idea of its salary system. Suppose he knows that the following prediction equation is satisfied:

$$y = \alpha_1 + \alpha_2 x + e \tag{6.2}$$

where y is the annual salary, x is the number of years employed with the company, and e is a random variable that is, say, normally distributed with mean 0 and variance σ^2. That is to say, the average salary is linearly related to the number of years employed. Suppose he wants to predict what his salary will be 10 years hence. From (6.2) he knows that the average salary of all employees who have been with the company 10 years is $\alpha_1 + 10\alpha_2$. However, he also knows that the salaries of all people who have been employed exactly 10 years with the company varies around the mean $\alpha_1 + 10\alpha_2$ with a normal frequency function whose variance is σ^2. Thus, there may be people who have been employed 10 years with this company and whose salaries might be extremely high or extremely low, but most of the salaries (approximately 95 per cent) are within 2σ units of $\alpha_1 + 10\alpha_2$. If σ is quite small, an employee can predict with a fairly high precision what his salary will be in 10 years. In general α_i will not be known; so he must take some observations of salaries within the company, estimate α_i, and then substitute this estimated value of α_i into Eq. (6.2) in order to predict. If σ^2 is so large that the prediction is not very precise, one might try a different prediction equation, say,

$$y = \beta_0 + \beta_1 x_1 + \beta_2 x_2 + f \tag{6.3}$$

where x_1 is the number of years employed, x_2 is the number of years of formal education, and the remaining symbols are as defined in (6.2) with the exception that the random normal variable f has variance σ_1^2. Thus, if $\sigma_1^2 < \sigma^2$, one can do a "better" job of predicting with Eq. (6.3) than with Eq. (6.2).

Now let us consider the general problem, where the model is a linear function of p variables as given in (6.1).

Since, in general, the β_i in Eq. (6.1) are not known, we shall have to estimate them in order to utilize the prediction equation. To estimate the β_i, a random sample of size n will be taken from the distribution $f(y;x;\beta)$. The sample will be denoted by $\mathbf{P}_1', \mathbf{P}_2', \ldots, \mathbf{P}_n'$, and the relations within the system of observations can be written

$$y_j = \sum_{i=1}^{p} \beta_i x_{ji} + e_j \qquad j = 1, 2, \ldots, n \tag{6.4}$$

or, in vector form,

$$\mathbf{Y} = \mathbf{X}\boldsymbol{\beta} + \mathbf{e} \tag{6.5}$$

where
$$\mathbf{Y} = \begin{pmatrix} y_1 \\ y_2 \\ \cdots \\ y_n \end{pmatrix} \qquad \mathbf{X} = \begin{pmatrix} x_{11} & x_{12} & \cdots & x_{1p} \\ x_{21} & x_{22} & \cdots & x_{2p} \\ \cdots & \cdots & \cdots & \cdots \\ x_{n1} & x_{n2} & \cdots & x_{np} \end{pmatrix} \qquad \boldsymbol{\beta} = \begin{pmatrix} \beta_1 \\ \beta_2 \\ \cdots \\ \beta_p \end{pmatrix}$$

$$\mathbf{e} = \begin{pmatrix} e_1 \\ e_2 \\ \cdots \\ e_n \end{pmatrix} \tag{6.6}$$

Examining $\mathbf{Y} = \mathbf{X}\boldsymbol{\beta} + \mathbf{e}$ in more detail, we see that we must first select (either at random or by design) a set of x's, say, $x_{11}, x_{12}, \ldots, x_{1p}$, and then randomly select an observation y_1 from the distribution $f(y; x_1 = x_{11}, x_2 = x_{12}, \ldots, x_p = x_{1p})$. Then we select another set of x's, say, $x_{21}, x_{22}, \ldots, x_{2p}$, and an observation y_2 at random from the distribution $f(y; x_1 = x_{21}, x_2 = x_{22}, \ldots, x_p = x_{2p})$. We repeat this process until n values of y are drawn. No stipulation has yet been put on the x_{ij}. They need not all be distinct; in fact, they might all be the same. However, some restriction will be put on the matrix of the x_{ij} in special instances. *Therefore, whenever the model* $\mathbf{Y} = \mathbf{X}\boldsymbol{\beta} + \mathbf{e}$ *appears, it will be assumed that it was constructed by the sampling process defined above.* On the basis of the observed matrix $\mathbf{X}$ and observed random variable $\mathbf{Y}$, estimators for the β_i and for σ^2 will be derived.

The preceding will be formulated in the following definition.

◆ **Definition 6.1** The model $\mathbf{Y} = \mathbf{X}\boldsymbol{\beta} + \mathbf{e}$ [where the quantities given in (6.6) are such that $\mathbf{Y}$ is a random observed vector, $\mathbf{e}$ is a random vector, $\mathbf{X}$ is an $n \times p$ matrix of known fixed quantities and $\boldsymbol{\beta}$ is a $p \times 1$ vector of unknown parameters] will be called model 1, the general-linear-hypothesis model of full rank, if the rank of $\mathbf{X}$ is equal to p where $p \leqslant n$.

Two cases concerning the distribution of the vector $\mathbf{e}$ will be examined:

Case A: $\mathbf{e}$ is distributed $N(\mathbf{0},\sigma^2\mathbf{I})$, where σ^2 is unknown.

Case B: $\mathbf{e}$ is a random vector such that $E(\mathbf{e}) = \mathbf{0}$ and $\text{cov}(\mathbf{e}) = E(\mathbf{ee}') = \sigma^2\mathbf{I}$, where σ^2 is unknown.

The definition of case A is equivalent to saying that each e_i is normally distributed with mean 0 and variance σ^2 and that the e_i are jointly independent. That of case B is equivalent to saying that the expected value of each e_i is zero, the e_i are uncorrelated, and the e_i have a common unknown variance σ^2. Case A will be referred to as the *normal-theory case*.

In using Eq. (6.1) as a model we shall be interested in ascertaining its many properties. Below is a list of the things we shall investigate:

1. The point estimation of σ^2, the β_i, and the linear functions of the β_i
2. The point estimation of $E(y)$
3. The interval estimation of σ^2, the β_i, and the linear functions of the β_i
4. The interval estimation of $E(y)$ and of a future observation y
5. The interval estimation of x for an observed y when $p = 2$, that is, in simple linear models
6. The test of the hypothesis that $\beta_1 = \beta_1^*, \beta_2 = \beta_2^*, \ldots, \beta_p = \beta_p^*$, where the β_i^* are given constants

7. The test of the hypothesis that the linear function $\mathbf{r}'\boldsymbol{\beta}$ is equal to r_0, where $\mathbf{r}$ is a known vector and r_0 is a known scalar

8. The test of the hypothesis that a given subset of the β_i is equal to a set of known constants, where the remaining β_i are unspecified; that is, a test of the hypothesis that $\beta_1 = \beta_1^*, \beta_2 = \beta_2^*, \ldots, \beta_k = \beta_k^*$, where $\beta_1^*, \beta_2^*, \ldots, \beta_k^*$ are known constants and $k < p$

9. The test of the hypothesis $\boldsymbol{\lambda}_1'\boldsymbol{\beta} = \boldsymbol{\lambda}_2'\boldsymbol{\beta} = \cdots = \boldsymbol{\lambda}_k'\boldsymbol{\beta} = 0$, where $\boldsymbol{\lambda}_1'\boldsymbol{\beta}, \boldsymbol{\lambda}_2'\boldsymbol{\beta}, \ldots, \boldsymbol{\lambda}_k'\boldsymbol{\beta}$ are linear functions of $\boldsymbol{\beta}$, where $\boldsymbol{\lambda}_1, \boldsymbol{\lambda}_2, \ldots, \boldsymbol{\lambda}_k$ are linearly independent vectors

For point estimation both case A and case B will be studied, but for interval estimation and testing of hypotheses we shall study the normal-theory situation, case A, only.

6.2 Point Estimation

The point estimates of the parameters in the general-linear-hypothesis model will be considered separately for the two cases.

6.2.1 Case A: Estimation of $\boldsymbol{\beta}$ and σ^2 under Normal Theory. Since we are considering the case in which the vector of errors $\mathbf{e}$ is normally distributed, the maximum-likelihood method will be used to estimate $\beta_1, \beta_2, \ldots, \beta_p$ and σ^2. The likelihood equation is

$$f(\mathbf{e}; \boldsymbol{\beta}, \sigma^2) = \frac{1}{(2\pi\sigma^2)^{n/2}} \exp(-\mathbf{e}'\mathbf{e}/2\sigma^2)$$

$$= \frac{1}{(2\pi\sigma^2)^{n/2}} \exp\left[-\frac{(\mathbf{Y} - \mathbf{X}\boldsymbol{\beta})'(\mathbf{Y} - \mathbf{X}\boldsymbol{\beta})}{2\sigma^2}\right]$$

Using logarithms, we get

$$\log f(\mathbf{e}; \boldsymbol{\beta}, \sigma^2) = -\frac{n}{2}\log 2\pi - \frac{n}{2}\log\sigma^2 - \frac{1}{2\sigma^2}(\mathbf{Y} - \mathbf{X}\boldsymbol{\beta})'(\mathbf{Y} - \mathbf{X}\boldsymbol{\beta})$$

The maximum-likelihood estimates of $\boldsymbol{\beta}$ and σ^2 are the solutions to the equations:

$$\frac{\partial}{\partial\beta_1}[\log f(\mathbf{e}; \boldsymbol{\beta}, \sigma^2)] = 0$$

$$\frac{\partial}{\partial\beta_2}[\log f(\mathbf{e}; \boldsymbol{\beta}, \sigma^2)] = 0$$

$$\cdots\cdots\cdots\cdots$$

$$\frac{\partial}{\partial\beta_p}[\log f(\mathbf{e}; \boldsymbol{\beta}, \sigma^2)] = 0$$

$$\frac{\partial}{\partial\sigma^2}[\log f(\mathbf{e}; \boldsymbol{\beta}, \sigma^2)] = 0$$

Taking the derivative of $\log f(\mathbf{e};\, \boldsymbol{\beta},\sigma^2)$ with respect to the vector $\boldsymbol{\beta}$, the above can be written

$$\frac{\partial}{\partial \boldsymbol{\beta}}[\log f(\mathbf{e};\, \boldsymbol{\beta},\sigma^2)] = \mathbf{0} \qquad \frac{\partial}{\partial \sigma^2}[\log f(\mathbf{e};\, \boldsymbol{\beta},\sigma^2)] = 0$$

which give $$\frac{\partial}{\partial \boldsymbol{\beta}}[\log f(\mathbf{e};\, \boldsymbol{\beta},\sigma^2)] = \frac{2}{2\sigma^2}(\mathbf{X}'\mathbf{Y} - \mathbf{X}'\mathbf{X}\boldsymbol{\beta}) = \mathbf{0}$$

$$\frac{\partial}{\partial \sigma^2}[\log f(\mathbf{e};\, \boldsymbol{\beta},\sigma^2)] = -\frac{n}{2\sigma^2} + \frac{(\mathbf{Y} - \mathbf{X}\boldsymbol{\beta})'(\mathbf{Y} - \mathbf{X}\boldsymbol{\beta})}{2\sigma^4} = 0$$

If $\tilde{\boldsymbol{\beta}}$ and $\tilde{\sigma}^2$ are the solutions to the resulting equations,[1] we get

$$\mathbf{X}'\mathbf{X}\tilde{\boldsymbol{\beta}} = \mathbf{X}'\mathbf{Y}$$

and $$\tilde{\sigma}^2 = \frac{1}{n}(\mathbf{Y} - \mathbf{X}\tilde{\boldsymbol{\beta}})'(\mathbf{Y} - \mathbf{X}\tilde{\boldsymbol{\beta}})$$

The matrix equations $\mathbf{X}'\mathbf{X}\tilde{\boldsymbol{\beta}} = \mathbf{X}'\mathbf{Y}$ are called the *normal equations*, and they will play an extremely important role in our theory. Since $\mathbf{X}$ is of rank p, $\mathbf{X}'\mathbf{X}$ is of rank p and, hence, has an inverse. Therefore, we get

$$\tilde{\boldsymbol{\beta}} = \begin{pmatrix} \tilde{\beta}_1 \\ \tilde{\beta}_2 \\ \cdots \\ \tilde{\beta}_p \end{pmatrix} = \mathbf{S}^{-1}\mathbf{X}'\mathbf{Y}$$

where $\mathbf{S} = \mathbf{X}'\mathbf{X}$. We shall let $\mathbf{C} = \mathbf{S}^{-1}$. Since $\tilde{\boldsymbol{\beta}}$ and $\tilde{\sigma}^2$ are maximum-likelihood estimators, they are consistent and efficient, but we must examine them to see whether they are unbiased.

To examine $\tilde{\boldsymbol{\beta}}$ for unbiasedness, we proceed as follows:

$$E(\tilde{\boldsymbol{\beta}}) = E(\mathbf{S}^{-1}\mathbf{X}'\mathbf{Y}) = \mathbf{S}^{-1}\mathbf{X}'E(\mathbf{Y}) = \mathbf{S}^{-1}\mathbf{X}'E(\mathbf{X}\boldsymbol{\beta} + \mathbf{e}) = \mathbf{S}^{-1}\mathbf{X}'\mathbf{X}\boldsymbol{\beta} = \boldsymbol{\beta}$$

So $\tilde{\boldsymbol{\beta}}$ is an unbiased estimate of $\boldsymbol{\beta}$. To examine $\tilde{\sigma}^2$ for unbiasedness we proceed in similar fashion and get $E(\tilde{\sigma}^2) = (1/n)E[(\mathbf{Y} - \mathbf{X}\tilde{\boldsymbol{\beta}})'(\mathbf{Y} - \mathbf{X}\tilde{\boldsymbol{\beta}})]$. If we substitute for the value of $\tilde{\boldsymbol{\beta}}$, we get, after some simplification,

$$E(\tilde{\sigma}^2) = \frac{1}{n}E[\mathbf{Y}'(\mathbf{I} - \mathbf{X}\mathbf{S}^{-1}\mathbf{X}')'(\mathbf{I} - \mathbf{X}\mathbf{S}^{-1}\mathbf{X}')\mathbf{Y}]$$

[1] In this article we shall use the notation that the symbol $\sim$ refers to a maximum-likelihood estimate. If it is unbiased we shall use the symbol $\frown$.

It is easily shown that $\mathbf{I} - \mathbf{X}\mathbf{S}^{-1}\mathbf{X}'$ is an idempotent matrix and, hence, that

$$E(\tilde{\sigma}^2) = \frac{1}{n} E[\mathbf{Y}'(\mathbf{I} - \mathbf{X}\mathbf{S}^{-1}\mathbf{X}')\mathbf{Y}]$$

$$E[\mathbf{Y}'(\mathbf{I} - \mathbf{X}\mathbf{S}^{-1}\mathbf{X}')\mathbf{Y}] = E[\mathbf{e}'(\mathbf{I} - \mathbf{X}\mathbf{S}^{-1}\mathbf{X}')\mathbf{e}]$$

and by Theorem 4.18, we find that this equals

$$\sigma^2 \operatorname{tr}(\mathbf{I} - \mathbf{X}\mathbf{S}^{-1}\mathbf{X}')$$

and
$$\operatorname{tr}(\mathbf{I} - \mathbf{X}\mathbf{S}^{-1}\mathbf{X}') = \operatorname{tr}(\mathbf{I}) - \operatorname{tr}(\mathbf{X}\mathbf{S}^{-1}\mathbf{X}')$$

where $\mathbf{I}$ is the $n \times n$ identity matrix. Therefore, $\operatorname{tr}(\mathbf{I}) = n$ and, by Theorem 1.46, $\operatorname{tr}(\mathbf{X}\mathbf{S}^{-1}\mathbf{X}') = \operatorname{tr}(\mathbf{X}'\mathbf{X}\mathbf{S}^{-1}) = \operatorname{tr}(\mathbf{I}) = p$, since $\mathbf{X}'\mathbf{X}\mathbf{S}^{-1}$ is a $p \times p$ identity matrix. So

$$E(\tilde{\sigma}^2) = \frac{n-p}{n} \sigma^2$$

Therefore, $\tilde{\sigma}^2$ is biased, but

$$\hat{\sigma}^2 = \frac{n}{n-p} \tilde{\sigma}^2 = \frac{(\mathbf{Y} - \mathbf{X}\hat{\boldsymbol{\beta}})'(\mathbf{Y} - \mathbf{X}\hat{\boldsymbol{\beta}})}{n-p}$$

is an unbiased estimate of σ^2.

Next we shall examine for sufficiency. The joint frequency of the e_i can be written

$$f(\mathbf{e}) = \frac{1}{(2\pi\sigma^2)^{n/2}} \exp\left[-\frac{1}{2\sigma^2}(\mathbf{Y} - \mathbf{X}\boldsymbol{\beta})'(\mathbf{Y} - \mathbf{X}\boldsymbol{\beta})\right]$$

The identity in $\boldsymbol{\beta}$

$$(\mathbf{Y} - \mathbf{X}\boldsymbol{\beta})'(\mathbf{Y} - \mathbf{X}\boldsymbol{\beta}) = (\mathbf{Y} - \mathbf{X}\hat{\boldsymbol{\beta}})'(\mathbf{Y} - \mathbf{X}\hat{\boldsymbol{\beta}}) + (\hat{\boldsymbol{\beta}} - \boldsymbol{\beta})'\mathbf{S}(\hat{\boldsymbol{\beta}} - \boldsymbol{\beta}) \quad (6.7)$$

can be readily established. Using this identity, we find that the joint frequency function is

$$f(\mathbf{e}) = \frac{1}{(2\pi\sigma^2)^{n/2}} \exp\left[-\frac{(\mathbf{Y} - \mathbf{X}\boldsymbol{\beta})'(\mathbf{Y} - \mathbf{X}\boldsymbol{\beta})}{2\sigma^2}\right]$$

$$= \frac{1}{(2\pi\sigma^2)^{n/2}} \exp\left[-\frac{(n-p)\hat{\sigma}^2 + (\hat{\boldsymbol{\beta}} - \boldsymbol{\beta})'\mathbf{S}(\hat{\boldsymbol{\beta}} - \boldsymbol{\beta})}{2\sigma^2}\right]$$

and, using the definition of sufficiency in Chap. 2, we see that $\hat{\sigma}^2, \hat{\beta}_1, \hat{\beta}_2, \ldots, \hat{\beta}_p$ form a set of estimators that are jointly sufficient for $\sigma^2, \beta_1, \beta_2, \ldots, \beta_p$. We shall not give the proof, but it can be shown that the estimators $\hat{\sigma}^2, \hat{\beta}_1, \hat{\beta}_2, \ldots, \hat{\beta}_p$ are complete.

Suppose, therefore, that it is desired to find an estimator of the function $g(\sigma^2, \beta_1, \beta_2, \ldots, \beta_p)$. Since $\hat{\sigma}^2, \hat{\beta}_1, \hat{\beta}_2, \ldots, \hat{\beta}_p$ form a set of jointly sufficient statistics for the parameters $\sigma^2, \beta_1, \beta_2, \ldots, \beta_p$ and since the estimators are complete, it follows that, if a function $h(\hat{\sigma}^2, \hat{\beta}_1, \hat{\beta}_2, \ldots, \hat{\beta}_p)$ can be found such that

$$E[h(\hat{\sigma}^2, \hat{\beta}_1, \hat{\beta}_2, \ldots, \hat{\beta}_p)] = g(\sigma^2, \beta_1, \ldots, \beta_p)$$

then $h(\hat{\sigma}^2, \hat{\beta}_1, \hat{\beta}_2, \ldots, \hat{\beta}_p)$ is an unbiased estimator of $g(\sigma^2, \beta_1, \beta_2, \ldots, \beta_p)$ and has a smaller variance for a given sample size than *any* other unbiased estimator of $g(\sigma^2, \beta_1, \beta_2, \ldots, \beta_p)$. For example, $\hat{\beta}_1$ is the minimum-variance unbiased estimator of β_1; $\hat{\beta}_1 - 2\hat{\beta}_2$ is the minimum-variance unbiased estimator of $\beta_1 - 2\beta_2$; etc.

Since $\hat{\boldsymbol{\beta}}$ is equal to the product of a constant matrix $\mathbf{S}^{-1}\mathbf{X}'$ and a normally distributed vector $\mathbf{Y}$, Theorem 3.22 can be used to show that $\hat{\boldsymbol{\beta}}$ has the p-variate normal distribution. We have already shown that the mean of $\hat{\boldsymbol{\beta}}$ is $\boldsymbol{\beta}$. The covariance matrix of $\hat{\boldsymbol{\beta}}$ is

$$\operatorname{cov}(\hat{\boldsymbol{\beta}}) = E[(\hat{\boldsymbol{\beta}} - \boldsymbol{\beta})(\hat{\boldsymbol{\beta}} - \boldsymbol{\beta})'] = E[(\mathbf{S}^{-1}\mathbf{X}'\mathbf{Y} - \boldsymbol{\beta})(\mathbf{S}^{-1}\mathbf{X}'\mathbf{Y} - \boldsymbol{\beta})']$$

If we substitute $\mathbf{X}\boldsymbol{\beta} + \mathbf{e}$ for $\mathbf{Y}$, we get

$$\begin{aligned}\operatorname{cov}(\hat{\boldsymbol{\beta}}) &= E\{[\mathbf{S}^{-1}\mathbf{X}'(\mathbf{X}\boldsymbol{\beta} + \mathbf{e}) - \boldsymbol{\beta}][\mathbf{S}^{-1}\mathbf{X}'(\mathbf{X}\boldsymbol{\beta} + \mathbf{e}) - \boldsymbol{\beta}]'\} \\ &= E[(\mathbf{S}^{-1}\mathbf{X}'\mathbf{e})(\mathbf{S}^{-1}\mathbf{X}'\mathbf{e})'] = E(\mathbf{S}^{-1}\mathbf{X}'\mathbf{e}\mathbf{e}'\mathbf{X}\mathbf{S}^{-1}) \\ &= \mathbf{S}^{-1}\mathbf{X}'[E(\mathbf{e}\mathbf{e}')]\mathbf{X}\mathbf{S}^{-1} = \sigma^2\mathbf{S}^{-1}\mathbf{X}'\mathbf{X}\mathbf{S}^{-1} = \sigma^2\mathbf{S}^{-1}\end{aligned}$$

So $\hat{\boldsymbol{\beta}}$ is distributed $N(\boldsymbol{\beta}, \sigma^2\mathbf{S}^{-1})$. Since $(n - p)\hat{\sigma}^2 = \mathbf{Y}'(\mathbf{I} - \mathbf{X}\mathbf{S}^{-1}\mathbf{X}')\mathbf{Y}$ and since $\mathbf{I} - \mathbf{X}\mathbf{S}^{-1}\mathbf{X}'$ is an idempotent matrix of rank $n - p$, we can use Corollary 4.7.1 to show that $(n - p)(\hat{\sigma}^2/\sigma^2)$ is distributed as $\chi'^2(n - p, \lambda)$, where $\lambda = (1/2\sigma^2)\boldsymbol{\beta}'\mathbf{X}'(\mathbf{I} - \mathbf{X}\mathbf{S}^{-1}\mathbf{X}')\mathbf{X}\boldsymbol{\beta}$. But $(\mathbf{I} - \mathbf{X}\mathbf{S}^{-1}\mathbf{X}')\mathbf{X} = \mathbf{0}$; so $\lambda = 0$, and $\chi'^2(n - p, \lambda = 0)$ reduces to the central chi-square distribution with $n - p$ degrees of freedom.

We shall now use Theorem 4.17 to show that $\hat{\sigma}^2$ is distributed independently of the vector $\hat{\boldsymbol{\beta}}$. The theorem states that the p linear forms $\hat{\boldsymbol{\beta}} = (\mathbf{S}^{-1}\mathbf{X}')\mathbf{Y}$ are independent of the quadratic form $\hat{\sigma}^2 = [1/(n - p)](\mathbf{Y} - \mathbf{X}\hat{\boldsymbol{\beta}})'(\mathbf{Y} - \mathbf{X}\hat{\boldsymbol{\beta}})$, which can be written $\hat{\sigma}^2 = [1/(n - p)]\mathbf{Y}'(\mathbf{I} - \mathbf{X}\mathbf{S}^{-1}\mathbf{X}')\mathbf{Y}$, if the product of the matrix of the linear forms and the quadratic form is the null matrix. That is to say, $\hat{\boldsymbol{\beta}}$ and $\hat{\sigma}^2$ are independent if $(\mathbf{S}^{-1}\mathbf{X}')(\mathbf{I} - \mathbf{X}\mathbf{S}^{-1}\mathbf{X}') = \mathbf{0}$. Simply by performing the indicated multiplication, we can easily show that this is true.

The foregoing discussion is summed up in the following theorem.

◆ **Theorem 6.1** If $\mathbf{Y} = \mathbf{X}\boldsymbol{\beta} + \mathbf{e}$ is a general-linear-hypothesis model of full rank and if e is distributed $N(\mathbf{0}, \sigma^2\mathbf{I})$, the estimators

$$\hat{\boldsymbol{\beta}} = \mathbf{S}^{-1}\mathbf{X}'\mathbf{Y}, \quad \hat{\sigma}^2 = \frac{\mathbf{Y}'(\mathbf{I} - \mathbf{X}\mathbf{S}^{-1}\mathbf{X}')\mathbf{Y}}{n - p}$$

have the following properties:

(1) Consistent
(2) Efficient
(3) Unbiased
(4) Sufficient
(5) $\hat{\boldsymbol{\beta}}$ is distributed $N(\boldsymbol{\beta},\sigma^2\mathbf{S}^{-1})$
(6) Complete
(7) Minimum variance unbiased
(8) $\frac{(n-p)\hat{\sigma}^2}{\sigma^2}$ is distributed as $\chi^2(n-p)$
(9) $\hat{\boldsymbol{\beta}}$ and $\hat{\sigma}^2$ are independent

6.2.2 Case B: Estimation of $\boldsymbol{\beta}$ and σ^2 and the Gauss-Markoff Theorem. In this article we shall assume that the random vector $\mathbf{e}$ has zero for its mean and $\sigma^2\mathbf{I}$ for its covariance matrix. The form of the frequency function of $\mathbf{e}$ will be unspecified; therefore, the principle of maximum likelihood cannot be used to obtain the estimators of the unknown parameters. Instead, we shall use the method of least squares; that is to say, we shall find the value of $\boldsymbol{\beta}$, say, $\hat{\boldsymbol{\beta}}$, such that the sum of squares (abbreviated SS) $\sum_{i=1}^{n} e_i^2$ is a minimum. This gives

$$\sum_{i=1}^{n} e_i^2 = \mathbf{e}'\mathbf{e} = (\mathbf{Y} - \mathbf{X}\boldsymbol{\beta})'(\mathbf{Y} - \mathbf{X}\boldsymbol{\beta})$$

The value of $\boldsymbol{\beta}$ that minimizes $\mathbf{e}'\mathbf{e}$ is given by the solution to

$$\frac{\partial}{\partial\boldsymbol{\beta}}(\mathbf{e}'\mathbf{e}) = \mathbf{0}$$

We get $\frac{\partial}{\partial\boldsymbol{\beta}}(\mathbf{e}'\mathbf{e}) = 2\mathbf{X}'\mathbf{Y} - 2\mathbf{X}'\mathbf{X}\hat{\boldsymbol{\beta}} = 0$. The least-squares estimate of $\boldsymbol{\beta}$ is, therefore,

$$\hat{\boldsymbol{\beta}} = \mathbf{S}^{-1}\mathbf{X}'\mathbf{Y}$$

which, of course, is the same as the maximum-likelihood estimate under normal theory. Minimizing the sum of squares $\mathbf{e}'\mathbf{e}$ does not provide an estimate of σ^2. However, the unbiased estimate of σ^2 based on the least-squares estimate of $\boldsymbol{\beta}$ is given by

$$\hat{\sigma}^2 = \frac{(\mathbf{Y} - \mathbf{X}\hat{\boldsymbol{\beta}})'(\mathbf{Y} - \mathbf{X}\hat{\boldsymbol{\beta}})}{n-p} = \frac{\mathbf{Y}'(\mathbf{I} - \mathbf{X}\mathbf{S}^{-1}\mathbf{X}')\mathbf{Y}}{n-p}$$

Next we shall investigate the properties of the least-squares estimators. Since the frequency form of the random vector $\mathbf{e}$ is unspecified, in general it will not be possible to examine the "goodness" of the estimator $\hat{\boldsymbol{\beta}}$ relative to all functions. Instead, we shall have to limit ourselves to a subset of functions: for example (since $\hat{\boldsymbol{\beta}}$ is a linear function of the y_i), to the set of all linear functions of the y_i. We have proved that when the vector $\mathbf{e}$ is normally distributed, $\hat{\boldsymbol{\beta}} = \mathbf{S}^{-1}\mathbf{X}'\mathbf{Y}$ has smaller variance than any other unbiased estimator of $\boldsymbol{\beta}$.

For the least-squares estimator we cannot make so broad a statement, but we can compare the "goodness" of $\hat{\boldsymbol{\beta}}$ with other estimators that are certain specified functions of the observations $\mathbf{Y}$. However, if we are interested in a general function $h(y_1,y_2, \ldots ,y_n)$ as an estimator for β_i, then, under quite general conditions, the function $h(y_1,y_2, \ldots ,y_n)$ can be expanded into a Taylor series and the linear term used as a first approximation. Although there are some good reasons for wanting to restrict ourselves to linear functions of the observations y_i as estimators of $\boldsymbol{\beta}$, often we do not want to so restrict ourselves. A theorem that asserts the "goodness" of least-squares estimates in the model $\mathbf{Y} = \mathbf{X}\boldsymbol{\beta} + \mathbf{e}$ is the Gauss–Markoff theorem. Although this theorem is often attributed to Markoff, it seems that Gauss was the first to give the proof. The theorem is as follows.

◆ **Theorem 6.2** If the general-linear-hypothesis model of full rank $\mathbf{Y} = \mathbf{X}\boldsymbol{\beta} + \mathbf{e}$ is such that the following two conditions on the random vector $\mathbf{e}$ are met:

(1) $$E(\mathbf{e}) = \mathbf{0}$$

(2) $$E(\mathbf{e}\mathbf{e}') = \sigma^2\mathbf{I}$$

the best (minimum-variance) linear (linear functions of the y_i) unbiased estimate of $\boldsymbol{\beta}$ is given by least squares; that is, $\hat{\boldsymbol{\beta}} = \mathbf{S}^{-1}\mathbf{X}'\mathbf{Y}$ is the best linear unbiased estimate of $\boldsymbol{\beta}$.

Proof: Let $\mathbf{A}$ be any $p \times n$ constant matrix and let $\boldsymbol{\beta}^* = \mathbf{A}\mathbf{Y}$; $\boldsymbol{\beta}^*$ is a general linear function of $\mathbf{Y}$, which we shall take as an estimate of $\boldsymbol{\beta}$. We must specify the elements of $\mathbf{A}$ so that $\boldsymbol{\beta}^*$ will be the best unbiased estimate of $\boldsymbol{\beta}$. Let $\mathbf{A} = \mathbf{S}^{-1}\mathbf{X}' + \mathbf{B}$. Since $\mathbf{S}^{-1}\mathbf{X}'$ is known, we must find $\mathbf{B}$ in order to be able to specify $\mathbf{A}$. For unbiasedness, we have

$$E(\boldsymbol{\beta}^*) = E(\mathbf{A}\mathbf{Y}) = E[(\mathbf{S}^{-1}\mathbf{X}' + \mathbf{B})\mathbf{Y}] = (\mathbf{S}^{-1}\mathbf{X}' + \mathbf{B})\mathbf{X}\boldsymbol{\beta} = \boldsymbol{\beta} + \mathbf{B}\mathbf{X}\boldsymbol{\beta}$$

But, to be unbiased, $E(\boldsymbol{\beta}^*)$ must equal $\boldsymbol{\beta}$, and this implies that $\mathbf{B}\mathbf{X}\boldsymbol{\beta} = \mathbf{0}$ for all $\boldsymbol{\beta}$. Thus, unbiasedness specifies that $\mathbf{B}\mathbf{X} = \mathbf{0}$.

For the property of "best" we must find the matrix $\mathbf{B}$ that minimizes $\text{var}(\beta_i^*)$, where $i = 1, 2, \ldots, p$, subject to the restriction $\mathbf{B}\mathbf{X} = \mathbf{0}$. To examine this, consider the covariance

$$\begin{aligned}\text{cov}(\boldsymbol{\beta}^*) &= E[(\boldsymbol{\beta}^* - \boldsymbol{\beta})(\boldsymbol{\beta}^* - \boldsymbol{\beta})'] \\ &= E\{[(\mathbf{S}^{-1}\mathbf{X}' + \mathbf{B})\mathbf{Y} - \boldsymbol{\beta}][(\mathbf{S}^{-1}\mathbf{X}' + \mathbf{B})\mathbf{Y} - \boldsymbol{\beta}]'\}\end{aligned}$$

Substituting $\mathbf{X\beta} + \mathbf{e}$ for $\mathbf{Y}$ and using $\mathbf{BX} = \mathbf{0}$, we get

$$\begin{aligned}\operatorname{cov}(\boldsymbol{\beta}^*) &= E(\mathbf{S}^{-1}\mathbf{X}'\mathbf{ee}'\mathbf{XS}^{-1} + \mathbf{Bee}'\mathbf{B}' + \mathbf{S}^{-1}\mathbf{X}'\mathbf{ee}'\mathbf{B}' + \mathbf{Bee}'\mathbf{XS}^{-1}) \\ &= \sigma^2(\mathbf{S}^{-1} + \mathbf{BB}')\end{aligned}$$

Let $\mathbf{BB}' = \mathbf{G} = (g_{ij})$. Then $\operatorname{cov}(\boldsymbol{\beta}^*) = \sigma^2(\mathbf{S}^{-1} + \mathbf{G})$. The diagonal elements of $\operatorname{cov}(\boldsymbol{\beta}^*)$ are the respective variances of the β_i^*. To minimize each $\operatorname{var}(\beta_i^*)$, we must, therefore, minimize each diagonal element of $\operatorname{cov}(\boldsymbol{\beta}^*)$. Since σ^2 and $\mathbf{S}^{-1}$ are constants, we must find a matrix $\mathbf{G}$ such that each diagonal element of $\mathbf{G}$ is a minimum. But $\mathbf{G} = \mathbf{BB}'$ is positive semidefinite; hence $g_{ii} \geqslant 0$. Thus the diagonal elements of $\operatorname{cov}(\boldsymbol{\beta}^*)$ will attain their minimum when $g_{ii} = 0$ for $i = 1, 2, \ldots, p$. But, if $\mathbf{B} = (b_{ij})$, then $g_{ii} = \sum_{j=1}^{n} b_{ij}^2$. Therefore, if g_{ii} is to equal 0 for all i, it must be true that $b_{ij} = 0$ for all i and all j. This implies that $\mathbf{B} = \mathbf{0}$. The condition $\mathbf{B} = \mathbf{0}$ is compatible with the condition of unbiasedness, $\mathbf{BX} = \mathbf{0}$. Therefore, $\mathbf{A} = \mathbf{S}^{-1}\mathbf{X}'$ and $\boldsymbol{\beta}^* = \hat{\boldsymbol{\beta}}$. This completes the proof. Following the method used for case A, it is easily shown that $\hat{\sigma}^2$ is an unbiased estimate of σ^2.

In the foregoing proof, the discussion has centered around the estimation of the β_i themselves. However, we are also often interested in estimating certain functions of the β_i. Because of the invariance property of maximum-likelihood estimators, it is a straightforward procedure to obtain the maximum-likelihood estimator of any one-to-one transformation of the β_i. The situation for least-squares estimators is not so simple. There is one important case, however, in which the invariance property holds for least-squares estimators, and this is the case of linear functions of the β_i. This is stated in the following theorem.

◆ **Theorem 6.3** Under the general-linear-hypothesis model given in Theorem 6.2, the best linear unbiased estimate of any linear combination of the β_i is the same linear combination of the best linear unbiased estimates of the β_i; that is, the best linear unbiased estimate of $\mathbf{t}'\boldsymbol{\beta}$ (where $\mathbf{t}$ is a $p \times 1$ known vector of constants) is $\mathbf{t}'\hat{\boldsymbol{\beta}}$, where $\hat{\boldsymbol{\beta}}$ is the best linear unbiased estimate of $\boldsymbol{\beta}$, that is, $\mathbf{t}'\hat{\boldsymbol{\beta}} = \mathbf{t}'\mathbf{S}^{-1}\mathbf{X}'\mathbf{Y}$.

The proof follows the line of Theorem 6.2 and will be left as an exercise for the reader.

For example, if the least-squares estimators of β_1, β_2, and β_3, respectively, are 3, 6, and -4, then the value of the best linear unbiased estimator of, say, $3\beta_1 - 2\beta_2 + 5\beta_3$ is $(3)(3) - (2)(6) + (5)(-4) = -23$.

This is not true for functions in general. For example, if we wish to estimate $u = (\beta_1 + \beta_2)/2\beta_3$, the value of the best linear unbiased estimate of u is *not* given by

$$\frac{\hat{\beta}_1 + \hat{\beta}_2}{2\hat{\beta}_3} = \frac{3 + 6}{(2)(-4)} = -\frac{9}{8}$$

6.2.3 Point Estimation of $E(y)$. To estimate the mean of y for a given set of $x_1, x_2, \ldots, x_p$, we can use the fact that $E(y)$ is a linear function of the β_i, that is, $E(y) = \sum_{i=1}^{p} \beta_i x_i$. Hence, invoking the theorem concerning estimators of linear functions of the β_i, we have

$$\widehat{E(y)} = \sum_{i=1}^{p} \hat{\beta}_i x_i$$

where $\widehat{E(y)}$ is the least-squares estimator of $E(y)$ for case B, and it is also the maximum-likelihood estimator of $E(y)$ for case A.

This article will be concluded with a theorem concerning the estimation of the vector $\boldsymbol{\beta}$ when the random variables are not uncorrelated and do not have a common variance.

◆ **Theorem 6.4** If the covariance matrix of the vector **e** in the general-linear-hypothesis model of full rank is equal to $\mathbf{V}\sigma^2$, where **V** is a known positive definite matrix, the maximum-likelihood estimator of $\boldsymbol{\beta}$ is $\hat{\boldsymbol{\beta}} = (\mathbf{X}'\mathbf{V}^{-1}\mathbf{X})^{-1}\mathbf{X}'\mathbf{V}^{-1}\mathbf{Y}$ and this has properties similar to those listed in Theorem 6.1.

The proof of this theorem will be left for the reader.

6.2.4 The Normal Equations. Early in this chapter we stated that the normal equations

$$\mathbf{X}'\mathbf{X}\hat{\boldsymbol{\beta}} = \mathbf{X}'\mathbf{Y}$$

would play an important part in estimation and testing hypotheses about the parameters in model 1. In the previous articles of this chapter the normal equations occur many times; this is some indication of their importance in point estimation. The student should become familiar with equations of this type and with various methods of solving them, since in most cases of estimation and testing hypotheses their solution will be needed. We shall not discuss computing procedures in this chapter but defer this discussion until Chap. 7.

The ijth element of $\mathbf{X}'\mathbf{X}$ is the quantity $\sum_k x_{ik}x_{jk}$, and the ith element of $\mathbf{X}'\mathbf{Y}$ is $\sum_k x_{ik}y_k$. These quantities can readily be found by using a desk calculator.

6.2.5 Choosing the X matrix. The x_{ij} values must be known before the y_i values are selected at random. In some cases x_{ij} values may be picked or chosen by the experimenter in any way he wishes; in other cases they cannot be so controlled. If the experimenter can choose the x_{ij} values, the question of how to select them arises. In general, it would seem that the best way to pick them is so that the variance of certain estimators will be as small as possible.

For example, the variance of the estimator of $\boldsymbol{\lambda}'\boldsymbol{\beta}$ is $\sigma^2\boldsymbol{\lambda}'\mathbf{S}^{-1}\boldsymbol{\lambda}$, and we might want to choose the x_{ij} that minimize $\boldsymbol{\lambda}'\mathbf{S}^{-1}\boldsymbol{\lambda}$. We cannot do this in general for all vectors $\boldsymbol{\lambda}$, but we might be able to for some. We shall discuss this further when we discuss specific examples.

6.2.6 Examples. A simple linear model is

$$y_i = \beta_1 + \beta_2 x_i + e_i \qquad i = 1, 2, \ldots, n$$

where β_1 and β_2 are unknown scalar constants and the x_i are known scalar constants. Therefore, referring to Eq. (6.6), we find that $(p = 2)$

$$\mathbf{X} = \begin{pmatrix} 1 & x_1 \\ 1 & x_2 \\ & \cdots \\ 1 & x_n \end{pmatrix} \qquad \boldsymbol{\beta} = \begin{pmatrix} \beta_1 \\ \beta_2 \end{pmatrix}$$

It is easy to see that

$$\mathbf{S} = \mathbf{X}'\mathbf{X} = \begin{pmatrix} n & \Sigma x_i \\ \Sigma x_i & \Sigma x_i^2 \end{pmatrix} \qquad \mathbf{S}^{-1} = \frac{1}{n\Sigma(x_i - \bar{x})^2}\begin{pmatrix} \Sigma x_i^2 & -\Sigma x_i \\ -\Sigma x_i & n \end{pmatrix}$$

and

$$\mathbf{X}'\mathbf{Y} = \begin{pmatrix} \Sigma y_i \\ \Sigma x_i y_i \end{pmatrix}$$

Thus,

$$\hat{\boldsymbol{\beta}} = \begin{pmatrix} \hat{\beta}_1 \\ \hat{\beta}_2 \end{pmatrix} = \mathbf{S}^{-1}\mathbf{X}'\mathbf{Y} = \frac{1}{n\Sigma(x_i - \bar{x})^2}\begin{pmatrix} \Sigma x_i^2 \Sigma y_i - \Sigma x_i \Sigma x_i y_i \\ -\Sigma x_i \Sigma y_i + n\Sigma y_i x_i \end{pmatrix}$$

or

$$\hat{\beta}_2 = \frac{\Sigma(x_i - \bar{x})(y_i - \bar{y})}{\Sigma(x_i - \bar{x})^2} \qquad \hat{\beta}_1 = \bar{y} - \hat{\beta}_2\bar{x}$$

β_2 is the slope of the line, or, in other words, it is the change in $E(y)$ per unit change in x, and β_1 is the value of $E(y)$ when $x = 0$.

Since the $\text{cov}(\hat{\boldsymbol{\beta}}) = \mathbf{S}^{-1}\sigma^2$, we see that

$$\text{cov}(\hat{\beta}_1, \hat{\beta}_2) = -\frac{\sigma^2 \Sigma x_i}{n\Sigma(x_i - \bar{x})^2} \qquad \text{var}(\hat{\beta}_1) = \frac{\sigma^2 \Sigma x_i^2}{n\Sigma(x_i - \bar{x})^2}$$

and

$$\text{var}(\hat{\beta}_2) = \frac{\sigma^2}{\Sigma(x_i - \bar{x})^2}$$

To minimize $\text{var}(\hat{\beta}_2)$, we choose our x_i such that $\Sigma(x_i - \bar{x})^2$ is as large as possible. To minimize $\text{var}(\hat{\beta}_1)$, we choose the x_i such that $\Sigma x_i^2/\Sigma(x_i - \bar{x})^2$ is as small as possible. Since $\Sigma(x_i - \bar{x})^2 \leqslant \Sigma x_i^2$, the $\text{var}(\hat{\beta}_1)$ is minimum if the x_i are chosen such that $\bar{x} = 0$. This also makes the $\text{cov}(\hat{\beta}_1,\hat{\beta}_2) = 0$. Note that we have assumed that n is fixed.

To estimate σ^2, we note that any of the following formulas can be used:

$$\hat{\sigma}^2 = \frac{1}{n-2}(\mathbf{Y}'\mathbf{Y} - \mathbf{Y}'\mathbf{X}\mathbf{S}^{-1}\mathbf{X}'\mathbf{Y}) = \frac{1}{n-2}(\mathbf{Y}'\mathbf{Y} - \hat{\boldsymbol{\beta}}'\mathbf{X}'\mathbf{Y})$$

$$= \frac{1}{n-2}(\mathbf{Y} - \mathbf{X}\hat{\boldsymbol{\beta}})'(\mathbf{Y} - \mathbf{X}\hat{\boldsymbol{\beta}})$$

$$= \frac{1}{n-2}(\mathbf{Y}'\mathbf{Y} - \hat{\boldsymbol{\beta}}'\mathbf{S}\hat{\boldsymbol{\beta}}) = \frac{1}{n-2}\left\{\Sigma(y_i - \bar{y})^2 - \frac{[\Sigma(x_i - \bar{x})(y_i - \bar{y})]^2}{\Sigma(x_i - \bar{x})^2}\right\}$$

As another example, let us suppose that we want to predict the distance s that a particle will travel in time t if the velocity is constant and equal to v and if the initial distance from a certain reference point is d_0. The relationship is

$$s = d_0 + vt$$

However, let us assume that we cannot measure s accurately but that there is a random error attached; i.e., we measure not s but d, where $d = s + e$, where e is a random error. Then our relationship is

$$d = d_0 + vt + e$$

Since d_0 and v are not known, we shall take a series of observations of d and t and estimate our unknown constants d_0 and v. If d_i represents the ith distance measured at time t_i and e_i is the corresponding error, we have

$$d_i = d_0 + vt_i + e_i \qquad i = 1, 2, \ldots, n$$

a general-linear-hypothesis model ($p = 2$). The observations taken are given below.

Distance d	9	15	19	20	45	55	78
Time t	1	2	3	4	10	12	18

The formulas above give

$$\mathbf{S} = \begin{pmatrix} 7 & 50 \\ 50 & 598 \end{pmatrix} \qquad \mathbf{S}^{-1} = \begin{pmatrix} .35469 & -.029656 \\ -.029656 & .0041518 \end{pmatrix}$$

$$\mathbf{Y}'\mathbf{Y} = 12{,}201 \qquad \mathbf{X}'\mathbf{Y} = \begin{pmatrix} 241 \\ 2{,}690 \end{pmatrix} \qquad \text{and} \qquad \hat{\boldsymbol{\beta}} = \begin{pmatrix} \hat{\beta}_1 \\ \hat{\beta}_2 \end{pmatrix} = \begin{pmatrix} 5.71 \\ 4.02 \end{pmatrix}$$

$$\hat{\boldsymbol{\beta}}'\mathbf{X}'\mathbf{Y} = 12{,}189.9$$

Therefore, $\hat{s} = \hat{d}_0 + \hat{v}t$, or $\hat{s} = 5.71 + 4.02t$, is the estimate of the distance the particle travels from the given reference in time t. We know that this prediction is best in the sense of Theorem 6.3. The estimate of σ^2 is

$$\tfrac{1}{5}(12{,}201.0 - 12{,}189.9) = 2.22 = \hat{\sigma}^2$$

6.3 Interval Estimation

6.3.1 Interval Estimation of β_i, σ^2, and a Linear Function of β_i. We shall now turn our attention to the problem of interval estimation of β_i and σ^2 for case A only. Since $(n - p)(\hat{\sigma}^2/\sigma^2) = u$ is distributed as $\chi^2(n - p)$, a confidence interval can be put about σ^2 as follows:

Let α_0 and α_1 be two constants such that

$$P\left[\alpha_0 \leqslant \frac{\hat{\sigma}^2(n-p)}{\sigma^2} \leqslant \alpha_1\right] = 1 - \alpha = \int_{\alpha_0}^{\alpha_1} g(u)\, du$$

where $1 - \alpha$ is the confidence-interval coefficient. After some algebraic manipulations, we arrive at

$$P\left[\frac{\hat{\sigma}^2(n-p)}{\alpha_1} \leqslant \sigma^2 \leqslant \frac{\hat{\sigma}^2(n-p)}{\alpha_0}\right] = 1 - \alpha \tag{6.8}$$

and the $1 - \alpha$ confidence interval about σ^2 is defined by the bracketed quantity. The width of the confidence interval is

$$\frac{\hat{\sigma}^2(n-p)}{\alpha_0} - \frac{\hat{\sigma}^2(n-p)}{\alpha_1} = \hat{\sigma}^2(n-p)\left(\frac{1}{\alpha_0} - \frac{1}{\alpha_1}\right)$$

To set a $1 - \alpha$ confidence interval on β_i, we use the fact that $\hat{\beta}_i$ is distributed $N(\beta_i, c_{ii}\sigma^2)$, where c_{ij} is the ijth element of $\mathbf{S}^{-1} = \mathbf{C}$. Therefore, $(\hat{\beta}_i - \beta_i)/\sigma\sqrt{c_{ii}}$ is distributed $N(0,1)$ and is independent of $(n - p)\hat{\sigma}^2/\sigma^2$, which is distributed as $\chi^2(n - p)$. It follows that

$$u = \frac{\hat{\beta}_i - \beta_i}{\sigma\sqrt{c_{ii}}}\sqrt{\frac{\sigma^2}{\hat{\sigma}^2}} = \frac{\hat{\beta}_i - \beta_i}{\sqrt{\hat{\sigma}^2 c_{ii}}}$$

is distributed as $t(n - p)$. Thus,

$$\int_{-t_{\alpha/2}}^{t_{\alpha/2}} t(u)\,du = P\left(-t_{\alpha/2} \leqslant \frac{\hat{\beta}_i - \beta_i}{\sqrt{\hat{\sigma}^2 c_{ii}}} \leqslant t_{\alpha/2}\right) = 1 - \alpha$$

where $t(u)$ is the density function of t. After some manipulations we get

$$P(\hat{\beta}_i - t_{\alpha/2}\sqrt{c_{ii}\hat{\sigma}^2} \leqslant \beta_i \leqslant \hat{\beta}_i + t_{\alpha/2}\sqrt{c_{ii}\hat{\sigma}^2}) = 1 - \alpha \tag{6.9}$$

and the quantity in parentheses defines a $1 - \alpha$ confidence interval on β_i. The width of the interval is $2t_{\alpha/2}\sqrt{c_{ii}\hat{\sigma}^2}$.

Frequently, an experimenter is interested in setting confidence limits on some function of the β_i. Exact methods for setting confidence intervals are not known, except for special functions. However, a method is available for one of the most important cases, i.e., for the case of a *linear* function of the β_i. Let $\mathbf{r}$ be a known $p \times 1$ vector of constants; then, to set confidence limits of size $1 - \alpha$ on $\mathbf{r}'\boldsymbol{\beta}$, we proceed as follows. Since $\mathbf{r}'\hat{\boldsymbol{\beta}}$ is distributed $N(\mathbf{r}'\boldsymbol{\beta},\sigma^2\mathbf{r}'\mathbf{S}^{-1}\mathbf{r})$, it is clear that $(\mathbf{r}'\hat{\boldsymbol{\beta}} - \mathbf{r}'\boldsymbol{\beta})/\sigma\sqrt{\mathbf{r}'\mathbf{S}^{-1}\mathbf{r}}$ is distributed $N(0,1)$. Proceeding as before, we arrive at the probability equation

$$P(\mathbf{r}'\hat{\boldsymbol{\beta}} - t_{\alpha/2}\sqrt{\hat{\sigma}^2\mathbf{r}'\mathbf{S}^{-1}\mathbf{r}} \leqslant \mathbf{r}'\boldsymbol{\beta} \leqslant \mathbf{r}'\hat{\boldsymbol{\beta}} + t_{\alpha/2}\sqrt{\hat{\sigma}^2\mathbf{r}'\mathbf{S}^{-1}\mathbf{r}}) = 1 - \alpha \tag{6.10}$$

and the quantity in parentheses defines the desired $1 - \alpha$ confidence interval. The width of the interval is $2t_{\alpha/2}\sqrt{\hat{\sigma}^2\mathbf{r}'\mathbf{S}^{-1}\mathbf{r}}$.

6.3.2 Interval Estimation of $E(y)$. If the β_i are known, then for a given set of $x_1, x_2, \ldots, x_p$ the quantity $E(y)$ can be accurately estimated, since $E(y) = \sum_{i=1}^{p} \beta_i x_i$. However, if the β_i are unknown, we must select a sample of size n, $\mathbf{P}_j' = (y_j, x_{j1}, x_{j2}, \ldots, x_{jp})$, where $j = 1, 2, \ldots, n$, and from these sample values obtain an estimate $\hat{\beta}_i$ of the β_i. Then we use $\sum_{i=1}^{p} \hat{\beta}_i x_i$ to obtain an estimate of $E(y)$ corresponding to the point $(x_1,x_2, \ldots ,x_p)$. Since the estimate of $E(y)$ is a random variable, subject to random fluctuations, we might desire a confidence-interval estimate of $E(y)$. To obtain this, let

$$z = \sum_{i=1}^{p} (\hat{\beta}_i - \beta_i)x_i = \mathbf{x}'(\hat{\boldsymbol{\beta}} - \boldsymbol{\beta})$$

where $\mathbf{x}$ is a $p \times 1$ vector whose ith element is x_i. Therefore, z is distributed $N(0,\sigma^2\mathbf{x}'\mathbf{S}^{-1}\mathbf{x})$, and $\sum_{i=1}^{p} \frac{(\hat{\beta}_i - \beta_i)x_i}{\sigma\sqrt{\mathbf{x}'\mathbf{S}^{-1}\mathbf{x}}}$ is distributed $N(0,1)$.

It is also true that $(n - p)\hat{\sigma}^2/\sigma^2 = w$ is distributed as $\chi^2(n - p)$ independently of z. Therefore,

$$u = \frac{z}{\hat{\sigma}\sqrt{\mathbf{x}'\mathbf{S}^{-1}\mathbf{x}}}$$

is distributed as $t(n - p)$, and a $1 - \alpha$ confidence interval is obtained from

$$1 - \alpha = \int_{-t_{\alpha/2}}^{t_{\alpha/2}} t(u)\,du = P(-t_{\alpha/2} \leqslant u \leqslant t_{\alpha/2})$$

Substituting for u, we get

$$P\left(-t_{\alpha/2} \leqslant \frac{\Sigma\hat{\beta}_i x_i - \Sigma\beta_i x_i}{\sqrt{\hat{\sigma}^2\mathbf{x}'\mathbf{S}^{-1}\mathbf{x}}} \leqslant t_{\alpha/2}\right) = 1 - \alpha$$

After some manipulations, we arrive at

$$P(\Sigma\hat{\beta}_i x_i - t_{\alpha/2}\sqrt{\hat{\sigma}^2\mathbf{x}'\mathbf{S}^{-1}\mathbf{x}} \leqslant \Sigma\beta_i x_i \leqslant \Sigma\hat{\beta}_i x_i + t_{\alpha/2}\sqrt{\hat{\sigma}^2\mathbf{x}'\mathbf{S}^{-1}\mathbf{x}}) = 1 - \alpha \tag{6.11}$$

and the quantity in parentheses defines the desired confidence interval on $E(y)$. This is a special case of the situation described in the previous section. The width of the interval is $2t_{\alpha/2}\sqrt{\hat{\sigma}^2\mathbf{x}'\mathbf{S}^{-1}\mathbf{x}}$.

6.3.3 Interval Estimate of the Mean of Future Observations. Let us suppose that the equation $w = \beta_0 + \beta_1 t$ describes the weight of chickens that have been fed a new ration for a given time t. The weight w is in grams, and t is in weeks. Suppose that t can be measured with no error but that there is either an equation error and/or an error in measuring the weight w. We observe y, where $y = w + e$. So the model will be written

$$y = \beta_0 + \beta_1 t + e$$

where β_0 is the initial weight of a chicken and β_1 is the growth rate. If we want to estimate the β_i by observing the weight of a few chickens, we can use the theorems developed in the previous articles. Suppose that a farmer plans to ship k chickens to market after feeding them for 12 weeks and that he wants to know what the average weight of these k chickens will be. The average weight of all chickens that have been fed for 12 weeks is $\beta_0 + 12\beta_1$. However, the farmer may not be interested in estimating this quantity; he may want instead a confidence interval on the mean of k values of y selected from the population with mean $\beta_0 + 12\beta_1$. This type of problem is the subject of this article. We shall derive formulas more general than those illustrated in the above example.

Let us suppose that a random sample of k values of y is drawn from the frequency function

$$f(y; x_{01},x_{02}, \ldots ,x_{0p})$$

where each of the k values of y is selected from the same frequency function, i.e., with the same x_i values.

Let these k values of y be denoted by $y_{01}, y_{02}, \ldots, y_{0k}$, and let the mean of these sample values be denoted by

$$\bar{y}_0 = \frac{1}{k} \sum_{j=1}^{k} y_{0j}$$

Suppose a confidence interval is desired on the mean $\bar{y}_0$ of these k values of y. That is to say, suppose we construct an interval such that the probability that the mean $\bar{y}_0$ will fall within it is equal to some preassigned value $1 - \alpha$. The interval will be called a *prediction interval of size* $1 - \alpha$, since the probability that the mean $\bar{y}_0$ of a future sample of y values selected at random from the density $f(y; x_{01},x_{02}, \ldots ,x_{0p})$ will fall within it is equal to $1 - \alpha$.

Since $\boldsymbol{\beta}$ and σ^2 are not known, the procedure is to select n sample values $\mathbf{P}_j' = (y_j,x_{1j}, \ldots ,x_{pj})$, where $j = 1, 2, \ldots, n$, use these values to estimate $\boldsymbol{\beta}$ and σ^2, and construct the prediction interval using the estimates $\hat{\boldsymbol{\beta}}$ and $\hat{\sigma}^2$. To do this we observe that, since $\bar{y}_0$ and $\hat{\boldsymbol{\beta}}$ are independent, the quantity

$$z = \bar{y}_0 - \hat{\boldsymbol{\beta}}'\mathbf{x}_0$$

is distributed $N\left[0, \sigma^2\left(\frac{1}{k} + \mathbf{x}_0'\mathbf{S}^{-1}\mathbf{x}_0\right)\right]$, where $\mathbf{x}_0$ is a $p \times 1$ vector whose ith element is x_{0i}. It follows that

$$w = \frac{\bar{y}_0 - \hat{\boldsymbol{\beta}}'\mathbf{x}_0}{\sigma\sqrt{\frac{1}{k} + \mathbf{x}_0'\mathbf{S}^{-1}\mathbf{x}_0}}$$

is distributed $N(0,1)$. Also, $v = (n - p)\hat{\sigma}^2/\sigma^2$ is distributed as $\chi^2(n - p)$ and is independent of w. Hence,

$$u = \frac{w\sqrt{n - p}}{\sqrt{v}}$$

is distributed as $t(n - p)$. Therefore, $1 - \alpha = P(-t_{\alpha/2} \leqslant u \leqslant t_{\alpha/2})$. Substituting for u, we obtain

$$1 - \alpha = P\left[-t_{\alpha/2} \leqslant \frac{\bar{y}_0 - \hat{\boldsymbol{\beta}}'\mathbf{x}_0}{\sqrt{\hat{\sigma}^2\left(\frac{1}{k} + \mathbf{x}_0'\mathbf{S}^{-1}\mathbf{x}_0\right)}} \leqslant t_{\alpha/2}\right]$$

After some manipulations we arrive at

$$P\left[\hat{\boldsymbol{\beta}}'\mathbf{x}_0 - t_{\alpha/2}\sqrt{\hat{\sigma}^2\left(\frac{1}{k} + \mathbf{x}_0'\mathbf{S}^{-1}\mathbf{x}_0\right)} \leqslant \bar{y}_0 \right.$$

$$\left. \leqslant \hat{\boldsymbol{\beta}}'\mathbf{x}_0 + t_{\alpha/2}\sqrt{\hat{\sigma}^2\left(\frac{1}{k} + \mathbf{x}_0'\mathbf{S}^{-1}\mathbf{x}_0\right)}\right] = 1 - \alpha \quad (6.12)$$

which gives the desired prediction interval.

We shall give an example to illustrate the above theory.

6.3.4 Example. Suppose we desire a 95 per cent confidence interval on β_1 and on $\beta_1 - 2\beta_2$ in the model $y = \beta_1 + \beta_2 x + e$. We shall assume that nine observations are available and that the following have been computed:

$$\mathbf{S} = \mathbf{X}'\mathbf{X} = \begin{pmatrix} 9 & 12 \\ 12 & 18 \end{pmatrix} \qquad \mathbf{X}'\mathbf{Y} = \begin{pmatrix} 21 \\ 27 \end{pmatrix} \qquad \mathbf{Y}'\mathbf{Y} = 54.0 \qquad n = 9 \qquad p = 2$$

Therefore $\mathbf{C} = \mathbf{S}^{-1} = \begin{pmatrix} 1.0 & -.67 \\ -.67 & .5 \end{pmatrix} \qquad \hat{\boldsymbol{\beta}} = \begin{pmatrix} \hat{\beta}_1 \\ \hat{\beta}_2 \end{pmatrix} = \begin{pmatrix} 3.0 \\ -.5 \end{pmatrix}$

and $\hat{\boldsymbol{\beta}}'\mathbf{X}'\mathbf{Y} = 49.5$. Further,

$$\hat{\sigma}^2 = \frac{1}{n-p}(\mathbf{Y}'\mathbf{Y} - \hat{\boldsymbol{\beta}}'\mathbf{X}'\mathbf{Y}) = \frac{1}{7}(54.0 - 49.5) = .64$$

To set a confidence interval on β_1, we use Eq. (6.9). Since $t_{\alpha/2} = 2.4$, we get $t_{\alpha/2}\sqrt{c_{11}\hat{\sigma}^2} = (2.4)\sqrt{(1.0)(.64)} = 1.9$, where $c_{11} = 1.0$ (the element in the first row and first column of $\mathbf{C}$). By substituting into (6.9), we get

$$3.0 - 1.9 \leqslant \beta_1 \leqslant 3.0 + 1.9$$

which gives the probability equation

$$P(1.1 \leqslant \beta_1 \leqslant 4.9) = .95$$

The width of the confidence interval is $4.9 - 1.1 = 3.8$.

To set a 95 per cent confidence interval on $\beta_1 - 2\beta_2$, we use Eq. (6.10), where $\mathbf{r}' = (1, -2)$. We get

$$\mathbf{r}'\hat{\boldsymbol{\beta}} = (1, -2)\begin{pmatrix} 3.0 \\ -.5 \end{pmatrix} = 4.0 \qquad \mathbf{r}'\mathbf{S}^{-1}\mathbf{r} = \tfrac{17}{3}$$

We substitute into Eq. (6.10) and get

$$t_{\alpha/2}\sqrt{\hat{\sigma}^2\mathbf{r}'\mathbf{S}^{-1}\mathbf{r}} = (2.4)\sqrt{(0.64)(\tfrac{17}{3})} = 4.56$$

The probability equation is

$$P(-.56 \leqslant \mathbf{r}'\boldsymbol{\beta} \leqslant 8.56) = .95$$

with width equal to 9.12.

6.3.5 Interval Estimate on x for an Observed y in a Simple Linear Model. Let us consider the relationship between growth of chickens and a given time interval that we discussed at the beginning of Art. 6.3.3. Suppose that the farmer is now interested in a somewhat different problem. He receives a group of k chickens that have been fed a standard ration for t (unknown) weeks. The farmer observes the weights of the chickens and wants to estimate the time t that they have been on this ration.

At first it may seem that we should solve the equation $w = \beta_1 + \beta_2 t$ for t, obtain $t = \alpha_1 + \alpha_2 w$, and use this as the prediction equation. However, if t is measured without error and there is error in measuring w, then, if $y = w + e$, we get $t = \alpha_1 + \alpha_2 y - \alpha_2 e$. This clearly does not fit the definition of any of the models in Chap. 5. Therefore, we shall use the model in the form $y = \beta_1 + \beta_2 t + e$ and follow an "inverse" method of estimation. This type of problem is the subject of this article.

Suppose we use the model $y_j = \beta_1 + \beta_2 x_j + e_j$ and select a sample of n pairs of values denoted by $\mathbf{P}'_1 = (y_1,x_1)$; $\mathbf{P}'_2 = (y_2,x_2)$; . . . ; $\mathbf{P}'_n = (y_n,x_n)$. Suppose, further, that we select k additional values of y, each with the same x value (say, x_0) which is unknown. In other words, $n + k$ sample points are selected as follows: $\mathbf{P}'_1, \mathbf{P}'_2, \ldots, \mathbf{P}'_n, \mathbf{P}'_{n+1}, \ldots, \mathbf{P}'_{n+k}$, where $\mathbf{P}'_{n+i} = (y_{n+i},x_0)$ for $i > 0$ and where x_0 is not known. Using these points, we want to estimate x_0. The likelihood of the sample is

$$f(e_1,e_2,\ldots,e_{n+k}) = \frac{1}{(2\pi\sigma^2)^{\frac{1}{2}(n+k)}} \exp\left\{-\frac{1}{2\sigma^2}\left[\sum_{j=1}^{n}(y_j - \beta_1 - \beta_2 x_j)^2 + \sum_{j=n+1}^{n+k}(y_j - \beta_1 - \beta_2 x_0)^2\right]\right\}$$

Taking the derivative with respect to σ^2, β_1, β_2, x_0 gives, after some simplification ($\hat{\sigma}^2$ is corrected for bias),

$$\begin{aligned}
\hat{x}_0 &= \frac{\bar{y}_0 - \hat{\beta}_1}{\hat{\beta}_2} \\
\hat{\beta}_2 &= \frac{\Sigma(y_i - \bar{y})(x_i - \bar{x})}{\Sigma(x_i - \bar{x})^2} \\
\hat{\beta}_1 &= \bar{y} - \hat{\beta}_2\bar{x} \\
\hat{\sigma}^2 &= \frac{1}{n+k-3}\left[\sum_{i=1}^{n}(y_i - \hat{\beta}_1 - \hat{\beta}_2 x_i) + \sum_{i=n+1}^{n+k}(y_i - \bar{y}_0)^2\right]
\end{aligned} \tag{6.13}$$

where $\quad \bar{y}_0 = \frac{1}{k}\sum_{j=n+1}^{n+k} y_j \qquad \bar{x} = \frac{1}{n}\sum_{i=1}^{n} x_i \qquad \bar{y} = \frac{1}{n}\sum_{i=1}^{n} y_i$

The second and third equations in (6.13) can be solved for $\hat{\beta}_1$ and $\hat{\beta}_2$ and these values substituted into the first equation to find $\hat{x}_0$. This gives a point estimate of x_0.

To obtain an interval estimate of x_0, we shall proceed in the following fashion. We shall let

$$u = \bar{y}_0 - \hat{\beta}_1 - \hat{\beta}_2 x_0$$

Clearly, u is distributed normally with mean 0 and variance

$$\sigma_u^2 = \sigma^2\left[\frac{1}{k} + \frac{\Sigma x_i^2}{n\Sigma(x_i - \bar{x})^2} + \frac{x_0^2}{\Sigma(x_i - \bar{x})^2} - \frac{2x_0\bar{x}}{\Sigma(x_i - \bar{x})^2}\right] \tag{6.14}$$

From this we obtain

$$\sigma_u^2 = \sigma^2\left[\frac{1}{k} + \frac{1}{n} + \frac{(x_0 - \bar{x})^2}{\Sigma(x_i - \bar{x})^2}\right] \tag{6.15}$$

Therefore u/σ_u is distributed $N(0,1)$. The quantity

$$\frac{v_1}{\sigma^2} = \sum_{i=1}^{n} \frac{(y_i - \hat{\beta}_1 - \hat{\beta}_2 x_i)^2}{\sigma^2}$$

is distributed as $\chi^2(n - 2)$ and is independent of u. Also, the quantity

$$\frac{v_2}{\sigma^2} = \sum_{i=n+1}^{n+k} \frac{(y_i - \bar{y}_0)^2}{\sigma^2}$$

is distributed as $\chi^2(k - 1)$ and is independent of u and v_1. Hence,

$$\frac{v}{\sigma^2} = \frac{v_1 + v_2}{\sigma^2}$$

is distributed as $\chi^2(n + k - 3)$ and is independent of u. Therefore, the quantity

$$\frac{u\sqrt{n + k - 3}}{\sigma_u\sqrt{v/\sigma^2}}$$

is distributed as $t(n + k - 3)$.

If we desire a $1 - \alpha$ confidence interval about x_0, we can write

$$P\left\{-t_{\alpha/2} \leqslant \frac{(\bar{y}_0 - \hat{\beta}_1 - \hat{\beta}_2 x_0)\sqrt{n + k - 3}}{v^{\frac{1}{2}}\left[\frac{1}{k} + \sum_{i=1}^{n}(x_i - x_0)^2 \Big/ n\sum_{i=1}^{n}(x_i - \bar{x})^2\right]^{\frac{1}{2}}} \leqslant t_{\alpha/2}\right\} = 1 - \alpha$$

We shall let $\hat{\sigma}^2 = v/(n + k - 3)$. Therefore, the probability is $1 - \alpha$ that the following inequality holds:

$$\frac{(\bar{y}_0 - \hat{\beta}_1 - \hat{\beta}_2 x_0)^2}{\hat{\sigma}^2\left[\dfrac{1}{k} + \dfrac{1}{n} + \dfrac{(\bar{x} - x_0)^2}{\Sigma(x_i - \bar{x})^2}\right]} \leqslant t^2_{\alpha/2} \tag{6.16}$$

Using the equality, we have a quadratic equation in x_0, the only unknown in (6.16). Solving for x_0 gives

$$x_0 = \bar{x} + \frac{\hat{\beta}_2(\bar{y} - \bar{y}_0)}{\lambda} \pm \frac{t_{\alpha/2}\hat{\sigma}}{\lambda}\sqrt{\lambda\left(\frac{1}{n} + \frac{1}{k}\right) + \frac{(\bar{y}_0 - \bar{y})^2}{\Sigma(x_i - \bar{x})^2}}$$

where

$$\lambda = \hat{\beta}_2^2 - \frac{t^2_{\alpha/2}\hat{\sigma}^2}{\Sigma(x_i - \bar{x})^2}$$

Therefore, a $1 - \alpha$ confidence interval on x_0 is given by

$$\bar{x} - \frac{\hat{\beta}_2(\bar{y} - \bar{y}_0)}{\lambda} - \frac{t_{\alpha/2}\hat{\sigma}}{\lambda}\sqrt{\lambda\left(\frac{1}{n} + \frac{1}{k}\right) + \frac{(\bar{y}_0 - \bar{y})^2}{\Sigma(x_i - \bar{x})^2}} \leqslant x_0$$

$$\leqslant \bar{x} - \frac{\hat{\beta}_2(\bar{y} - \bar{y}_0)}{\lambda} + \frac{t_{\alpha/2}\hat{\sigma}}{\lambda}\sqrt{\lambda\left(\frac{1}{n} + \frac{1}{k}\right) + \frac{(\bar{y}_0 - \bar{y})^2}{\Sigma(x_i - \bar{x})^2}} \quad \text{if } \lambda > 0 \tag{6.17}$$

If $\lambda < 0$, it can be seen by examining (6.16) that confidence limits on x_0 may not exist. Even if confidence limits do exist when $\lambda < 0$, they will in general be too wide to be of any practical use.

6.3.6 Width of Confidence Intervals. When an experimenter decides to use a confidence interval, he should decide what width is necessary in order to make a decision. We must, therefore, investigate the width to see what is involved. We shall not discuss this here, but shall devote a chapter in Vol. II to this important subject.

6.3.7 Simultaneous Confidence Intervals. The frequency interpretation of the foregoing results on confidence intervals is that, if many samples are taken and an interval such as that given by (6.9) is constructed, then, on the average, $100(1 - \alpha)$ per cent of these intervals will cover the true (unknown) value β_i. This is true for one particular value of i only; i.e., a sample can be selected and a confidence interval set on β_1, and the above frequency interpretation will hold. But if the same data are used to set confidence intervals on β_1 and β_2, the probability is *not* equal to $1 - \alpha$ that the resulting confidence intervals will include *both* β_1 and β_2. *For each set of observations only one confidence statement can be made using the above formulas.* The same is true for Eqs. (6.8), (6.10), (6.11), and (6.12).

This procedure does not seem to be realistic—setting a confidence interval on only one β_i. It seems that an experimenter will want to set a confidence interval on each β_i and to know the probability that all these intervals contain their respective β_i. If the intervals were independent, this probability would be the product of the individual probabilities; i.e., in simple linear models we should calculate two intervals similar to (6.9), one corresponding to β_1 and one for β_2, and the probability that both intervals were correct would be $(1 - \alpha)^2$. Unfortunately, the intervals are not independent, and $(1 - \alpha)^2$ is not the correct probability. The problem is more complicated than this and will be dealt with in some detail in Vol. II.

6.4 Tests of Hypotheses

6.4.1 Testing the Hypothesis $\boldsymbol{\beta} = \boldsymbol{\beta}^*$. To test the hypothesis $\boldsymbol{\beta} = \boldsymbol{\beta}^*$ ($\boldsymbol{\beta}^*$ is a known vector) in the general-linear-hypothesis model of full rank, we shall assume that the vector $\mathbf{e}$ is distributed $N(\mathbf{0},\sigma^2\mathbf{I})$ unless it is specifically stated otherwise.

Testing $\boldsymbol{\beta} = \boldsymbol{\beta}^*$ ($\boldsymbol{\beta}^*$ known) is equivalent to testing simultaneously that each coefficient β_i equals a given constant β_i^*. It is quite important for an experimenter to have such a test available. In the model $\mathbf{Y} = \mathbf{X}\boldsymbol{\beta} + \mathbf{e}$, for example, if $\boldsymbol{\beta} = \mathbf{0}$, a knowledge of the factors that correspond to the x_i does not aid in the prediction of $E(y)$. However, if $\boldsymbol{\beta} \neq \mathbf{0}$, the x_i factors will be valuable in predicting $E(y)$.

To test the hypothesis H_0: $\boldsymbol{\beta} = \boldsymbol{\beta}^*$ we must devise a test function, say, $u = g(y_1,y_2, \ldots ,y_n,x_{11},x_{12}, \ldots ,x_{np})$, that is a function of the observations y_i and x_{ij} such that the distribution of u is known when $\boldsymbol{\beta} = \boldsymbol{\beta}^*$. In order to evaluate the power of the test, the distribution must also be known when the alternative hypothesis H_1: $\boldsymbol{\beta} \neq \boldsymbol{\beta}^*$ is true.

We shall use the likelihood ratio L as the test function. The likelihood equation is

$$f(\mathbf{e}; \boldsymbol{\beta},\sigma^2) = f(e_1,e_2, \ldots ,e_n; \boldsymbol{\beta},\sigma^2) = \frac{1}{(2\pi)^{n/2}\sigma^n} e^{-\mathbf{e}'\mathbf{e}/2\sigma^2}$$

$$= \frac{1}{(2\pi\sigma^2)^{n/2}} \exp - \frac{(\mathbf{Y} - \mathbf{X}\boldsymbol{\beta})'(\mathbf{Y} - \mathbf{X}\boldsymbol{\beta})}{2\sigma^2} \tag{6.18}$$

The test criterion is $L = L(\hat{\omega})/L(\hat{\Omega})$, where $L(\hat{\omega})$ and $L(\hat{\Omega})$ are as defined below.

The likelihood function is regarded as a function of the $p + 1$ parameters $\beta_1, \beta_2, \ldots, \beta_p$ and σ^2. The parameter space Ω is the $p + 1$-dimensional space defined by the inequalities $0 < \sigma^2 < \infty$; $-\infty < \beta_i < \infty$ $(i = 1, 2, \ldots, p)$. Let $\hat{\Omega}$ be the point (that is, the particular set of values of the parameters $\sigma^2, \beta_1, \beta_2, \ldots, \beta_p$) in Ω that maximizes the likelihood function, and let $L(\hat{\Omega})$ be the maximum value. $L(\hat{\omega})$ has a similar definition in the parameter space restricted by H_0. That is to say, ω is the space of $\sigma^2, \beta_1, \beta_2, \ldots, \beta_p$ defined by the inequality $0 < \sigma^2 < \infty$ and by $\beta_1 = \beta_1^*, \beta_2 = \beta_2^*, \ldots, \beta_p = \beta_p^*$. Actually, ω is a one-dimensional space. To find $L(\hat{\omega})$ and $L(\hat{\Omega})$ we shall work with the logarithm of the likelihood function.

To find $L(\hat{\omega})$ we proceed as follows:

$$\log f(\mathbf{e}; \sigma^2, \boldsymbol{\beta}^*) = -\frac{n}{2}\log 2\pi - \frac{n}{2}\log \sigma^2 - \frac{(\mathbf{Y} - \mathbf{X}\boldsymbol{\beta}^*)'(\mathbf{Y} - \mathbf{X}\boldsymbol{\beta}^*)}{2\sigma^2}$$

Since the β_i^* are fixed, this equation is a function of σ^2 only, and the value of σ^2 that maximizes it is the solution to

$$\frac{d}{d\sigma^2}[\log f(\mathbf{e}; \sigma^2, \boldsymbol{\beta}^*)] = \frac{(\mathbf{Y} - \mathbf{X}\boldsymbol{\beta}^*)'(\mathbf{Y} - \mathbf{X}\boldsymbol{\beta}^*)}{2\hat{\sigma}^4} - \frac{n}{2\hat{\sigma}^2} = 0$$

The solution is

$$\hat{\sigma}^2 = \frac{(\mathbf{Y} - \mathbf{X}\boldsymbol{\beta}^*)'(\mathbf{Y} - \mathbf{X}\boldsymbol{\beta}^*)}{n}$$

and

$$L(\hat{\omega}) = \frac{n^{n/2}e^{-n/2}}{(2\pi)^{n/2}[(\mathbf{Y} - \mathbf{X}\boldsymbol{\beta}^*)'(\mathbf{Y} - \mathbf{X}\boldsymbol{\beta}^*)]^{n/2}}$$

Proceeding in similar fashion to obtain $L(\hat{\Omega})$, we see that the point where $\log f(\mathbf{e}; \sigma^2, \boldsymbol{\beta})$ attains its maximum is the solution to the equations

$$\frac{\partial}{\partial \boldsymbol{\beta}}[\log f(\mathbf{e}; \sigma^2, \boldsymbol{\beta})] = \frac{\mathbf{X}'\mathbf{Y}}{\hat{\sigma}^2} - \frac{\mathbf{X}'\mathbf{X}\hat{\boldsymbol{\beta}}}{\hat{\sigma}^2} = \mathbf{0}$$

$$\frac{\partial}{\partial \sigma^2}[\log f(\mathbf{e}; \sigma^2, \boldsymbol{\beta})] = \frac{(\mathbf{Y} - \mathbf{X}\hat{\boldsymbol{\beta}})'(\mathbf{Y} - \mathbf{X}\hat{\boldsymbol{\beta}})}{2\hat{\sigma}^4} - \frac{n}{2\hat{\sigma}^2} = 0$$

We get

$$L(\hat{\Omega}) = \frac{n^{n/2}e^{-n/2}}{(2\pi)^{n/2}[(\mathbf{Y} - \mathbf{X}\hat{\boldsymbol{\beta}})'(\mathbf{Y} - \mathbf{X}\hat{\boldsymbol{\beta}})]^{n/2}}$$

where $\hat{\boldsymbol{\beta}} = \mathbf{S}^{-1}\mathbf{X}'\mathbf{Y}$. Therefore,

$$L = \left[\frac{(\mathbf{Y} - \mathbf{X}\hat{\boldsymbol{\beta}})'(\mathbf{Y} - \mathbf{X}\hat{\boldsymbol{\beta}})}{(\mathbf{Y} - \mathbf{X}\boldsymbol{\beta}^*)'(\mathbf{Y} - \mathbf{X}\boldsymbol{\beta}^*)}\right]^{n/2}$$

If $g(L; \boldsymbol{\beta}^*)$ is the distribution of L under H_0: $\boldsymbol{\beta} = \boldsymbol{\beta}^*$, the critical region is $0 \leqslant L \leqslant A$, where A is such that

$$\int_0^A g(L; \boldsymbol{\beta}^*)\, dL = \alpha \qquad (6.19)$$

where α is the probability of type I error. To determine A from Eq. (6.19) we must find $g(L;\, \boldsymbol{\beta}^*)$, the distribution of L when $\boldsymbol{\beta} = \boldsymbol{\beta}^*$. Then L is determined from collected data, and H_0 is rejected if $L \leq A$.

Before determining the distribution of L, we shall study the quantities that are involved, that is, $(\mathbf{Y} - \mathbf{X}\boldsymbol{\beta}^*)'(\mathbf{Y} - \mathbf{X}\boldsymbol{\beta}^*)$ and $(\mathbf{Y} - \mathbf{X}\hat{\boldsymbol{\beta}})'(\mathbf{Y} - \mathbf{X}\hat{\boldsymbol{\beta}})$.

From (6.18) we obtain the identity

$$(\mathbf{Y} - \mathbf{X}\boldsymbol{\beta}^*)'(\mathbf{Y} - \mathbf{X}\boldsymbol{\beta}^*) = (\mathbf{Y} - \mathbf{X}\hat{\boldsymbol{\beta}})'(\mathbf{Y} - \mathbf{X}\hat{\boldsymbol{\beta}}) + (\hat{\boldsymbol{\beta}} - \boldsymbol{\beta}^*)'\mathbf{X}'\mathbf{X}(\hat{\boldsymbol{\beta}} - \boldsymbol{\beta}^*) \tag{6.20}$$

If we substitute $\mathbf{S}^{-1}\mathbf{X}'\mathbf{Y}$ for $\hat{\boldsymbol{\beta}}$, Eq. (6.20) becomes

$$\begin{aligned}(\mathbf{Y} - \mathbf{X}\boldsymbol{\beta}^*)'(\mathbf{Y} - \mathbf{X}\boldsymbol{\beta}^*) &= (\mathbf{Y} - \mathbf{X}\boldsymbol{\beta}^*)'(\mathbf{I} - \mathbf{X}\mathbf{S}^{-1}\mathbf{X}')(\mathbf{Y} - \mathbf{X}\boldsymbol{\beta}^*) \\ &\quad + (\mathbf{Y} - \mathbf{X}\boldsymbol{\beta}^*)'\mathbf{X}\mathbf{S}^{-1}\mathbf{X}'(\mathbf{Y} - \mathbf{X}\boldsymbol{\beta}^*)\end{aligned} \tag{6.21}$$

If we let $\mathbf{Z} = \mathbf{Y} - \mathbf{X}\boldsymbol{\beta}^*$, we get

$$\mathbf{Z}'\mathbf{Z} = (\mathbf{Y} - \mathbf{X}\boldsymbol{\beta}^*)'(\mathbf{Y} - \mathbf{X}\boldsymbol{\beta}^*) = \mathbf{Z}'\mathbf{A}_1\mathbf{Z} + \mathbf{Z}'\mathbf{A}_2\mathbf{Z}$$

where $\mathbf{A}_1 = \mathbf{I} - \mathbf{X}\mathbf{S}^{-1}\mathbf{X}'$ and $\mathbf{A}_2 = \mathbf{X}\mathbf{S}^{-1}\mathbf{X}'$ are idempotent matrices. Since $\mathbf{Y}$ is distributed $N(\mathbf{X}\boldsymbol{\beta}, \sigma^2\mathbf{I})$, it follows that $\mathbf{Z}$ is distributed $N(\mathbf{X}\boldsymbol{\beta} - \mathbf{X}\boldsymbol{\beta}^*,\, \sigma^2\mathbf{I})$. And, since $\mathbf{A}_1$ and $\mathbf{A}_2$ are idempotent matrices, we note (see Theorem 4.16) that

(a) $\dfrac{\mathbf{Z}'\mathbf{A}_1\mathbf{Z}}{\sigma^2}$ is distributed as $\chi^2(n - p)$.

(b) $\dfrac{\mathbf{Z}'\mathbf{A}_2\mathbf{Z}}{\sigma^2}$ is distributed as $\chi'^2(p,\, \lambda)$, where

$$\lambda = \frac{(\boldsymbol{\beta} - \boldsymbol{\beta}^*)'\mathbf{X}'\mathbf{A}_2\mathbf{X}(\boldsymbol{\beta} - \boldsymbol{\beta}^*)}{2\sigma^2} = \frac{(\boldsymbol{\beta} - \boldsymbol{\beta}^*)'\mathbf{S}(\boldsymbol{\beta} - \boldsymbol{\beta}^*)}{2\sigma^2} \tag{6.22}$$

(c) $\dfrac{\mathbf{Z}'\mathbf{A}_1\mathbf{Z}}{\sigma^2}$ and $\dfrac{\mathbf{Z}'\mathbf{A}_2\mathbf{Z}}{\sigma^2}$ are independent.

In order to examine statements (6.22*a*), (6.22*b*), (6.22*c*) more closely and to see how they follow from Theorem 4.16, we note that $\mathbf{Z}'\mathbf{A}_1\mathbf{Z}/\sigma^2$ is distributed as $\chi'^2(k_1, \lambda_1)$, where k_1 is the rank of $\mathbf{A}_1$ and

$$\lambda_1 = \frac{(\boldsymbol{\beta} - \boldsymbol{\beta}^*)'\mathbf{X}'\mathbf{A}_1\mathbf{X}(\boldsymbol{\beta} - \boldsymbol{\beta}^*)}{2\sigma^2}$$

But, since $\mathbf{A}_1$ is idempotent, the rank equals the trace, and so $k_1 = n - p$. Also, it is evident that $\lambda_1 = 0$. Thus $\chi'^2(n - p,\, \lambda_1 = 0)$ is equivalent to $\chi^2(n - p)$; hence we have (6.22*a*).

To see how the theorem implies (6.22*b*) we continue as follows: $\mathbf{Z}'\mathbf{A}_2\mathbf{Z}/\sigma^2$ is distributed as $\chi'^2(k_2,\lambda)$, where $k_2 = \text{rank of } \mathbf{A}_2 = \text{tr}(\mathbf{A}_2) = p$ and

$$\lambda = \frac{(\boldsymbol{\beta} - \boldsymbol{\beta}^*)'\mathbf{X}'(\mathbf{X}\mathbf{S}^{-1}\mathbf{X}')\mathbf{X}(\boldsymbol{\beta} - \boldsymbol{\beta}^*)}{2\sigma^2} = \frac{(\boldsymbol{\beta} - \boldsymbol{\beta}^*)'\mathbf{S}(\boldsymbol{\beta} - \boldsymbol{\beta}^*)}{2\sigma^2}$$

Therefore, since $\mathbf{S}$ is positive definite, $\mathbf{Z}'\mathbf{A}_2\mathbf{Z}/\sigma^2$ has a central chi-square distribution if and only if $\boldsymbol{\beta} - \boldsymbol{\beta}^* = \mathbf{0}$; that is, if and only if $\text{H}_0\colon \boldsymbol{\beta} = \boldsymbol{\beta}^*$ is true.

Statement (6.22*c*) follows immediately, by Theorem 4.16.

Because of the distributional properties of $\mathbf{Z}'\mathbf{A}_1\mathbf{Z}/\sigma^2$ and $\mathbf{Z}'\mathbf{A}_2\mathbf{Z}/\sigma^2$ given in (6.22) it follows that

$$u = \frac{\mathbf{Z}'\mathbf{A}_2\mathbf{Z}}{\mathbf{Z}'\mathbf{A}_1\mathbf{Z}} \frac{n-p}{p} \tag{6.23}$$

is distributed as $F'(p, n-p, \lambda)$ and reduces to Snedecor's F distribution if and only if H_0 is true. It also follows that

$$v = \frac{pu}{(n-p) + pu} = \frac{\mathbf{Z}'\mathbf{A}_2\mathbf{Z}}{\mathbf{Z}'\mathbf{Z}} \tag{6.24}$$

is distributed as $E^2(p, n-p, \lambda)$.

Returning now to the equation for the likelihood ratio L, we have found that

$$L = \left[\frac{(\mathbf{Y} - \mathbf{X}\hat{\boldsymbol{\beta}})'(\mathbf{Y} - \mathbf{X}\hat{\boldsymbol{\beta}})}{(\mathbf{Y} - \mathbf{X}\boldsymbol{\beta}^*)'(\mathbf{Y} - \mathbf{X}\boldsymbol{\beta}^*)}\right]^{n/2} = (1 - v)^{n/2} = \left\{\frac{1}{1 + [p/(n-p)]u}\right\}^{n/2} \tag{6.25}$$

We observed from (6.25) that L is a monotonic function of v, and L is also a monotonic function of u. From these facts we see that either v or u can be used as a test function for the hypothesis $\text{H}_0\colon \boldsymbol{\beta} = \boldsymbol{\beta}^*$. For the distribution of u (Snedecor's F) the critical region that corresponds to $0 \leqslant L \leqslant A$ is $F_\alpha \leqslant u < \infty$, where F_α is a constant such that

$$\int_{F_\alpha}^{\infty} F(u; p, n-p, \lambda = 0)\, du = \alpha$$

Similarly, for the distribution of v (Tang's E^2) the critical region is $E_\alpha^2 \leqslant v \leqslant 1$ where E_α^2 is such that

$$\int_{E_\alpha{}^2}^{1} E^2(v; p, n-p, \lambda = 0)\, dv = \alpha$$

The power of the test is given by

$$\beta(\lambda) = \int_{E_\alpha^2}^{1} E^2(v; p, n-p, \lambda)\, dv$$

where
$$\lambda = \frac{(\boldsymbol{\beta} - \boldsymbol{\beta}^*)'\mathbf{S}(\boldsymbol{\beta} - \boldsymbol{\beta}^*)}{2\sigma^2}$$

and may be evaluated for different values of p, $n - p$, and λ by using Tang's tables.

TABLE 6.1 ANALYSIS OF VARIANCE FOR TESTING $\boldsymbol{\beta} = \boldsymbol{\beta}^*$

SV	DF	SS	MS
Total	n	$Q = (\mathbf{Y} - \mathbf{X}\boldsymbol{\beta}^*)'(\mathbf{Y} - \mathbf{X}\boldsymbol{\beta}^*)$	
Due to $\boldsymbol{\beta}$	p	$Q_2 = (\mathbf{Y} - \mathbf{X}\boldsymbol{\beta}^*)'\mathbf{X}\mathbf{S}^{-1}\mathbf{X}'(\mathbf{Y} - \mathbf{X}\boldsymbol{\beta}^*)$	$\frac{Q_2}{p}$
Error	$n - p$	$Q_1 = \mathbf{Y}'(\mathbf{I} - \mathbf{X}\mathbf{S}^{-1}\mathbf{X}')\mathbf{Y}$	$\frac{Q_1}{n-p}$

The evaluation of u is generally put in the form shown in Table 6.1; this is an *analysis-of-variance* (AOV) *table*, in which the degrees of freedom (DF), sum of squares (SS), and mean square (MS) corresponding to the various sources of variation (SV) are listed. The procedure for calculating u is to obtain Q and Q_2 by direct computation and to get Q_1 by using the identity $Q_1 = Q - Q_2$.

By observing the likelihood equation it is easily shown that L can be obtained by evaluating the following [where min $(\mathbf{Y} - \mathbf{X}\boldsymbol{\beta})'(\mathbf{Y} - \mathbf{X}\boldsymbol{\beta})$ means the minimum value of $(\mathbf{Y} - \mathbf{X}\boldsymbol{\beta})'(\mathbf{Y} - \mathbf{X}\boldsymbol{\beta})$ with respect to $\boldsymbol{\beta}$]:

$$\begin{aligned} Q &= (\mathbf{Y} - \mathbf{X}\boldsymbol{\beta}^*)'(\mathbf{Y} - \mathbf{X}\boldsymbol{\beta}^*) \\ &= \min\,[(\mathbf{Y} - \mathbf{X}\boldsymbol{\beta})'(\mathbf{Y} - \mathbf{X}\boldsymbol{\beta}) \text{ when } \mathrm{H}_0\text{: } \boldsymbol{\beta} = \boldsymbol{\beta}^* \text{ is true}] \\ Q_1 &= (\mathbf{Y} - \mathbf{X}\hat{\boldsymbol{\beta}})'(\mathbf{Y} - \mathbf{X}\hat{\boldsymbol{\beta}}) \\ &= \min\,[(\mathbf{Y} - \mathbf{X}\boldsymbol{\beta})'(\mathbf{Y} - \mathbf{X}\boldsymbol{\beta}) \text{ with no restriction on } \boldsymbol{\beta}] \\ L &= (Q_1/Q)^{n/2} \end{aligned}$$

We now state the following important theorem.

◆ **Theorem 6.5** In the general-linear-hypothesis model given in Definition 6.1 where $\mathbf{e}$ is distributed $N(\mathbf{0},\sigma^2\mathbf{I})$, the quantity

$$u = \frac{n-p}{p}\frac{Q_2}{Q_1}$$

is distributed as $F'(p, n - p, \lambda)$, where $Q_1 + Q_2$ is the minimum with respect to $\boldsymbol{\beta}$ of $(\mathbf{Y} - \mathbf{X}\boldsymbol{\beta})'(\mathbf{Y} - \mathbf{X}\boldsymbol{\beta})$ when H_0 is true, Q_1 is the minimum with respect to $\boldsymbol{\beta}$ when there is no restriction on $\boldsymbol{\beta}$, and

$$\lambda = \frac{(\boldsymbol{\beta} - \boldsymbol{\beta}^*)'\mathbf{S}(\boldsymbol{\beta} - \boldsymbol{\beta}^*)}{2\sigma^2}$$

One of the most important quantities in the AOV table is the error mean square $Q_1/n - p$. (It should be noted that the quantity $\hat{\sigma}^2$ that appears in the confidence-interval equations of Sec. 6.3 is equal to $Q_1/n - p$.) To calculate Q_1 we can let $\boldsymbol{\beta}^* = \mathbf{0}$; then the total sum of squares Q is $\mathbf{Y}'\mathbf{Y}$. Because of its importance we repeat the fact that $Q_1 = \mathbf{Y}'\mathbf{Y} - \hat{\boldsymbol{\beta}}'\mathbf{X}'\mathbf{Y}$, where $\hat{\boldsymbol{\beta}}$ is the solution of the normal equations. We shall discuss methods for computing these quantities in a later chapter.

6.4.2 A Test of the Hypothesis That a Given Linear Function $\mathbf{r}'\boldsymbol{\beta}$ (r Is a Known Vector) Is Equal to r_0. In deriving Eq. (6.10), Art. 6.3.1, we noticed that u was distributed as Student's t with $n - p$ degrees of freedom, where

$$u = \frac{\mathbf{r}'\hat{\boldsymbol{\beta}} - \mathbf{r}'\boldsymbol{\beta}}{\sqrt{(\mathbf{r}'\mathbf{S}^{-1}\mathbf{r})\hat{\sigma}^2}} \tag{6.26}$$

To test H_0: $\mathbf{r}'\boldsymbol{\beta} = r_0$, we simply replace $\mathbf{r}'\boldsymbol{\beta}$ by the known scalar r_0 in (6.26) and compare this value of u with the tabulated value of the Student's t distribution with $n - p$ degrees of freedom. In deriving Eq. (6.10) we noted that

$$v = \frac{\mathbf{r}'\hat{\boldsymbol{\beta}} - r_0}{\sqrt{\sigma^2\mathbf{r}'\mathbf{S}^{-1}\mathbf{r}}}$$

is distributed $N(\mu,1)$, where $\mu = (\mathbf{r}'\boldsymbol{\beta} - r_0)(\sigma^2\mathbf{r}'\mathbf{S}^{-1}\mathbf{r})^{-\frac{1}{2}}$; therefore, v^2 is distributed as $\chi'^2(1, \lambda)$, where

$$\lambda = \frac{(\mathbf{r}'\boldsymbol{\beta} - r_0)^2}{2\sigma^2\mathbf{r}'\mathbf{S}^{-1}\mathbf{r}}$$

Thus,

$$z = \frac{v^2}{\hat{\sigma}^2/\sigma^2} = \frac{(\mathbf{r}'\hat{\boldsymbol{\beta}} - r_0)^2}{\hat{\sigma}^2\mathbf{r}'\mathbf{S}^{-1}\mathbf{r}}$$

is distributed as $F'(1, n - p, \lambda)$ and reduces to Snedecor's F if and only if H_0: $(\mathbf{r}'\boldsymbol{\beta} = r_0)$ is true.

6.4.3 Testing a Subhypothesis, i.e., the Hypothesis $\beta_1 = \beta_1^*$, $\beta_2 = \beta_2^*, \ldots, \beta_r = \beta_r^*$ with the Remaining β_i Unspecified. In this article we shall examine the problem of testing the hypothesis $\beta_1 = \beta_1^*, \beta_2 = \beta_2^*, \ldots, \beta_r = \beta_r^*$ $(r < p)$ in the model $\mathbf{Y} = \mathbf{X}\boldsymbol{\beta} + \mathbf{e}$ in Definition 6.1. This is a very useful and very important test.

For the case $p = 2$ the equation is

$$y = x_1\beta_1 + x_2\beta_2 + e \tag{6.27}$$

Now suppose an experimenter wants to know if he can replace Eq. (6.27) with the model

$$y = x_1\beta_1 + e \tag{6.28}$$

In other words, he wants to test the hypothesis $\beta_2 = 0$ in the model given in (6.27) with no stipulation about the value of β_1. This test is clearly different from the test discussed in the preceding article; there a procedure was devised to test whether β_1 and β_2, for example, were simultaneously equal to zero. Here, if the conclusion is drawn that $\beta_2 \neq 0$ in model (6.27), it follows that the two factors x_1 and x_2 together are better for predicting y than the factor x_1 alone. On the other hand, if $\beta_2 = 0$, x_1 and x_2 together are *no better* than x_1 alone.

A test of this type is extremely valuable in simple linear models, e.g., when $x_1 = 1$ in model (6.27). In this case β_1 is the intercept of the line and β_2 the slope. In many cases it is extremely desirable to be able to test whether the slope β_2 is zero regardless of the value of the intercept β_1, or to test whether the intercept β_1 is zero regardless of the slope.

In the model

$$\mathbf{Y} = \mathbf{X}\boldsymbol{\beta} + \mathbf{e}$$

partition the matrix $\mathbf{X}$ and the vector $\boldsymbol{\beta}$ so that

$$\mathbf{X} = (\mathbf{X}_1, \mathbf{X}_2) \qquad \boldsymbol{\beta} = \begin{pmatrix} \boldsymbol{\gamma}_1 \\ \boldsymbol{\gamma}_2 \end{pmatrix}$$

where $\mathbf{X}_1$ has dimension $n \times r$ and $\boldsymbol{\gamma}_1$ has dimension $r \times 1$. Then the model can be written

$$\mathbf{Y} = \mathbf{X}\boldsymbol{\beta} + \mathbf{e} = (\mathbf{X}_1, \mathbf{X}_2)\begin{pmatrix} \boldsymbol{\gamma}_1 \\ \boldsymbol{\gamma}_2 \end{pmatrix} + \mathbf{e}$$

or

$$\mathbf{Y} = \mathbf{X}_1\boldsymbol{\gamma}_1 + \mathbf{X}_2\boldsymbol{\gamma}_2 + \mathbf{e}$$

We desire a test of the hypothesis $\boldsymbol{\gamma}_1 = \boldsymbol{\gamma}_1^*$ ($\boldsymbol{\gamma}_1^*$ known) with no stipulations on $\boldsymbol{\gamma}_2$. Since

$$\boldsymbol{\gamma}_1 = \begin{pmatrix} \beta_1 \\ \beta_2 \\ \cdots \\ \beta_r \end{pmatrix} \qquad \text{and} \qquad \boldsymbol{\gamma}_1^* = \begin{pmatrix} \beta_1^* \\ \beta_2^* \\ \cdots \\ \beta_r^* \end{pmatrix}$$

this test is equivalent to testing the hypothesis $\beta_1 = \beta_1^*, \beta_2 = \beta_2^*, \ldots, \beta_r = \beta_r^*$ $(r < p)$ in the model

$$y_j = \sum_{i=1}^{p} x_{ji}\beta_i + e_j$$

The test of the hypothesis $H_0: \gamma_1 = \gamma_1^*$ and its construction are stated in the following.

◆ **Theorem 6.6** Let the model $\mathbf{Y} = \mathbf{X}\boldsymbol{\beta} + \mathbf{e}$ given in Definition 6.1 be partitioned so that

$$\mathbf{Y} = \mathbf{X}_1\boldsymbol{\gamma}_1 + \mathbf{X}_2\boldsymbol{\gamma}_2 + \mathbf{e}$$

where $\boldsymbol{\gamma}_1$ is of dimension $r \times 1$, and let $\mathbf{e}$ be distributed $N(\mathbf{0},\sigma^2\mathbf{I})$. Then to test the hypothesis $H_0: \boldsymbol{\gamma}_1 = \boldsymbol{\gamma}_1^*$ by the likelihood ratio, the procedure is:

(1) Obtain the minimum value of $\mathbf{e}'\mathbf{e}$ with respect to the unknown parameters in $\boldsymbol{\beta}$ in the model $\mathbf{Y} = \mathbf{X}\boldsymbol{\beta} + \mathbf{e}$, and denote this minimum value by Q_0.

(2) Obtain the minimum value of $\mathbf{e}'\mathbf{e}$ with respect to the unknown parameters in $\boldsymbol{\gamma}_2$ in the model $\mathbf{Y} = \mathbf{X}_1\boldsymbol{\gamma}_1^* + \mathbf{X}_2\boldsymbol{\gamma}_2 + \mathbf{e}$ (this will be called the *model restricted by the hypothesis* H_0). Denote this minimum value by $Q_0 + Q_1$.

(3) Let $Q = Q_0 + Q_1 + Q_2$, where $Q = (\mathbf{Y} - \mathbf{X}_1\boldsymbol{\gamma}_1^*)'(\mathbf{Y} - \mathbf{X}_1\boldsymbol{\gamma}_1^*)$. Then the quantity u given by

$$u = \frac{n-p}{r}\frac{Q_1}{Q_0}$$

is distributed as $F'(r, n - p, \lambda)$, where

$$\lambda = \frac{(\boldsymbol{\gamma}_1 - \boldsymbol{\gamma}_1^*)'\mathbf{B}(\boldsymbol{\gamma}_1 - \boldsymbol{\gamma}_1^*)}{2\sigma^2}$$

and where $\mathbf{B} = \mathbf{X}_1'\mathbf{X}_1 - \mathbf{X}_1'\mathbf{X}_2(\mathbf{X}_2'\mathbf{X}_2)^{-1}\mathbf{X}_2'\mathbf{X}_1$

Since $\mathbf{B}$ is positive definite, u is distributed as $F(r, n - p)$ if and only if H_0 is true, i.e., if and only if $\boldsymbol{\gamma}_1 = \boldsymbol{\gamma}_1^*$.

Proof: The likelihood function is

$$f(\mathbf{Y}; \boldsymbol{\gamma}_1,\boldsymbol{\gamma}_2,\sigma^2) = \frac{1}{(2\pi\sigma^2)^{n/2}} \exp\left[-\frac{1}{2\sigma^2}(\mathbf{Y} - \mathbf{X}_1\boldsymbol{\gamma}_1 - \mathbf{X}_2\boldsymbol{\gamma}_2)'(\mathbf{Y} - \mathbf{X}_1\boldsymbol{\gamma}_1 - \mathbf{X}_2\boldsymbol{\gamma}_2)\right] \tag{6.29}$$

The parameter space Ω for the unrestricted model has dimension

$p + 1$ and is defined by the inequalities $-\infty < \beta_i < \infty$ ($i = 1, 2, \ldots, p$), $0 < \sigma^2 < \infty$. Since this situation is identical with that in Art. 6.4.1, we see that

$$L(\hat{\Omega}) = \frac{n^{n/2}e^{-n/2}}{(2\pi)^{n/2}[(\mathbf{Y} - \mathbf{X}\hat{\boldsymbol{\beta}})'(\mathbf{Y} - \mathbf{X}\hat{\boldsymbol{\beta}})]^{n/2}}$$

where $\hat{\boldsymbol{\beta}} = \mathbf{S}^{-1}\mathbf{X}'\mathbf{Y}$ is the value of $\boldsymbol{\beta}$ that maximizes (6.29). If we examine $\mathbf{e}'\mathbf{e} = (\mathbf{Y} - \mathbf{X}\boldsymbol{\beta})'(\mathbf{Y} - \mathbf{X}\boldsymbol{\beta})$, we see that the minimum value of the $\mathbf{e}'\mathbf{e}$ with respect to $\boldsymbol{\beta}$ (denoted by Q_0) is

$$Q_0 = (\mathbf{Y} - \mathbf{X}\hat{\boldsymbol{\beta}})'(\mathbf{Y} - \mathbf{X}\hat{\boldsymbol{\beta}})$$

Therefore,
$$L(\hat{\Omega}) = \frac{n^{n/2}e^{-n/2}}{(2\pi)^{n/2}} Q_0^{-n/2} \tag{6.30}$$

The parameter space ω for the restricted model $\mathbf{Y} = \mathbf{X}_1\boldsymbol{\gamma}_1^* + \mathbf{X}_2\boldsymbol{\gamma}_2 + \mathbf{e}$ has dimension $p - r + 1$ and is defined by $\beta_1 = \beta_1^*$, $\beta_2 = \beta_2^*, \ldots, \beta_r = \beta_r^*$, $-\infty < \beta_i < \infty$ ($i = r + 1, r + 2, \ldots, p$), $0 < \sigma^2 < \infty$. To obtain $L(\hat{\omega})$ we must find the maximum value of (6.29) after substituting $\boldsymbol{\gamma}_1^*$ for $\boldsymbol{\gamma}_1$. If we let

$$\mathbf{T} = \mathbf{Y} - \mathbf{X}_1\boldsymbol{\gamma}_1^*$$

Eq. (6.29) can be written

$$f(\mathbf{T}; \boldsymbol{\gamma}_2, \sigma^2) = \frac{1}{(2\pi\sigma^2)^{n/2}} \exp\left[-\frac{1}{2\sigma^2}(\mathbf{T} - \mathbf{X}_2\boldsymbol{\gamma}_2)'(\mathbf{T} - \mathbf{X}_2\boldsymbol{\gamma}_2)\right] \tag{6.31}$$

If we let $\hat{\boldsymbol{\gamma}}_2(\boldsymbol{\gamma}_1^*)$ and $\hat{\sigma}^2(\boldsymbol{\gamma}_1^*)$ denote respectively the values of $\boldsymbol{\gamma}_2$ and σ^2 that maximize Eq. (6.31), we get

$$\begin{aligned} \hat{\boldsymbol{\gamma}}_2(\boldsymbol{\gamma}_1^*) &= (\mathbf{X}_2'\mathbf{X}_2)^{-1}\mathbf{X}_2'\mathbf{T} = (\mathbf{X}_2'\mathbf{X}_2)^{-1}\mathbf{X}_2'(\mathbf{Y} - \mathbf{X}_1\boldsymbol{\gamma}_1^*) \\ \hat{\sigma}^2(\boldsymbol{\gamma}_1^*) &= \frac{1}{n}[\mathbf{T} - \mathbf{X}_2\hat{\boldsymbol{\gamma}}_2(\boldsymbol{\gamma}_1^*)]'[\mathbf{T} - \mathbf{X}_2\hat{\boldsymbol{\gamma}}_2(\boldsymbol{\gamma}_1^*)] \end{aligned} \tag{6.32}$$

Substituting the values in (6.32) into the likelihood function (6.31), we have

$$L(\hat{\omega}) = \frac{n^{n/2}e^{-n/2}}{(2\pi)^{n/2}\{[\mathbf{T} - \mathbf{X}_2\hat{\boldsymbol{\gamma}}_2(\boldsymbol{\gamma}_1^*)]'[\mathbf{T} - \mathbf{X}_2\hat{\boldsymbol{\gamma}}_2(\boldsymbol{\gamma}_1^*)]\}^{n/2}}$$

If we examine $\mathbf{e}'\mathbf{e}$ in the model restricted by $\boldsymbol{\gamma}_1 = \boldsymbol{\gamma}_1^*$, we see that

$$\begin{aligned} \mathbf{e}'\mathbf{e} &= (\mathbf{Y} - \mathbf{X}_1\boldsymbol{\gamma}_1^* - \mathbf{X}_2\boldsymbol{\gamma}_2)'(\mathbf{Y} - \mathbf{X}_1\boldsymbol{\gamma}_1^* - \mathbf{X}_2\boldsymbol{\gamma}_2) \\ &= (\mathbf{T} - \mathbf{X}_2\boldsymbol{\gamma}_2)'(\mathbf{T} - \mathbf{X}_2\boldsymbol{\gamma}_2) \end{aligned} \tag{6.33}$$

and that the minimum of $\mathbf{e}'\mathbf{e}$ with respect to the unknown parameter $\boldsymbol{\gamma}_2$ is

$$[\mathbf{T} - \mathbf{X}_2\hat{\boldsymbol{\gamma}}_2(\boldsymbol{\gamma}_1^*)]'[\mathbf{T} - \mathbf{X}_2\hat{\boldsymbol{\gamma}}_2(\boldsymbol{\gamma}_1^*)] = Q_0 + Q_1$$

Therefore,
$$L = \frac{L(\hat{\omega})}{L(\hat{\Omega})} = \left(\frac{Q_0 + Q_1}{Q_0}\right)^{-n/2}$$

To find the distribution of L, let us examine the quantities Q_0, Q_1, and Q_2. In (6.32) we have stated in effect that $\mathbf{X}_2'\mathbf{X}_2$ has an inverse. This follows from the fact that $\mathbf{X}_2'\mathbf{X}_2$ is a principal minor of the positive definite matrix $\mathbf{X}'\mathbf{X}$. Hence, $\mathbf{X}_2'\mathbf{X}_2$ is also positive definite. From the fact that

$$\mathbf{X}'(\mathbf{I} - \mathbf{X}\mathbf{S}^{-1}\mathbf{X}') = \mathbf{0}$$

we get
$$\begin{pmatrix}\mathbf{X}_1' \\ \mathbf{X}_2'\end{pmatrix}(\mathbf{I} - \mathbf{X}\mathbf{S}^{-1}\mathbf{X}') = \mathbf{0}$$

and finally the two equations

$$\begin{aligned}\mathbf{X}_1'(\mathbf{I} - \mathbf{X}\mathbf{S}^{-1}\mathbf{X}') &= \mathbf{0}\\ \mathbf{X}_2'(\mathbf{I} - \mathbf{X}\mathbf{S}^{-1}\mathbf{X}') &= \mathbf{0}\end{aligned} \tag{6.34}$$

Substituting for $\mathbf{T}$ and $\hat{\boldsymbol{\gamma}}_2(\boldsymbol{\gamma}_1^*)$, we get

$$Q_0 + Q_1 = (\mathbf{Y} - \mathbf{X}_1\boldsymbol{\gamma}_1^*)'[\mathbf{I} - \mathbf{X}_2(\mathbf{X}_2'\mathbf{X}_2)^{-1}\mathbf{X}_2'](\mathbf{Y} - \mathbf{X}_1\boldsymbol{\gamma}_1^*)$$

From (6.34), Q_0 can be written

$$Q_0 = (\mathbf{Y} - \mathbf{X}_1\boldsymbol{\gamma}_1^*)'(\mathbf{I} - \mathbf{X}\mathbf{S}^{-1}\mathbf{X}')(\mathbf{Y} - \mathbf{X}_1\boldsymbol{\gamma}_1^*)$$

Using the notation

$$\begin{aligned}\mathbf{T} &= \mathbf{Y} - \mathbf{X}_1\boldsymbol{\gamma}_1^*\\ \mathbf{A} &= \mathbf{I} - \mathbf{X}\mathbf{S}^{-1}\mathbf{X}'\\ \mathbf{A}_2 &= \mathbf{I} - \mathbf{X}_2(\mathbf{X}_2'\mathbf{X}_2)^{-1}\mathbf{X}_2'\end{aligned}$$

we shall study the identity

$$\mathbf{T}'\mathbf{T} = \mathbf{T}'\mathbf{A}\mathbf{T} + \mathbf{T}'(\mathbf{A}_2 - \mathbf{A})\mathbf{T} + \mathbf{T}'(\mathbf{I} - \mathbf{A}_2)\mathbf{T} = Q_0 + Q_1 + Q_2 \tag{6.35}$$

It follows that $\mathbf{T}$ is distributed $N[\mathbf{X}(\boldsymbol{\beta} - \boldsymbol{\beta}^*), \sigma^2\mathbf{I}]$, where $\boldsymbol{\beta}^* = \begin{pmatrix}\boldsymbol{\gamma}_1^* \\ \mathbf{0}\end{pmatrix}$, since

$$E(\mathbf{T}) = E(\mathbf{Y} - \mathbf{X}_1\boldsymbol{\gamma}_1^*) = \mathbf{X}\boldsymbol{\beta} - \mathbf{X}_1\boldsymbol{\gamma}_1^* = \mathbf{X}\boldsymbol{\beta} - \mathbf{X}\begin{pmatrix}\boldsymbol{\gamma}_1^* \\ \mathbf{0}\end{pmatrix}$$

In Eq. (6.35) the identity

$$\mathbf{I} = \mathbf{A} + (\mathbf{A}_2 - \mathbf{A}) + (\mathbf{I} - \mathbf{A}_2)$$

is obtained. Using (6.34) it is easily verified that $\mathbf{A}$, $(\mathbf{A}_2 - \mathbf{A})$, and $(\mathbf{I} - \mathbf{A}_2)$ are each idempotent matrices. Hence, using Theorem 4.16, we see that:

1. $\mathbf{T}'\mathbf{A}\mathbf{T}/\sigma^2 = Q_0/\sigma^2$ is distributed as $\chi^2(n - p)$, since, by (6.34), the noncentrality parameter is zero.

2. $\mathbf{T}'(\mathbf{A}_2 - \mathbf{A})\mathbf{T}/\sigma^2 = Q_1/\sigma^2$ is distributed as $\chi'^2(r, \lambda)$ where r is the rank of $\mathbf{A}_2 - \mathbf{A}$ and $\lambda = (1/2\sigma^2)[E(\mathbf{T}')](\mathbf{A}_2 - \mathbf{A})[\mathrm{E}(\mathbf{T})]$.

3. Q_0/σ^2 and Q_1/σ^2 are independent. Therefore, the quantity $u = (Q_1/Q_0)(n - p)/r$ is distributed as $F'(r, n - p, \lambda)$.

To show that the rank of $\mathbf{A}_2 - \mathbf{A}$ equals r we can use the fact that $\mathbf{A}_2 - \mathbf{A}$ is idempotent and, hence, that $\rho(\mathbf{A}_2 - \mathbf{A}) = \mathrm{tr}(\mathbf{A}_2 - \mathbf{A}) = \mathrm{tr}(\mathbf{A}_2) - \mathrm{tr}(\mathbf{A}) = n - (p - r) - (n - p) = r$. To examine λ in more detail, substitute $\mathbf{X}(\boldsymbol{\beta} - \boldsymbol{\beta}^*)$ for $E(\mathbf{T})$ and get

$$2\sigma^2\lambda = (\boldsymbol{\beta} - \boldsymbol{\beta}^*)'\mathbf{X}'(\mathbf{A}_2 - \mathbf{A})\mathbf{X}(\boldsymbol{\beta} - \boldsymbol{\beta}^*) = [(\boldsymbol{\gamma}_1 - \boldsymbol{\gamma}_1^*)'\mathbf{X}_1' + \boldsymbol{\gamma}_2'\mathbf{X}_2']$$
$$\times [\mathbf{X}\mathbf{S}^{-1}\mathbf{X}' - \mathbf{X}_2(\mathbf{X}_2'\mathbf{X}_2)^{-1}\mathbf{X}_2']\mathbf{X}[_1(\boldsymbol{\gamma}_1 - \boldsymbol{\gamma}_1^*) + \mathbf{X}_2\boldsymbol{\gamma}_2]$$

or $$\lambda = \frac{1}{2\sigma^2}(\boldsymbol{\gamma}_1 - \boldsymbol{\gamma}_1^*)'[\mathbf{X}_1'\mathbf{X}_1 - \mathbf{X}_1'\mathbf{X}_2(\mathbf{X}_2'\mathbf{X}_2)^{-1}\mathbf{X}_2'\mathbf{X}_1](\boldsymbol{\gamma}_1 - \boldsymbol{\gamma}_1^*)$$

But, by Theorem 1.49, $\mathbf{X}_1'\mathbf{X}_1 - \mathbf{X}_1'\mathbf{X}_2(\mathbf{X}_2'\mathbf{X}_2)^{-1}\mathbf{X}_2'\mathbf{X}_1$ is positive definite; hence, $\lambda = 0$ if and only if $\boldsymbol{\gamma}_1 - \boldsymbol{\gamma}_1^* = \mathbf{0}$, that is, if and only if H_0 is true. This completes the proof of the theorem.

We see that $L = \{1 + [r/(n - p)]u\}^{-n/2}$ and that use of the test function u is equivalent to the likelihood-ratio test, since L is a monotonic function of u.

6.4.4 The Analysis-of-variance Table. Since the test of a subhypothesis given in the previous article is so important, we shall investigate it further. Since it is often desired to test the hypothesis $\mathrm{H}: \boldsymbol{\gamma}_1 = \mathbf{0}$, we shall examine this test. The quantities needed are

$$Q_0 = \mathbf{Y}'(\mathbf{I} - \mathbf{X}\mathbf{S}^{-1}\mathbf{X}')\mathbf{Y}$$

$$Q_1 = \mathbf{Y}'[\mathbf{X}\mathbf{S}^{-1}\mathbf{X}' - \mathbf{X}_2(\mathbf{X}_2'\mathbf{X}_2)^{-1}\mathbf{X}_2']\mathbf{Y}$$

We can write $$Q_0 = \mathbf{Y}'\mathbf{Y} - \hat{\boldsymbol{\beta}}'\mathbf{X}'\mathbf{Y}$$

$\mathbf{Y}'\mathbf{Y}$ will be called the *total sum of squares*; this quantity is very easy to compute. $\hat{\boldsymbol{\beta}}'\mathbf{X}'\mathbf{Y}$ will be called the *reduction due to* $\boldsymbol{\beta}$ and will be written $R(\boldsymbol{\beta})$. Again we see the importance of the normal equations $\mathbf{X}'\mathbf{X}\hat{\boldsymbol{\beta}} = \mathbf{X}'\mathbf{Y}$, since $R(\boldsymbol{\beta})$ comprises the elements of the vector $\hat{\boldsymbol{\beta}}$, which is the solution of the normal equations, multiplied by the corresponding elements of the right-hand side $\mathbf{X}'\mathbf{Y}$. Now $Q_1 = \mathbf{Y}'\mathbf{X}\,\mathbf{S}^{-1}\mathbf{X}'\mathbf{Y} - \mathbf{Y}'\mathbf{X}_2(\mathbf{X}_2'\mathbf{X}_2)^{-1}\mathbf{X}_2'\mathbf{Y} = \hat{\boldsymbol{\beta}}'\mathbf{X}'\mathbf{Y} - \tilde{\boldsymbol{\gamma}}_2'\mathbf{X}_2'\mathbf{Y}$, where $\tilde{\boldsymbol{\gamma}}_2$ is the solution to the reduced normal equations $\mathbf{X}_2'\mathbf{X}_2\tilde{\boldsymbol{\gamma}}_2 = \mathbf{X}_2'\mathbf{Y}$.

The quantity $\tilde{\boldsymbol{\gamma}}_2'\mathbf{X}_2'\mathbf{Y}$ will be called the *reduction due to* $\boldsymbol{\gamma}_2$ *ignoring* $\boldsymbol{\gamma}_1$ and will be written $R(\boldsymbol{\gamma}_2)$.

If we look again to our normal equations, we partition $\mathbf{X}$ and $\boldsymbol{\beta}$ to get

$$\begin{pmatrix} \mathbf{X}_1'\mathbf{X}_1 & \mathbf{X}_1'\mathbf{X}_2 \\ \mathbf{X}_2'\mathbf{X}_1 & \mathbf{X}_2'\mathbf{X}_2 \end{pmatrix} \begin{pmatrix} \boldsymbol{\gamma}_1 \\ \boldsymbol{\gamma}_2 \end{pmatrix} = \begin{pmatrix} \mathbf{X}_1'\mathbf{Y} \\ \mathbf{X}_2'\mathbf{Y} \end{pmatrix}$$

So from the normal equations we can get the reduced normal equations $\mathbf{X}_2'\mathbf{X}_2\tilde{\boldsymbol{\gamma}}_2 = \mathbf{X}_2'\mathbf{Y}$, from which we get $\tilde{\boldsymbol{\gamma}}_2'\mathbf{X}_2'\mathbf{Y} = R(\boldsymbol{\gamma}_2)$. If we so desire, we can get the reduced normal equations from the model $\mathbf{Y} = \mathbf{X}_2\boldsymbol{\gamma}_2 + \mathbf{e}$, which is obtained from the original model $\mathbf{Y} = \mathbf{X}\boldsymbol{\beta} + \mathbf{e}$ with $\mathbf{0}$ substituted for $\boldsymbol{\gamma}_1$. $\hat{\boldsymbol{\beta}}'\mathbf{X}'\mathbf{Y} - \tilde{\boldsymbol{\gamma}}_2'\mathbf{X}_2'\mathbf{Y}$ will be called *reduction due to* $\boldsymbol{\gamma}_1$ *adjusted for* $\boldsymbol{\gamma}_2$ and will be denoted by $R(\boldsymbol{\gamma}_1 \mid \boldsymbol{\gamma}_2)$. Therefore, $Q_1 = R(\boldsymbol{\gamma}_1 \mid \boldsymbol{\gamma}_2) = R(\boldsymbol{\beta}) - R(\boldsymbol{\gamma}_2)$ and $Q_0 = \mathbf{Y}'\mathbf{Y} - R(\boldsymbol{\beta})$.

These quantities can be put into an AOV table, shown in Table 6.2.

TABLE 6.2 TEST OF A SUBHYPOTHESIS: $\boldsymbol{\gamma}_1 = \mathbf{0}$

SV	DF	SS	MS	F
Total	n	$\mathbf{Y}'\mathbf{Y}$		
Due to $\boldsymbol{\beta}$	p	$\hat{\boldsymbol{\beta}}'\mathbf{X}'\mathbf{Y}$		
Due to $\boldsymbol{\gamma}_2$ (unadj)	$p - r$	$\tilde{\boldsymbol{\gamma}}_2'\mathbf{X}_2'\mathbf{Y}$		
Due to $\boldsymbol{\gamma}_1$ (adj)	r	$\hat{\boldsymbol{\beta}}'\mathbf{X}'\mathbf{Y} - \tilde{\boldsymbol{\gamma}}_2'\mathbf{X}_2'\mathbf{Y} = Q_1$	$\dfrac{Q_1}{r}$	$\dfrac{n-p}{r}\dfrac{Q_1}{Q_0}$
Error	$n - p$	$\mathbf{Y}'\mathbf{Y} - \hat{\boldsymbol{\beta}}'\mathbf{X}'\mathbf{Y} = Q_0$	$\dfrac{Q_0}{n-p}$	

The notation in this table should be studied in detail since it will be used often in subsequent sections. The things to notice are as follows:

1. $R(\boldsymbol{\beta})$ is $\hat{\boldsymbol{\beta}}'\mathbf{X}'\mathbf{Y}$, where all the elements in $\boldsymbol{\beta}$ are included in the normal equations.

2. $R(\boldsymbol{\gamma}_2)$ is $\tilde{\boldsymbol{\gamma}}_2'\mathbf{X}_2'\mathbf{Y}$, where only the elements in $\boldsymbol{\gamma}_2$ are used in the model from which we obtain the normal equations.

3. $R(\boldsymbol{\gamma}_1 \mid \boldsymbol{\gamma}_2)$ is $R(\boldsymbol{\beta}) - R(\boldsymbol{\gamma}_2)$, where $\boldsymbol{\gamma}_1$ and $\boldsymbol{\gamma}_2$ comprise all the elements of $\boldsymbol{\beta}$.

That is to say, if we want to test the hypothesis H_0: $\boldsymbol{\gamma}_1 = \mathbf{0}$, we have two models:

1. The unrestricted model $\mathbf{Y} = \mathbf{X}\boldsymbol{\beta} + \mathbf{e}$. From this model we compute the normal equations $\mathbf{X}'\mathbf{X}\hat{\boldsymbol{\beta}} = \mathbf{X}'\mathbf{Y}$ and obtain $R(\boldsymbol{\beta}) = \hat{\boldsymbol{\beta}}'\mathbf{X}'\mathbf{Y}$.

2. The model restricted by the hypothesis $\boldsymbol{\gamma}_1 = \mathbf{0}$, that is, $\mathbf{Y} = \mathbf{X}_2\boldsymbol{\gamma}_2 + \mathbf{e}$. From this model we compute the normal equations $\mathbf{X}_2'\mathbf{X}_2\tilde{\boldsymbol{\gamma}}_2 = \mathbf{X}_2'\mathbf{Y}$ and obtain $R(\boldsymbol{\gamma}_2) = \tilde{\boldsymbol{\gamma}}_2'\mathbf{X}_2'\mathbf{Y}$.

After the total sum of squares $\mathbf{Y}'\mathbf{Y}$, $R(\boldsymbol{\beta})$, and $R(\boldsymbol{\gamma}_2)$ are obtained, the rest of the quantities in SS can be obtained by subtraction.

6.4.5 Orthogonality. We can use the sum of squares $R(\boldsymbol{\beta})$ and the error in Table 6.2 to test the hypothesis $\boldsymbol{\beta} = \mathbf{0}$. This can be seen by observing that these are exactly the quantities used in Art. 6.4.3.

If we want to test the hypothesis $\boldsymbol{\gamma}_2 = \mathbf{0}$, we can get this test from the quantities in Table 6.2 if and only if the relationship $\mathbf{X}_1'\mathbf{X}_2 = \mathbf{0}$ holds. If $\mathbf{X}_1'\mathbf{X}_2 = \mathbf{0}$, we see that the $\mathbf{X}'\mathbf{X}$ matrix is diagonal in blocks, and we get

$$\begin{pmatrix} \mathbf{X}_1'\mathbf{X}_1 & \mathbf{0} \\ \mathbf{0} & \mathbf{X}_2'\mathbf{X}_2 \end{pmatrix}\begin{pmatrix} \hat{\boldsymbol{\gamma}}_1 \\ \hat{\boldsymbol{\gamma}}_2 \end{pmatrix} = \begin{pmatrix} \mathbf{X}_1'\mathbf{Y} \\ \mathbf{X}_2'\mathbf{Y} \end{pmatrix}$$

From this we obtain the equations

$$\begin{pmatrix} \hat{\boldsymbol{\gamma}}_1 \\ \hat{\boldsymbol{\gamma}}_2 \end{pmatrix} = \begin{pmatrix} (\mathbf{X}_1'\mathbf{X}_1)^{-1} & \mathbf{0} \\ \mathbf{0} & (\mathbf{X}_2'\mathbf{X}_2)^{-1} \end{pmatrix}\begin{pmatrix} \mathbf{X}_1'\mathbf{Y} \\ \mathbf{X}_2'\mathbf{Y} \end{pmatrix}$$

and

$$\hat{\boldsymbol{\beta}} = \begin{pmatrix} \hat{\boldsymbol{\gamma}}_1 \\ \hat{\boldsymbol{\gamma}}_2 \end{pmatrix} = \begin{pmatrix} (\mathbf{X}_1'\mathbf{X}_1)^{-1}\mathbf{X}_1'\mathbf{Y} \\ (\mathbf{X}_2'\mathbf{X}_2)^{-1}\mathbf{X}_2'\mathbf{Y} \end{pmatrix}$$

Also,

$$R(\boldsymbol{\beta}) = \mathbf{Y}'\mathbf{X}_1(\mathbf{X}_1'\mathbf{X}_1)^{-1}\mathbf{X}_1'\mathbf{Y} + \mathbf{Y}'\mathbf{X}_2(\mathbf{X}_2'\mathbf{X}_2)^{-1}\mathbf{X}_2'\mathbf{Y}$$

and

$$R(\boldsymbol{\gamma}_2) = \mathbf{Y}'\mathbf{X}_2(\mathbf{X}_2'\mathbf{X}_2)^{-1}\mathbf{X}_2'\mathbf{Y}$$

so

$$R(\boldsymbol{\gamma}_1 \mid \boldsymbol{\gamma}_2) = R(\boldsymbol{\beta}) - R(\boldsymbol{\gamma}_2) = \mathbf{Y}'\mathbf{X}_1(\mathbf{X}_1'\mathbf{X}_1)^{-1}\mathbf{X}_1'\mathbf{Y} = R(\boldsymbol{\gamma}_1)$$

If we wanted to test the hypothesis $\boldsymbol{\gamma}_2 = \mathbf{0}$, then, by a method similar to that used to obtain Table 6.2, we would need $R(\boldsymbol{\gamma}_2 \mid \boldsymbol{\gamma}_1) = R(\boldsymbol{\beta}) - R(\boldsymbol{\gamma}_1) = R(\boldsymbol{\beta}) - \tilde{\boldsymbol{\gamma}}_1'\mathbf{X}_1'\mathbf{Y}$. If $\mathbf{X}_1'\mathbf{X}_2 = \mathbf{0}$, we see that $R(\boldsymbol{\gamma}_1 \mid \boldsymbol{\gamma}_2) = R(\boldsymbol{\gamma}_1)$ and $R(\boldsymbol{\gamma}_2 \mid \boldsymbol{\gamma}_1) = R(\boldsymbol{\gamma}_2)$, and we can use Table 6.2 to test any one of the three hypotheses $\boldsymbol{\beta} = \mathbf{0}$, $\boldsymbol{\gamma}_1 = \mathbf{0}$, and $\boldsymbol{\gamma}_2 = \mathbf{0}$.

The noncentrality parameter in the test $\boldsymbol{\gamma}_1 = 0$ given in Art. 6.4.3 is

$$\lambda = \frac{\boldsymbol{\gamma}_1'[\mathbf{X}_1'\mathbf{X}_1 - \mathbf{X}_1'\mathbf{X}_2(\mathbf{X}_2'\mathbf{X}_2)^{-1}\mathbf{X}_2'\mathbf{X}_1]\boldsymbol{\gamma}_1}{2\sigma^2}$$

This can be written

$$\lambda = \frac{1}{2\sigma^2}\,\boldsymbol{\gamma}_1'\mathbf{X}_1'\mathbf{X}_1\boldsymbol{\gamma}_1 - \frac{1}{2\sigma^2}\,\boldsymbol{\gamma}_1'\mathbf{X}_1'\mathbf{X}_2(\mathbf{X}_2'\mathbf{X}_2)^{-1}\mathbf{X}_2'\mathbf{X}_1\boldsymbol{\gamma}_1$$

Since the power function $\beta(\lambda)$ is an increasing function of λ, we want λ to be as large as possible. If $\mathbf{X}_1'\mathbf{X}_1$ is fixed, λ is a maximum if $\mathbf{X}_1'\mathbf{X}_2 = \mathbf{0}$. Therefore, if possible, we choose the $\mathbf{X}$ matrix such that $\mathbf{X}_1'\mathbf{X}_2 = \mathbf{0}$.

We now formulate the definition of orthogonality.

◆ **Definition 6.2** If $\mathbf{X}_1'\mathbf{X}_2 = \mathbf{0}$ in the model $\mathbf{Y} = \mathbf{X}_1\boldsymbol{\gamma}_1 + \mathbf{X}_2\boldsymbol{\gamma}_2 + \mathbf{e}$, then $\boldsymbol{\gamma}_1$ will be said to be orthogonal to $\boldsymbol{\gamma}_2$.

Suppose we write the model $\mathbf{Y} = \mathbf{X}\boldsymbol{\beta} + \mathbf{e}$ as

$$\mathbf{Y} = \mathbf{X}_1\beta_1 + \mathbf{X}_2\beta_2 + \cdots + \mathbf{X}_p\beta_p + \mathbf{e}$$

where $\mathbf{X}_i$ is an $n \times 1$ vector and β_i is a scalar. Suppose further that we wish to make the following tests (there could be many others):

1. $\beta_1 = 0$
2. $\beta_2 = 0$
3. $\beta_p = 0$
4. $\beta_1 = \beta_3 = \beta_5 = 0$
5. $\beta_2 = \beta_3 = \beta_p = 0$; etc.

We need the quantities $\hat{\boldsymbol{\beta}}'\mathbf{X}'\mathbf{Y}$ and $\mathbf{Y}'\mathbf{Y}$, which remain the same for all five tests. And we need the following for the respective tests:

1. $R(\beta_2,\beta_3, \ldots ,\beta_p)$ that is, the reduction due to $\beta_2, \ldots , \beta_p$ ignoring β_1
2. $R(\beta_1,\beta_3, \ldots ,\beta_p)$
3. $R(\beta_1,\beta_2, \ldots ,\beta_{p-1})$
4. $R(\beta_2,\beta_4,\beta_6, \ldots ,\beta_p)$
5. $R(\beta_1,\beta_4, \ldots ,\beta_{p-1})$, etc.

Most of these tests require that the appropriate β_i be set equal to zero in the normal equations and that the remaining equations be solved. This is a tedious and time-consuming job. However, if $\beta_1, \beta_2, \ldots , \beta_p$ are all *orthogonal*, then all tests similar to those listed above are easily made. If $\beta_1, \ldots , \beta_p$ are orthogonal, then $\mathbf{X}'\mathbf{X}$ is a diagonal matrix, and the reduction due to any set of the β_i adjusted for any other set is simply the reduction due to that particular set of the β_i ignoring the other set.

For example, let us suppose in the model $\mathbf{Y} = \mathbf{X}_1\beta_1 + \mathbf{X}_2\beta_2 + \mathbf{X}_3\beta_3 + \mathbf{e}$ that the normal equations are

$$\begin{aligned} 2\hat{\beta}_1 \qquad\qquad &= 4 \\ 3\hat{\beta}_2 \qquad &= 3 \\ \hat{\beta}_3 &= 5 \end{aligned}$$

Since $\mathbf{X}'\mathbf{X}$ is diagonal, the β_i are orthogonal. We get $\hat{\beta}_1 = 2, \hat{\beta}_2 = 1$, $\hat{\beta}_3 = 5$, and $R(\boldsymbol{\beta}) = \hat{\boldsymbol{\beta}}'\mathbf{X}'\mathbf{Y} = 36$. Suppose $\mathbf{Y}'\mathbf{Y} = 48$; then the error sum of squares is 12.

Suppose we wish to test the following

1. $\beta_1 = 0$
2. $\beta_2 = 0$
3. $\beta_1 = \beta_2 = 0$

For 1 we need $R(\beta_1 \mid \beta_2,\beta_3)$, which equals $R(\boldsymbol{\beta}) - R(\beta_2,\beta_3)$. But

$R(\beta_2,\beta_3)$ is obtained from the normal equations by setting $\hat{\beta}_1 = 0$ and ignoring the first equation. We get $\tilde{\beta}_2 = 1$, $\tilde{\beta}_3 = 5$; so $R(\beta_2,\beta_3) = 28$. Therefore, $R(\boldsymbol{\beta}) - R(\beta_2,\beta_3) = 8$. But this is $R(\beta_1)$.

For test 2 we need $R(\beta_2 \mid \beta_1,\beta_3)$, which equals $R(\boldsymbol{\beta}) - R(\beta_1,\beta_3)$. By a procedure similar to that given above we can show that $R(\beta_2 \mid \beta_1,\beta_3) = R(\beta_2) = 3$.

Now, for test 3 we need $R(\beta_1,\beta_2 \mid \beta_3)$, which equals $R(\boldsymbol{\beta}) - R(\beta_3)$. But, by setting $\hat{\beta}_1$ and $\hat{\beta}_2$ zero and using the last of the normal equations, we get $R(\beta_3) = 25$. So $R(\beta_1,\beta_2 \mid \beta_3) = 11$. But this equals $R(\beta_1,\beta_2)$, which is $R(\beta_1) + R(\beta_2)$.

6.4.6 Example. The distance a particle travels from a given reference point is given theoretically by the curve

$$D^* = \beta_0 + \beta_1 T_1 + \beta_2 T_2$$

where D^* is the distance, T_1 is the time the particle moves, and T_2 is the temperature of the medium through which the particle moves. The time and temperature can be measured without error, but, instead of observing D^*, we observe $y = D^* + e$, where e is a normal random error with mean zero. The following set of measurements were taken:

y	6.0	13.0	13.0	29.2	33.1	32.0	46.2	117.5
T_1	1	2	3	4	5	6	8	20
T_2	10	10	12	11	14	15	18	30

1. Find $\hat{\boldsymbol{\beta}}$, $\hat{\sigma}^2$.
2. Find $\hat{D}^*$.
3. Test the hypothesis $\beta_1 = \beta_2 = 0$.

We get

$$\mathbf{X'X} = \begin{pmatrix} 8 & 49 & 120 \\ 49 & 555 & 1{,}014 \\ 120 & 1{,}014 & 2{,}110 \end{pmatrix}$$

The normal equations are

$$\begin{aligned} 8\hat{\beta}_0 + \phantom{1{,}0}49\hat{\beta}_1 + \phantom{2{,}}120\hat{\beta}_2 &= 290.0 \\ 49\hat{\beta}_0 + \phantom{1{,}}555\hat{\beta}_1 + 1{,}014\hat{\beta}_2 &= 3{,}264.9 \\ 120\hat{\beta}_0 + 1{,}014\hat{\beta}_1 + 2{,}110\hat{\beta}_2 &= 5{,}967.2 \end{aligned}$$

and

$$\mathbf{X'Y} = \begin{pmatrix} 290.0 \\ 3{,}264.9 \\ 5{,}967.2 \end{pmatrix}$$

Solving the equations gives

$$\hat{\boldsymbol{\beta}} = \begin{pmatrix} 22.79 \\ 8.79 \\ -2.69 \end{pmatrix} \qquad \hat{D}^* = 22.79 + 8.79T_1 - 2.69T_2$$

To estimate σ^2 and to test the hypothesis $\beta_1 = \beta_2 = 0$, we shall construct an AOV table. The quantities needed are given below.

First, $R(\boldsymbol{\beta}) = \hat{\boldsymbol{\beta}}'\mathbf{X}'\mathbf{Y} = (22.79)(290.0) + (8.79)(3{,}264.9) - (2.69) \times (5{,}967.2) = 19{,}240.5$.

If we set $\boldsymbol{\gamma}_1 = \begin{pmatrix} \beta_1 \\ \beta_2 \end{pmatrix}$ and $\boldsymbol{\gamma}_2 = \beta_0$, the reduced normal equation is obtained from the model $y = \beta_0 + e$; that is, we set $\beta_1 = \beta_2 = 0$. We get the normal equation $8\tilde{\beta}_0 = 290.0$ and $\tilde{\boldsymbol{\gamma}}_2 = \tilde{\beta}_0 = 290.0/8 = 36.3$ and $R(\boldsymbol{\gamma}_2) = \tilde{\boldsymbol{\gamma}}_2'\mathbf{X}_2'\mathbf{Y} = (290.0)^2/8 = 10{,}512.5$: $R(\boldsymbol{\gamma}_1 \mid \boldsymbol{\gamma}_2) = R(\boldsymbol{\beta}) - R(\boldsymbol{\gamma}_2) = 8{,}743.3$.

The AOV is given in Table 6.3.

TABLE 6.3 ANALYSIS OF VARIANCE

SV	DF	SS	MS	F
Total	8	19,286.9		
Due to $\boldsymbol{\beta}$	3	19,240.5		
Due to $\boldsymbol{\gamma}_2$ (unadj)	1	10,512.5		
Due to $\boldsymbol{\gamma}_1$ (adj)	2	8,728.0	4,364.0	470.3
Error	5	46.4	9.28	

From this we get $\hat{\sigma}^2 = 9.28$. Using Eqs. (6.8) and (6.9) we can set confidence intervals on the β_i and on σ^2.

6.4.7 Test of the Hypothesis $\boldsymbol{\lambda}_1'\boldsymbol{\beta} = \boldsymbol{\lambda}_2'\boldsymbol{\beta} = \cdots = \boldsymbol{\lambda}_r'\boldsymbol{\beta} = 0$. In this test we assume that the $r \times p$ matrix

$$\mathbf{G}_1 = \begin{pmatrix} \boldsymbol{\lambda}_1' \\ \boldsymbol{\lambda}_2' \\ \cdots \\ \boldsymbol{\lambda}_r' \end{pmatrix}$$

has rank r. Let $\mathbf{G}$ be a $p \times p$ matrix of rank p such that $\mathbf{G} = \begin{pmatrix} \mathbf{G}_1 \\ \mathbf{G}_2 \end{pmatrix}$.

Let $\mathbf{G}^{-1} = \boldsymbol{\Delta} = (\boldsymbol{\Delta}_1, \boldsymbol{\Delta}_2)$, where $\boldsymbol{\Delta}_1$ has dimension $p \times r$. Now $\mathbf{G}$ is a known matrix, hence $\mathbf{G}^{-1} = \boldsymbol{\Delta}$ is also known. Let

$$\mathbf{X}\mathbf{G}^{-1} = \mathbf{X}\boldsymbol{\Delta} = (\mathbf{X}\boldsymbol{\Delta}_1, \mathbf{X}\boldsymbol{\Delta}_2) = (\mathbf{Z}_1, \mathbf{Z}_2) = \mathbf{Z}$$

Let

$$\mathbf{G}\boldsymbol{\beta} = \begin{pmatrix} \mathbf{G}_1\boldsymbol{\beta} \\ \mathbf{G}_2\boldsymbol{\beta} \end{pmatrix} = \begin{pmatrix} \boldsymbol{\alpha}_1 \\ \boldsymbol{\alpha}_2 \end{pmatrix}$$

We can write the model $\mathbf{Y} = \mathbf{X}\boldsymbol{\beta} + \mathbf{e}$ as $\mathbf{Y} = \mathbf{X}\mathbf{G}^{-1}\mathbf{G}\boldsymbol{\beta} + \mathbf{e}$ or as $\mathbf{Y} = \mathbf{Z}\boldsymbol{\alpha} + \mathbf{e}$. This becomes $\mathbf{Y} = \mathbf{Z}_1\boldsymbol{\alpha}_1 + \mathbf{Z}_2\boldsymbol{\alpha}_2 + \mathbf{e}$. Now $\boldsymbol{\alpha}_1 = \mathbf{0}$ is equivalent to $\mathbf{G}_1\boldsymbol{\beta} = \mathbf{0}$, which is the test we desire. We now follow the procedure given in Art. 6.4.4 to find $R(\boldsymbol{\alpha})$ and $R(\boldsymbol{\alpha}_2)$. $R(\boldsymbol{\alpha})$ is the same as $R(\boldsymbol{\beta})$, and we set $\boldsymbol{\alpha}_1 = \mathbf{0}$ and compute $R(\boldsymbol{\alpha}_2)$. From these quantities we get an AOV table.

To sum up, we have the following instructions: To test $\boldsymbol{\lambda}_1'\boldsymbol{\beta} = \boldsymbol{\lambda}_2'\boldsymbol{\beta} = \cdots = \boldsymbol{\lambda}_r'\boldsymbol{\beta} = 0$ we have two models:

1. The unrestricted model $\mathbf{Y} = \mathbf{X}\boldsymbol{\beta} + \mathbf{e}$, from which we obtain the normal equations $\mathbf{X}'\mathbf{X}\hat{\boldsymbol{\beta}} = \mathbf{X}'\mathbf{Y}$ and get $R(\boldsymbol{\beta}) = \hat{\boldsymbol{\beta}}'\mathbf{X}'\mathbf{Y}$

2. The model restricted by the hypothesis $\boldsymbol{\lambda}_1'\boldsymbol{\beta} = \boldsymbol{\lambda}_2'\boldsymbol{\beta} = \cdots = \boldsymbol{\lambda}_r'\boldsymbol{\beta} = 0$. This we shall write as $\mathbf{Y} = \mathbf{Z}_2\boldsymbol{\alpha}_2 + \mathbf{e}$; from it we obtain the normal equations $\mathbf{Z}_2'\mathbf{Z}_2\tilde{\boldsymbol{\alpha}}_2 = \mathbf{Z}_2'\mathbf{Y}$ and get $R(\boldsymbol{\alpha}_2) = \tilde{\boldsymbol{\alpha}}_2'\mathbf{Z}_2'\mathbf{Y}$.

We compute $\mathbf{Y}'\mathbf{Y}$ and obtain the AOV of Table 6.4. For example, suppose we have the model

$$y_j = \beta_0 + \beta_1 x_{1j} + \beta_2 x_{2j} + \beta_3 x_{3j} + \beta_4 x_{4j} + e_j$$

TABLE 6.4 ANALYSIS OF VARIANCE FOR TESTING
$\boldsymbol{\lambda}_1'\boldsymbol{\beta} = \boldsymbol{\lambda}_2'\boldsymbol{\beta} = \cdots = \boldsymbol{\lambda}_r'\boldsymbol{\beta} = 0$

SV	DF	SS	MS	F
Total	n	$\mathbf{Y}'\mathbf{Y}$		
Due to $\boldsymbol{\beta}$	p	$\hat{\boldsymbol{\beta}}'\mathbf{X}'\mathbf{Y}$		
Due to $\boldsymbol{\alpha}_2$ (unadj)	$p - r$	$\tilde{\boldsymbol{\alpha}}_2'\mathbf{Z}_2'\mathbf{Y}$		
Due to $\boldsymbol{\alpha}_1$ (adj)	r	$\hat{\boldsymbol{\beta}}'\mathbf{X}'\mathbf{Y} - \tilde{\boldsymbol{\alpha}}_2\mathbf{Z}_2'\mathbf{Y}$		
Error	$n - p$	$\mathbf{Y}'\mathbf{Y} - \hat{\boldsymbol{\beta}}'\mathbf{X}'\mathbf{Y}$		

and we want to test $\beta_1 - \beta_2 = 0$, $\beta_1 - 2\beta_3 = 0$. We can substitute $\beta_2 = \beta_1$ and $\beta_3 = \frac{1}{2}\beta_1$ and get

$$y_j = \beta_0 + \beta_1(x_{1j} + x_{2j} + \tfrac{1}{2}x_{3j}) + \beta_4 x_{4j} + e_j$$

for the model $\mathbf{Y} = \mathbf{Z}_2\boldsymbol{\alpha}_2 + \mathbf{e}$ restricted by the hypothesis.

As another example, suppose we wanted to test the hypothesis $\beta_3 = \beta_4$, $\beta_1 = \beta_2$. Substituting, we obtain

$$y_j = \beta_0 + \beta_1(x_{1j} + x_{2j}) + \beta_3(x_{3j} + x_{4j}) + e_j$$

for the model $\mathbf{Y} = \mathbf{Z}_2\boldsymbol{\alpha}_2 + \mathbf{e}$ restricted by the hypothesis.

6.4.8 Example. Suppose that, in the example of Art. 6.4.6, we wish to test simultaneously the two hypotheses $\beta_1 = 2\beta_2$ and $\beta_0 = \beta_1$.

Here $r = 2$ and $\boldsymbol{\lambda}_1' = (0, 1, -2)$; $\boldsymbol{\lambda}_2' = (1, -1, 0)$; the only new quantity we need to compute is $R(\boldsymbol{\alpha}_2)$. In the model we put $\beta_1 = \beta_0 = 2\beta_2$. This gives us

$$y = 2\beta_2 + 2\beta_2 T_1 + \beta_2 T_2 + e$$

or
$$y = \beta_2(2 + 2T_1 + T_2) + e$$

which we write as
$$y = \beta_2 x_2 + e$$

where $x_2 = 2 + 2T_1 + T_2$. The reduced normal equation is $9{,}290\tilde{\beta}_2 = 13{,}077$, which gives $\tilde{\beta}_2 = 1.41$. $R(\boldsymbol{\alpha}_2) = \tilde{\beta}_2'\mathbf{X}_2'\mathbf{Y} = 18{,}407.7$, and $R(\boldsymbol{\beta}) - R(\boldsymbol{\alpha}_2) = R(\boldsymbol{\alpha}_1 \mid \boldsymbol{\alpha}_2) = 848.1$. The F value is $848.1/9.28 = 91.39$. This is larger than the tabulated F value for 1 and 5 degrees of freedom at the 1 per cent level.

Problems

6.1 Prove that, for simple linear models such as those of Art. 6.2.6,

$$\frac{1}{n-2}\left[\Sigma(y_i - \bar{y})^2 - \frac{[\Sigma(x_i - \bar{x})(y_i - \bar{y})]^2}{\Sigma(x_i - \bar{x})^2}\right] = \frac{1}{n-2}[(\mathbf{Y} - \mathbf{X}\hat{\boldsymbol{\beta}})'(\mathbf{Y} - \mathbf{X}\hat{\boldsymbol{\beta}})]$$

6.2 An experimenter has theoretical reasons to believe that, in storing ice cream at low temperatures, the average weight loss $E(y) = L$ of the ice cream (in pint containers) is linearly related to the storage time. He, therefore, assumes that the linear model

$$L = E(y) = \beta_1 t$$

relates average weight loss to storage time. He knows there are many other factors besides time that affect weight loss, but it seems feasible to assume that these factors together act as a random error e. Hence the model can be written

$$y = \beta_1 t + e$$

To estimate β_1, he conducts an experiment in which he measures the weight loss of pints of ice cream over an extended period. The results (t is in weeks; y is in grams) are

t_i	1	2	3	4	5	6	7	8
y_i	.15	.21	.30	.41	.49	.59	.72	.83

Estimate β_1 and σ^2.

6.3 Find $\hat{\beta}_1$, $\hat{\beta}_2$, $\hat{\beta}_3$, $\hat{\sigma}^2$, and $\hat{\beta}_1 - \hat{\beta}_2 + 2\hat{\beta}_3$ for the data below, assuming that the linear model $y_i = \beta_1 + \beta_2 x_{1i} + \beta_3 x_{2i} + e_i$ fits Definition 6.1.

y	1	5	0	4	4	-1
x_1	1	2	1	3	3	3
x_2	1	1	2	1	2	3

6.4 If $\mathbf{X} = \begin{pmatrix} 1 & 2 \\ 2 & 2 \\ 3 & 2 \end{pmatrix}$, find $\mathbf{X}'\mathbf{X}$, $(\mathbf{X}'\mathbf{X})^{-1}$, $\mathbf{X}(\mathbf{X}'\mathbf{X})^{-1}$, $\mathbf{X}(\mathbf{X}'\mathbf{X})^{-1}\mathbf{X}'$, and $\mathbf{I} - \mathbf{X}(\mathbf{X}'\mathbf{X})^{-1}\mathbf{X}'$.

6.5 In simple linear models, as illustrated in Art. 6.2.5, prove that, if the x_i can be selected anywhere in the interval (a,b) and if n is an even integer, then the variance of $\hat{\beta}_2$ is minimized if we selected $n/2$ values of x_i equal to a and $n/2$ values equal to b.

6.6 Prove the identity in Eq. (6.7).

6.7 Use the $\mathbf{X}$ matrix in Prob. 6.4 and find $\text{tr}(\mathbf{X}\mathbf{S}^{-1}\mathbf{X}')$ and $\text{tr}(\mathbf{I} - \mathbf{X}\mathbf{S}^{-1}\mathbf{X}')$.

6.8 Let k_1^2 and k_2^2 be positive constants. Prove that

$$\mathbf{V} = k_1^2\mathbf{X}\mathbf{S}^{-1}\mathbf{X}' + k_2^2(\mathbf{I} - \mathbf{X}\mathbf{S}^{-1}\mathbf{X}')$$

is positive definite, where $\mathbf{X}$ is an $n \times p$ matrix of rank $p \leqslant n$.

6.9 In Prob. 6.8, show that

$$\mathbf{V}^{-1} = \frac{1}{k_1^2}\mathbf{X}\mathbf{S}^{-1}\mathbf{X}' + \frac{1}{k_2^2}(\mathbf{I} - \mathbf{X}\mathbf{S}^{-1}\mathbf{X}')$$

6.10 In Theorem 6.4, if $\mathbf{V} = k_1^2\mathbf{X}\mathbf{S}^{-1}\mathbf{X}' + k_2^2(\mathbf{I} - \mathbf{X}\mathbf{S}^{-1}\mathbf{X}')$, show that $\hat{\boldsymbol{\beta}}$ is the same as it is if $\mathbf{V} = \mathbf{I}$.

6.11 Prove the following statement, made in Art. 6.2.6: $\text{var}(\hat{\beta}_1)$ is a minimum if the x_i are chosen so that $\bar{x} = 0$.

6.12 Prove Theorem 6.3.

6.13 Prove Theorem 6.4.

6.14 In the model $\mathbf{Y} = \mathbf{X}\boldsymbol{\beta} + \mathbf{e}$, where $n = 10$ and $p = 3$, the following normal equations were computed ($\mathbf{Y}'\mathbf{Y} = 58$):

$$3\hat{\beta}_1 + \hat{\beta}_2 - 2\hat{\beta}_3 = 1$$

$$\hat{\beta}_1 + 2\hat{\beta}_2 + \hat{\beta}_3 = 7$$

$$-2\hat{\beta}_1 + \hat{\beta}_2 + 4\hat{\beta}_3 = 9$$

(*a*) Find $\hat{\boldsymbol{\beta}}$.
(*b*) Find $R(\boldsymbol{\beta}) = \hat{\boldsymbol{\beta}}'\mathbf{X}'\mathbf{Y}$.
(*c*) Set confidence limits on σ^2, β_1, β_2, β_3, and $\beta_1 - \beta_2$.
(*d*) Find $R(\boldsymbol{\gamma}_1)$, where $\boldsymbol{\gamma}_1 = \beta_1$.
(*e*) Find $R(\boldsymbol{\gamma}_2 \mid \boldsymbol{\gamma}_1)$, where $\boldsymbol{\gamma}_2' = (\beta_2, \beta_3)$.

6.15 In the model $y_j = \beta_0 + \beta_1 x_j + e_j$ $(j = 1, 2, \ldots, 20)$, the following normal equations were computed ($\Sigma y_j^2 = 90$):

$$20\hat{\beta}_0 + 10\hat{\beta}_1 = 40$$

$$10\hat{\beta}_0 + 6\hat{\beta}_1 = 22$$

(*a*) Set a 95 per cent confidence interval on $E(y)$, when $X = 3$.
(*b*) If we select a sample of 30 values of y from the above model when $X = 10$, find a 95 per cent confidence interval on the mean of these y values. Use Eq. (6.12).

6.16 In the model $\mathbf{Y} = \mathbf{X}\boldsymbol{\beta} + \mathbf{e}$, $n = 8$, the values shown in Table 6.5 were recorded:

TABLE 6.5

y	X_0	X_1	X_2	X_3	X_4	X_5	X_6	X_7
4	1	1	1	1	1	1	1	1
5	1	1	−1	1	−1	1	−1	−1
3	1	1	1	−1	1	−1	−1	−1
2	1	1	−1	−1	−1	−1	1	1
6	1	−1	−1	1	1	−1	1	−1
3	1	−1	1	1	−1	−1	−1	1
4	1	−1	−1	−1	1	1	−1	1
8	1	−1	1	−1	−1	1	1	−1

(*a*) Find $\mathbf{X}'\mathbf{X}$.
(*b*) Find $R(\boldsymbol{\beta})$, $R(\beta_1 \mid \beta_0,\beta_2, \ldots ,\beta_7)$, $R(\beta_3)$.
(*c*) Show that $R(\beta_2,\beta_3 \mid \beta_0,\beta_1,\beta_4, \ldots ,\beta_7) = R(\beta_2) + R(\beta_3)$.
(*d*) Find the reduction due to H_0 (adj), where H_0 is $\beta_1 = \beta_2 = \beta_3$.
(*e*) Show that there is no unbiased estimate of σ^2.

6.17 In the proof of Theorem 6.6, show that the matrices $\mathbf{A}$, $(\mathbf{A}_2 - \mathbf{A})$, and $\mathbf{I} - \mathbf{A}_2$ are each idempotent.

6.18 In Prob. 6.17, show that the products $\mathbf{A}(\mathbf{A}_2 - \mathbf{A})$, $\mathbf{A}(\mathbf{I} - \mathbf{A}_2)$ and $(\mathbf{A}_2 - \mathbf{A})(\mathbf{I} - \mathbf{A}_2)$ are each equal to the null matrix.

6.19 In Prob. 6.15, suppose a value of y equal to 5.3 is selected from an unknown X value. Use Eq. (6.17) to set a 95 per cent confidence interval on the unknown X.

6.20 In the example of Art. 6.4.6, set a 90 per cent confidence interval on $\beta_1 - 2\beta_2$.

6.21 If $n = 20$, $p = 6$, $\alpha = .05$ in Eq. (6.8), find the expected width of the confidence interval.

6.22 Repeat Prob. 6.21 for $n = 40$, $p = 6$, $\alpha = .05$ in Eq. (6.9).

6.23 In Prob. 6.21, show that $\lim_{n\to\infty} E(\omega) = 0$, where $E(\omega)$ is the expected width.

6.24 In Eq. (6.9), find the distribution of the width ω.

6.25 In Prob. 6.24, find $E(\omega)$ and $\text{var}(\omega)$.

6.26 Prove that the coefficient of σ^2 in Eq. (6.15) is equal to

$$\frac{1}{k} + \frac{\Sigma(x_i - x_0)^2}{n\Sigma(x_i - \bar{x})^2}$$

6.27 In Eq. (6.12), find $\mathbf{x}_0'\mathbf{S}^{-1}\mathbf{x}_0$ for the simple linear model

$$y_j = \beta_0 + \beta_1 x_j + e_j,$$

where

$$\mathbf{x}_0 = \begin{pmatrix} 1 \\ x \end{pmatrix}$$

Further Reading

1 O. Kempthorne: "Design and Analysis of Experiments," John Wiley & Sons, Inc., New York, 1952.

2 R. L. Anderson and T. A. Bancroft: "Statistical Theory in Research," McGraw-Hill Book Company, Inc., New York, 1952.

3 C. R. Rao: "Advanced Statistical Methods in Biometric Research," John Wiley & Sons, Inc., New York, 1952.

4 H. B. Mann: "Analysis and Design of Experiments," Dover Publications, New York, 1949.

5 F. N. David and J. Neyman: Extension of the Markoff Theorem on Least Squares, *Statist. Research Mem.*, vol. 2, pp. 105–116, 1938.

6 C. R. Rao: A Theorem in Least Squares, *Sankhyā*, vol. 11, pp. 9–12, 1951.

7 C. R. Rao: On Transformations Useful in the Distribution Problems of Least Squares, *Sankhyā*, vol. 12, pp. 339–346, 1952–1953.

8 C. R. Rao: Generalization of Markoff's Theorem and Tests of Linear Hypotheses, *Sankhyā*, vol. 7, pp. 9–19, 1945–1946.

9 C. R. Rao: On the Linear Combination of Observations and the General Theory of Least Squares, *Sankhyā*, vol. 7, pp. 237–256, 1945–1946.

10 R. L. Plackett: Some Theorems in Least Squares, *Biometrika*, vol. 37, pp. 149–157, 1950.

11 P. C. Tang: The Power Function of AOV Tests, *Statist. Research Mem.*, vol. 2, 1938.

12 A. Wald: On the Power Function of the Analysis of Variance Test, *Ann. Math. Statist.*, vol. 13, pp. 434–439, 1942.

13 A. Wald: On the Efficient Design of Statistical Investigations, *Ann. Math. Statist.*, vol. 14, pp. 134–140, 1943.

14 S. W. Nash: Note on Power of the *F* Test, *Ann. Math. Statist.*, vol. 19, p. 434, 1948.

15 J. Wolfowitz: The Power of the Classical Tests Associated with the Normal Distribution, *Ann. Math. Statist.*, vol. 20, pp. 540–551, 1949.

16 J. E. Maxfield and R. S. Gardner: Note on Linear Hypotheses with Prescribed Matrix of Normal Equations, *Ann. Math. Statist.*, vol. 26, pp. 149–150, 1955.

17 R. A. Fisher: The Goodness of Fit and Regression Formulae, and the Distribution of Regression Coefficients, *J. Roy. Statist. Soc.*, vol. 85, part IV, 1922; reprinted in "Contributions to Mathematical Statistics," John Wiley & Sons, Inc., New York, 1950.

18 S. Kolodziejczyk: On an Important Class of Statistical Hypotheses, *Biometrika*, vol. 27, 1935.

7

Computing Techniques

7.1 Introduction

It was shown in the preceding chapter that, to estimate $\boldsymbol{\beta}$ or any component β_1 from the model $\mathbf{Y} = \mathbf{X}\boldsymbol{\beta} + \mathbf{e}$, we must find the solution $\hat{\boldsymbol{\beta}}$ of the set of equations $\mathbf{X}'\mathbf{X}\hat{\boldsymbol{\beta}} = \mathbf{X}'\mathbf{Y}$. We can solve for $\hat{\boldsymbol{\beta}}$ by finding the inverse of $\mathbf{X}'\mathbf{X}$, and get $\hat{\boldsymbol{\beta}} = (\mathbf{X}'\mathbf{X})^{-1}\mathbf{X}'\mathbf{Y}$. If we want only $\hat{\boldsymbol{\beta}}$, we can solve the system of equations $\mathbf{X}'\mathbf{X}\hat{\boldsymbol{\beta}} = \mathbf{X}'\mathbf{Y}$ without actually finding the inverse of $\mathbf{X}'\mathbf{X}$. But, if we want also the covariance matrix of $\hat{\boldsymbol{\beta}}$, we *must* find the inverse of $\mathbf{X}'\mathbf{X}$, since $\text{cov}(\hat{\boldsymbol{\beta}}) = \sigma^2(\mathbf{X}'\mathbf{X})^{-1}$. Both ways will be discussed. Section 7.2 will be devoted to a method of solving a system of symmetric equations without actually finding the inverse of a matrix, and Sec. 7.3 will be devoted to finding the inverse of a symmetric matrix.

There are many ways to invert a matrix, and the method to use in a particular instance depends upon what type of calculator or computer is available. We shall present only one method, the Doolittle method, which is especially adaptable to ordinary desk calculators.

7.2 Solving a System of Symmetric Equations

7.2.1 The Doolittle Method. To find the solution to the set of equations

$$\mathbf{X}'\mathbf{X}\hat{\boldsymbol{\beta}} = \mathbf{X}'\mathbf{Y}$$

where $\mathbf{X}'\mathbf{X}$ is a known symmetric matrix and $\mathbf{X}'\mathbf{Y}$ is a known vector, we shall employ a technique known as the *forward solution* of the *Doolittle method.* We shall illustrate this method by a simple example.

Suppose we want to solve the system

$$\begin{aligned}
&(1) \qquad & 2\hat\beta_1 + 4\hat\beta_2 + 2\hat\beta_3 &= 6 \\
&(2) \qquad & 4\hat\beta_1 + 10\hat\beta_2 + 2\hat\beta_3 &= 18 \\
&(3) \qquad & 2\hat\beta_1 + 2\hat\beta_2 + 12\hat\beta_3 &= -16
\end{aligned} \tag{7.1}$$

By methods of elementary algebra we can solve the three simultaneous equations by the following steps:

a. Multiply Eq. (7.1.1) by $\frac{1}{2}$.
b. Subtract 2 times Eq. (7.1.1) from Eq. (7.1.2).
c. Subtract Eq. (7.1.1) from Eq. (7.1.3).

This gives

$$\begin{aligned}
&(a) \qquad & \hat\beta_1 + 2\hat\beta_2 + \hat\beta_3 &= 3 \\
&(b) \qquad & 2\hat\beta_2 - 2\hat\beta_3 &= 6 \\
&(c) \qquad & -2\hat\beta_2 + 10\hat\beta_3 &= -22
\end{aligned}$$

Now let us continue:

d. Rewrite Eq. (*a*).
e. Multiply Eq. (*b*) by $\frac{1}{2}$.
f. Add Eq. (*b*) to Eq. (*c*).

This gives

$$\begin{aligned}
&(d) \qquad & \hat\beta_1 + 2\hat\beta_2 + \hat\beta_3 &= 3 \\
&(e) \qquad & \hat\beta_2 - \hat\beta_3 &= 3 \\
&(f) \qquad & 8\hat\beta_3 &= -16
\end{aligned}$$

If we write (*d*) and (*e*) and divide Eq. (*f*) by 8, we get the triangular system

$$\begin{aligned}
&(g) \qquad & \hat\beta_1 + 2\hat\beta_2 + \hat\beta_3 &= 3 \\
&(h) \qquad & \hat\beta_2 - \hat\beta_3 &= 3 \\
&(i) \qquad & \hat\beta_3 &= -2
\end{aligned}$$

It is now quite simple to find the solutions

$$\hat\beta_3 = -2 \qquad \hat\beta_2 = 1 \qquad \hat\beta_1 = 3$$

We can solve the set of equations (7.1) in fewer steps than we have indicated; we do not, for example, need to write $\hat\beta_1, \hat\beta_2, \hat\beta_3$ in Eqs. (*a*) to (*i*), but we can let position indicate which element is involved. Equations (7.1) can be put into the matrix form:

$$\begin{pmatrix} 2 & 4 & 2 \\ 4 & 10 & 2 \\ 2 & 2 & 12 \end{pmatrix} \begin{pmatrix} \hat\beta_1 \\ \hat\beta_2 \\ \hat\beta_3 \end{pmatrix} = \begin{pmatrix} 6 \\ 18 \\ -16 \end{pmatrix}$$

If we write this

$$\begin{pmatrix} 2 & 4 & 2 & \vline & 6 \\ 4 & 10 & 2 & \vline & 18 \\ 2 & 2 & 12 & \vline & -16 \end{pmatrix}$$

and reduce the 3×3 matrix left of the vertical line to a triangular matrix by row operations a to i, we get the results shown in Table 7.1.

TABLE 7.1 REDUCTION OF EQ. (7.1) TO TRIANGULAR FORM

Instruction		C_1	C_2	C_3	C_0	Check (sum)
	R_1	2	4	2	6	14
	R_2	4	10	2	18	34
	R_3	2	2	12	−16	0
$\frac{1}{2}R_1$	R_4	1	2	1	3	7
$R_2 - 2R_1$	R_5	0	2	−2	6	6
$R_3 - R_1$	R_6	0	−2	10	−22	−14
R_4	R_7	1	2	1	3	7
$\frac{1}{2}R_5$	R_8	0	1	−1	3	3
$R_6 + R_5$	R_9	0	0	8	−16	−8
R_7	R_{10}	1	2	1	3	7
R_8	R_{11}	0	1	−1	3	3
$\frac{1}{8}R_9$	R_{12}	0	0	1	−2	−1

If we sum the elements of each row, we get the corresponding elements in each row of the check column. The instructions each refer to an operation on each element of the corresponding row including the element in the check column. The rows R_{10}, R_{11}, and R_{12} complete the *forward* solution of the Doolittle method. We can solve for the $\hat{\beta}_i$: that is, using R_{12}, we get $\hat{\beta}_3 = -2$; substituting this into R_{11}, we get $\hat{\beta}_2 = 1$; continuing into R_{10}, we get $\hat{\beta}_1 = 3$.

We can further reduce the labor involved in solving the system (7.1) by a technique known as the *abbreviated Doolittle method*, as follows: Write the original matrix, the C_0 column, and the check column, i.e., rows R'_1, R'_2, and R'_3 in Table 7.2, omitting the elements below the main diagonal, since these elements are known owing to symmetry. Then reduce the matrix to triangular form by following the instruction column. The symbols R_{ij}, R'_{ij}, and r_{ij} represent the elements in the jth columns of the rows R_i, R'_i, and r_i, respectively. The row corresponding to r_i is obtained by dividing every element in the row R_i by the first nonzero element. The spaces where an asterisk (*) appears need not be filled, since the elements in them would be zero if they were completed via the instruction. After each row is completed, it should be totaled to see whether it equals the element in the check column.

TABLE 7.2 ABBREVIATED DOOLITTLE METHOD FOR SOLVING EQ. (7.1)

Instruction	Row	C_1	C_2	C_3	C_0	Check
	R'_1	2	4	2	6	14
	R'_2		10	2	18	34
	R'_3			12	−16	0
R'_1	R_1	2	[4]	(2)	6	14
$\frac{1}{2}R_1$	r_1	1	2	1	3	7
$R'_{2j} - R_{12}r_{1j}$	R_2	*	2	(−2)	6	6
$\frac{1}{2}R_2$	r_2	*	1	−1	3	3
$R'_{3j} - R_{13}r_{1j} - R_{23}r_{2j}$	R_3	*	*	8	−16	−8
$\frac{1}{8}R_3$	r_3	*	*	1	−2	−1

After the matrix is reduced to triangular form, the rows r_3, r_2, and r_1 (with check omitted) can be used to obtain the values of $\hat{\beta}_1$, $\hat{\beta}_2$, and $\hat{\beta}_3$ that satisfy Eq. (7.1). The things to notice here are the pivotal elements. For example, to get the jth element in row R_2, we need only the jth element in row R'_2, the jth element in row r_1, and the pivotal element R_{12}, which is marked with a square for emphasis. To calculate the fourth element in row R_2: it is equal to the fourth element in R'_2 (which is 18) minus the pivotal element (which is 4) times the fourth element in row r_1 (which is 3); that is, $18 - (4)(3) = 6$. To find the values of the quantities in the jth column of R_3 we do a similar operation, except that we have two pivotal elements, one for each row of the matrix that has been computed (this pair of pivotal elements is circled for emphasis). To compute the fifth element of row R_3: it is equal to the fifth element of R'_3 (which is 0) minus the fifth element of r_1 (which is 7) times the pivotal element in R_1 (which is 2) minus the fifth element of r_2 (which is 3) times the pivotal element in R_2 (which is -2); that is, $0 - (7)(2) - (3)(-2) = -8$.

7.2.2 The Application of the Doolittle Method to Analysis-of-variance Problems for Model 1. If the system (7.1) is a set of normal equations, then C_0 is $\mathbf{X'Y}$, and the reduction due to $\beta_1, \beta_2, \beta_3$ is $\hat{\boldsymbol{\beta}}'\mathbf{X'Y}$, which in this example gives $R(\boldsymbol{\beta}) = \hat{\boldsymbol{\beta}}'\mathbf{X'Y} = (3)(6) + (1)(18) + (-2)(-16) = 68$.

However, the reduction due to $\boldsymbol{\beta}$ can be obtained very easily from Table 7.2. It is

$$R(\boldsymbol{\beta}) = \hat{\boldsymbol{\beta}}'\mathbf{X}'\mathbf{Y} = R_{10}r_{10} + R_{20}r_{20} + R_{30}r_{30}$$

which is the sum of the product of the three pairs of values in the C_0 column.

Suppose we wanted to test the hypothesis $\beta_3 = 0$. We know from Theorem 6.5 and Table 6.2 that we need to obtain:

1. The reduction due to β_1, β_2, and β_3, which is $R(\boldsymbol{\beta}) = \hat{\boldsymbol{\beta}}'\mathbf{X}'\mathbf{Y}$, where $\hat{\boldsymbol{\beta}}$ is the solution to Eq. (7.1).

2. The reduction due to β_1, β_2 (when $\beta_3 = 0$), which is $R(\boldsymbol{\gamma}_1) = \tilde{\boldsymbol{\gamma}}_1'\mathbf{X}_1'\mathbf{Y}$, where $\tilde{\boldsymbol{\gamma}}_1 = \begin{pmatrix}\tilde{\boldsymbol{\beta}}_1\\ \tilde{\boldsymbol{\beta}}_2\end{pmatrix}$ is the solution to the normal equations in (7.1) except with $\hat{\beta}_3 = 0$ and the β_3 equation (7.1.3) omitted. That is, we solve

$$\begin{aligned} 2\tilde{\beta}_1 + 4\tilde{\beta}_2 &= 6 \\ 4\tilde{\beta}_1 + 10\tilde{\beta}_2 &= 18 \end{aligned} \tag{7.2}$$

and get
$$R(\boldsymbol{\gamma}_1) = \tilde{\boldsymbol{\gamma}}_1'\mathbf{X}_1'\mathbf{Y} = \tilde{\beta}_1(6) + \tilde{\beta}_2(18)$$

Then the sum of squares for β_3 adjusted for β_1 and β_2 is

$$R(\boldsymbol{\gamma}_2 \mid \boldsymbol{\gamma}_1) = R(\boldsymbol{\beta}) - R(\boldsymbol{\gamma}_1) = \hat{\boldsymbol{\beta}}'\mathbf{X}'\mathbf{Y} - \tilde{\boldsymbol{\gamma}}_1'\mathbf{X}_1'\mathbf{Y}$$

Now these values can be picked immediately from Table 7.2.

The reduced system of equations (7.2) is represented in Table 7.2 by rows R_1' and R_2' with column C_3 omitted. Therefore, the solution can be obtained from r_1 and r_2 if column C_3 and the check column are omitted.

The solution is

$$\tilde{\beta}_2 = 3 \qquad \tilde{\beta}_1 = -3$$

and, since $\mathbf{X}_1'\mathbf{Y}$ is the corresponding element in the C_0 column, we get

$$R(\boldsymbol{\gamma}_1) = \tilde{\boldsymbol{\gamma}}_1'\mathbf{X}_1'\mathbf{Y} = (-3)(6) + (3)(18) = 36$$

However, as explained before, this reduction is immediately available from Table 7.2. It is as follows: $R_{10}r_{10} + R_{20}r_{20} = R(\boldsymbol{\gamma}_1) = R(\beta_1, \beta_2)$ = reduction due to β_1, β_2 ignoring β_3. Therefore, the reduction due to β_3 adjusted for β_1 and β_2 is $R(\boldsymbol{\gamma}_2 \mid \boldsymbol{\gamma}_1) = R(\boldsymbol{\beta}) - R(\boldsymbol{\gamma}_1) = \hat{\boldsymbol{\beta}}'\mathbf{X}'\mathbf{Y} - \tilde{\boldsymbol{\gamma}}_1'\mathbf{X}_1'\mathbf{Y} = (R_{10}r_{10} + R_{20}r_{20} + R_{30}r_{30}) - (R_{10}r_{10} + R_{20}r_{20}) = R_{30}r_{30}$, which is very simple to find from Table 7.2. It is equal to $(-16)(-2) = 32$.

To illustrate further, suppose we wish to test the hypothesis $\beta_2 = \beta_3 = 0$. We need the reduction due to β_2 and β_3 adjusted for β_1, that is, $R(\beta_2,\beta_3 \mid \beta_1)$. This is equal to

$$r_{20}R_{20} + r_{30}R_{30}$$

or, in our problem, $(3)(6) + (-2)(-16) = 50$.

If we desire to test the hypothesis $\beta_1 = 0$, we must obtain the reduction due to β_1 adjusted for β_2 and β_3. This cannot be readily obtained from Table 7.2 as it stands. It could have been obtained if we had interchanged β_1 and β_3 and the β_1 and β_3 equations in the system (7.1) so that, when the system was reduced to triangular form, the last row r_3 would give a solution for β_1. The β_1 equation in the normal equations would have to be in row R_3'. To generalize, we have

◆ **Theorem 7.1** To test $\boldsymbol{\gamma}_2 = \mathbf{0}$ in the general-linear-hypothesis model $\mathbf{Y} = \mathbf{X}_1\boldsymbol{\gamma}_1 + \mathbf{X}_2\boldsymbol{\gamma}_2 + \mathbf{e}$, where $\boldsymbol{\gamma}_1$ is $k \times 1$ and the normal equations are $\mathbf{X}'\mathbf{X}\hat{\boldsymbol{\beta}} = \mathbf{X}'\mathbf{Y}$, put the normal equations in the form of Table 7.2 with the coefficients pertaining to $\boldsymbol{\gamma}_2$ in the right-hand columns of the matrix and reduce to triangular form by the abbreviated Doolittle technique. Then (see Theorem 6.5 and Table 6.2) the error sum of squares is

$$\mathbf{Y}'\mathbf{Y} - \sum_{i=1}^{p} R_{i0} r_{i0}$$

and the reduction due to $\boldsymbol{\gamma}_2$ adjusted for $\boldsymbol{\gamma}_1$ is

$$R(\boldsymbol{\gamma}_2 \mid \boldsymbol{\gamma}_1) = \sum_{i=k+1}^{p} R_{i0} r_{i0}$$

We shall not prove this theorem, but it is extremely important. The reader should become well acquainted with it if he is going to use the abbreviated Doolittle technique to obtain the quantities needed to test hypotheses in model 1.

7.2.3 Alternative Method of Presenting the Analysis-of-variance Table. Many times the $\mathbf{X}$ matrix in model 1 has unity for every element in the first column, and the equation takes the form

$$y_j = \mu + \sum_{i=1}^{p-1} x_{ji}\beta_i + e_j \qquad j = 1, 2, \ldots, n \tag{7.3}$$

If we add and subtract $\sum_{i=1}^{p-1} \beta_i \bar{x}_i$, we can write this

$$y_j = (\mu + \Sigma\beta_i\bar{x}_i) + \sum_i (x_{ji} - \bar{x}_i)\beta_i + e_j$$

where $\bar{x}_i = \frac{1}{n}\sum_j x_{ji}$. If we let $x_{ji} - \bar{x}_i = z_{ji}$, $z_{j0} = 1$, and $\mu + \Sigma\beta_i\bar{x}_i = \beta_0$, we get

$$y_j = \sum_{i=0}^{p-1} \beta_i z_{ji} + e_j$$

In matrix form this could be written

$$\mathbf{Y} = \mathbf{Z}\boldsymbol{\beta} + \mathbf{e}$$

where we must remember that $\boldsymbol{\beta}' = (\beta_0,\beta_1, \ldots ,\beta_{p-1})$ and that $\beta_0 = \mu + \sum_{i=1}^{p-1} \beta_i \bar{x}_i$. If we write the model as we did in Art. 6.4.3, we get $\mathbf{Y} = \mathbf{Z}_1\boldsymbol{\gamma}_1 + \mathbf{Z}_2\boldsymbol{\gamma}_2 + \mathbf{e}$, where $\boldsymbol{\gamma}_1 = \beta_0$ and $\mathbf{Z}_1$ is an $n \times 1$ matrix with each element equal to unity. It is clearly true that $\boldsymbol{\gamma}_1$ and $\boldsymbol{\gamma}_2$ are orthogonal according to Definition 6.2 (Art. 6.4.5), since $\mathbf{Z}_1'\mathbf{Z}_2 = \mathbf{0}$. Therefore, whenever the model can be written as (7.3), and we use deviations of each x_{ji}, β_0 is orthogonal to the other β_i. Also, $R(\beta_0,\beta_1, \ldots ,\beta_{p-1}) = R(\beta_0) + R(\beta_1, \ldots ,\beta_{p-1})$. We shall call $R(\beta_0)$ the *reduction due to the mean*; it is clearly equal to $(\Sigma y_j)^2/n$.

If we want to test the hypothesis $\beta_{p-k} = \beta_{p-k+1} = \cdots = \beta_{p-1} = 0$, the analysis of variance can be written in the form of Table 7.3.

TABLE 7.3 TEST OF THE HYPOTHESIS
$\beta_{p-k} = \beta_{p-k+1} = \cdots = \beta_{p-1} = 0$

SV	DF	SS
Total	$n - 1$	$\sum_{j=1}^{n} (y_j - \bar{y})^2$
$R(\beta_1,\beta_2, \ldots ,\beta_{p-1})$	$p - 1$	$\sum_{i=1}^{p-1} r_{i0}^* R_{i0}^*$
$R(\beta_1,\beta_2, \ldots ,\beta_{p-k-1})$	$p - k - 1$	$\sum_{i=1}^{p-k-1} r_{i0}^* R_{i0}^*$
$R(\beta_{p-k}, \ldots ,\beta_{p-1} \mid \beta_1,\beta_2, \ldots ,\beta_{p-k-1})$	k	$\sum_{i=p-k}^{p-1} r_{i0}^* R_{i0}^*$
Error	$n - p$	$\sum_{j=1}^{n} (y_j - \bar{y})^2 - \sum_{i=1}^{p-1} r_{i0}^* R_{i0}^*$

The values r_{i0}^* and R_{i0}^* are values taken from a table similar to Table 7.2 except that the matrix entered in the table is $(p - 1) \times (p - 1)$ and the elements are the corrected (deviations from the means) sums of squares and cross products. The row and column corresponding to μ are omitted. The elements of the C_0 column are the corrected cross products of the x and y values.

7.3 Computing the Inverse of a Symmetric Matrix

7.3.1 The Inverse. If a confidence interval is desired on any element of $\boldsymbol{\beta}$ or on any linear combination of β_i, then the elements of the inverse of $\mathbf{X'X}$ are needed [see Eqs. (6.9) and (6.10)]. The inverse can be found by using the *abbreviated* Doolittle technique.

The theory can be explained by supposing that $\mathbf{A}$ is a symmetric matrix whose inverse is desired.

Let $\mathbf{B} = \mathbf{A}^{-1}$; then $\mathbf{AB} = \mathbf{I}$.

Let $\mathbf{B} = (\mathbf{B}_1, \mathbf{B}_2, \ldots, \mathbf{B}_p)$, where $\mathbf{B}_i$ is the ith column of $\mathbf{A}^{-1}$. Then $\mathbf{A}(\mathbf{B}_1, \mathbf{B}_2, \ldots, \mathbf{B}_p) = (\mathbf{E}_1, \mathbf{E}_2, \ldots, \mathbf{E}_p)$, where $\mathbf{E}_i$ is the ith column of the identity matrix $\mathbf{I}$.

We can then obtain systems of p equations $\mathbf{AB}_i = \mathbf{E}_i$ $(i = 1, 2, \ldots, p)$. We must solve these p systems and then obtain the elements of $\mathbf{A}^{-1}$. By using a table such as Table 7.2, we can solve all p of the systems by reducing $\mathbf{A}$ to triangular form. We shall illustrate by finding the inverse of the matrix in (7.1).

$$\mathbf{X'X} = \begin{pmatrix} 2 & 4 & 2 \\ 4 & 10 & 2 \\ 2 & 2 & 12 \end{pmatrix}$$

We can find $\mathbf{B}_1$ by solving

$$\left(\begin{array}{ccc|c} 2 & 4 & 2 & 1 \\ 4 & 10 & 2 & 0 \\ 2 & 2 & 12 & 0 \end{array}\right)$$

by the abbreviated Doolittle method and $\mathbf{B}_2$ by solving

$$\left(\begin{array}{ccc|c} 2 & 4 & 2 & 0 \\ 4 & 10 & 2 & 1 \\ 2 & 2 & 12 & 0 \end{array}\right)$$

and similarly for $\mathbf{B}_3$. We can save work by putting the matrix in the form

$$\left(\begin{array}{ccc|ccc} 2 & 4 & 2 & 1 & 0 & 0 \\ 4 & 10 & 2 & 0 & 1 & 0 \\ 2 & 2 & 12 & 0 & 0 & 1 \end{array}\right)$$

and reducing to triangular form. This is illustrated in Table 7.4, where all the operations except those of the last three rows are exactly the same as in Table 7.2.

Using columns C_1, C_2, C_3, and E_1 of rows r_1, r_2, and r_3, we find the elements of $\mathbf{B}_1$ to be

$$\mathbf{B}_1 = \begin{pmatrix} \frac{29}{8} \\ -\frac{11}{8} \\ -\frac{3}{8} \end{pmatrix}$$

TABLE 7.4 THE INVERSE OF X′X IN EQ. (7.1)

Instruction	Row	C_1	C_2	C_3	C_0	E_1	E_2	E_3	Check
	R'_1	2	4	2	6	1	0	0	15
	R'_2		10	2	18	0	1	0	35
	R'_3			12	−16	0	0	1	1
R'_1	R_1	2	4	2	6	1	0	0	15
$\frac{1}{2}R_1$	r_1	1	2	1	3	$\frac{1}{2}$	0	0	$\frac{15}{2}$
$R'_{2j} - R_{12}r_{1j}$	R_2	*	2	−2	6	−2	1	0	5
$\frac{1}{2}R_2$	r_2	*	1	−1	3	−1	$\frac{1}{2}$	0	$\frac{5}{2}$
$R'_{3j} - R_{13}r_{1j} - R_{23}r_{2j}$	R_3	*	*	8	−16	−3	1	1	−9
$\frac{1}{8}R_3$	r_3	*	*	1	−2	$-\frac{3}{8}$	$\frac{1}{8}$	$\frac{1}{8}$	$-\frac{9}{8}$
$R_{15}r_{1j} + R_{25}r_{2j} + R_{35}r_{3j}$	B_1					$\frac{29}{8}$	$-\frac{11}{8}$	$-\frac{3}{8}$	
$R_{26}r_{2j} + R_{36}r_{3j}$	B_2						$\frac{5}{8}$	$\frac{1}{8}$	
$R_{37}r_{37}$	B_3							$\frac{1}{8}$	

Using columns C_1, C_2, C_3, and E_2 of rows r_1, r_2, and r_3 we get

$$\mathbf{B}_2 = \begin{pmatrix} -\frac{11}{8} \\ \frac{5}{8} \\ \frac{1}{8} \end{pmatrix}$$

and, by a similar process, we get

$$\mathbf{B}_3 = \begin{pmatrix} -\frac{3}{8} \\ \frac{1}{8} \\ \frac{1}{8} \end{pmatrix}$$

Therefore, $$\mathbf{B} = (\mathbf{B}_1, \mathbf{B}_2, \mathbf{B}_3) = \frac{1}{8}\begin{pmatrix} 29 & -11 & -3 \\ -11 & 5 & 1 \\ -3 & 1 & 1 \end{pmatrix}$$

There is a short-cut method for obtaining **B** by using the instructions for the last three rows of Table 7.4 and using only columns E_1, E_2, and E_3. This gives the elements of **B** immediately. In computing the rows $\mathbf{B}_1$, $\mathbf{B}_2$, $\mathbf{B}_3$ notice the pivotal elements R_{15}, R_{25}, etc.

We might point out that the determinant of **A** is equal to the product of the first nonzero elements in rows R_1, R_2, and R_3. In this particular problem it is equal to $(2)(2)(8) = 32$.

TABLE 7.5 ABBREVIATED DOOLITTLE FORMAT FOR A $p \times p$ SYMMETRIC MATRIX

Instruction	Row	$C_1C_2 \cdots C_pC_0E_1 \cdots E_p$
	R_1' . . . R_p'	
R_1' A_1	R_1 r_1	
$R_{2j}' - R_{12}r_{1j}$ A_2	R_2 r_2	
. . .	. . .	
$R_{tj}' - \sum_{q=1}^{t-1} R_{qt}r_{qj}$ A_t	R_t r_t	
. . .	. . .	
$R_{pj}' - \sum_{q=1}^{p-1} R_{qp}r_{qj}$ A_p	R_p r_p	
$\sum_{i=1}^{p} R_{i,p+2}r_{ij}$	B_1	
$\sum_{i=2}^{p} R_{i,p+3}r_{ij}$	B_2	
. . .	. . .	
$\sum_{i=t}^{p} R_{i,p+t+1}r_{ij}$	B_t	
. . .	. . .	
$R_{p,2p+1}r_{p,2p+1}$	B_p	

7.3.2 The General Case. We shall look at the solution for a general $p \times p$ symmetric matrix. Everything proceeds as for the 3×3 case except that we shall have more rows. The instruction and row columns for a $p \times p$ symmetric matrix are given in Table 7.5. The instruction A_t means: Divide the elements in the row by the first nonzero element of row R_t.

7.3.3 Example. We shall demonstrate the abbreviated Doolittle technique by an example. In measuring the various constituents of cow's milk it would be desirable to find a function relating the quantities: solids nonfat, fat, and protein. Suppose we wish to estimate protein by using a knowledge of factors fat x_1 and solids nonfat x_2 by the equation

$$y_j = \mu + \beta_1 x_{ji} + \beta_2 x_{j2} + e_j$$

By taking samples from 16 cows the values shown in Table 7.6 were obtained.

TABLE 7.6 DATA FOR EXAMPLE OF ART. 7.3.3

Protein y	Fat x_1	Solid nonfat x_2	Protein y	Fat x_1	Solid nonfat x_2
3.75	4.74	9.50	3.16	3.36	8.86
3.19	3.66	8.56	3.65	3.64	9.21
2.99	4.27	8.54	3.36	3.92	8.93
3.46	4.03	8.62	3.60	2.99	9.16
3.27	3.51	9.35	3.87	3.28	8.45
3.27	3.97	8.39	3.14	3.23	9.09
2.78	3.23	7.87	3.00	3.65	8.36
3.59	3.79	9.33	3.18	4.23	9.28

To test the hypothesis $\beta_1 = \beta_2 = 0$, we shall use the *abbreviated* Doolittle technique with both the *corrected* and *uncorrected* sum of squares, and we shall find the inverse of $\mathbf{X}'\mathbf{X}$ by both methods (we shall use the uncorrected sum of squares first). The normal equations are

$$\begin{pmatrix} 16.00 & 59.50 & 141.50 \\ 59.50 & 224.47 & 527.01 \\ 141.50 & 527.01 & 1{,}254.57 \end{pmatrix} \begin{pmatrix} \mu \\ \beta_1 \\ \beta_2 \end{pmatrix} = \begin{pmatrix} 53.26 \\ 198.28 \\ 472.10 \end{pmatrix}$$

The equations are reduced by the abbreviated Doolittle method and the desired quantities computed. This is illustrated in Table 7.7. The AOV is given in Table 7.8. Solving for the regression coefficients gives $\hat{\beta}_2 = .342566$, $\hat{\beta}_1 = -.017799$, $\hat{\mu} = .365340$. Using Theorem 7.1, we get

$$R(\boldsymbol{\beta}) = \hat{\boldsymbol{\beta}}'\mathbf{X}'\mathbf{Y} = 177.65 \qquad \text{and} \qquad R(\mu) = 177.29$$

Table 7.7 Abbreviated Doolittle Format for Data of Table 7.6

	C_1	C_2	C_3	C_0	E_1	E_2	E_3	Check
R'_1	16.00	59.50	141.50	53.26	1.00000	.00000	.00000	271.26
R'_2		224.47	527.01	198.28	.00000	1.00000	.00000	1,010.26
R'_3			1,254.57	472.10	.00000	.00000	1.00000	2,396.18
R_1	16.00	59.50	141.50	53.26	1.00000	.00000	.00000	271.26
r_1	1.00	3.71875	8.84375	3.32875	.0625	.00000	.00000	16.95375
R_2	*	3.204375	.806875	.219375	−3.71875	1.00000	.00000	1.511875
r_2	*	1.000000	.2518041	.068461	−1.160523	.3120733	.00000	.471816
R_3	*	*	2.976170	1.019534	−7.907353	−.2518335	1.00000	−3.163395
r_3	*	*	1.000000	.342566	−2.656892	−.0846170	.3360027	−1.062909
B_1					25.387143	−.419390	−2.656892	
B_2						.334380	−.084162	
B_3							.336003	

The F value found in Table 7.8 is less than the tabulated F value at the 5 per cent level. A test of the hypothesis μ equal to zero (or to some other constant) can be made by the t test, using the elements from $(\mathbf{X}'\mathbf{X})^{-1}$ given in lines B_1, B_2, and B_3 of Table 7.7. Similarly for testing

TABLE 7.8 ANALYSIS OF VARIANCE FOR DATA OF TABLE 7.6

SV	DF	SS	MS	F
Total	16	178.66		
$R(\mu,\beta_1,\beta_2)$	3	177.65		
$R(\mu)$	1	177.29		
$R(\beta_1,\beta_2 \mid \mu)$	2	.36	.18	2.30
Error	13	1.01	.077	

β_1 or β_2. Also, the inverse elements can be used to test a linear combination of μ, β_1, β_2 equal to some constant and for setting confidence limits on μ, β_1, or β_2 and on a linear combination of the parameters. We shall refer to this example again later in the chapter.

Now we shall test the hypothesis $\beta_1 = \beta_2 = 0$ using the abbreviated Doolittle method on the *corrected* sums of squares and cross products. The corrected sums of squares and cross products are entered in a table like Table 7.9 and reduced as demonstrated there. The corresponding

TABLE 7.9 ABBREVIATED DOOLITTLE FORMAT FOR DATA OF TABLE 7.6, CORRECTED SUM OF SQUARES

	C_1	C_2	C_0	E_1	E_2	Check
R_1'	3.2018	.80158	.22198	1	0	5.22536
R_2'		3.1822	1.08558	0	1	6.06936
R_1	3.2018	.80158	.22198	1	0	5.22536
r_1	1.0000	.25035	.069330	.31234	.00000	1.63200
R_2	*	2.98152	1.03001	−.25035	1.00000	4.76118
r_2	*	1.00000	.34546	−.08397	.33540	1.59689
B_1				.33334	−.08397	
B_2					.33540	

AOV is given in Table 7.10. From Table 7.9 we see that the values of $\hat{\beta}_1$ and $\hat{\beta}_2$ are the same as when they are computed in Table 7.7, except for rounding errors. The analysis of variance in Table 7.10 gives the same mean square for error and for the reduction due to β_1 and β_2 adjusted for μ as Table 7.8 and, hence, the same F value. The work is probably reduced somewhat by using the corrected sums of squares and cross products.

7.3.4 Rounding Errors. It is noticeable in Tables 7.7 and 7.9 that rounding errors can become quite troublesome. It is essential to carry at least two more significant figures in the computation than are needed in the final result.

TABLE 7.10 ANALYSIS OF VARIANCE OF CORRECTED SUMS OF SQUARES AND CROSS PRODUCTS

SV	DF	SS	MS	F
Total	15	1.37		
Due to β_1, β_2 (adj)	2	.37	.19	2.30
Error	13	1.00	.077	

Rounding errors can also become quite large in the abbreviated method of finding the inverse elements. However, there is a method for improving the inverse of a matrix, which is due to Hotelling; this is given in the following theorem.

◆ **Theorem 7.2** If $\mathbf{B}$ is an approximate inverse of the matrix $\mathbf{A}$, then an improvement on $\mathbf{B}$ is given by $\mathbf{B}^*$, where

$$\mathbf{B}^* = \mathbf{B}(2\mathbf{I} - \mathbf{AB})$$

Suppose we want the inverse of the matrix

$$\mathbf{A} = \begin{pmatrix} 2 & -3 \\ -3 & 5 \end{pmatrix}$$

The inverse is
$$\mathbf{A}^{-1} = \begin{pmatrix} 5 & 3 \\ 3 & 2 \end{pmatrix}$$

Suppose, however, that we have only an approximate inverse

$$\mathbf{B} = \begin{pmatrix} 4.989 & 3.011 \\ 3.012 & 1.001 \end{pmatrix}$$

Then
$$2\mathbf{I} - \mathbf{AB} = \begin{pmatrix} 1.058 & -.049 \\ -.093 & 1.078 \end{pmatrix}$$

and, therefore,
$$\mathbf{B}^* = \mathbf{B}(2\mathbf{I} - \mathbf{AB}) = \begin{pmatrix} 4.908 & 3.001 \\ 3.002 & 1.999 \end{pmatrix}$$

If still further accuracy is desired, the operation can be repeated on $\mathbf{B}^*$.

Problems

7.1 Use the abbreviated Doolittle method to solve the system of equations

$$\begin{aligned} 3\hat\beta_1 + 2\hat\beta_2 + \hat\beta_3 &= 7 \\ 2\hat\beta_1 + 2\hat\beta_2 + \hat\beta_3 &= 5 \\ \hat\beta_1 + \hat\beta_2 + 4\hat\beta_3 &= -1 \end{aligned}$$

and find the inverse of the matrix.

7.2 Repeat Prob. 7.1 for the system

$$\begin{aligned} 2\hat\beta_1 + \hat\beta_2 + 3\hat\beta_3 - \hat\beta_4 &= 8 \\ \hat\beta_1 + \hat\beta_2 - \hat\beta_3 + \hat\beta_4 &= -1 \\ 3\hat\beta_1 - \hat\beta_2 + 3\hat\beta_3 - 2\hat\beta_4 &= 2 \\ -\hat\beta_1 + \hat\beta_2 - 2\hat\beta_3 + 4\hat\beta_4 &= 1 \end{aligned}$$

7.3 If the approximate inverse of

$$\begin{pmatrix} 2 & 1 & 1 \\ 1 & 1 & 1 \\ 1 & 1 & 3 \end{pmatrix} \quad \text{is} \quad \begin{pmatrix} .998 & -.987 & .011 \\ -.986 & 2.521 & -.507 \\ .010 & -.501 & .503 \end{pmatrix}$$

use Theorem 7.2 to improve the approximation.

7.4 In Art. 7.2.3, prove that the maximum-likelihood estimate of β_0 is the same whether we use the model in the z_j or x_j.

7.5 In Art. 7.2.3, prove that $R(\beta_0) = (\Sigma y)^2/n$.

7.6 Work the problem in the example of Art. 6.4.6 using the Doolittle method.

7.7 Prove that the total corrected sum of squares divided by σ^2 is distributed as $\chi^2(n-1)$; i.e., prove that $\Sigma(y_i - \bar y)^2/\sigma^2$ is distributed as $\chi^2(n-1)$ if $\mathbf{Y}$ is distributed $N(\mathbf{0},\sigma^2\mathbf{I})$.

7.8 Prove that, if the $p \times p$ matrix

$$\mathbf{B} = \begin{pmatrix} 1 & -\bar x_1 & -\bar x_2 & \cdots & -\bar x_{p-1} \\ 0 & 1 & 0 & \cdots & 0 \\ 0 & 0 & 1 & \cdots & 0 \\ \cdots & \cdots & \cdots & \cdots & \cdots \\ 0 & 0 & 0 & \cdots & 1 \end{pmatrix}$$

then

$$\mathbf{B}^{-1} = \begin{pmatrix} 1 & \bar x_1 & \bar x_2 & \cdots & \bar x_{p-1} \\ 0 & 1 & 0 & \cdots & 0 \\ 0 & 0 & 1 & \cdots & 0 \\ \cdots & \cdots & \cdots & \cdots & \cdots \\ 0 & 0 & 0 & \cdots & 1 \end{pmatrix}$$

7.9 If $\mathbf{X}'\mathbf{X}\hat{\boldsymbol{\beta}} = \mathbf{X}'\mathbf{Y}$ is a system of normal equations, where all the elements in the first column of $\mathbf{X}$ are equal to unity and where $\hat{\boldsymbol{\beta}}' = (\hat{\mu},\hat{\beta}_1, \ldots ,\hat{\beta}_{p-1})$, prove the following:

(*a*) $\mathbf{B}'\mathbf{X}'\mathbf{X}\mathbf{B} = \begin{pmatrix} n & 0 \\ 0 & \mathbf{Z}'\mathbf{Z} \end{pmatrix}$, where the rsth element of $\mathbf{Z}'\mathbf{Z}$ is equal to

$$\sum_j (x_{jr} - x_{.r})(x_{js} - x_{.s}) \qquad r \neq 1; \qquad s \neq 1$$

and where $\mathbf{B}$ is as given in Prob. 7.8.

(*b*) $(\mu,\beta_1, \ldots ,\beta_{p-1})\mathbf{B}'^{-1} = (\beta_0,\beta_1, \ldots ,\beta_{p-1})$, where $\beta_0 = \mu + \sum_{i=1}^{p-1} \beta_i\bar{x}_i$.

(*c*) $\mathbf{B}'\mathbf{X}'\mathbf{Y}$ is a vector whose rth element is

$$\Sigma(y_j - y_.)(x_{jr} - x_{.r}) \qquad r \neq 1$$

(*d*) Find the set of equations $\mathbf{B}'\mathbf{X}'\mathbf{X}\mathbf{B}\mathbf{B}^{-1}\hat{\boldsymbol{\beta}} = \mathbf{B}'\mathbf{X}'\mathbf{Y}$.

Further Reading

1 P. S. Dwyer and F. V. Waugh: On Errors in Martix Inversion, *J. Am. Statist. Assoc.*, vol. 48, pp. 289–319, 1953.

2 P. S. Dwyer: The Doolittle Technique, *Ann. Math. Statist.*, vol. 12, pp. 419–458, 1940–1941.

3 J. Ullman: The Probability of Convergence of an Iterative Process of Inverting a Matrix, *Ann. Math. Statist.*, vol. 15, pp. 205–213, 1944.

4 P. S. Dwyer: A Matrix Presentation of Least Squares and Correlation Theory with Matrix Justification of Improved Methods of Solution, *Ann. Math. Statist.*, vol. 15, pp. 82–89, 1944.

5 P. L. Hsu: On the Power Functions of the E^2-Test and the T^2-Test, *Ann. Math. Statist.*, vol. 16, pp. 278–286, 1945.

6 W. L. Ferrar: "Algebra," Oxford University Press, New York, 1946.

7 M. J. Weiss: "Higher Algebra for the Undergraduate," John Wiley & Sons Inc., New York, 1949.

8 S. Perlis: "Theory of Matrices," Addison-Wesley Publishing Company, Cambridge, Mass., 1952.

9 R. L. Anderson and T. A. Bancroft: "Statistical Theory in Research," McGraw-Hill Book Company, Inc., New York, 1952.

10 C. R. Rao: "Advanced Statistical Methods in Biometric Research," John Wiley & Sons, Inc., New York, 1952.

11 R. A. Fisher: "Statistical Methods for Research Workers," Oliver and Boyd, Ltd., London, 1946.

8

Polynomial or Curvilinear Models

8.1 Introduction

In this chapter we shall consider a special case of model 1, commonly called a *curvilinear* or *polynomial* model. The model referred to can be written

$$y_j = \beta_0 + \beta_1 x_j + \beta_2 x_j^2 + \cdots + \beta_p x_j^p + e_j$$

There are two situations in which an experimenter may want to use a polynomial model:

1. Where he knows, theoretically or otherwise, that his data fit a polynomial of degree p or less, and he wishes to find the maximum-likelihood or least-squares estimate of the β_i, set confidence intervals on the β_i, or test hypotheses about the β_i.

2. Where it is not known what function fits the data; so a search must be made to find a polynomial of low degree that adequately describes them.

8.2 Estimating and Testing Coefficients in a Polynomial Model

Consider the model

$$y_j = \beta_0 + \beta_1 x_{1j} + \beta_2 x_{2j} + \beta_3 x_{3j} + e_j \qquad j = 1, 2, \ldots, n$$

If we let $x_{1j} = x_j$, $x_{2j} = x_j^2$, $x_{3j} = x_j^3$, we get the particular model

$$y_j = \beta_0 + \beta_1 x_j + \beta_2 x_j^2 + \beta_3 x_j^3 + e_j$$

This model is exactly model 1. The pertinent matrices are

$$\mathbf{X} = \begin{pmatrix} 1 & x_1 & x_1^2 & x_1^3 \\ 1 & x_2 & x_2^2 & x_2^3 \\ 1 & x_3 & x_3^2 & x_3^3 \\ \cdot & \cdot & \cdot & \cdot \\ 1 & x_n & x_n^2 & x_n^3 \end{pmatrix} \qquad \mathbf{X'X} = \begin{pmatrix} n & \Sigma x_i & \Sigma x_i^2 & \Sigma x_i^3 \\ \Sigma x_i & \Sigma x_i^2 & \Sigma x_i^3 & \Sigma x_i^4 \\ \Sigma x_i^2 & \Sigma x_i^3 & \Sigma x_i^4 & \Sigma x_i^5 \\ \Sigma x_i^3 & \Sigma x_i^4 & \Sigma x_i^5 & \Sigma x_i^6 \end{pmatrix}$$

$$\mathbf{X'Y} = \begin{pmatrix} \Sigma y_i \\ \Sigma x_i y_i \\ \Sigma x_i^2 y_i \\ \Sigma x_i^3 y_i \end{pmatrix}$$

The theory in Chap. 6 is applicable.

8.3 Finding the Degree of a Polynomial That Describes a Given Set of Data

8.3.1 Theory. Let us suppose that we have evidence that a quantity y is functionally related to a quantity x by $y = f(x)$. Up to now we have been discussing prediction models where the functional form of $f(x)$ is known but contains unknown parameters. The general procedure is to collect data (various values of y and x) and estimate or test hypotheses about the parameters in $f(x)$.

We shall now turn our attention to a different problem, that of *finding the functional form of $f(x)$ by using a small amount of data.* In general, this problem is very complex; in most cases satisfactory solutions do not exist.

There are various ways in which a scientist may determine the functional form of $f(x)$. He may determine it from a knowledge of more fundamental facts, such as differential equations; he may reason from a related fundamental law that is known; he may collect such a vast amount of data that he can perceive certain trends; he may evolve a function by intuitive or visionary inspiration.

We shall discuss the problem faced by a scientist who takes a few observations and, having no function at his disposal, wants a systematic way of determining a function that fits his observations. We shall not attempt here to prove or test that the data satisfy certain functional relationships. Rather we shall attempt to find a function that *fits the data well* and make decisions accordingly.

We shall assume that $f(x)$ can be expanded into a Taylor series, so that we obtain

$$y = f(x) = \theta_0 + \theta_1 x + \theta_2 x^2 + \cdots + \theta_k x^k + \cdots$$

This can be written $y = \alpha_0 + \alpha_1 x + e_1$, where e_1 is the remainder in a Taylor series, which we assume for our purposes is random error (see the equation-error model in Chap. 5). If the fit to a linear equation is not satisfactory, we assume a new model

$$y = \beta_0 + \beta_1 x + \beta_2 x^2 + e_2$$

where e_2 is the remainder after a quadratic is fitted to the data. We can continue to any degree polynomial we desire. Clearly, if the true functional relationship is $y = f(x)$ and if $f(x)$ is not a polynomial, the above procedure will not determine the true form $f(x)$, but will simply aid the experimenter in finding a polynomial of low degree that will help describe the data. As we mentioned in Sec. 8.2, this model actually satisfies Definition 6.1 (we are assuming that the x's are fixed and that y is a random variable).

The procedure is first to fit the linear polynomial $y = \alpha_0 + \alpha_1 x + e_1$, then to fit the quadratic polynomial $y = \beta_0 + \beta_1 x + \beta_2 x^2 + e_2$, then to fit the cubic $y = \gamma_0 + \gamma_1 x + \gamma_2 x^2 + \gamma_3 x^3 + e_3$, and so forth, until we find the polynomial that fits the data "best." (We note that the coefficient of x in the first-degree polynomial is not necessarily the same as the coefficient of x in the quadratic polynomial. Similarly for other terms.) Polynomials of degree $n - 1$ will go through every point and, hence, fit the data perfectly. However, in most cases, this is not what we want. We desire a low-degree polynomial that *represents* the data.

We shall begin by fitting a first-degree polynomial $y = \alpha_0 + \alpha_1 x + e_1$ and test the hypothesis $\alpha_1 = 0$.

The normal equations are

$$\begin{pmatrix} n & \Sigma x_i \\ \Sigma x_i & \Sigma x_i^2 \end{pmatrix} \begin{pmatrix} \hat{\alpha}_0 \\ \hat{\alpha}_1 \end{pmatrix} = \begin{pmatrix} \Sigma y_i \\ \Sigma x_i y_i \end{pmatrix} \tag{8.1}$$

and $R(\alpha_0,\alpha_1) = \hat{\alpha}_0 \Sigma y_i + \hat{\alpha}_1 \Sigma y_i x_i$. The term $R(\alpha_0) = \tilde{\alpha}_0 \Sigma y_i$, where $\tilde{\alpha}_0$ is the solution to $n\tilde{\alpha}_0 = \Sigma y_i$; that is, $\tilde{\alpha}_0 = \bar{y}$. This equation is obtained from (8.1) by striking out the last row and column of the matrix and the last row of the vectors. The AOV is given in Table 8.1.

If, by using the test, we decide that $\alpha_1 = 0$, we conclude that the line $y = \alpha_0 + e$ fits the data adequately. If instead we decide that $\alpha_1 \neq 0$, we fit the second-degree polynomial $y = \beta_0 + \beta_1 x + \beta_2 x^2 + e_2$, as in Table 8.2.

TABLE 8.1 ANALYSIS OF VARIANCE FOR LINEAR POLYNOMIAL

SV	DF	SS	MS	F
Total	n	$\mathbf{Y}'\mathbf{Y}$		
Due to α_0, α_1	2	$R(\alpha_0,\alpha_1)$		
Due to α_0	1	$R(\alpha_0)$		
Due to α_1 (adj)	1	$R(\alpha_1 \mid \alpha_0)$	A_1	$\dfrac{A_1}{E_1}$
Error	$n - 2$	$\mathbf{Y}'\mathbf{Y} - R(\alpha_0,\alpha_1)$	E_1	

To get the quantities in Table 8.2, we find the normal equations, which are

$$\begin{pmatrix} n & \Sigma x_i & \Sigma x_i^2 \\ \Sigma x_i & \Sigma x_i^2 & \Sigma x_i^3 \\ \Sigma x_i^2 & \Sigma x_i^3 & \Sigma x_i^4 \end{pmatrix} \begin{pmatrix} \hat{\beta}_0 \\ \hat{\beta}_1 \\ \hat{\beta}_2 \end{pmatrix} = \begin{pmatrix} \Sigma y_i \\ \Sigma x_i y_i \\ \Sigma x_i^2 y_i \end{pmatrix} \tag{8.2}$$

and $R(\beta_0,\beta_1,\beta_2) = \hat{\beta}_0 \Sigma y_i + \hat{\beta}_1 \Sigma y_i x_i + \hat{\beta}_2 \Sigma y_i x_i^2$. $R(\beta_0,\beta_1) = \tilde{\beta}_0 \Sigma y_i + \tilde{\beta}_1 \Sigma y_i x_i$, where $\tilde{\beta}_0$ and $\tilde{\beta}_1$ are solutions to the equations

$$\begin{pmatrix} n & \Sigma x_i \\ \Sigma x_i & \Sigma x_i^2 \end{pmatrix} \begin{pmatrix} \tilde{\beta}_0 \\ \tilde{\beta}_1 \end{pmatrix} = \begin{pmatrix} \Sigma y_i \\ \Sigma x_i y_i \end{pmatrix} \tag{8.3}$$

TABLE 8.2 ANALYSIS OF VARIANCE FOR QUADRATIC POLYNOMIAL

SV	DF	SS	MS	F
Total	n			
Due to β_0, β_1, β_2	3	$R(\beta_0,\beta_1,\beta_2)$		
Due to β_0, β_1 (unadj)	2	$R(\beta_0,\beta_1) = R(\alpha_0,\alpha_1)$		
Due to β_2 (adj)	1	$R(\beta_2 \mid \beta_0,\beta_1)$	A_2	$\dfrac{A_2}{E_2}$
Error	$n - 3$	$\mathbf{Y}'\mathbf{Y} - R(\beta_0,\beta_1,\beta_2)$	E_2	

These are obtained from (8.2) by striking out the last row and column from the 3×3 matrix and the last element from each of the vectors. Clearly, $\tilde{\beta}_0$ and $\tilde{\beta}_1$ in (8.3) are the same as $\hat{\alpha}_0$ and $\hat{\alpha}_1$ in (8.1) and, hence, do not need to be computed again. Also, $R(\beta_0,\beta_1) = R(\alpha_0,\alpha_1)$.

In Table 8.2 we test $\beta_2 = 0$ without reference to β_1. If we conclude from Table 8.1 that $\alpha_1 \neq 0$ and from Table 8.2 that $\beta_2 = 0$, we then conclude that $y = \alpha_0 + \alpha_1 x + e_1$ fits the data adequately.

If we conclude that $\beta_2 \neq 0$, we next fit the cubic $y = \gamma_0 + \gamma_1 x + \gamma_2 x^2 + \gamma_3 x^3 + e_3$, as in Table 8.3.

To get the quantities in Table 8.3, we work with the normal equations

$$\begin{pmatrix} n & \Sigma x_i & \Sigma x_i^2 & \Sigma x_i^3 \\ \Sigma x_i & \Sigma x_i^2 & \Sigma x_i^3 & \Sigma x_i^4 \\ \Sigma x_i^2 & \Sigma x_i^3 & \Sigma x_i^4 & \Sigma x_i^5 \\ \Sigma x_i^3 & \Sigma x_i^4 & \Sigma x_i^5 & \Sigma x_i^6 \end{pmatrix} \begin{pmatrix} \hat{\gamma}_0 \\ \hat{\gamma}_1 \\ \hat{\gamma}_2 \\ \hat{\gamma}_3 \end{pmatrix} = \begin{pmatrix} \Sigma y_i \\ \Sigma x_i y_i \\ \Sigma x_i^2 y_i \\ \Sigma x_i^3 y_i \end{pmatrix} \tag{8.4}$$

$R(\gamma_0,\gamma_1,\gamma_2,\gamma_3) = \hat{\gamma}_0\Sigma y_i + \hat{\gamma}_1\Sigma y_i x_i + \hat{\gamma}_2\Sigma y_i x_i^2 + \hat{\gamma}_3\Sigma y_i x_i^3$, and $R(\gamma_0,\gamma_1,\gamma_2) = \tilde{\gamma}_0\Sigma y_i + \tilde{\gamma}_1\Sigma y_i x_i + \tilde{\gamma}_2\Sigma y_i x_i^2$, where $\tilde{\gamma}_0$, $\tilde{\gamma}_1$, $\tilde{\gamma}_2$ are obtained from (8.4) by striking out the last row and column of the 4×4 matrix and

TABLE 8.3 ANALYSIS OF VARIANCE FOR CUBIC POLYNOMIAL

SV	DF	SS	MS	F
Total	n			
Due to $\gamma_0, \gamma_1, \gamma_2, \gamma_3$	4	$R(\gamma_0,\gamma_1,\gamma_2,\gamma_3)$		
Due to $\gamma_0, \gamma_1, \gamma_2$ (unadj)	3	$R(\gamma_0,\gamma_1,\gamma_2) = R(\beta_0,\beta_1,\beta_2)$		
Due to γ_3 (adj)	1	$R(\gamma_3 \mid \gamma_0,\gamma_1,\gamma_2)$	C_3	$\frac{C_3}{E_3}$
Error	$n - 4$	$\mathbf{Y}'\mathbf{Y} - R(\gamma_0,\gamma_1,\gamma_2,\gamma_3)$	E_3	

omitting the last element in the two vectors. Clearly, $\tilde{\gamma}_0$, $\tilde{\gamma}_1$, $\tilde{\gamma}_2$ are respectively equal to $\hat{\beta}_0$, $\hat{\beta}_1$, $\hat{\beta}_2$ in Eq. (8.2), and $R(\gamma_0,\gamma_1,\gamma_2) = R(\beta_0,\beta_1,\beta_2)$.

If F is not significant, we conclude that $\gamma_3 = 0$ and that a second-degree polynomial adequately fits the data. If F is significant, we conclude that $\gamma_3 \neq 0$ and proceed to fit a fourth-degree polynomial to the data. We continue in this fashion until we arrive at the first nonsignificant result; if we get significance for linear, quadratic, . . . , up to a $k - 1$-degree polynomial and then get nonsignificance for a polynomial of degree k, we conclude that a $k - 1$-degree polynomial fits the data.

Whenever we conclude that a certain-degree polynomial adequately fits the data, we can then estimate the coefficients and thus determine a least-squares curve.

There are two things in the above procedure that need comment:

1. The nonsignificance of a result does not imply that the data actually came from any specified degree of polynomial. It is merely a procedural criterion for establishing what polynomial is adequate.

Suppose it is decided that the quadratic model $y = \beta_0 + \beta_1 x + \beta_2 x^2 + e_2$ is an adequate representation of the data. When we examined the linear model $y = \alpha_0 + \alpha_1 x + e_1$, the term e_1, which we assumed took on the aspect of a random variable, has a strong component of x^2 in it. This may introduce a bias in the error sum of squares for linear. Similar remarks hold when we decide that a pth-degree polynomial fits the data. The remainder sum of squares for the lower degrees may be biased. If an independent estimate of the variance of the random variable e is available, a better method is the one described in Sec. 8.5.

Some statisticians recommend that two consecutive nonsignificant results appear before a decision is made on the degree of polynomial. Suppose, for example, that the first two consecutive nonsignificant results are found when testing $\beta_2 = 0$ and when testing $\gamma_3 = 0$; then we conclude that a linear polynomial fits the data.

2. The computing involved is not difficult for low-degree polynomials. Table 8.1 requires no difficult computations; Table 8.2 requires that we solve a system of three equations with three unknowns, to obtain the sum of squares due to $\beta_0, \beta_1, \beta_2$. Table 8.3 requires that we solve a system of four equations and four unknowns, to obtain the sum of squares due to $\gamma_0, \gamma_1, \gamma_2, \gamma_3$. If, however, the data fit a polynomial of degree 3 or more, the computation can become quite troublesome.

If the x values are in arithmetic progression, we can greatly reduce the computations by using orthogonal polynomials, which will be discussed in Sec. 8.4. First we shall illustrate the above ideas with an example.

8.3.2 Example. Given the following data, we want to find what degree of polynomial adequately represents them.

y	24.0	20.0	10.0	13.0	12.0	6.0	5.0	1.0	1.0	.0
x	.8	1.0	1.2	1.4	1.6	1.8	2.0	2.2	2.4	2.6

First we shall fit a linear model $y = \alpha_0 + \alpha_1 x + e_1$.

$$\mathbf{X'X} = \begin{pmatrix} 10.0 & 17.0 \\ 17.0 & 32.2 \end{pmatrix} \qquad \mathbf{X'Y} = \begin{pmatrix} 92.0 \\ 114.0 \end{pmatrix}$$

So we get

$$\begin{pmatrix} \hat{\alpha}_0 \\ \hat{\alpha}_1 \end{pmatrix} = \begin{pmatrix} 31.0424 \\ -12.8480 \end{pmatrix} \qquad \mathbf{Y'Y} = 1{,}452.0$$

$$R(\alpha_0, \alpha_1) = (31.0424)(92.0) + (-12.8480)(114.0) = 1391.16$$

$$R(\alpha_0) = (9.20)(92.0) = 846.0$$

The AOV is given in Table 8.4.

TABLE 8.4 ANALYSIS OF VARIANCE FOR LINEAR POLYNOMIAL MODEL

SV	DF	SS	MS	F	Pr
Total	10	1,452.00			
Due to α_0, α_1	2	1,391.16			
Due to α_0	1	846.40			
Due to α_1 (adj)	1	544.76	544.76	71.58	$<5\%$
Error	8	60.84	7.61		

Since we conclude that $\alpha_1 \neq 0$, we next fit a quadratic. We get

$$\mathbf{X}'\mathbf{X} = \begin{pmatrix} 10.00 & 17.00 & 32.20 \\ 17.00 & 32.20 & 65.96 \\ 32.20 & 65.96 & 142.68 \end{pmatrix} \qquad \mathbf{X}'\mathbf{Y} = \begin{pmatrix} 92.0 \\ 114.0 \\ 156.0 \end{pmatrix}$$

We get $\hat{\beta}_0 = 42.96$, $\hat{\beta}_1 = -28.68$, $\hat{\beta}_2 = 4.66$, and $R(\beta_0,\beta_1,\beta_2) = 1{,}409.35$. $R(\beta_0,\beta_1) = R(\alpha_0,\alpha_1) = 1{,}391.16$.

The AOV is given in Table 8.5.

TABLE 8.5 ANALYSIS OF VARIANCE FOR QUADRATIC POLYNOMIAL MODELS

SV	DF	SS	MS	F	Pr
Total	10	1,452.00			
Due to β_0, β_1, β_2	3	1,409.35			
Due to β_0, β_1 (unadj)	2	1,391.16			
Due to β_2 (adj)	1	18.19	18.19	2.99	$>5\%$
Error	7	42.65	6.09		

This F value is not significant at the 5 per cent level; so we assume that $\beta_2 = 0$ and that the linear polynomial adequately fits the data. These two tables can be summarized in one, as illustrated in Table 8.6.

TABLE 8.6 ANALYSIS OF VARIANCE FOR LINEAR AND QUADRATIC POLYNOMIAL MODELS

SV	DF	SS	MS	F	Pr
Total	10	1,452.00			
Mean	1	846.40			
Linear term	1	544.76	544.76	71.58	$<5\%$
Error for linear	8	60.84	7.61		
Quadratic term	1	18.19	18.19	2.99	$>5\%$
Error for quadratic	7	42.65	6.09		

8.4 Orthogonal Polynomials

8.4.1 Introduction. The central idea involved in using *orthogonal polynomials* in a curvilinear model is that the computing is made easier by a transformation made on the x values so that the matrix $\mathbf{X}'\mathbf{X}$ becomes diagonal. Since $\mathbf{X}'\mathbf{X}$ is diagonal, the parameters are orthogonal, according to Art. 6.4.5.

Orthogonal polynomials can be used whether the x values are equally or unequally spaced. However, they simplify computation greatly when the x values are in equal steps; so we shall discuss that case only.

8.4.2 Orthogonal Polynomials for Linear Models. We shall discuss the ideas for a linear model first and then point out some generalizations.

Suppose we have some data (y and x values) to which we wish to fit a polynomial. Suppose the x values are in equal steps; i.e., suppose $x_1 = a + h$; $x_2 = a + 2h$; . . . ; $x_i = a + ih$; . . . ; $x_n = a + nh$. For example, if the x_i are 4.3, 4.5, 4.7, 4.9, 5.1, then $a = 4.1$, $h = .2$, and $n = 5$. First we shall fit the linear model $y_i = \beta_0 + \beta_1 x_i + e_i$; but, instead of writing it this way, we shall write it

$$y_i = \alpha_0 + \alpha_1 P_1(i - \bar{\imath}) + e_i$$

where $P_1(i - \bar{\imath})$ is a first-degree polynomial in $i - \bar{\imath}$, where $\bar{\imath}$ is the mean of the i values. That is to say, since i takes on the values $1, \ldots, n$, $\bar{\imath} = (1/n)\,\Sigma i = n(n + 1)/2n = (n + 1)/2$. Therefore,

$$P_1(i - \bar{\imath}) = P_1\left(i - \frac{n+1}{2}\right) = c_0 + c_1\left(i - \frac{n+1}{2}\right)$$

is a first-degree polynomial in $i - \bar{\imath}$, where c_0 and c_1 are constants. If we set the polynomial in x_i equal to the polynomial in $i - \bar{\imath}$, we get

$$\beta_0 + \beta_1 x_i = \alpha_0 + \alpha_1\left[c_0 + c_1\left(i - \frac{n+1}{2}\right)\right]$$

Since this is an identity in the x_i, we can equate coefficients (remembering $x_i = a + ih$) and get

$$\begin{aligned} &\textit{Coefficient of } i^0\textit{:} \quad \alpha_0 + \left[c_0 - \frac{c_1(n+1)}{2}\right]\alpha_1 = \beta_0 + a\beta_1 \\ &\textit{Coefficient of } i\textit{:} \qquad\qquad\qquad\qquad c_1\alpha_1 = h\beta_1 \end{aligned} \tag{8.5}$$

Whatever the values of c_0 and c_1 ($h \neq 0$), we can solve for the β's if the α's are known. Therefore, we shall use the model

$$y_i = \alpha_0 + \alpha_1\left[c_0 + c_1\left(i - \frac{n+1}{2}\right)\right] + e_i$$

which we shall write as

$$\mathbf{Y} = \mathbf{X}\boldsymbol{\alpha} + \mathbf{e}$$

to estimate α_0 and α_1, and then use these estimates and solve for β_0 and β_1 by (8.5). Let us look at the $\mathbf{X}$ vector and $\mathbf{X}'\mathbf{X}$ matrix (remembering $i = 1, 2, \ldots, n$):

$$\mathbf{X} = \begin{pmatrix} 1 & c_0 + \dfrac{(1-n)c_1}{2} \\ 1 & c_0 + \dfrac{(3-n)c_1}{2} \\ \cdots & \cdots \\ 1 & c_0 + \dfrac{(n-1)c_1}{2} \end{pmatrix} \qquad \mathbf{X}'\mathbf{X} = \begin{pmatrix} n & nc_0 \\ nc_0 & \Sigma c_0^2 + c_1^2 \Sigma i\left(i - \dfrac{n+1}{2}\right) \end{pmatrix}$$

Since the constants c_0 and c_1 are at our disposal, we shall choose $c_0 = 0$ such that $\mathbf{X}'\mathbf{X}$ is diagonal and c_1 such that fractions are eliminated in the model. If n is an odd number, $c_1 = 1$; if n is even, $c_1 = 2$. The model becomes

$$y_i = \alpha_0 + \alpha_1 P_1\left(i - \frac{n+1}{2}\right) + e_i = \alpha_0 + \alpha_1 c_1\left(\frac{2i - n - 1}{2}\right) + e_i$$

where

$$P_1\left(i - \frac{n+1}{2}\right) = \frac{2i - n - 1}{2} \qquad \text{if } n \text{ is an odd integer}$$

$$P_1\left(i - \frac{n+1}{2}\right) = 2i - n - 1 \qquad \text{if } n \text{ is an even integer} \tag{8.6}$$

This gives

$$\mathbf{X}'\mathbf{X} = \begin{pmatrix} n & 0 \\ 0 & c_1^2 \dfrac{n(n+1)(n-1)}{12} \end{pmatrix} = \begin{pmatrix} n & 0 \\ 0 & \Sigma\left[P_1\left(i - \dfrac{n+1}{2}\right)\right]^2 \end{pmatrix} \tag{8.7}$$

Also,

$$\mathbf{X}'\mathbf{Y} = \begin{pmatrix} \Sigma y_i \\ \Sigma y_i P_1\left(i - \dfrac{n+1}{2}\right) \end{pmatrix} \qquad \hat{\boldsymbol{\alpha}} = \begin{pmatrix} \hat{\alpha}_0 \\ \hat{\alpha}_1 \end{pmatrix}$$

So the normal equations $\mathbf{X}'\mathbf{X}\hat{\boldsymbol{\alpha}} = \mathbf{X}'\mathbf{Y}$ are easily solved. We get

$$\hat{\boldsymbol{\alpha}} = \begin{pmatrix} \dfrac{1}{n}\Sigma y_i \\ \dfrac{\Sigma y_i P_1\left(i - \dfrac{n+1}{2}\right)}{\Sigma\left[P_1\left(i - \dfrac{n+1}{2}\right)\right]^2} \end{pmatrix} \tag{8.8}$$

Substituting into Eq. (8.5), we can easily obtain the $\hat{\beta}_j$.

If we wish to test the hypothesis $\alpha_1 = 0$ we need the following quantities: *reduction due to* α_1 *adjusted for* α_0, denoted by $R(\alpha_1 \mid \alpha_0)$, and *reduction due to* α_0 *and* α_1, denoted by $R(\alpha_0,\alpha_1)$. Because of the orthogonality of α_0 and α_1, these quantities are

$$R(\alpha_1 \mid \alpha_0) = R(\alpha_1) = \hat{\alpha}_1\mathbf{X}_1\mathbf{Y} = \frac{\{\Sigma y_i P_1[i - (n+1)/2]\}^2}{\Sigma\{P_1[i - (n+1)/2]\}^2}$$

and

$$R(\alpha_0,\alpha_1) = R(\alpha_0) + R(\alpha_1) = \hat{\alpha}_0\mathbf{X}_0'\mathbf{Y} + \hat{\alpha}_1\mathbf{X}_1'\mathbf{Y} = \frac{(\Sigma y_i)^2}{n} + \frac{\{\Sigma y_i P_1[i - (n+1)/2]\}^2}{\Sigma\{P_1[i - (n+1)/2]\}^2}$$

We see, therefore, that the analysis of the model is very simple provided only that the polynomial

$$P_1\left(i - \frac{n+1}{2}\right) = c_1\left(\frac{2i - n - 1}{2}\right)$$

can be evaluated for various values of n. The values of this polynomial have been tabulated [1] up to $n = 104$. The computations for a few values of n are given in Table 8.7 [Eq. (8.6) is used].

TABLE 8.7

n	1	2	3	4	5	6	ΣP_1^2
3	−1	0	1				2
4	−3	−1	1	3			20
5	−2	−1	0	1	2		10
6	−5	−3	−1	1	3	5	70

8.4.3 Example. Given the data

y	1.0	2.0	4.0	3.0	5.0
x_i	5.1	5.4	5.7	6.0	6.3

we see that the x_i are equally spaced, and $a = 4.8$, $h = .3$, $n = 5$. By using Table 8.7 for $n = 5$ we get the table below.

y_i	1.0	2.0	4.0	3.0	5.0	$15 = \Sigma y_i$
$P_1(i - \bar{i})$	−2	−1	0	1	2	$9 = \Sigma y_i P_1\left(i - \frac{n+1}{2}\right)$

$\Sigma y_i^2 = 55 \qquad \Sigma[P_1(i - \bar{i})]^2 = \Sigma P_1^2 = 10$

From Eq. (8.8) we get

$$\hat{\boldsymbol{\alpha}} = \begin{pmatrix} 3.0 \\ .9 \end{pmatrix}$$

From Eq. (8.7),

$$\mathbf{X}'\mathbf{X} = \begin{pmatrix} 5 & 0 \\ 0 & 10 \end{pmatrix}$$

From Eq. (8.5),

$$\hat{\alpha}_0 - 3\hat{\alpha}_1 = \hat{\beta}_0 + 4.8\hat{\beta}_1$$

$$\hat{\alpha}_1 = .3\hat{\beta}_1$$

which gives $\hat{\beta}_1 = 3.33\hat{\alpha}_1$, $\hat{\beta}_0 = \hat{\alpha}_0 - 19\hat{\alpha}_1$. We get

$$\hat{\boldsymbol{\beta}} = \begin{pmatrix} -14.1 \\ 3.0 \end{pmatrix}$$

Also, $R(\alpha_1 \mid \alpha_0) = 8.1$; $R(\alpha_0,\alpha_1) = 53.1$; $\mathbf{Y}'\mathbf{Y} = 55.0$. So the AOV is as shown in Table 8.8.

TABLE 8.8 ANALYSIS OF VARIANCE FOR LINEAR POLYNOMIAL

SV	DF	SS	MS	F	Pr
Total	5	55.0			
$R(\alpha_0)$	1	45.0			
$R(\alpha_1 \mid \alpha_0)$	1	8.1	8.1	13.5	<5%
Error	3	1.9	.6		

The central features to notice in using orthogonal polynomials to fit a first-degree polynomial when the x_i are equally spaced are as follows:

1. Since, by Eq. (8.5), β_1 is zero if and only if α_1 is zero, we can test $\beta_1 = 0$ by using the test for $\alpha_1 = 0$.
2. We can obtain point estimates and interval estimates of β_0 and β_1 by using Eq. (8.5).
3. The AOV on the α_i is very easy to compute, since the quantities in Eqs. (8.7) and (8.8) are easily obtained.

If only a linear polynomial is to be fitted, orthogonal polynomials are not worth while. But if we want to fit first a linear, then a quadratic, etc., until we find a "best" fit and if the x_i are equally spaced, then orthogonal polynomials are a great help. In the next article we shall use them to fit a second-degree polynomial.

8.4.4 Fitting a Quadratic by Orthogonal Polynomials. The model we shall fit is

$$y_i = \gamma_0 + \gamma_1 x_i + \gamma_2 x_i^2 + e_i \tag{8.9}$$

which we shall write as

$$y_i = \delta_0 + \delta_1 P_1\left(i - \frac{n+1}{2}\right) + \delta_2 P_2\left(i - \frac{n+1}{2}\right) + e_i \tag{8.10}$$

where $P_1\left(i - \frac{n+1}{2}\right) = d_0 + d_1\left(i - \frac{n+1}{2}\right) \qquad d_1 \neq 0$

and $P_2\left(i - \frac{n+1}{2}\right) = f_0 + f_1\left(i - \frac{n+1}{2}\right) + f_2\left(i - \frac{n+1}{2}\right)^2 \qquad f_2 \neq 0$

The quantities d_0, d_1, f_0, f_1, f_2 are at our disposal to determine as we please. The two polynomials (8.9) and (8.10) can be rewritten (since $x_i = a + ih$)

$$y_i = \gamma_0 + \gamma_1(a + ih) + \gamma_2(a + ih)^2 + e_i \tag{8.11}$$

$$y_i = \delta_0 + \delta_1\left[d_0 + d_1\left(i - \frac{n+1}{2}\right)\right] + \delta_2\left[f_0 + f_1\left(i - \frac{n+1}{2}\right) + f_2\left(i - \frac{n+1}{2}\right)^2\right] + e_i \tag{8.12}$$

If we equate coefficients of i^p ($p = 0, 1, 2$), we get

Coefficient of i^0: $\delta_0 + \left(d_0 - d_1\frac{n+1}{2}\right)\delta_1 + \left[f_0 - f_1\frac{n+1}{2} + f_2\left(\frac{n+1}{2}\right)^2\right]\delta_2$

$$= \gamma_0 + a\gamma_1 + a^2\gamma_2 \tag{8.13}$$

Coefficient of i: $d_1\delta_1 + (f_1 - (n+1)f_2)\delta_2 = \gamma_1 h + 2ah\gamma_2$

Coefficient of i^2: $f_2\delta_2 = \gamma_2 h^2$

If we can estimate the δ's we can use Eq. (8.13) to estimate the γ's. The quantities d_0, d_1, f_0, f_1, f_2 are at our disposal, and we shall set them equal to quantities such that the $\mathbf{X}'\mathbf{X}$ matrix of the model given in (8.12) becomes diagonal. If (8.12) is the ith term of

$$\mathbf{Y} = \mathbf{X}\boldsymbol{\delta} + \mathbf{e}$$

we have

$$\mathbf{X} = \begin{pmatrix} 1 & P_1\left(1 - \frac{n+1}{2}\right) & P_2\left(1 - \frac{n+1}{2}\right) \\ 1 & P_1\left(2 - \frac{n+1}{2}\right) & P_2\left(2 - \frac{n+1}{2}\right) \\ \cdots & \cdots & \cdots \\ 1 & P_1\left(n - \frac{n+1}{2}\right) & P_2\left(n - \frac{n+1}{2}\right) \end{pmatrix}$$

and $\mathbf{X}'\mathbf{X} =$

$$\begin{pmatrix} n & \sum_i P_1\left(i - \frac{n+1}{2}\right) & \sum_i P_2\left(i - \frac{n+1}{2}\right) \\ \sum_i P_1\left(i - \frac{n+1}{2}\right) & \sum_i \left[P_1\left(i - \frac{n+1}{2}\right)\right]^2 & \sum_i P_1\left(i - \frac{n+1}{2}\right)P_2\left(i - \frac{n+1}{1}\right) \\ \sum_i P_2\left(i - \frac{n+1}{2}\right) & \sum_i P_1\left(i - \frac{n+1}{2}\right)P_2\left(i - \frac{n+1}{2}\right) & \sum_i \left[P_2\left(i - \frac{n+1}{2}\right)\right]^2 \end{pmatrix}$$

If $\mathbf{X}'\mathbf{X}$ is to be diagonal, we must have $\Sigma P_1 = \Sigma P_2 = \Sigma P_1 P_2 = 0$. This gives us $d_0 = 0$; $f_0 = -f_2[(n+1)(n-1)]/12$; and $f_1 = 0$, since $d_1 \neq 0$.

With these values the polynomials can be written

$$\begin{aligned} P_1\left(i - \frac{n+1}{2}\right) &= d_1\left(i - \frac{n+1}{2}\right) \\ P_2\left(i - \frac{n+1}{2}\right) &= f_2\left[-\frac{(n-1)(n+1)}{12} + \left(i - \frac{n+1}{2}\right)^2\right] \end{aligned} \tag{8.14}$$

where d_1 and f_2 are chosen such that each polynomial has integral coefficients. We can write Eq. (8.12) as

$$y_i = \delta_0 + \delta_1 d_1\left(i - \frac{n+1}{2}\right) + \delta_2 f_2\left[-\frac{(n-1)(n+1)}{12} + \left(i - \frac{n+1}{2}\right)^2\right] + e_i$$

Then

$$\mathbf{X}'\mathbf{Y} = \begin{pmatrix} \Sigma y_i \\ \Sigma y_i P_1\left(i - \frac{n+1}{2}\right) \\ \Sigma y_i P_2\left(i - \frac{n+1}{2}\right) \end{pmatrix}$$

$$\hat{\boldsymbol{\delta}} = (\mathbf{X}'\mathbf{X})^{-1}\mathbf{X}'\mathbf{Y} = \begin{pmatrix} \dfrac{\Sigma y_i}{n} \\ \dfrac{\Sigma y_i P_1\left(i - \frac{n+1}{2}\right)}{\Sigma\left[P_1\left(i - \frac{n+1}{2}\right)\right]^2} \\ \dfrac{\Sigma y_i P_2\left(i - \frac{n+1}{2}\right)}{\Sigma\left[P_2\left(i - \frac{n+1}{2}\right)\right]^2} \end{pmatrix}$$

An important point to notice here is that the coefficients of δ_0 and δ_1 are the same as those of the polynomial when only a linear is fitted; that is, $\hat{\delta}_0 = \hat{\alpha}_0$ and $\hat{\delta}_1 = \hat{\alpha}_1$ (where $\hat{\alpha}_0$ and $\hat{\alpha}_1$ are as in Art. 8.4.2). That is to say, when we fit the quadratic, the only coefficient we need to compute is $\hat{\delta}_2$. *The other coefficients remain the same.* Also,

$$R(\delta_2 \mid \delta_0,\delta_1) = R(\delta_2) = \frac{\left[\Sigma y_i P_2\left(i - \frac{n+1}{2}\right)\right]^2}{\Sigma\left[P_2\left(i - \frac{n+1}{2}\right)\right]^2}$$

The reason that the computations are so easy is that $P_1\left(i - \frac{n+1}{2}\right)$ and $P_2\left(i - \frac{n+2}{2}\right)$ have been tabulated [1].

Table 8.9 gives the polynomials for a few values of n. They are computed from the quantities in Eq. (8.14).

TABLE 8.9

n	P_1					P_2					ΣP_2^2
	1	2	3	4	5	1	2	3	4	5	
3	−1	0	1			1	−2	1			6
4	−3	−1	1	3		1	−1	−1	1		4
5	−2	−1	0	1	2	2	−1	−2	−1	2	14

The values for P_1 are the same as those given in Table 8.7.

8.4.5 Example. Suppose we want to fit a quadratic polynomial to the data in Art. 8.4.3. We get

y_i	1	2	4	3	5	$\Sigma y_i = 15$
P_1	−2	−1	0	1	2	$\Sigma P_1^2 = 10$
P_2	2	−1	−2	−1	2	$\Sigma P_2^2 = 14$

$$\Sigma P_1 y_i = 9 \qquad \Sigma P_2 y_i = -1 \qquad \Sigma y_i^2 = 55$$

$$\hat{\delta}_0 = \tfrac{15}{5} = 3.0 \qquad \hat{\delta}_1 = \tfrac{9}{10} = .9 \qquad \hat{\delta}_2 = \tfrac{1}{14} = -.07$$

$$R(\delta_0 \mid \delta_1,\delta_2) = R(\delta_0) = 45 \quad R(\delta_1 \mid \delta_0,\delta_2) = R(\delta_1) = 8.1 \quad R(\delta_2 \mid \delta_0,\delta_1) = .07$$

The AOV is given in Table 8.10.

TABLE 8.10 ANALYSIS OF VARIANCE FOR QUADRATIC POLYNOMIAL

SV	DF	SS	MS	F	Pr
Total	5	55.0			
Mean $= R(\delta_0)$	1	45.0			
Linear $= R(\delta_1)$	1	8.1	8.1		
Quadratic $= R(\delta_2)$	1	.07	.07	<1	$>5\%$
Error	2	1.83	.92		

All the entries in Table 8.10 are the same as in Table 8.8 except the quadratic and the new error term, which is obtained by subtraction. The estimates of δ_0 and δ_1 are the same as those of α_0 and α_1, respectively, in Art. 8.4.2. If we want to estimate the γ_0, γ_1, and γ_2 in Eq. (8.9), we use Eq. (8.13). We get

$$\hat{\gamma}_2 = \frac{\hat{\delta}_2}{.09} = -\frac{.07}{.09} = -.79$$

$$\hat{\gamma}_1 = 12.0$$

$$\hat{\gamma}_0 = -39.6$$

8.4.6 The General *p*th-degree Orthogonal Polynomial. In the previous articles we have explained in detail the construction and use of orthogonal polynomials of degree 1 and degree 2. We shall now discuss pth-degree orthogonal polynomials.

Let a pth-degree polynomial

$$y_i = \beta_0 + \beta_1 x_i + \beta_2 x_i^2 + \cdots + \beta_p x_i^p + e_i \tag{8.15}$$

where the x_i are equally spaced ($x_i = a + ih$), be represented by the polynomial

$$y_i = \alpha_0 + \alpha_1 P_1(i - \bar{\imath}) + \alpha_2 P_2(i - \bar{\imath}) + \cdots + \alpha_p P_p(i - \bar{\imath}) + e_i \qquad i = 1, 2, \ldots, n \tag{8.16}$$

$P_t(i - \bar{\imath})$ is a tth-degree polynomial in $i - \bar{\imath}$, where $\bar{\imath} = (n + 1)/2$. We can represent this as

$$P_t = P_t(i - \bar{\imath}) = P_t\left(i - \frac{n+1}{2}\right) = a_{0t} + a_{1t}\left(i - \frac{n+1}{2}\right) + a_{2t}\left(i - \frac{n+1}{2}\right)^2 + \cdots + a_{tt}\left(i - \frac{n+1}{2}\right)^t \tag{8.17}$$

where the $a_{0t}, a_{1t}, \ldots, a_{tt}$ are constants, which we shall determine so that the $\mathbf{X'X}$ matrix is diagonal and so that all the elements in P_t are whole numbers.

If we substitute $a + ih$ for x_i in (8.15) and equate coefficients of i^t in (8.15) and (8.16), we get a set of equations that we can solve for the α_i in terms of the β_j. The equations are

Coefficient of i^0:

$$A_{00}\alpha_0 + A_{01}\alpha_1 + \cdots + A_{0p}\alpha_p = B_{00}\beta_0 + B_{01}\beta_1 + \cdots + B_{0p}\beta_p$$

Coefficient of i:

$$A_{11}\alpha_1 + \cdots + A_{1p}\alpha_p = B_{11}\beta_1 + \cdots + B_{1p}\beta_p \qquad (8.18)$$

. . . .

Coefficient of i^p: $\quad A_{pp}\alpha_p = B_{pp}\beta_p$

where A_{st} is the coefficient of i^s in the polynomial $P_t[i - (n + 1)/2]$ and B_{st} is the coefficient of i^s in the quantity $(a + ih)^t$.

We can write $\mathbf{A\alpha} = \mathbf{B\beta}$, where $\boldsymbol{\alpha}$ is the vector of α_i, $\boldsymbol{\beta}$ is the vector of β_i, and $\mathbf{A}$ and $\mathbf{B}$ are $(p + 1) \times (p + 1)$ matrices whose ijth elements are A_{ij} and B_{ij}, respectively. Since $A_{ii} \neq 0$ ($i = 0, 1, \ldots, p$) and $B_{ii} \neq 0$ ($i = 0, 1, \ldots, p$) and since $\mathbf{A}$ and $\mathbf{B}$ are triangular matrices, the inverses $\mathbf{A}^{-1}$ and $\mathbf{B}^{-1}$ both exist. We can write $\boldsymbol{\beta} = \mathbf{B}^{-1}\mathbf{A}\boldsymbol{\alpha} = \mathbf{C}\boldsymbol{\alpha}$. By the theorems in Chap. 6 we know that $\hat{\boldsymbol{\beta}} = \mathbf{C}\hat{\boldsymbol{\alpha}}$. We also know that we can test hypotheses about the elements of $\boldsymbol{\beta}$ or set confidence limits on elements in $\boldsymbol{\beta}$ by using the fact that the β_i are linear combinations of the α_i. In using orthogonal polynomials we are estimating or testing hypotheses only about the α_i. If we want to say something about the β_j we can use Eq. (8.18).

One very important thing to observe is that $\alpha_p = 0$ if and only if $\beta_p = 0$, where p is the highest-order term; this follows from the last equation in (8.18). It means that, if we fit a linear equation in x, the coefficient of the linear term in x is zero if and only if α_1 is zero. If we fit a quadratic in x, the coefficient of the quadratic term is zero if and only if α_2 is zero; and so forth. We must remember that each time we change the degree of the equation in x we change all the coefficients. This is not true of the α_j; the coefficient of the first-degree orthogonal polynomial is α_1 no matter how many more terms we fit. A similar situation holds for the coefficient of any degree orthogonal polynomial.

Therefore, the procedure is to use orthogonal polynomials only so that we can easily add a higher-degree term until we find out what *degree* polynomial adequately represents the data. If all we are looking for is the *degree* of the polynomial, then we are finished. If, however, we want to estimate this polynomial (its coefficients), we can use either Eq. (8.18) or the original model to fit the polynomial of the desired degree.

If we turn back to Eq. (8.16) and write the model as $\mathbf{Y} = \mathbf{X}\boldsymbol{\alpha} + \mathbf{e}$, the pertinent matrices are:

$$\mathbf{X} = \begin{pmatrix} 1 & P_1\left(1 - \frac{n+1}{2}\right) & \cdots & P_p\left(1 - \frac{n+1}{2}\right) \\ 1 & P_1\left(2 - \frac{n+1}{2}\right) & \cdots & P_p\left(2 - \frac{n+1}{2}\right) \\ \cdots & \cdots & \cdots & \cdots \\ 1 & P_1\left(n - \frac{n+1}{2}\right) & \cdots & P_p\left(n - \frac{n+1}{2}\right) \end{pmatrix}$$

If we use the notation $\Sigma P_r P_s$ to mean $\sum_{i=1}^{p} P_r\left(i - \frac{n+1}{2}\right)P_s\left(i - \frac{n+1}{2}\right)$, we have

$$\mathbf{X'X} = \begin{pmatrix} n & \Sigma P_1 & \Sigma P_2 & \cdots & \Sigma P_p \\ \Sigma P_1 & \Sigma P_1^2 & \Sigma P_1 P_2 & \cdots & \Sigma P_1 P_p \\ \cdots & \cdots & \cdots & \cdots & \cdots \\ \Sigma P_p & \Sigma P_p P_1 & \Sigma P_p P_2 & \cdots & \Sigma P_p^2 \end{pmatrix}$$

So we must choose the coefficients a_{tt} in (8.17) so as to make $\mathbf{X'X}$ diagonal. These coefficients have been calculated [1] for n up to 104. If we use these tables, the calculations are extremely easy. Since

$$\mathbf{X'Y} = \begin{pmatrix} \Sigma y_i \\ \Sigma y_i P_1 \\ \Sigma y_i P_2 \\ \cdots \\ \Sigma y_i P_p \end{pmatrix}$$

we get $\qquad \mathbf{X'X}\hat{\boldsymbol{\alpha}} = \mathbf{X'Y} \qquad$ and $\qquad \hat{\boldsymbol{\alpha}} = (\mathbf{X'X})^{-1}\mathbf{X'Y}$

which is

$$\hat{\boldsymbol{\alpha}} = \begin{pmatrix} \frac{1}{n}\Sigma y_i \\ \frac{\Sigma y_i P_1}{\Sigma P_1^2} \\ \frac{\Sigma y_i P_2}{\Sigma P_2^2} \\ \cdots \\ \frac{\Sigma y_i P_p}{\Sigma P_p^2} \end{pmatrix}$$

Also, $$R(\alpha_q \mid \alpha_0,\alpha_1,\alpha_{q-1},\alpha_{q+1}, \ldots ,\alpha_p) = R(\alpha_q) = \frac{(\Sigma y_i P_q)^2}{\Sigma P_q^2}$$

Also, $\text{var}(\hat{\alpha}_q) = \sigma^2/\Sigma P_q^2$. The quantities ΣP_q^2 are also tabulated in Reference [1]. The only quantities that need to be computed are Σy_i^2, Σy_i, and $\Sigma P_q y_i$ $(q = 1, 2, \ldots, p)$.

8.4.7 Example. The following rainfall was recorded at the indicated year. Suppose we desire to find the polynomial that describes the data.

Year	1944	1945	1946	1947	1948	1949	1950	1951	1952	1953	1954	1955
Rainfall	30.2	32.2	35.1	34.2	39.1	41.3	36.1	30.1	30.5	26.1	24.8	28.2

From [1] we record the linear, quadratic, cubic, and quartic polynomial for $n = 12$. We shall use the 2 per cent F point as the index for adequate fit. The AOV is given in Table 8.12.

TABLE 8.11 POLYNOMIALS FOR RAINFALL DATA

i (year coded)	1	2	3	4	5	6	7	8	9	10	11	12	ΣP^2
Rainfall	30.2	32.2	35.1	34.2	39.1	41.3	36.1	30.1	30.5	26.1	24.8	28.2	
P_1	−11	−9	−7	−5	−3	−1	1	3	5	7	9	11	572
P_2	55	25	1	−17	−29	−35	−35	−29	−17	1	25	55	12,012
P_3	−33	3	21	25	19	7	−7	−19	−25	−21	−3	33	5,148
P_4	33	−27	−33	−13	12	28	28	12	−13	−33	−27	33	8,008

$\bar{y} = 32.325$ $\Sigma P_1 y_i = -202.30$ $\Sigma P_2 y_i = -1{,}117.5$ $\Sigma P_3 y_i = 445.10$
$\Sigma P_4 y_i = 525.10$

Table 8.12 tells us that the quadratic polynomial adequately describes the data if we use the 2 per cent value of F as the index.

The polynomial can, therefore, be written as

$$Y = \beta_0 + \beta_1 x + \beta_2 x^2 + e$$

The abbreviated Doolittle technique can be used to estimate β_i and obtain standard errors for β_i.

8.4.8 Summary. If the x's are equally spaced and it is desired to find the degree polynomial that adequately represents a given set of data, pick the desired significance level and use the method of orthogonal polynomials presented in this section. It is advisable to obtain two nonsignificant results before deciding on the degree polynomial to use. This is because, if a cubic fits the data, a linear is quite likely to make a significant reduction, but a quadratic may contribute nothing. In fact, when any even-degree polynomial fits the data, any odd-degree polynomial is likely to contribute nothing; hence, if we stop at the first nonsignificant result we may miss the important results.

TABLE 8.12 ANALYSIS OF VARIANCE OF RAINFALL DATA

SV	DF	SS	MS	F	Pr
Total	12	12,815.99			
Reduction for mean	1	12,538.87			
Remainder from mean	11	277.12			
Linear	1	71.55	71.55	3.48	>2%
Error for linear	10	205.57	20.56		
Quadratic	1	103.96	103.96	9.21	<2%
Error for quadratic	9	101.61	11.29		
Cubic	1	38.48	38.48	4.88	>2%
Error for cubic	8	63.13	7.89		
Quartic	1	34.43	34.43	8.40	>2%
Error for quartic	7	28.70	4.10		

If we desire to find the *degree* polynomial that represents a given set of data and to *estimate* the parameters in the polynomial, the procedure is (1) to use the results of this section to find the desired degree polynomial, and (2) to use the results of Sec. 8.2 (the results of Chap. 6) to estimate the parameters in the polynomial or to test hypotheses about them.

8.5 Repeated Observations for Each X

Suppose that the observed random variables y have the following structure:

$$y_{ij} = f(x_i) + e_{ij} \qquad \begin{array}{l} j = 1, 2, \ldots, m; m > 1 \\ i = 1, 2, \ldots, k \end{array}$$

where we postulate that $f(x)$ is a function which can be expanded into a Taylor series, and e_{ij} are uncorrelated normal random variables with means zero and variances σ^2. The problem is to find a polynomial that adequately represents $f(x)$. This model may be realistic in a situation where the experimenter can control the points x_i where the observations y_{ij} are taken. The model assumes that at each x_i point, m ($m > 1$) values of y_{ij} are obtained. Thus we can get an unbiased estimate of σ^2 by using the quantity

$$\hat{\sigma}^2 = \frac{1}{k(m-1)} \sum_{i=1}^{k} \sum_{j=1}^{m} (y_{ij} - y_{i.})^2 = \frac{1}{k(m-1)} \sum_{i=1}^{k} \sum_{j=1}^{m} (e_{ij} - \bar{e}_{i.})^2$$

Also $\dfrac{k(m-1)\hat{\sigma}^2}{\sigma^2}$ is distributed as $\chi^2[k(m-1)]$ regardless of what function $f(x)$ is.

The computations proceed as outlined in this chapter except that in each table the error sum of squares can now be broken into two parts: one term called the lack of fit, the other the error sum of squares $k(m-1)\hat{\sigma}^2$. The error sum of squares is easily computed, and so it is subtracted from the remainder to obtain the lack of fit term.

If the model is actually linear, then the mean square lack of fit term, which is the error sum of squares for linear minus $k(m-1)\hat{\sigma}^2$ all divided by the d.f., is an unbiased estimate of σ^2. If the model is *not* linear, then this term is on the average larger than σ^2. The lack of fit mean square divided by $\hat{\sigma}^2$ is distributed as F if the model is linear. If the hypothesis is rejected, then the lack of fit term for quadratic can be used in the numerator and $\hat{\sigma}^2$ in the denominator of the F to test for quadratic. This process is continued until the proper fit is concluded.

Problems

8.1 It is known that the following data fit a cubic model. Set a 95 per cent confidence interval on each of the coefficients γ_0, γ_1, γ_2, and γ_3.

y	.5	3.0	1.6	1.3	.2	.5	−.1	1.2
x	1.0	1.5	2.0	2.5	3.0	3.5	4.0	4.5

8.2 In the example of Art. 8.3.2, fit a cubic and quartic polynomial by using the abbreviated Doolittle method, and interpret the results.

8.3 In the model $y_i = \beta_0 + \beta_1 x_i + \beta_2 x_i^2 + e_i$, find the maximum-likelihood estimate of the maximum or minimum value of $E(y)$ (assume $i = 1, 2, \ldots, n$).

8.4 In an experiment on corrosion of metal the following data were collected:

Voltage applied	1.5	2.0	2.5	3.0	3.5	4.0	4.5	5.0
Corrosion, %	1.10	1.43	2.11	3.12	2.50	2.21	2.50	5.90

(*a*) Find the degree of polynomial that describes the data.

(*b*) Estimate the coefficients in this polynomial by using the abbreviated Doolittle method.

(*c*) Find the standard errors of the coefficients.

8.5 Use the results of this chapter to find the first three orthogonal polynomials for $n = 5$.

8.6 In a quadratic polynomial model $y = \beta_0 + \beta_1 x + \beta_2 x^2 + e$, find the coefficients γ_i in terms of the coefficients β_i if we use the model $y = \gamma_0 + \gamma_1 z + \gamma_2 z^2 + e$, where z is a coded value of x; that is, $z = x - h$, where h is known constant.

Further Reading

1 R. L. Anderson and E. E. Houseman: Tables of Orthogonal Polynomials Values Extended to $N = 104$, *Iowa State Coll. Agri. Exp. Sta. Bul. No.* 297, April, 1942.

2 A. M. Mood: "Introduction to the Theory of Statistics," McGraw-Hill Book Company, Inc., New York, 1950.

3 M. G. Kendall: "The Advanced Theory of Statistics," vols. I, II, Charles Griffin & Co., Ltd., London, 1946.

4 R. L. Anderson and T. A. Bancroft: "Statistical Theory in Research," McGraw-Hill Book Company, Inc., New York, 1952.

9

Model 2: Functional Relationships

9.1 Introduction and Definitions

In this chapter we shall consider model 2, in which there exists a functional relationship among mathematical variables, variables that cannot be observed owing to errors of measurement. For example, suppose distance S and time T are related by the model $S = \alpha + \beta T$, where α is the distance at time $T = 0$ and β is the velocity. Now suppose that S and T are not observable but that s and t can be observed, where $s = S + e$ and $t = T + d$, where d and e are measuring errors. We can write the relationship as

$$s = \alpha + \beta t + (e - \beta d)$$

It may appear at first sight that we can set the random variable $e - \beta d$ equal to b, write the model as $s = \alpha + \beta t + b$, and obtain a case of model 1. However, in model 1 the quantity analogous to t is specified to be a mathematical variable, whereas here t is a random variable and t is not independent of the error b; therefore, this relation does not fit into the framework of model 1.

The fact that b and t are correlated introduces a great many complexities. In most of the discussion of model 2 we shall consider only the case of one independent and one dependent variable. The results will generalize. There may be many special cases of model 2, but we shall discuss only those listed below.

Throughout this chapter the quantities Y, X, U, and V will be *unobservable* mathematical quantities, and y and x will be *observable* random variables, where $y = Y + e$ and $x = X + d$. Also, a, b, d, and e will be *unobservable* random errors with means 0 and variances σ_a^2, σ_b^2, σ_d^2, and σ_e^2, respectively. The functional relationship between the unobservable mathematical variables X and Y will be given by $Y = \alpha + \beta X$.

Given a sample of n pairs of observed values $(x_1, y_1), (x_2, y_2), \ldots, (x_n, y_n)$, the problem is to estimate α, β, σ_e^2, and σ_d^2 and to test certain hypotheses about these parameters.

We shall assume that the e_i and d_i are independent normal variables, and we shall assume further that we have one of the following two cases:

1. The case in which the ratio $\lambda = \sigma_d^2/\sigma_e^2$ is known
2. A controlled-independent-variable model

We must make some such assumption because otherwise the problems associated with estimating the parameters are very difficult; a complete solution does not seem to have been obtained up to the present time. In other words, although we may prefer not to assume either case 1 or 2, we find that we must do so, since the theory is not yet sufficiently developed to do without them. There are other assumptions for which solutions have been obtained, but we shall discuss only the two listed above.

Let us consider first the assumption of case 1. An experimenter may never know the exact value of λ; however, he may feel quite strongly that the variance of the measurement error in Y is of the same order of magnitude as the variance of the measurement error in X. If this is so, then λ can be taken equal to 1. Or the experimenter may have a "good" estimate of λ from previous theory or experience.

We consider next the controlled-independent-variable model, case 2. This concept is due to Berkson [2] and serves to make the problem tractable. Essentially, the conditions for this type of model are that the experimenter must decide beforehand what values of x_i he is going to observe and then obtain them. In other words, in the illustration given at the beginning of this chapter, the experimenter might, for example, decide that he will measure his distance s at times $t = 6, 8$, and 10, respectively. Clearly, the true times T will not equal 6, 8, and 10, respectively. This does not matter. It is, however, important that readings for s be taken when his instrument reads $t = 6, 8$, and 10. This is why it is called the *controlled-independent-variable* model.

These assumptions seem to be quite distinct from each other, and in fact they are. The experimenter must decide which (if either) seems plausible in each particular situation. We shall notice that the inferences drawn will in general be different for the two cases.

9.2 Point Estimation

9.2.1 The General Procedure. The problem is as follows: There is a functional relationship $Y_i = \alpha + \beta X_i$ relating the mathematical quantities Y_i and X_i. However, Y_i and X_i cannot be observed, because of measurement errors; y_i and x_i are observed, where $y_i = Y_i + e_i$ and $x_i = X_i + d_i$. The e_i are independent normal variables with mean 0 and variance σ_e^2. The d_i are similarly distributed with variance σ_d^2. We shall attempt to find maximum-likelihood estimates of α, β, σ_d^2, and σ_e^2. The likelihood function is

$$L = f(e_1,e_2, \ldots ,e_n,d_1,d_2, \ldots ,d_n) = \frac{1}{(\sigma_e\sigma_d 2\pi)^n} \exp\left[-\frac{1}{2}\Sigma\left(\frac{e_i^2}{\sigma_e^2} + \frac{d_i^2}{\sigma_d^2}\right)\right] \tag{9.1}$$

Substituting for e_i and d_i, we get

$$L = \frac{1}{(\sigma_e\sigma_d 2\pi)^n} \exp\left\{-\frac{1}{2}\left[\frac{\Sigma(y_i - Y_i)^2}{\sigma_e^2} + \frac{\Sigma(x_i - X_i)^2}{\sigma_d^2}\right]\right\}$$

Substituting $\alpha + \beta X_i$ for Y_i and taking logarithms,

$$\log L = -\frac{n}{2}\log \sigma_e^2 - \frac{n}{2}\log \sigma_d^2 - n \log 2\pi - \frac{1}{2}\left[\frac{\Sigma(y_i - \alpha - \beta X_i)^2}{\sigma_e^2} + \frac{\Sigma(x_i - X_i)^2}{\sigma_d^2}\right]$$

There are $n + 4$ unknown parameters, namely, $X_1, X_2, \ldots, X_n, \alpha, \beta, \sigma_e^2$, and σ_d^2. Taking the derivative with respect to these unknown parameters,

$$\frac{\partial \log L}{\partial X_t} = \frac{(y_t - \hat{\alpha} - \hat{\beta}\hat{X}_t)\hat{\beta}}{\hat{\sigma}_e^2} + \frac{x_t - \hat{X}_t}{\hat{\sigma}_d^2} = 0 \qquad t = 1, 2, \ldots, n \tag{9.2}$$

$$\frac{\partial \log L}{\partial \beta} = \frac{\Sigma(y_i - \hat{\alpha} - \hat{\beta}\hat{X}_i)\hat{X}_i}{\hat{\sigma}_e^2} = 0 \tag{9.3}$$

$$\frac{\partial \log L}{\partial \alpha} = \frac{\Sigma(y_i - \hat{\alpha} - \hat{\beta}\hat{X}_i)}{\hat{\sigma}_e^2} = 0 \tag{9.4}$$

$$\frac{\partial \log L}{\partial \sigma_e^2} = \frac{-n}{2\hat{\sigma}_e^2} + \frac{1}{2\hat{\sigma}_e^4}\sum (y_i - \hat{\alpha} - \hat{\beta}\hat{X}_i)^2 = 0 \tag{9.5}$$

$$\frac{\partial \log L}{\partial \sigma_d^2} = \frac{-n}{2\hat{\sigma}_d^2} + \frac{1}{2\hat{\sigma}_d^4}\Sigma(x_i - \hat{X}_i)^2 = 0 \tag{9.6}$$

Solving Eqs. (9.5) and (9.6), we obtain

$$\hat{\sigma}_e^2 = \frac{1}{n}\Sigma(y_i - \hat{\alpha} - \hat{\beta}\hat{X}_i)^2 \tag{9.7}$$

$$\hat{\sigma}_d^2 = \frac{1}{n}\Sigma(x_i - \hat{X}_i)^2 \tag{9.8}$$

If we sum the n equations in (9.2), we get

$$\hat{\beta}\,\frac{\Sigma(y_t - \hat{\alpha} - \hat{\beta}\hat{X}_t)}{\hat{\sigma}_e^2} + \frac{\Sigma(x_t - \hat{X}_t)}{\hat{\sigma}_d^2} = 0 \tag{9.9}$$

If we substitute Eq. (9.4) into the first term of Eq. (9.9), then (9.9) becomes

$$\Sigma(x_t - \hat{X}_t) = 0 \tag{9.10}$$

If we transfer the second term of (9.2) to the right-hand side of the equality sign, we get

$$\frac{(y_t - \hat{\alpha} - \hat{\beta}\hat{X}_t)\hat{\beta}}{\hat{\sigma}_e^2} = -\frac{x_t - \hat{X}_t}{\hat{\sigma}_d^2} \tag{9.11}$$

Squaring both sides and then summing,

$$\frac{\hat{\beta}^2}{\hat{\sigma}_e^4}\Sigma(y_t - \hat{\alpha} - \hat{\beta}\hat{X}_t)^2 = \frac{\Sigma(x_t - \hat{X}_t)^2}{\hat{\sigma}_d^4} \tag{9.12}$$

Substituting (9.7) and (9.8) into (9.12),

$$\frac{n\hat{\beta}^2\hat{\sigma}_e^2}{\hat{\sigma}_e^4} = \frac{n\hat{\sigma}_d^2}{\hat{\sigma}_d^4} \tag{9.13}$$

or

$$\hat{\sigma}_d^2\hat{\beta}^2 = \hat{\sigma}_e^2 \tag{9.14}$$

This result is not satisfactory, since it does not seem reasonable that the ratio of the variances should be equal to β^2. Apparently, we must have some information about the unknown parameters if we are to get a reasonable solution to the likelihood equations. The one assumption that an experimenter may be willing to make is that the ratio of the two variances is known; i.e., that λ is known, where $\lambda = \sigma_d^2/\sigma_e^2$. This is case 1 and will be discussed in the next article.

9.2.2 Case 1: $\lambda = \sigma_d^2/\sigma_e^2$ Known. If we assume that λ is known, we can substitute $\sigma_d^2 = \lambda\sigma_e^2$ in the likelihood equation (9.1). The likelihood equation now becomes

$$\log L = -\frac{n}{2}\log \sigma_e^2 - \frac{n}{2}\log \lambda\sigma_e^2 - n\log 2\pi - \frac{1}{2}\left[\frac{\Sigma(y_i - \alpha - \beta X_i)^2}{\sigma_e^2} + \frac{\Sigma(x_i - X_i)^2}{\lambda\sigma_e^2}\right] \tag{9.15}$$

We now have $n + 3$ unknown parameters $X_1, X_2, \ldots, X_n, \alpha, \beta$, and

σ_e^2. Taking the partial derivatives with respect to each of these $n + 3$ parameters, we obtain

$$\frac{\partial \log L}{\partial X_t} = \frac{(y_t - \hat{\alpha} - \hat{\beta}\hat{X}_t)\hat{\beta}}{\hat{\sigma}_e^2} + \frac{x_t - \hat{X}_t}{\lambda \hat{\sigma}_e^2} = 0 \qquad t = 1, 2, \ldots, n \tag{9.16}$$

$$\frac{\partial \log L}{\partial \sigma_e^2} = -\frac{n}{\hat{\sigma}_e^2} + \frac{1}{2\hat{\sigma}_e^4}\left[\Sigma(y_i - \hat{\alpha} - \hat{\beta}\hat{X}_i)^2 + \frac{\Sigma(x_i - \hat{X}_i)^2}{\lambda}\right] = 0 \tag{9.17}$$

$$\frac{\partial \log L}{\partial \beta} = \frac{\Sigma(y_i - \hat{\alpha} - \hat{\beta}\hat{X}_i)\hat{X}_i}{\hat{\sigma}_e^2} = 0 \tag{9.18}$$

$$\frac{\partial \log L}{\partial \alpha} = \frac{\Sigma(y_i - \hat{\alpha} - \hat{\beta}\hat{X}_i)}{\hat{\sigma}_e^2} = 0 \tag{9.19}$$

From (9.19) we get

$$\bar{y} - \hat{\alpha} - \frac{\hat{\beta}}{n}\Sigma\hat{X}_i = 0$$

and from (9.16)

$$\left(\hat{\alpha} + \frac{\hat{\beta}}{n}\Sigma\hat{X}_i - \bar{y}\right)\lambda\hat{\beta} = \frac{\Sigma(x_i - \hat{X}_i)}{n}$$

which give $\Sigma(x_i - \hat{X}_i) = 0$ and, finally,

$$\hat{\alpha} = \bar{y} - \hat{\beta}\bar{x} \tag{9.20}$$

Solving (9.16) for $\hat{X}_t$ gives

$$\hat{X}_t = \frac{x_t + \lambda\hat{\beta}y_t - \lambda\hat{\beta}\hat{\alpha}}{1 + \lambda\hat{\beta}^2} \tag{9.21}$$

From (9.18) we get

$$\hat{\beta} = \frac{\Sigma(y_i - \hat{\alpha})\hat{X}_i}{\Sigma(\hat{X}_i)^2}$$

Substituting (9.20) and (9.21) into this last equation, after some simplification, gives

$$\hat{\beta}^2[\lambda\Sigma(y_i - \bar{y})(x_i - \bar{x})] + \hat{\beta}[\Sigma(x_i - \bar{x})^2 - \lambda\Sigma(y_i - \bar{y})^2] - \Sigma(y_i - \bar{y})(x_i - \bar{x}) = 0$$

This is a quadratic in $\hat{\beta}$ and, when solved, gives

$$\hat{\beta} = \pm\left[U^2 + \frac{1}{\lambda}\right]^{\frac{1}{2}} + U \tag{9.22}$$

where
$$U = \frac{\Sigma(y_i - \bar{y})^2 - (1/\lambda)\Sigma(x_i - \bar{x})^2}{2\Sigma(x_i - \bar{x})(y_i - \bar{y})}$$

and where the sign used for the first term is that which will maximize the likelihood function. If we substitute into (9.17), we get

$$\hat{\sigma}_e^2 = \frac{1}{2n} \Sigma(y_i - \bar{y})[(y_i - \bar{y}) - \hat{\beta}(x_i - \bar{x})] \tag{9.23}$$

It can be shown that this is not a consistent estimate of σ_e^2. If it is to be consistent, the denominator must be of the order of n instead of $2n$. Accordingly, we let the denominator be equal to $2n$, the number of degrees of freedom, minus $n + 2$, the number of unknown parameters (that is, α, β, and the X_i), and we get $2n - (n + 2) = n - 2$. Using this denominator, we obtain

$$\hat{\sigma}_e^2 = \frac{1}{n-2}[\Sigma(y_i - \bar{y})^2 - \hat{\beta}\Sigma(x_i - \bar{x})(y_i - \bar{y})]$$

We have the following theorem:

◆ **Theorem 9.1** If y_i, x_i, Y_i, X_i satisfy the conditions in Sec. 9.1 and if $\lambda = \sigma_d^2/\sigma_e^2$ is known, then the maximum-likelihood estimates of σ_e^2, α, and β (the divisor of $\hat{\sigma}_e^2$ is changed) are

$$\hat{\beta} = \pm\left(U^2 + \frac{1}{\lambda}\right)^{\frac{1}{2}} + U \qquad \text{where } U = \frac{\Sigma(y_i - \bar{y})^2 - (1/\lambda)\Sigma(x_i - \bar{x})^2}{2\Sigma(y_i - \bar{y})(x_i - \bar{x})}$$

$$\hat{\alpha} = \bar{y} - \hat{\beta}\bar{x}$$

$$\hat{\sigma}_e^2 = \frac{1}{n-2}[\Sigma(y_i - \bar{y})^2 - \hat{\beta}\Sigma(y_i - \bar{y})(x_i - \bar{x})]$$

These estimates are consistent estimates of their respective parameters.

9.2.3 Example. Suppose the following values of y_i, x_i are recorded under the assumptions of the preceding article (assume $\lambda = 1$):

y_i	6.3	7.9	8.2	9.4	10.8	11.5	12.9
x_i	11.1	9.9	8.2	7.3	6.1	6.0	4.2

We get $\Sigma(x_i - \bar{x})^2 = 34.337$; $\Sigma(y_i - \bar{y})^2 = 31.714$; $\Sigma(y_i - \bar{y})(x_i - \bar{x}) = -32.311$; $n = 7$; $\bar{x} = 7.543$; $\bar{y} = 9.571$; $U = .040589$; $\hat{\beta} = -.960$; $\hat{\alpha} = 16.81$; $\hat{\sigma}_e^2 = .139$.

9.2.4 Case 2: Controlled-independent-variable Model. Let the functional relationship be $V = \alpha + \beta U$, where capital letters stand for mathematical quantities and lower-case letters for random variables. We assume that V and U cannot be observed but that we can observe x and y, where

$$y = V + e \qquad x = U + d \tag{9.24}$$

where d and e are errors of measurement. In case 1 we assumed that e and d were normal and independent; since U is a mathematical variable, this would make d and x functionally related. For case 2, however, we shall assume that x is set equal to predetermined quantities X_i and, hence, that x takes the role of a mathematical variable. This forces U to be a random variable; hence V also is a random variable. We shall rewrite (9.24)

$$y = v + e \qquad X = u + d \qquad v = \alpha + \beta u \tag{9.25}$$

and assume that e is independent of u, v, and d but that v, u, and d are functionally related. The physical meaning of this assumption is this: We must decide beforehand what values x will take (we denote these values by X) and then bring the recording instrument to those values and measure y. Accordingly, X is termed a *controlled variable*.

We can now rewrite $v = \alpha + \beta u$

$$y - e = \alpha + \beta(X - d) \qquad \text{or} \qquad y = \alpha + \beta X + (e - \beta d) \tag{9.26}$$

If we let $f = e - \beta d$, we get

$$y = \alpha + \beta X + f$$

and this fits the definition of model 1 in Chap. 6. Hence the theory in Chap. 6 applies.

If we have the more general functional relationship

$$\begin{aligned} V &= \alpha + \beta_1 U_1 + \beta_2 U_2 + \cdots + \beta_k U_k \\ \text{and if} \qquad y &= V + e \qquad x_i = U_i + d_i \qquad i = 1, 2, \ldots, k \end{aligned} \tag{9.27}$$

we can control x_i (i.e., set $x_i = X_i$) and get

$$\begin{aligned} v &= \alpha + \beta_1 u_1 + \cdots + \beta_k u_k \\ y &= v + e \qquad X_i = u_i + d_i \qquad i = 1, 2, \ldots, k \end{aligned} \tag{9.28}$$

Substituting, we get

$$y = \alpha + \beta_1 X_1 + \beta_2 X_2 + \cdots + \beta_k X_k + g \tag{9.29}$$

where g is a random error such that

$$g = e - \beta_1 d_1 - \beta_2 d_2 - \cdots - \beta_k d_k$$

The theory in Chap. 6 now applies to this model.

If the U_i are not linear, the theory in Chap. 6 may not apply. For example, if the functional relationship is

$$V = \alpha + \beta_1 U + \beta_2 U^2$$

we substitute $X - d$ for U, $y - e$ for V, and get

$$y - e = \alpha + \beta_1(X - d) + \beta_2(X - d)^2$$

which becomes

$$y = \alpha + \beta_1 X + \beta_2 X^2 + (e - \beta_1 d + \beta_2 d^2 - 2\beta_2 X d)$$

But this does not fit the definition of model 1, since the error term $e - \beta_1 d + \beta_2 d^2 - 2\beta_2 X d$ depends on X.

9.2.5 Example. Suppose an experimenter has a resistor that he wants to measure. He assumes that the law $E = RI$ holds, where E is the voltage of the circuit and I is the current measured in amperes. He cannot measure E and I exactly, but he can observe $y = E + e$ and $x = I + d$. To use the controlled-independent-variable method, he decides he will bring his ammeter to the following values: $X = 10, 15, 20, 25$, and 30, and read the corresponding voltages y. He observes a voltage for each of the ampere readings and gets a set of values:

X	10.0	15.0	20.0	25.0	30
y	3.1	4.8	7.2	8.4	9.7

From these he computes $\hat{R} = \Sigma y_i X_i / \Sigma X_i^2 = .33$; this estimate of R has all the properties listed in the consequent of Theorem 6.1.

9.3 Interval Estimation and Tests of Hypotheses

The only case we shall discuss for interval estimation and tests of hypotheses is the controlled-variable model, case 2. The results of Chap. 6 are applicable. It should be noticed that the variance of y in this case is equal to $\sigma_e^2 + \beta^2 \sigma_d^2$. This appears in the noncentrality parameter in tests of hypotheses, and the power function is affected.

Problems

9.1 The functional relationship $Y = \alpha + \beta X$ is known to hold. The mathematical variables Y and X cannot be observed, but the following values of y_i and x_i are observed, where $y_i = Y_i + e_i$ and $x_i = X_i + d_i$:

y_i	10.3	12.6	11.2	15.3	18.1	20.0
x_i	3.2	3.3	5.1	7.9	10.3	12.0

Assuming $\lambda = \sigma_d^2/\sigma_e^2 = 1$, find the maximum-likelihood estimates of α, β, and σ_e^2.

9.2 Give an example of a functional relationship from some scientific field where there are errors in measuring all variables.

9.3 Discuss the physical interpretation to be put on the *controlled-variable* model.

9.4 In an experiment designed to measure the density of gold, the following values were obtained (w is weight, v is volume):

w	19.0	44.0	60.7	74.0	82.0	90.3	109.0	156.0
v	1.0	2.2	3.1	3.8	4.1	4.7	5.8	8.0

The functional relationship is $W = \delta V$, where W is weight in grams and V is volume in cubic centimeters. However, since there are errors of measurement, w and v were recorded. Assuming $\lambda = 1$, find the maximum-likelihood estimate of the density δ.

9.5 In Prob. 9.4, assume that there is no error in measuring V, and find the maximum-likelihood estimate of δ. Notice the difference between the two estimates.

9.6 Prove that $\hat{\sigma}_e^2$ is greater than or equal to zero in Theorem 9.1.

9.7 For the quantities in Art. 9.2.2, prove that $\sum_{i=1}^{n} (x_i - \hat{X}_i)\hat{X}_i = 0$.

9.8 For the quantities in Art. 9.2.2, prove that $\Sigma(y_i - \hat{\alpha} - \hat{\beta}x_i)\hat{X}_i = 0$.

9.9 For the quantities in Art. 9.2.2, prove that Eq. (9.23) can be written

$$\hat{\sigma}_e^2 = \frac{1}{2n} \frac{\Sigma[(y_i - \bar{y}) - \hat{\beta}(x_i - \bar{x})]^2}{1 + \lambda\hat{\beta}^2}$$

Further Reading

1 O. L. Davies: "Design and Analysis of Industrial Experiments," Oliver and Boyd, Ltd., London, 1954.

2 J. Berkson: Are There Two Regressions?, *J. Am. Statist. Assoc.*, vol. 45, pp. 164–180, 1950.

3 E. S. Keeping: Note on Wald's Method of Fitting a Straight Line when Both Variables Are Subject to Error, *Biometrics*, vol. 12, pp. 445–448, 1956.

4 M. S. Bartlett: Fitting a Straight Line when Both Variables Are Subject to Error, *Biometrics*, vol. 5, pp. 207–212, 1949.

5 E. L. Scott: Note on Consistent Estimates of the Linear Structural Relation between Two Variables, *Ann. Math. Statist.*, vol. 21, pp. 284–288, 1950.

6 T. C. Koopmans and O. Reiersol: The Identification of Structural Characteristics, *Ann. Math. Statist.*, vol. 27, pp. 165–181, 1956.

7 M. G. Kendall: Regression, Structure and Functional Relationship, *Biometrika*, parts I and II, vol. 39, pp. 96–108, 1952.

8 R. C. Geary: Non Linear Functional Relationship between Two Variables when One Variable Is Controlled, *J. Am. Statist. Assoc.*, vol. 48, pp. 94–103, 1953.

9 A. Wald: The Fitting of Straight Lines if Both Variables Are Subject to Error, *Ann. Math. Statist.*, vol. 11, pp. 284–300, 1940–1941.

10 D. V. Lindley: Estimation of a Functional Relationship, *Biometrika*, vol. 40, pp. 47–49, 1953.

10

Model 3: Regression Models

10.1 Introduction

In this chapter we shall discuss what is often called *multiple regression*. If $y, x_1, x_2, \ldots, x_k$ are random variables whose joint distribution is given by $f(y, x_1, x_2, \ldots, x_p)$ and if the expected value of y in the conditional distribution $f(y \mid x_1 = X_1, x_2 = X_2, \ldots, x_k = X_k)$ is equal to $\beta_0 + \beta_1 X_1 + \beta_2 X_2 + \cdots + \beta_k X_k$, we shall say that there is a *linear regression of y on* $x_1, x_2, \ldots, x_k$ and call this model 3. We can write

$$\tilde{y} = \beta_0 + \beta_1 X_1 + \cdots + \beta_k X_k + e$$

where the β_i are unknown parameters, where X_i are particular (known) values of the random variables x_i, and where e is a random variable with mean zero.

This model is similar to model 1, the difference being that in model 1 the X_i are mathematical variables and not particular values of random variables. This difference may seem slight at first, but it is quite important when we consider its consequences. For example, in model 1 there are np values of X_i involved. The probability of the confidence statements and tests of hypotheses was based on the fact that repeated values of y_i were to be drawn from *these same np* values of X_i. We shall now extend this inference to other X_i by using model 3. Also, correlation was not defined for model 1 (correlation is not defined for mathematical variables), but it will be defined and used for the random variables in model 3.

In the functional form $s = vt$, for example, it is difficult to imagine the time t playing the part of a random variable. This would be possible, but it seems as if we should want to calculate the distance s for preselected values of t. Also, it is clear that, if t is a random variable, then s must also be a random variable, and the correlation between the two quantities is equal to unity. On the other hand, consider the following examples, which seem to be different in character from the above:

1. The relationship among height, weight, arm length, etc., of 12-year-old children
2. The relationship among maximum temperature, maximum humidity, etc., for each hour of the year
3. The relationship among per cent calcium, per cent protein, per cent butterfat, etc., of the milk of a certain cow for each day of the year

In these examples there would in general be no attempt to find *functional* relationships among the various factors, as in the equation $s = vt$. In these examples we would be interested rather in *predicting* one factor from the other, realizing that there is available no function that exactly relates the various quantities. We shall return later to this problem of *prediction* versus *functional form*. It may seem reasonable to assume that the quantities in 1, 2, or 3 above have a multivariate distribution and that one of the factors has a linear regression on the others. If this is true, then we can use model 3 and the results of this chapter to find this linear prediction function.

It seems worth repeating that in model 3 there is in general no *functional* relationship among the factors but there is a *linear regression*. This regression function can be used for *predicting*.

10.1.1 Definitions. Although model 3 would ordinarily be defined for $k + 1$ random variables $y, x_1, x_2, \ldots, x_k$, we shall study only the two-variable case, denoted simply by y, x. All the results will generalize to the $k + 1$ situation, and we shall state some general theorems but omit the proofs. Therefore, throughout this chapter, lower-case letters y_i, x_i will denote random variables from a bivariate distribution. The linear regression function of y on x will be denoted by $\tilde{y}$, and $E(\tilde{y})$ will be denoted by Y; so $Y = \beta_0 + \beta_1 X$, where capital letters Y and X are *not* random variables.

We shall consider the following three special cases of model 3:

Case 1. The random variables y and x have a bivariate *normal* distribution.

Case 2. The random variables y and x have a bivariate *nonnormal* distribution, but the conditional distribution $g(y \mid x = X)$ is

$$\frac{1}{\sigma\sqrt{2\pi}} e^{-(1/2\sigma^2)(y-\beta_0-\beta_1 X)^2}$$

and the marginal distribution of x does not depend on the unknown parameters β_0, β_1, and σ^2. This means that the joint distribution of y and x can be written

$$f(y,x) = h(x)\frac{1}{\sigma\sqrt{2\pi}} e^{-(1/2\sigma^2)(y-\beta_0-\beta_1 x)^2} \qquad -\infty < y < \infty; \quad -\infty < x < \infty$$

where $h(x)$ does not depend on σ^2, β_0, β_1.

Case 3. The random variables x and y have a bivariate normal distribution, but there also are errors of measurement involved. Hence we cannot observe y and x, but we do observe u and v, where $u = x + d$ and $v = y + e$, where e and d are independent normal variables which are also independent of y and x.

We shall discuss these three cases in turn, but first we shall prove some theorems concerning the multivariate normal distribution. These will be an extension of the theory in Chap. 3.

10.2 Case 1: The Multivariate Normal

10.2.1 Introduction. We shall use the notation introduced in Chap. 3. In this article we shall find point and interval estimates of the parameters in the multivariate normal and give certain interesting tests of hypotheses. In most cases the proofs will be given only for the bivariate case.

10.2.2 Point Estimation. Let us assume that the vector $\begin{pmatrix} x \\ y \end{pmatrix}$ is distributed as a bivariate normal distribution with means $\begin{pmatrix} \mu_1 \\ \mu_2 \end{pmatrix}$ and covariance matrix

$$\mathbf{V} = \begin{pmatrix} \sigma_{11} & \sigma_{12} \\ \sigma_{21} & \sigma_{22} \end{pmatrix}$$

If we select n vectors $\begin{pmatrix} x_i \\ y_i \end{pmatrix}$ for $i = 1, 2, \ldots, n$ from this distribution, the likelihood equation is

$$f = f(y_1,x_1,y_2,x_2, \ldots ,y_n,x_n) = \frac{e^{-Q/2(1-\rho^2)}}{[2\pi\sqrt{(1-\rho^2)\sigma_{11}\sigma_{22}}]^n}$$

$$\text{where } Q = \frac{\Sigma(x_i - \mu_1)^2}{\sigma_{11}} - 2\sigma_{12}\frac{\Sigma(x_i - \mu_1)(y_i - \mu_2)}{\sigma_{11}\sigma_{22}} + \frac{\Sigma(y_i - \mu_2)^2}{\sigma_{22}}$$

where $\rho^2 = \sigma_{12}^2/\sigma_{11}\sigma_{22}$. Solving the equations

$$\frac{\partial f}{\partial \mu_1} = 0 \qquad \frac{\partial f}{\partial \mu_2} = 0 \qquad \frac{\partial f}{\partial \sigma_{11}} = 0 \qquad \frac{\partial f}{\partial \sigma_{22}} = 0 \qquad \frac{\partial f}{\partial \sigma_{12}} = 0$$

gives the maximum-likelihood estimators (the last three quantities are divided by $n - 1$ instead of n, to correct for bias)

$$\hat{\mu}_1 = \frac{1}{n} \Sigma x_i = \bar{x}$$

$$\hat{\mu}_2 = \frac{1}{n} \Sigma y_i = \bar{y}$$

$$\hat{\sigma}_{11} = \frac{1}{n-1} \Sigma(x_i - \bar{x})^2 \tag{10.1}$$

$$\hat{\sigma}_{22} = \frac{1}{n-1} \Sigma(y_i - \bar{y})^2$$

$$\hat{\sigma}_{12} = \frac{1}{n-1} \Sigma(x_i - \bar{x})(y_i - \bar{y})$$

From Theorem 3.11 we know that

$$E(y \mid x = X) = \mu_2 - \frac{\sigma_{12}}{\sigma_{11}} \mu_1 + \frac{\sigma_{12}}{\sigma_{11}} X$$

which we shall write as $Y = \beta_0 + \beta_1 X$. So we shall let

$$\hat{\beta}_1 = \frac{\hat{\sigma}_{12}}{\hat{\sigma}_{11}} \qquad \text{and} \qquad \hat{\beta}_0 = \hat{\mu}_2 - \frac{\hat{\sigma}_{12}}{\hat{\sigma}_{11}} \hat{\mu}_1 \tag{10.2}$$

since

$$\beta_1 = \frac{\sigma_{12}}{\sigma_{11}} \qquad \text{and} \qquad \beta_0 = \mu_2 - \frac{\sigma_{12}}{\sigma_{11}} \mu_1$$

Since $\hat{\mu}_1$, $\hat{\mu}_2$, $\hat{\sigma}_{11}$, $\hat{\sigma}_{22}$, and $\hat{\sigma}_{12}$ are maximum-likelihood estimators, they are consistent and efficient. It can also be shown that they form a set of sufficient estimators and they are complete and unbiased. We shall sum up this information in the following theorem.

◆ **Theorem 10.1** If (x_i, y_i) for $i = 1, 2, \ldots, n$ represents a random sample from the bivariate normal distribution, the maximum-likelihood estimates (corrected for bias) of $\mu_1, \mu_2, \sigma_{11}, \sigma_{22}, \sigma_{12}$ given in (10.1) have the following properties:

(1) Consistency
(2) Efficiency
(3) Completeness
(4) Minimum variance unbiased
(5) Sufficiency
(6) That $\hat{\mu}_1$ be independent of $\hat{\sigma}_{11}$, $\hat{\sigma}_{22}$, $\hat{\sigma}_{12}$

(7) That $\hat{\mu}_2$ be independent of $\hat{\sigma}_{11}$, $\hat{\sigma}_{22}$, $\hat{\sigma}_{12}$

(8) That $\hat{\mu}_1$, $\hat{\mu}_2$ be jointly distributed as the bivariate normal with mean μ_1 and μ_2 and covariance matrix $(1/n)\mathbf{V}$

We shall examine $\hat{\beta}_0$ and $\hat{\beta}_1$ to show that they are unbiased. To do this we shall prove the following lemma.

◆ **Lemma 10.1** Let $f(x,y)$ be the joint distribution of two random variables x and y. Then

$$E[h(x,y)] = E_x\{E_{y|x}[h(x,y)]\}$$

where $h(x,y)$ is any function such that $E[h(x,y)]$ exists and where the notation $E_{y|x}$ denotes the expected value in the *conditional* distribution of y given x and E_x the expected value in the *marginal* distribution of x.

Proof: Let $f(x,y) = k(y \mid x)g(x)$.

$$E[h(x,y)] = \int_{-\infty}^{\infty}\int_{-\infty}^{\infty} h(x,y)f(x,y)\,dx\,dy$$

$$= \int_{-\infty}^{\infty}\int_{-\infty}^{\infty} h(x,y)k(y \mid x)g(x)\,dy\,dx$$

$$= \int_{-\infty}^{\infty}\left[\int_{-\infty}^{\infty} h(x,y)k(y \mid x)\,dy\right]g(x)\,dx$$

The term in the brackets is $E_{y|x}[h(x,y)]$, which is a function of x only. Substituting gives

$$E[h(x,y)] = \int_{-\infty}^{\infty}\{E_{y|x}[h(x,y)]\}g(x)\,dx = E_x\{E_{y|x}[h(x,y)]\}$$

and the lemma is proved.

Now, $$E(\hat{\beta}_1) = E_x[E_{y|x}(\hat{\beta}_1)] = E_x\left\{E_{y|x}\left[\frac{\Sigma(y_i - \bar{y})(x_i - \bar{x})}{\Sigma(x_i - \bar{x})^2}\right]\right\}$$

$$= E_x\left\{\frac{\Sigma(x_i - \bar{x})}{\Sigma(x_i - \bar{x})^2}\,E_{y|x}(y_i - \bar{y})\right\}$$

But $$E_{y|x}(y_i) = \mu_2 - \frac{\sigma_{12}}{\sigma_{11}}\,\mu_1 + \frac{\sigma_{12}}{\sigma_{11}}\,x_i$$

and $$E_{y|x}(\bar{y}) = \mu_2 - \frac{\sigma_{12}}{\sigma_{11}}\,\mu_1 + \frac{\sigma_{12}}{\sigma_{11}}\,\bar{x}$$

so $$E(\hat{\beta}_1) = E_x\left[\frac{\Sigma(x_i - \bar{x})}{\Sigma(x_i - \bar{x})^2}\,\frac{\sigma_{12}}{\sigma_{11}}\,(x_i - \bar{x})\right] = \frac{\sigma_{12}}{\sigma_{11}} = \beta_1$$

Also, $$E(\hat{\beta}_0) = \mu_2 - \frac{\sigma_{12}}{\sigma_{11}}\,\mu_1 = \beta_0$$

So $\hat{\beta}_0$ and $\hat{\beta}_1$ are unbiased estimators of β_0 and β_1, respectively, but they are *not* normally distributed as when the x_i are fixed. We have established the following theorem.

◆ **Theorem 10.2** The maximum-likelihood estimates of β_0 and β_1 given in (10.2) have the following properties:

(1) Consistency
(2) Efficiency
(3) Minimum variance unbiased

10.2.3 The Wishart Distribution. We shall now find the joint distribution of the quantities $(n-1)\hat{\sigma}_{11}$, $(n-1)\hat{\sigma}_{12}$, $(n-1)\hat{\sigma}_{22}$. First we shall prove the following theorem.

◆ **Theorem 10.3** Let $\mathbf{U}_i = \begin{pmatrix} x_i \\ y_i \end{pmatrix}$ for $i = 1, 2, \ldots, n$; $n \geqslant 2$ be a random sample from the distribution $N(\mathbf{0},\mathbf{V})$. Let $b_{11} = \Sigma x_i^2$; $b_{12} = b_{21} = \Sigma x_i y_i$; $b_{22} = \Sigma y_i^2$. Then the joint distribution of b_{11}, b_{12}, b_{22} is

$$f(b_{11},b_{12},b_{22}) = \begin{cases} \dfrac{e^{-\frac{1}{2}\mathrm{tr}(\mathbf{BV}^{-1})}|\mathbf{B}|^{\frac{1}{2}(n-3)}}{2^n|\mathbf{V}|^{n/2}\pi^{1/2}\Gamma\left(\dfrac{n}{2}\right)\Gamma\left(\dfrac{n-1}{2}\right)} & \text{if } \mathbf{B} \text{ is positive definite} \\ 0 & \text{if } \mathbf{B} \text{ is not positive definite} \end{cases} \tag{10.3}$$

where $b_{11} \geqslant 0 \qquad b_{22} \geqslant 0 \qquad -\infty < b_{12} < \infty \qquad \mathbf{B} = \begin{pmatrix} b_{11} & b_{12} \\ b_{21} & b_{22} \end{pmatrix}$

Proof: It is clear that $\Sigma \mathbf{U}_i\mathbf{U}_i' = \mathbf{B}$. We shall find the joint moment-generating function of Σx_i^2, $\Sigma x_i y_i$, Σy_i^2, show that it is equal to the moment-generating function of b_{11}, b_{12}, b_{22}, where the distribution of b_{11}, b_{12}, b_{22} is given by (10.3), and use Theorem 2.1. It can be shown by direct integration that

$$\int_0^\infty \int_0^\infty \int_{-\infty}^\infty f(b_{11},b_{22},b_{12})\, db_{11}\, db_{22}\, db_{12} = 1$$

and hence
$$\int_0^\infty \int_0^\infty \int_{-\infty}^\infty e^{-\frac{1}{2}\mathrm{tr}(\mathbf{BV}^{-1})}|\mathbf{B}|^{\frac{1}{2}(n-3)}\, db_{11}\, db_{22}\, db_{12}$$

$$= 2^n|\mathbf{V}|^{n/2}\pi^{\frac{1}{2}}\Gamma\left(\frac{n}{2}\right)\Gamma\left(\frac{n-1}{2}\right) = K \tag{10.4}$$

We shall use this result to find the moment-generating function. The joint moment-generating function of b_{11}, b_{22}, b_{12} is equal to

$$M(t_1,t_2,t_3) = E(e^{t_1b_{11}+t_2b_{22}+t_3b_{12}})$$

If we let $$\mathbf{T} = \begin{pmatrix} t_1 & \frac{1}{2}t_3 \\ \frac{1}{2}t_3 & t_2 \end{pmatrix}$$

then $$\operatorname{tr}(\mathbf{BT}) = t_1 b_{11} + t_2 b_{22} + t_3 b_{12}$$

So $$M(t_1,t_2,t_3) = E[e^{\operatorname{tr}(\mathbf{BT})}] = \iiint \frac{e^{\operatorname{tr}(\mathbf{BT})} e^{-\frac{1}{2}\operatorname{tr}(\mathbf{BV}^{-1})} |\mathbf{B}|^{\frac{1}{2}(n-3)}}{K} \, db_{11}\, db_{22}\, db_{12}$$

But $$e^{\operatorname{tr}(\mathbf{BT})} e^{-\frac{1}{2}\operatorname{tr}(\mathbf{BV}^{-1})} = e^{-\frac{1}{2}\operatorname{tr}(\mathbf{BV}^{-1} - 2\mathbf{BT})} = e^{-\frac{1}{2}\operatorname{tr}[\mathbf{B}(\mathbf{V}^{-1} - 2\mathbf{T})]}$$

The integral exists, since we can choose t_i small enough so that $\mathbf{V}^{-1} - 2\mathbf{T}$ is positive definite. Therefore, if we let $\mathbf{V}^{-1} - 2\mathbf{T} = \mathbf{C}$, we get

$$M(t_1,t_2,t_3) = \iiint \frac{e^{-\frac{1}{2}\operatorname{tr}(\mathbf{BC})} |\mathbf{B}|^{\frac{1}{2}(n-3)}}{K} \, db_{11}\, db_{22}\, db_{12}$$

Using the result of Eq. (10.4), we get

$$M(t_1,t_2,t_3) = \frac{|\mathbf{C}|^{-n/2}}{|\mathbf{V}|^{n/2}} = \frac{1}{|\mathbf{V}(\mathbf{V}^{-1} - 2\mathbf{T})|^{n/2}} = |\mathbf{I} - 2\mathbf{VT}|^{-n/2}$$

Now, from the distribution of $\mathbf{U}_i$ in the theorem, we shall find the moment-generating function of $\Sigma x_i^2\ \Sigma x_i y_i,\ \Sigma y_i^2$. Let us designate it by

$$m(t_1,t_2,t_3) = E(e^{t_1 \Sigma x_i^2 + t_2 \Sigma y_i^2 + t_3 \Sigma x_i y_i}) = E[e^{\operatorname{tr}(\mathbf{T}\Sigma \mathbf{U}_i \mathbf{U}_i')}]$$

$$= \int_{-\infty}^{\infty} \cdots \int_{-\infty}^{\infty} e^{\operatorname{tr}(\mathbf{T}\Sigma \mathbf{U}_i \mathbf{U}_i')} g(x_1,x_2, \ldots ,y_n)\, dx_1\, dx_2 \cdots dy_n$$

By definition of a random sample of vectors,

$$g(x_1,x_2, \ldots ,y_n) = f(x_1,y_1) f(x_2,y_2) \cdots f(x_n,y_n)$$

so $$m(t_1,t_2,t_3) = \int_{-\infty}^{\infty} \cdots \int_{-\infty}^{\infty} \frac{e^{\operatorname{tr}(\mathbf{T}\Sigma \mathbf{U}_i \mathbf{U}_i')}}{|\mathbf{V}|^{n/2}(2\pi)^n} e^{-\frac{1}{2}\Sigma \mathbf{U}_i' \mathbf{V}^{-1} \mathbf{U}_i}\, dx_1\, dx_2 \cdots dy_n$$

$$= \frac{1}{|\mathbf{V}|^{n/2}(2\pi)^n} \int_{-\infty}^{\infty} \cdots \int_{-\infty}^{\infty} e^{\operatorname{tr}[(\Sigma \mathbf{U}_i \mathbf{U}_i')\mathbf{T}] - \frac{1}{2}\Sigma \mathbf{U}_i' \mathbf{V}^{-1} \mathbf{U}_i}\, du$$

where $du = dx_1\, dx_2 \cdots dy_n$. Using Theorem 1.46, we see that $\operatorname{tr}(\Sigma \mathbf{U}_i \mathbf{U}_i' \mathbf{T}) = \operatorname{tr}[\Sigma \mathbf{U}_i (\mathbf{U}_i' \mathbf{T})] = \operatorname{tr}(\Sigma \mathbf{U}_i' \mathbf{T} \mathbf{U}_i) = \Sigma \mathbf{U}_i' \mathbf{T} \mathbf{U}_i$, since $\Sigma \mathbf{U}_i' \mathbf{T} \mathbf{U}_i$ is a scalar. The exponent can now be simplified:

$$e^{-\frac{1}{2}\Sigma \mathbf{U}_i'(\mathbf{V}^{-1} - 2\mathbf{T})\mathbf{U}_i} = e^{-\frac{1}{2}\Sigma \mathbf{U}_i' \mathbf{C} \mathbf{U}_i}$$

We get $$m(t_1,t_2,t_3) = \frac{|\mathbf{C}|^{-n/2}}{|\mathbf{V}|^{n/2}} = |\mathbf{I} - 2\mathbf{VT}|^{-n/2}$$

We see that the two moment-generating functions $m(t_1,t_2,t_3)$ and $M(t_1,t_2,t_3)$ are equal; hence, with the application of Theorem 2.1 the proof is complete.

The distribution given in Theorem 10.3 is known in statistical literature as the *Wishart distribution.* We shall generalize the theorem to a p-variate normal but we shall not give the proof.

◆ **Theorem 10.4** If a vector $\mathbf{Y}$ is distributed $N(\mathbf{0},\mathbf{V})$, where $\mathbf{Y}$ is $p \times 1$, and if $\mathbf{Y}_j$ $(j = 1, 2, \ldots, n)$ is a set of n random vectors $(n \geqslant p)$ from this distribution, the elements of $\mathbf{B} = \Sigma \mathbf{Y}_j\mathbf{Y}_j' = (b_{ij})$ have the joint distribution

$$f(b_{11},b_{12}, \ldots ,b_{pp}) = \begin{cases} \dfrac{e^{-\frac{1}{2}\mathrm{tr}(\mathbf{BV}^{-1})} \, |\mathbf{B}|^{\frac{1}{2}(n-p-1)}}{\pi^{\frac{1}{4}p(p-1)} 2^{np/2} |\mathbf{V}|^{n/2} \prod_{i=1}^{p} \Gamma\left(\dfrac{n+1-i}{2}\right)} & \text{if } \mathbf{B} \text{ is positive definite} \\ 0 & \text{if } \mathbf{B} \text{ is not positive definite} \end{cases}$$

When a set of variables b_{ij} has this distribution, we shall say that *the matrix* $\mathbf{B}$ *is distributed as* $W(p,n,\mathbf{V})$.

◆ **Theorem 10.5** If $\mathbf{B}_1$ is distributed as $W(p,n_1,\mathbf{V})$, if $\mathbf{B}_2$ is distributed as $W(p,n_2,\mathbf{V})$, and if the elements of $\mathbf{B}_1$ and $\mathbf{B}_2$ are mutually independent, then $\mathbf{B} = \mathbf{B}_1 + \mathbf{B}_2$ is distributed as $W(p, n_1 + n_2, \mathbf{V})$.

This reproductive property of the Wishart distribution can be extended to a finite number of independent matrices $\mathbf{B}_i$. The theorem can be proved by using the moment-generating function of the distribution.

◆ **Theorem 10.6** If $\mathbf{Y}_i$ is a random vector from the distribution $N(\boldsymbol{\mu},\mathbf{V})$, then

$$\mathbf{A} = \sum_{i=1}^{n} (\mathbf{Y}_i - \overline{\mathbf{Y}})(\mathbf{Y}_i - \overline{\mathbf{Y}})'$$

is distributed as $W(p, n - 1, \mathbf{V})$, where $\overline{\mathbf{Y}} = (1/n)\Sigma\mathbf{Y}_i$ for $n \geqslant p + 1$.

Proof: It is clear that

$$\sum_{i=1}^{n} (\mathbf{Y}_i - \overline{\mathbf{Y}}) = 0$$

and, hence, that the $\mathbf{Y}_i - \overline{\mathbf{Y}}$ are not linearly independent. Let us make the following transformation on the vectors $\mathbf{Y}_i$:

$$\mathbf{W}_1 = \frac{1}{\sqrt{2}}(\mathbf{Y}_1 - \mathbf{Y}_2)$$

$$\mathbf{W}_2 = \frac{1}{\sqrt{6}}(\mathbf{Y}_1 + \mathbf{Y}_2 - 2\mathbf{Y}_3)$$

. . . .

$$\mathbf{W}_{n-1} = \frac{1}{\sqrt{n(n-1)}}[\mathbf{Y}_1 + \cdots + \mathbf{Y}_{n-1} - (n-1)\mathbf{Y}_n]$$

The $\mathbf{W}_i$ are independent, and $\mathbf{W}_i$ is distributed $N(\mathbf{0},\mathbf{V})$. Also, it is easy to see that

$$\sum_{i=1}^{n-1} \mathbf{W}_i\mathbf{W}_i' = \sum_{i=1}^{n} (\mathbf{Y}_i - \overline{\mathbf{Y}})(\mathbf{Y}_i - \overline{\mathbf{Y}})' = \mathbf{A}$$

Hence we have, by Theorem 10.4, that $\mathbf{A}$ is distributed as $W(p, n-1, \mathbf{V})$.

We have shown that, if x_i and y_i are observations from $N(\boldsymbol{\mu},\mathbf{V})$, then, letting

$$\hat{\mathbf{V}} = \begin{pmatrix} \hat{\sigma}_{11} & \hat{\sigma}_{12} \\ \hat{\sigma}_{21} & \hat{\sigma}_{22} \end{pmatrix}$$

where $\hat{\sigma}_{ij}$ is defined in (10.1), we find that $(n-1)\hat{\mathbf{V}}$ is distributed as $W(2, n-1, \mathbf{V})$.

10.2.4 Example. The widths x_i and surface area y_i of the leaves of a tree are assumed to follow a bivariate normal distribution. Since surface area is quite important in many investigations and since its measurement is a lengthy process, it is desired to find an equation that can be used to predict the surface area of a leaf from knowledge of its width. Ten leaves were examined; the values (in centimeters) are given below:

y_i	8.1	12.3	8.0	15.4	8.3	8.8	16.2	15.8	18.1	16.2
x_i	2.2	2.5	2.0	2.8	2.2	2.3	2.7	2.7	2.8	2.6

The following quantities were calculated:

$$\bar{y} = \hat{\mu}_2 = 12.72 \qquad \hat{\sigma}_{22} = 16.50 \qquad \hat{\beta}_0 = \hat{\mu}_2 - \frac{\hat{\sigma}_{12}}{\hat{\sigma}_{11}}\hat{\mu}_1 = -20.60$$

$$\bar{x} = \hat{\mu}_1 = 2.48 \qquad \hat{\sigma}_{12} = 1.10 \qquad \hat{\beta}_1 = \frac{\hat{\sigma}_{12}}{\hat{\sigma}_{11}} = 13.40$$

$$\hat{\rho}^2 = \frac{\hat{\sigma}_{12}^2}{\hat{\sigma}_{11}\hat{\sigma}_{22}} = .90 \qquad \hat{\sigma}_{11} = .082$$

The regression equation is $Y = -20.60 + 13.40X$.

10.2.5 Interval Estimation. In the bivariate normal distribution it might be desirable to be able to set confidence limits on the parameters μ_1, μ_2, β_0, and β_1. Let $z_i = ax_i + by_i$; then, by Theorem 3.22, the z_i are distributed independently $N(a\mu_1 + b\mu_2, \sigma^2)$. Therefore, Student's distribution can be used to set confidence limits on $a\mu_1 + b\mu_2$. If we let $a = 1$ and $b = 0$, we get confidence limits on μ_1, and similarly for μ_2.

The situation looks more complex if we desire to set confidence limits on β_0 or β_1. However, we shall show that the theory of Art. 6.3.1 can be used. To demonstrate this, suppose we consider the general-linear-hypothesis model of Definition 6.1, and in particular let us consider setting a confidence interval on $\mathbf{r}'\boldsymbol{\beta}$, as in Eq. (6.10). This probability statement as we derived it is valid for repeated random samples of y_i from the same set of X_i. However, if we change the X_i values and draw a random sample of y_i from the new X_i, the probability statement is still true. Hence it is true for any set of X_i, since the confidence coefficient does not depend on the X_i. Therefore, we can also select the X_i at random from any distribution, so long as the conditional distribution of y given x is normal and satisfies Definition 6.1.

To demonstrate further, the quantity

$$u = \frac{\mathbf{r}'\hat{\boldsymbol{\beta}} - \mathbf{r}'\boldsymbol{\beta}}{\sqrt{\hat{\sigma}^2\mathbf{r}'\mathbf{S}^{-1}\mathbf{r}}}$$

is distributed as Student's t when the X_i are fixed. Let this t distribution be written $g(t \mid x_1, \ldots, x_n)$. The probability statement of Eq. (6.10) can be written

$$1 - \alpha = P\left[-t_{\alpha/2} \leqslant \frac{\mathbf{r}'\hat{\boldsymbol{\beta}} - \mathbf{r}'\boldsymbol{\beta}}{(\hat{\sigma}^2\mathbf{r}'\mathbf{S}^{-1}\mathbf{r})^{\frac{1}{2}}} \leqslant t_{\alpha/2}\right]$$

$$= P(-t_{\alpha/2} \leqslant u \leqslant t_{\alpha/2}) = \int_{-t_{\alpha/2}}^{t_{\alpha/2}} g(t \mid x_1, \ldots, x_n)\, dt$$

If $h(x_1, \ldots, x_n)$ is the density function of the X_i, then

$$\begin{aligned} P(-t_{\alpha/2} \leqslant u \leqslant t_{\alpha/2}) &= 1 - \alpha \\ &= \int_{-\infty}^{\infty} \cdots \int_{-\infty}^{\infty} (1 - \alpha) h(x_1, \ldots, x_n)\, dx_1 \cdots dx_n \\ &= \int_{-\infty}^{\infty} \cdots \int_{-\infty}^{\infty} \left[\int_{-t_{\alpha/2}}^{t_{\alpha/2}} g(t \mid x_1, \ldots, x_n) h(x_1, \ldots, x_n)\, dt\right] dx_1 \cdots dx_n \\ &= \int_{-\infty}^{\infty} \cdots \int_{-\infty}^{\infty} \int_{-t_{\alpha/2}}^{t_{\alpha/2}} f(t, x_1, \ldots, x_n)\, dt\, dx_1 \cdots dx_n \end{aligned}$$

where $f(t,x_1, \ldots ,x_n)$ is the joint density of t and the X_i. Hence, the probability statement in Eq. (6.10) is true when the X_i are random variables, i.e., when we sample from a multivariate instead of from a conditional distribution. The joint distribution, however, must be such that the conditional is normal and satisfies Definition 6.1.

We selected only one type of confidence-interval statement to demonstrate the result; however, the same result holds for any confidence statement in Art. 6.3.1.

The expected width of the confidence interval is different in the two situations: (1) when the X_i are fixed and (2) when the X_i are random.

10.2.6 Example. We shall set a 95 per cent confidence interval on β_1 for the example of Art. 10.2.5. Regarding these data as coming from model 1, the x_i fixed, we can use Eq. (6.9) and obtain the 95 per cent confidence interval

$$9.8 \leqslant \beta_1 \leqslant 17.1$$

According to the theory of Art. 10.2.4, this statement means that, if we take repeated sets of y measurements from these x_i and calculate 95 per cent confidence intervals for each set, then, on the average, 95 per cent of all intervals will contain the true unknown parameter β_1. This type of inference does not seem to be too useful. However, by using the theory in the preceding article, we can select repeated sets of *leaves* (the x_i values may change from set to set) and set a confidence interval on β_1 for each set. We then know that 95 per cent of all intervals contain the true unknown parameter β_1.

10.2.7 Tests of Hypotheses. There are various hypotheses that are interesting to test in the bivariate normal distribution. Those we shall discuss are:

1. That the vector of means $\boldsymbol{\mu}$ is equal to a known vector $\boldsymbol{\mu}^*$
2. That the regression coefficient β_1 is equal to a known constant β_1^*; and similarly for β_0

There are other tests that are important (on the elements of $\mathbf{V}$ for example), but we shall not discuss them.

The hypothesis H_0: $\boldsymbol{\mu} = \boldsymbol{\mu}^*$. To test H_0: $\boldsymbol{\mu} = \boldsymbol{\mu}^*$, where $\boldsymbol{\mu}^*$ is a known vector, we shall use the following theorems, which we shall give without proof.

◆ **Theorem 10.7** If the $p \times 1$ vector $\mathbf{Z}$ is distributed $N(\boldsymbol{\mu},\mathbf{V})$, if the $p \times p$ matrix $\mathbf{A}$ is distributed $W(p,n,\mathbf{V})$, and if the elements of $\mathbf{Z}$ are independent of the elements of $\mathbf{A}$, the quantity

$$u = \mathbf{Z}'\mathbf{A}^{-1}\mathbf{Z}\,\frac{n-p+1}{p}$$

is distributed $F'(p, n-p+1, \lambda)$, where $\lambda = \frac{1}{2}\boldsymbol{\mu}'\mathbf{V}^{-1}\boldsymbol{\mu}$.

The proof of this theorem was given first by Hotelling. The quantity $[np/(n - p + 1)]u$ is generally referred to as *Hotelling's* T^2.

◆ **Theorem 10.8** If the n vectors $\mathbf{Y}_i$ $(i = 1, 2, \ldots, n)$ are random samples from the p-variate $N(\boldsymbol{\mu},\mathbf{V})$ and if

$$\mathbf{S} = \frac{1}{n-1} \sum_{i=1}^{n} (\mathbf{Y}_i - \overline{\mathbf{Y}})(\mathbf{Y}_i - \overline{\mathbf{Y}})'$$

then $\mathbf{S}$ and $\overline{\mathbf{Y}}$ are independent.

This theorem is an extension of the independence of $\bar{y}$ and s^2 in the univariate case.

We shall now state the theorem that will be used to test the hypothesis $\boldsymbol{\mu} = \boldsymbol{\mu}^*$.

◆ **Theorem 10.9** If $\mathbf{Y}_i$ $(i = 1, 2, \ldots, n)$ is a set of random vectors from the p-variate $N(\boldsymbol{\mu},\mathbf{V})$, the quantity

$$u = (\overline{\mathbf{Y}} - \boldsymbol{\mu}^*)'\mathbf{S}^{-1}(\overline{\mathbf{Y}} - \boldsymbol{\mu}^*) \frac{n(n-p)}{p(n-1)} \qquad n > p$$

is distributed $F(p, n - p)$ if $\boldsymbol{\mu} = \boldsymbol{\mu}^*$. The test function u is equivalent to the likelihood-ratio test of the hypothesis $\boldsymbol{\mu} = \boldsymbol{\mu}^*$.

To compute the test function u, the inverse of $\mathbf{S}$ need not be calculated. The Doolittle procedure outlined in Chap. 7 can be used. If we set up the equations $\mathbf{S}\hat{\boldsymbol{\beta}} = (\overline{\mathbf{Y}} - \boldsymbol{\mu}^*)$ in a form like Table 7.1, where $\hat{\boldsymbol{\beta}}$ is an unknown $p \times 1$ vector, we find that $\hat{\boldsymbol{\beta}} = \mathbf{S}^{-1}(\overline{\mathbf{Y}} - \boldsymbol{\mu}^*)$ and $\hat{\boldsymbol{\beta}}'(\overline{\mathbf{Y}} - \boldsymbol{\mu}^*) = (\overline{\mathbf{Y}} - \boldsymbol{\mu}^*)'\mathbf{S}^{-1}(\overline{\mathbf{Y}} - \boldsymbol{\mu}^*)$. But this latter quantity is the sum of products of the elements in the C_0 column of Table 7.1.

The hypothesis $\mathrm{H}_0\colon \beta_1 = \beta_1^*$. Any test of hypothesis that we derived in Chap. 6 was for a conditional distribution of y given $x = X$. However, by methods similar to the interval estimation of Art. 10.2.5, it can be shown that the probability of a type I error is exactly the same if the sampling is from a joint distribution of y and x, so long as the conditional distribution of y given $x = X$ meets the conditions of Definition 6.1.

It seems that the two situations are not identical (the X_i fixed, and the x_i random). In fact, the difference shows itself in the powers of the tests, which are different when X_i are fixed and when x_i are random.

10.3 Correlation

10.3.1 Simple Correlation. A simple correlation coefficient is defined as a certain operation on two *random* variables. Hence, in the general-linear-hypothesis model given in Definition 6.1, the correlation between y and X is not defined. In the model of Sec. 10.2, however, correlation does have meaning, as it did in the multivariate normal of Chap. 3.

As we saw in Theorem 3.14, the square of the correlation coefficient gives the percentage decrease in the variance of one random variable due to use of information available on a related variable. Also, if the correlation between two random normal variables is zero, the two variables are statistically independent. However, if the random variables are not jointly normal, then zero correlation does not necessarily imply independence. In fact, the correlation coefficient between two random variables can be zero and they can still be functionally related. For example, suppose we have four points (x, y), each with probability $\frac{1}{4}$, that is, $(1, 1)$, $(2, 4)$, $(-1, 1)$, and $(-2, 4)$. Now,

$$E(x) = (1)(\tfrac{1}{4}) + (2)(\tfrac{1}{4}) + (-1)(\tfrac{1}{4}) + (-2)(\tfrac{1}{4}) = 0$$

and $$E(xy) = (1)(\tfrac{1}{4}) + (8)(\tfrac{1}{4}) + (-1)(\tfrac{1}{4}) + (-8)(\tfrac{1}{4}) = 0$$

and so $$\rho_{xy} = 0$$

But y and x are functionally related, since $y = x^2$. This demonstrates the fact that, unless random variables are normal, it is difficult to draw conclusions on independence when the correlation is zero.

However, when the correlation coefficient between two random variables is equal to plus or minus unity, this does give valuable information, as follows.

◆ **Theorem 10.10** If x and y are random variables and if the correlation between x and y is equal to plus or minus unity, then y and x are functionally related in a linear manner; that is, $y = \alpha + \beta x$, where α and β are constants.

Proof: Suppose $\rho_{xy} = +1$. Let $y^* = (y - \mu_y)/\sigma_y$ and $x^* = (x - \mu_x)/\sigma_x$. Now,

$$E(y^*)^2 + E(x^*)^2 - 2E[(x^*)(y^*)] = 0$$

since, by hypothesis, $$E[(x^*)(y^*)] = \rho_{xy} = +1$$

Thus $$E(y^* - x^*)^2 = 0 \qquad \text{and} \qquad E(y^* - x^*) = 0$$

so the density function of $x^* - y^*$ is a point density with its total probability at $y^* - x^* = 0$; hence, $y^* = x^*$. Substituting gives

$$\frac{y - \mu_y}{\sigma_y} = \frac{x - \mu_x}{\sigma_x} \qquad \text{or} \qquad y = \alpha + \beta x$$

where $$\alpha = \mu_y - \frac{\sigma_y}{\sigma_x}\mu_x \qquad \text{and} \qquad \beta = \frac{\sigma_y}{\sigma_x}$$

A similar proof holds if $\rho_{xy} = -1$.

For example, if x and y are random variables and the correlation between y and $f(x)$ is equal to unity, then $y = \alpha + \beta f(x)$. In this respect correlation coefficients can be useful in obtaining the relationships among random variables.

If either of the variables is not a random variable, then a correlation coefficient merely describes the relationship between the sets of points observed, and it might be extremely dangerous to draw inferences as if both variables were random. However, when the two variables are indeed random, the square of the correlation coefficient is an estimate of the percentage reduction in the variance of one of the variables by using the other.

We shall state some theorems (some without proof) on the correlation coefficient. In Theorems 10.11 to 10.21 we shall assume that $\mathbf{U}_i = \begin{pmatrix} x_i \\ y_i \end{pmatrix}$ for $i = 1, 2, \ldots, n$ is a random sample from the bivariate normal distribution with correlation coefficient equal to ρ and that

$$\hat{\rho} = \frac{\Sigma(y_i - \bar{y})(x_i - \bar{x})}{\sqrt{\Sigma(y_i - \bar{y})^2 \Sigma(x_i - \bar{x})^2}}$$

◆ **Theorem 10.11** The density function of $\hat{\rho}$ is

$$f(\hat{\rho}; \rho) = \frac{(n-2)(1-\rho^2)^{\frac{1}{2}(n-1)}}{\pi}(1-\hat{\rho}^2)^{\frac{1}{2}(n-4)} \int_0^\infty \frac{dw}{(\cosh w - \hat{\rho}\rho)^{n-1}} \qquad -1 \leqslant \hat{\rho} \leqslant 1$$

The proof of this theorem will be omitted. The theorem can be obtained by using Theorem 10.3 with $p = 2$, substituting $\hat{\rho}$ for $\hat{\sigma}_{12}/\sqrt{\hat{\sigma}_{11}\hat{\sigma}_{22}}$, and integrating out $\hat{\sigma}_{11}$ and $\hat{\sigma}_{22}$.

The density function $f(\hat{\rho})$ is difficult to work with. The parameters in $f(\hat{\rho})$ are ρ, which is unknown, and n, which is known. To test hypotheses about ρ and set confidence intervals on ρ, it is desirable to have

$$\int_{-1}^{\rho_0} f(\hat{\rho};\rho)\, d\hat{\rho} = P(\hat{\rho} \leqslant \rho_0 \mid \rho)$$

tabulated for various values of ρ_0, ρ, and n. This has been done by F. N. David for the following values: $\rho_0 = -1.00, -.95, -.90, \ldots, .95, 1.00$; $n = 3, 4, 5, \ldots, 25, 50, 100, 200, 400$; $\rho = .0, .1, .2, \ldots, .9$. By using David's tables we can compute the power of the test for various hypotheses about ρ. The use of the tables will be demonstrated later. First we shall state some useful theorems about the distribution of $\hat{\rho}$.

◆ **Theorem 10.12** If $\rho = 0$ in the bivariate normal, then

$$f(\hat{\rho};0) = \frac{\Gamma\left(\frac{n-1}{2}\right)(1-\hat{\rho}^2)^{\frac{1}{2}(n-4)}}{\Gamma\left(\frac{n-2}{2}\right)\Gamma\left(\frac{1}{2}\right)} \qquad -1 \leqslant \hat{\rho} \leqslant 1$$

Proof: The proof is immediate from Theorem 10.11.

◆ **Theorem 10.13** If $\rho = 0$, then

$$f(\hat{\rho}^2;0) = \frac{\Gamma\left(\frac{n-1}{2}\right)(1-\hat{\rho}^2)^{\frac{1}{2}(n-4)}}{\Gamma\left(\frac{n-2}{2}\right)\Gamma\left(\frac{1}{2}\right)(\hat{\rho}^2)^{\frac{1}{2}}} \qquad 0 \leqslant \hat{\rho}^2 \leqslant 1$$

◆ **Theorem 10.14** If $\rho = 0$, $E(\hat{\rho}^2) = 1/(n-1)$.

◆ **Theorem 10.15** If $\rho = 0$, the quantity

$$t = \frac{\hat{\rho}\sqrt{n-2}}{\sqrt{1-\hat{\rho}^2}}$$

is distributed as Student's t with $n - 2$ degrees of freedom.

To test the hypothesis H_0: $\rho = 0$ we can use Student's distribution, but to test the hypothesis H_0: $\rho = \rho^*$ ($\rho^* \neq 0$) we must know the distribution of ρ when $\rho \neq 0$. If $n \leqslant 25$, David's tables can be used. If $n \geqslant 25$, interpolation in David's tables can be used, but with some sacrifice in accuracy.

It may be desirable to test the hypothesis that two or more correlation coefficients are equal. This type of problem can be handled quite easily by using a transformation due to R. A. Fisher. This is the substance of the next few theorems.

Fisher has shown that, if we make the transformation

$$Z = \tfrac{1}{2}\log_e \frac{1+\hat{\rho}}{1-\hat{\rho}} = \operatorname{arctanh} \hat{\rho}$$

and let

$$\delta = \tfrac{1}{2}\log_e \frac{1+\rho}{1-\rho} = \operatorname{arctanh} \rho$$

then $Z - \delta$ has the remarkable property of approximating the normal distribution even for fairly small n. We shall state this in the following.

◆ **Theorem 10.16** The quantity $Z = \operatorname{arctanh} \hat{\rho}$ is approximately normally distributed with mean equal to $\delta = \operatorname{arctanh} \rho$ and variance $(n-3)^{-1}$. A slightly more accurate mean of Z is $\operatorname{arctanh} \rho - \rho/2(n-1)$.

◆ **Theorem 10.17** To test the hypothesis $\rho = \rho^*$ (ρ^* known) David's tables [9] can be used if $n \leqslant 25$. If $n \geqslant 25$, then

$$u = (\text{arctanh } \hat{\rho} - \text{arctanh } \rho^*) \sqrt{n-3}$$

can be considered to be approximately distributed $N(0, 1)$.

◆ **Theorem 10.18** Suppose that k-bivariate normal populations have correlation coefficients $\rho_1, \rho_2, \ldots, \rho_k$ and that $\hat{\rho}_1, \hat{\rho}_2, \ldots, \hat{\rho}_k$ are the estimates based on samples of size $n_1, n_2, \ldots, n_k$, respectively. To test the hypothesis H_0: $\rho_1 = \rho_2 = \cdots = \rho_k = \rho^*$ (ρ^* known), the quantity,

$$U = \sum_{i=1}^{k} (\text{arctanh } \hat{\rho}_i - \text{arctanh } \rho^*)^2 (n_i - 3)$$

has an approximate chi-square distribution with k degrees of freedom. Reject H_0 with probability α if $U \geqslant \chi^2_\alpha(k)$ where $\chi^2_\alpha(k)$ is the appropriate chi-square value for k degrees of freedom and probability α.

◆ **Theorem 10.19** In the preceding theorem, suppose it is desired to test the hypothesis that all the correlation coefficients are equal without stating what they are equal to; that is, H_0: $\rho_1 = \rho_2 = \cdots = \rho_k$. Then we can use the quantity

$$W = \Sigma(n_i - 3)(Z_i - \bar{Z})^2$$

as a chi-square variate with $k - 1$ degrees of freedom, where

$$Z_i = \text{arctanh } \hat{\rho}_i \qquad \text{and} \qquad \bar{Z} = \frac{\Sigma(n_i - 3)Z_i}{\Sigma(n_i - 3)}$$

◆ **Theorem 10.20** If $\rho_1 = \rho_2 = \cdots = \rho_k$ in Thereom 10.19, then the "best" linear combined estimate $\hat{\rho}$ of the common correlation ρ ("best" means weighted with inverse variance) is given by $\hat{\rho} = \tanh Z^*$, where

$$Z^* = \bar{Z} - \frac{m\rho^*}{2}$$

where $\bar{Z}$ is defined as in Theorem 10.19 and

$$\rho^* = \frac{1}{k} \Sigma\hat{\rho}_i$$

$$m = \frac{\Sigma[(n_i - 3)/(n_i - 1)]}{\Sigma(n_i - 3)}$$

The improved estimate given in Theorem 10.16 is used to obtain the pooled estimate.

◆ **Theorem 10.21** To set an approximate $1 - \alpha$ confidence limit on ρ, we use the fact that $(n - 3)^{1/2}$ (arctanh $\hat{\rho}$ − arctanh ρ) is approximately distributed $N(0,1)$. A $1 - \alpha$ confidence interval on ρ is

$$\tanh\left(\operatorname{arctanh}\hat{\rho} - \frac{z_{\alpha/2}}{\sqrt{n-3}}\right) \leqslant \rho \leqslant \tanh\left(\operatorname{arctanh}\hat{\rho} + \frac{z_{\alpha/2}}{\sqrt{n-3}}\right)$$

where $z_{\alpha/2}$ is the appropriate value for the normal distribution.

R. A. Fisher has tabulated $\frac{1}{2}\log_e (1 + x)/(1 - x)$ for various values of x, and many of the approximate tests can be readily applied by using his tables. If $n \leqslant 25$, one should use the tables prepared by David. If $n > 25$, then the Z transformation is satisfactory.

10.3.2 Example. We shall illustrate some of the theorems with an example. Let us consider the correlation between protein and fat for four breeds of dairy cows. The two constituents protein and fat are regarded as elements in a bivariate normal distribution. Suppose the computed correlation coefficients between protein and fat for four breeds of dairy cows are

$\hat{\rho}_i$	.028	.054	.407	.381
n_i	24	28	24	20

We shall first test the following hypotheses with the probability of type I error equal to .05:

1. $\rho_1 = 0$; alternative hypothesis is $\rho_1 > 0$; that is, this is a one-sided test. Calculate the power if $\rho_1 = .2, .3$.

2. $\rho_3 = .5$; alternative hypothesis is $\rho_3 \neq .5$; that is, this is a two-sided test. Calculate the power if $\rho_3 = .2, .3$.

3. $\rho_1 = \rho_2 = 0$; alternative hypothesis is that one of the equalities does not hold.

4. $\rho_1 = \rho_2 = \rho_3 = \rho_4$; alternative hypothesis is that at least one equality does not hold.

1. We can use Theorem 10.15 and have an exact test. We get

$$t = \frac{.028\sqrt{22}}{\sqrt{1 - (.028)^2}} = .131$$

which is not significant at the 5 per cent level. Hence we cannot reject H_0. The power function for $\rho = \rho^*$ is

$$\beta(\rho^*) = \int_{\rho_x}^{1} f(\hat{\rho};\rho^*)\, d\hat{\rho}$$

where ρ_α is such that

$$.95 = \int_{-1}^{\rho_\alpha} f(\hat{\rho};0)\, d\hat{\rho}$$

From David's tables for $n = 24$ and $\rho = 0$ we get $\rho_\alpha = .3447$. From David's tables (using this value of ρ_α) we obtain $\beta(\rho^* = .2) = .241$; $\beta(\rho^* = .3) = .421$. The power using the Z transformation gives $\beta(.2) = .239$; $\beta(.3) = .413$.

2. We can use David's tables and have an exact test, or we can use Fisher's Z transformation and Theorem 10.17. Using the Z transformation we get

$$Z = \text{arctanh } .407 = .432$$

$$\delta = \text{arctanh } .500 = .549$$

$$\sqrt{\text{var}(Z)} = \sqrt{\frac{1}{21}} = \frac{1}{4.58} = .218$$

and

$$u = (.432 - .549)(4.58) = -.536$$

which is not significant at the 5 per cent level. Since u is distributed $N(0,1)$, the critical region is the two intervals $-\infty < u < -1.96$ and $1.96 < u < \infty$. The value $u = -.536$ does not fall in the critical region; hence, we shall not reject the hypothesis. When $\rho = .2$, we get u distributed $N(-1.587, 1)$. Integrating this over the critical region gives $\beta(.2)$ as approximately .350. For $\rho = .3$ the same procedure gives $\beta(.3)$ approximately .195.

The exact test from David's tables gives the critical region as the two intervals $-1.0 \leqslant x \leqslant .130$ and $.758 \leqslant x \leqslant 1.0$; since $\hat{\rho}$ is not in this region, the hypothesis cannot be rejected. The power function is

$$\beta(\rho^*) = 1 - \int_{+\rho_0}^{\rho_1} f(\hat{\rho};\rho^*)\, d\hat{\rho}$$

where ρ_0 and ρ_1 are such that

$$\int_{-1}^{+\rho_0} f(\hat{\rho};.5)\, d\hat{\rho} = \tfrac{1}{2}(1 - .95) = .025$$

and

$$\int_{\rho_1}^{1} f(\hat{\rho};.5)\, d\hat{\rho} = .025$$

Notice that an equal two-tail test has been used. Using this, we get $\rho_0 = .13$; $\rho_1 = .758$. This gives $\beta(.2) = .363$; $\beta(.3) = .198$.

3. We can use Theorem 10.18 with $\rho^* = 0$; we get

arctanh $\hat{\rho}_i$	.028	.054
$n_i - 3$	21	25

Arctanh $\rho^* = \text{arctanh } 0 = 0$. So $U = .0896$, and we compare this with the chi-square with 2 degrees of freedom. It is not significant at the 5 per cent level; hence, we shall not reject H_0.

4. We can use Theorem 10.19, which is applied by computing the entries in Table 10.1.

TABLE 10.1 TABLE FOR TESTING $\rho_1 = \rho_2 = \rho_3 = \rho_4$

	1	2	3	4	Sum
$\hat{\rho}_i$	.0280	.0540	.4070	.3810	
Z_i	.0282	.0540	.4320	.4012	
$n_i - 3$	21	25	21	17	84
$Z_i(n_i - 3)$	.5920	1.3500	9.0720	6.8204	17.8344
$Z_i^2(n_i - 3)$	.0167	.0729	3.9191	2.7360	6.7447

From the table we compute

$$\bar{Z} = \frac{\Sigma(n_i - 3)Z_i}{\Sigma(n_i - 3)} = \frac{17.8344}{84} = .2123$$

and

$$W = \Sigma(n_i - 3)(Z_i - \bar{Z})^2 = \Sigma(n_i - 3)Z_i^2 - \bar{Z}^2\Sigma(n_i - 3)$$

$$= 6.7447 - (.2123)^2(84) = 2.959.$$

If we compare this value of W with a chi-square variate with 3 degrees of freedom at the 5 per cent level of significance, we conclude that we have no reason to reject the hypothesis that these $\hat{\rho}_i$ are all estimates of the same ρ.

In addition to the above tests, suppose that we want (1) the 95 per cent confidence interval on ρ_4 and (2) to pool the estimates of $\rho_1, \rho_2, \rho_3, \rho_4$, assuming that $\rho_1 = \rho_2 = \rho_3 = \rho_4$.

The 95 per cent Confidence Interval on ρ_4. We shall use Theorem 10.21. For a 95 per cent confidence interval, $z_\alpha = 1.96$, and

$$.4012 - \frac{1.96}{\sqrt{17}} \leqslant \text{arctanh } \rho \leqslant .4012 + \frac{1.96}{\sqrt{17}}$$

which gives

$$-.0745 \leqslant \text{arctanh } \rho \leqslant .8769$$

and, finally,

$$-.0743 \leqslant \rho \leqslant .7049$$

Pooling the Estimates of $\rho_1, \rho_2, \rho_3,$ *and* ρ_4. To pool the estimates of ρ_i into a combined estimate $\hat{\rho}$ of ρ, we can use Theorem 10.20. We get $\rho^* = .87/4 = .2175$, $m = .0434$, $\bar{Z} = .2123$, and $Z^* = .2076$; hence, $\hat{\rho} = .205$.

10.3.3 Partial Correlation. To gain a better insight into relationships among random variables in a multivariate normal population, it may be desirable to stratify the population into subpopulations in which one or more of the variables is fixed and then determine the correlations among the other variables.

This was the case in observations about how average yearly temperature x_1 and average yearly rainfall x_2 affected the yield y of a certain agricultural crop. The simple correlations recorded were

$$\hat{\rho}_{y1} = .90 \qquad \hat{\rho}_{y2} = -.45 \qquad \hat{\rho}_{12} = -.55$$

It is surprising that there is a negative correlation between yield and rainfall ($-.45$), since rainfall is expected to increase the yield. Further investigation reveals that temperature and rainfall are negatively correlated ($-.55$), as they should be. The correlation between rainfall and yield in the subpopulation derived by holding temperature constant clarified the picture greatly. This implied that, for constant temperature, rainfall tends to increase yield, as expected.

In Chap. 3 the partial correlation coefficient was defined as the simple correlation between two random variables when certain other random variables are held constant. Let the $p \times 1$ vector $\mathbf{Y}$ be distributed $N(\boldsymbol{\mu},\mathbf{V})$. Further, let us partition $\mathbf{Y}$ into a $q \times 1$ vector $\mathbf{Y}^{(1)}$ and a $(p - q) \times 1$ vector $\mathbf{Y}^{(2)}$ such that

$$\mathbf{Y} = \begin{pmatrix} \mathbf{Y}^{(1)} \\ \mathbf{Y}^{(2)} \end{pmatrix}$$

From Theorem 3.10 we know that the conditional distribution of $\mathbf{Y}^{(1)}$ given $\mathbf{Y}^{(2)}$ is normal with mean $\mathbf{U}_1 + \mathbf{V}_{12}\mathbf{V}_{22}^{-1}(\mathbf{Y}^{(2)} - \mathbf{U}_2)$ and with covariance matrix $\mathbf{V}_{11} - \mathbf{V}_{12}\mathbf{V}_{22}^{-1}\mathbf{V}_{21}$. We shall use the symbol $\mathbf{V}_{11.2}$ to denote $\mathbf{V}_{11} - \mathbf{V}_{12}\mathbf{V}_{22}^{-1}\mathbf{V}_{21}$. The partial correlation between y_i and y_j[y_i and y_j are in $\mathbf{Y}^{(1)}$] holding $y_{q+1}, y_{q+2}, \ldots, y_p$ constant is equal to

$$\rho_{ij\cdot q+1,q+2,\ldots,p} = \frac{\sigma_{ij\cdot q+1,\ldots,p}}{\sqrt{\sigma_{ii\cdot q+1,\ldots,p}\sigma_{jj\cdot q+1,\ldots,p}}}$$

where $\sigma_{ij\cdot q+1,\ldots,p}$ is the ijth element in the covariance matrix $\mathbf{V}_{11.2}$.

Suppose a random sample of n vectors $\mathbf{Y}_1, \mathbf{Y}_2, \ldots, \mathbf{Y}_n$ is taken from the normal distribution described above. It was proved in preceding sections that the elements of the matrix

$$(n - 1)\mathbf{S} = \sum_{i=1}^{n} (\mathbf{Y}_i - \overline{\mathbf{Y}})(\mathbf{Y}_i - \overline{\mathbf{Y}})' = \mathbf{B}$$

are distributed as $W(p, n - 1, \mathbf{V})$. On the basis of this fact theorems on simple correlation coefficients were derived. We shall now state without proof a similar theorem that will enable us to estimate and test hypotheses about the partial correlation coefficients. Let us partition the matrix $\mathbf{B}$ as we partitioned $\mathbf{V}$. We obtain

$$\mathbf{B} = \begin{pmatrix} \mathbf{B}_{11} & \mathbf{B}_{12} \\ \mathbf{B}_{21} & \mathbf{B}_{22} \end{pmatrix}$$

Define $\mathbf{B}_{11.2}$ to be $(\mathbf{B}_{11} - \mathbf{B}_{12}\mathbf{B}_{22}^{-1}\mathbf{B}_{21})$, and let $b_{ij.2}$ denote the ijth element of $\mathbf{B}_{11.2}$.

◆ **Theorem 10.22** Under the conditions explained above, the maximum-likelihood estimate of the partial correlation coefficient $\rho_{ij\cdot q+1,\ldots,p}$ is $\hat{\rho}_{ij\cdot q+1,\ldots,p} = b_{ij.2}/\sqrt{b_{ii.2}b_{jj.2}}$.

Proof: This theorem follows immediately from the invariant property of the maximum-likelihood estimates.

◆ **Theorem 10.23** Under the conditions above, the elements of the matrix $\mathbf{B}_{11.2}$ are distributed as $W[p, (n-1)-(p-q), \mathbf{V}_{11.2}]$.

We shall not prove this theorem, but a proof can be found in [6]. We are now in a position to state a very important theorem.

◆ **Theorem 10.24** If the $p \times 1$ vector $\mathbf{Y}$ is partitioned into $\begin{pmatrix} \mathbf{Y}^{(1)} \\ \mathbf{Y}^{(2)} \end{pmatrix}$, where $\mathbf{Y}^{(1)}$ is a $q \times 1$ vector, and if $\mathbf{Y}$ is distributed $N(\boldsymbol{\mu},\mathbf{V})$, then all the theorems 10.11 to 10.21 are valid for the distribution of $\hat{\rho}_{ij\cdot q+1,\ldots,p}$ if n is replaced by $n - p + q$ and ρ is replaced by $\rho_{ij\cdot q+1,\ldots,p}$.

10.3.4 Example. It is assumed that the 4×1 vector $\mathbf{Y}$ is distributed $N(\boldsymbol{\mu},\mathbf{V})$. On the basis of 24 observations the following matrix was computed:

$$\mathbf{S} = \begin{pmatrix} 10 & 9 & -1 & -16 \\ 9 & 20 & -3 & -16 \\ -1 & -3 & 5 & 3 \\ -16 & -16 & 3 & 27 \end{pmatrix}$$

It is desired

1. To find $\hat{\rho}_{12\cdot 34}$
2. To test the hypothesis $\rho_{12\cdot 34} = 0$ at the 5 per cent level against the alternative $\rho_{12\cdot 34} \neq 0$
3. To set 95 per cent confidence limits on $\rho_{12\cdot 34}$

1. We find $\mathbf{S}_{11\cdot 2}$ to be equal to

$$\frac{1}{126}\begin{pmatrix} 49 & -35 \\ -35 & 1{,}285 \end{pmatrix}$$

and from this we get $\hat{\rho}_{12\cdot 34} = -.139$.

2. Using David's tables with parameter $n - (p - q) = 24 - (4 - 2) = 22$, we find that the two-tailed critical region is $-1.00 \leqslant x \leqslant -.423$ and $.426 \leqslant x \leqslant 1.00$. Since $\hat{\rho}_{12\cdot 34} = -.139$ is not in this critical region, we do not reject H_0.

3. The 95 per cent confidence is, by Fisher's transformations, $-.53 \leqslant \rho_{12\cdot 34} \leqslant .30$.

10.3.5 Multiple Correlation. The conditional distribution of y_1 given $y_2, y_3, \ldots, y_p$ is defined in Chap. 3. If the $p \times 1$ vector $\mathbf{Y}$ is distributed $N(\boldsymbol{\mu},\mathbf{V})$, the conditional distribution of y_1 given $y_2, \ldots, y_p$ is distributed normally with mean $\mu_1 + \mathbf{V}_{12}\mathbf{V}_{22}^{-1}(\mathbf{Y}_2 - \mathbf{U}_2)$ and variance $\mathbf{V}_{11} - \mathbf{V}_{12}\mathbf{V}_{22}^{-1}\mathbf{V}_{21}$. If we let $z = \mu_1 + \mathbf{V}_{12}\mathbf{V}_{22}^{-1}(\mathbf{Y}_2 - \mathbf{U}_2)$, then the multiple correlation coefficient is defined as the simple correlation between y_1 and z. The importance of this coefficient is explained in Chap. 3. It is defined as

$$\rho_1 = \left(1 - \frac{\sigma_{11\cdot 23\cdots p}}{\sigma_{11}}\right)^{\frac{1}{2}}$$

so the maximum likelihood estimate of ρ_1 is

$$\hat{\rho}_1 = \left(1 - \frac{\hat{\sigma}_{11\cdot 23\cdots p}}{\hat{\sigma}_{11}}\right)^{\frac{1}{2}}$$

We shall next state a theorem concerning the distribution of $\hat{\rho}_1$ when ρ_1 is zero.

◆ **Theorem 10.25** If $\mathbf{Y}_1, \mathbf{Y}_2, \ldots, \mathbf{Y}_n$ are independent $p \times 1$ vectors from the distribution $N(\boldsymbol{\mu},\mathbf{V})$, then the quantity

$$w = \frac{\hat{\rho}_1^2}{1 - \hat{\rho}_1^2}\,\frac{n-p}{p-1}$$

is distributed as $F(p - 1, n - p)$ if $\rho_1 = 0$, that is, if $\sigma_{12} = \cdots = \sigma_{1p} = 0$.

10.4 Case 2, Model 3

10.4.1 Definitions. Case 2 of regression models is defined by the following density function of the random variables y, x:

$$f(y,x) = \frac{h(x)e^{-(1/2\sigma^2)(y-\alpha-\beta x)^2}}{\sigma\sqrt{2\pi}}$$

where $h(x)$ is the marginal distribution of x and does not contain the parameters α, β, or σ. From this it follows that the conditional distribution of y given $x = X$ is normal with mean $\alpha + \beta X$ and variance σ^2; that is,

$$f(y \mid x = X) = \frac{1}{\sigma\sqrt{2\pi}} e^{-(1/2\sigma^2)(y-\alpha-\beta X)^2}$$

10.4.2 Point Estimation. If n pairs of sample values are selected at random from the distribution defined in the previous article, the likelihood function is

$$L = f(y_1,x_1,y_2,x_2, \ldots ,y_n,x_n) = h(x_1)h(x_2) \cdots h(x_n) \frac{e^{-(1/2\sigma^2)\Sigma(y_i-\alpha-\beta x_i)^2}}{(2\pi\sigma^2)^{n/2}}$$

To find the maximum-likelihood estimates of α, β, and σ^2, the derivatives of log L with respect to the three parameters are set equal to zero. This gives ($\hat{\sigma}^2$ is corrected for bias)

$$\begin{aligned} \hat{\beta} &= \frac{\Sigma(y_i - \bar{y})(x_i - \bar{x})}{\Sigma(x_i - \bar{x})^2} \\ \hat{\alpha} &= \bar{y} - \hat{\beta}\bar{x} \\ \hat{\sigma}^2 &= \frac{1}{n-2} \Sigma(y_i - \hat{\alpha} - \hat{\beta}x_i)^2 \end{aligned} \qquad (10.5)$$

By methods exactly analogous to those used to prove Theorem 10.1, we have

◆ **Theorem 10.26** The maximum-likelihood estimates for the parameters α, β, and σ^2 in the density defined in Art. 10.4.1 are given in Eq. (10.5). These estimates are
(1) Efficient
(2) Sufficient
(3) Unbiased
(4) Consistent

10.4.3 Interval Estimation. By the same argument as that used for interval estimation for case 1, model 3, in Sec. 10.2, it follows that all the confidence-interval statements in Art. 6.3.1 hold for α, β, and σ^2 in case 2, model 3.

10.4.4 Tests of Hypotheses. Again by the same argument as that used for testing hypotheses for case 1, model 3, in Sec. 10.2, it follows that tests of hypotheses on α, β, and σ^2 as defined in Chap. 6 have the same probability of a type I error if the model is case 2, model 3, as defined in Art. 10.4.1.

10.5 Case 3, Model 3

10.5.1 Definitions. Let x and y have a bivariate normal distribution with means μ_1 and μ_2, variances σ_1^2 and σ_2^2, respectively, and covariance σ_{12}. Suppose, however, that we cannot observe y and x, but that we observe u and v where $u = x + d$, $v = y + e$, where e and d are measurement errors that are independent normal variables with means 0 and variances σ_e^2 and σ_d^2, respectively. Also, let e and d be independent of y and x.

The main objective in a model such as this is to use the regression function of y and x to predict y from a knowledge of x. However, because of measurement errors, y and x cannot be observed. This was the case in an extensive set of measurements on the constituents in cow's milk. It was desired to predict the per cent protein in cow's milk by measuring the per cent butterfat, since the per cent butterfat is fairly inexpensive to measure. However, in a given day's milk from a cow, the per cent butterfat and per cent protein cannot be measured accurately. A few samples of the total day's production of milk is taken and measured; so u and v are observed instead of y and x.

In one sense there is always measurement error present when continuous variables are involved. The experimenter must decide whether these measurement errors are large enough to be of importance or can be ignored.

Returning to the above definitions of x, y, u, and v, we have, by Theorem 3.22, the following properties:

1. u and v have a bivariate normal distribution with means μ_1 and μ_2, variances $\sigma_1^2 + \sigma_d^2$ and $\sigma_2^2 + \sigma_e^2$, respectively, and covariance σ_{12}.

2. u and y have a bivariate normal distribution with means μ_1 and μ_2, variances $\sigma_1^2 + \sigma_d^2$ and σ_2^2, respectively, and covariance σ_{12}.

3. v and x have a bivariate normal distribution with means μ_1 and μ_2, variances $\sigma_2^2 + \sigma_e^2$ and σ_1^2, respectively, and covariance σ_{12}.

4. x and y have a bivariate normal distribution with means μ_1 and μ_2, variances σ_1^2 and σ_2^2, respectively, and covariance σ_{12}.

10.5.2 The Regression Functions. In functional relationships, say $Y = \alpha + \beta X$, the problem is to estimate α and β. If $\hat{\alpha}$ and $\hat{\beta}$ are unbiased estimates, $\hat{Y} = \hat{\alpha} + \hat{\beta}X$ is an unbiased estimate of Y.

If there is no functional relationship between two random variables y and x but if they have a bivariate normal distribution, then the regression function of y on x can be used as a predictor. The regression function is the expected value of y in the conditional distribution of y given $x = X$. We shall write this

$$E(y \mid x = X) = Y_x = \beta_0 + \beta_1 X$$

where $$\beta_0 = \mu_2 - \frac{\sigma_{12}}{\sigma_1^2}\mu_1 \qquad \text{and} \qquad \beta_1 = \frac{\sigma_{12}}{\sigma_1^2}$$

The problem then reduces to taking a set of n pairs of values y_i, x_i and estimating β_0 and β_1. We could then write $\hat{Y}_x = \hat{\beta}_0 + \hat{\beta}_1 X$, and

$\hat{Y}_x$ would be an unbiased estimate of Y_x if $\hat{\beta}_0$ and $\hat{\beta}_1$ were unbiased estimates of β_0 and β_1, respectively. Unfortunately, x_i cannot be measured; so we must use u in place of x to estimate y. That is to say, we must use the regression of y on u, which we shall write as $Y_u = \alpha_0 + \alpha_1 U$, where

$$\alpha_0 = \mu_2 - \frac{\sigma_{12}}{\sigma_1^2 + \sigma_d^2}\mu_1 \qquad \text{and} \qquad \alpha_1 = \frac{\sigma_{12}}{\sigma_1^2 + \sigma_d^2}$$

Therefore, $\hat{Y}_u = \hat{\alpha}_0 + \hat{\alpha}_1 U$ is an unbiased estimate of Y_u if $\hat{\alpha}_0$ and $\hat{\alpha}_1$ are unbiased estimates of α_0 and α_1, respectively. Again there is difficulty, because we cannot take pairs of values u_i, y_i in order to obtain $\hat{\alpha}_0$ and $\hat{\alpha}_1$. The only pairs of values that can be observed are u_i, v_i. Therefore, we shall examine the regression of v on u, denoted by $V_u = \gamma_0 + \gamma_1 U$, where

$$\gamma_0 = \mu_2 - \frac{\sigma_{12}}{\sigma_1^2 + \sigma_d^2}\mu_1 \qquad \text{and} \qquad \gamma_1 = \frac{\sigma_{12}}{\sigma_1^2 + \sigma_d^2}$$

Therefore, $\gamma_0 = \alpha_0$ and $\gamma_1 = \alpha_1$; so we observe n pairs of values u_i, v_i and get

$$\hat{\gamma}_0 = \hat{\alpha}_0 = \bar{v} - \hat{\alpha}_1\bar{u} \qquad \text{and} \qquad \hat{\gamma}_1 = \hat{\alpha}_1 = \frac{\Sigma(u_i - \bar{u})(v_i - \bar{v})}{\Sigma(u_i - \bar{u})^2}$$

Then $\hat{Y}_u = \hat{\gamma}_0 + \hat{\gamma}_1 U$ can be used to predict y. This is the solution to the problem that we set out to obtain.

10.5.3 Example. The ideas in the previous articles concerning case 3, model 3, will be expanded by continuing the example given in Art. 10.5.1. Let x, y represent respectively the per cent butterfat content and per cent protein of one day's production of milk of one cow. If we consider all the x, y values for all cows in a given dairy herd for a period of a year or more, we can imagine x, y as having a bivariate normal distribution with parameters as given in Art. 10.5.1. It would be desirable to know the per cent protein of each day's production of each cow. This seems impossible to predict for future days. One alternative is to choose one number and use it to represent each future day's production for each cow. The number that would be selected is the mean of the y values, that is, μ_2. Since μ_2 is the mean of a normal population with variance σ_2^2, it follows that, regardless of the true value y of a given quantity of milk, we would know that about 95 per cent of the time, the true value y will be within $2\sigma_2$ units of μ_2. Since μ_2 is not known, a few samples would be taken and an estimate $\hat{\mu}_2 = \bar{y}$ obtained. Therefore, $\bar{y}$ would be used to represent μ_2, which is the best number available to represent each cow's daily production of per cent protein. The variance of $\bar{y}$ is σ_2^2/n. Since a day's production of milk for each cow is (say) in the neighborhood of 3 or 4 gallons, the exact per cent protein for this amount could not be measured. A few samples would be secured and the per cent protein ascertained. Accordingly, we observe not y but v. So we cannot use $\bar{y}$ to represent μ_2; we must use instead $\bar{v}$ obtained from a sample of n. However, since $E(\bar{v}) = \mu_2$, $\bar{v}$, like $\bar{y}$, is an unbiased estimate of μ_2. The variance of $\bar{v}$ is $(\sigma_2^2 + \sigma_e^2)/n$, which is larger than the variance of $\bar{y}$. We must pay a price for not being able to measure y exactly, and this price is the increased variance of the estimator of μ_2.

The experimenter may feel that μ_2 is a poor number to represent per cent protein of a day's production of a given cow, and he may prefer to find something better. Since x, y are distributed as a bivariate normal, the regression of y on x can be used to represent per cent protein if the per cent butterfat x can be measured. Thus $\beta_0 + \beta_1 X$ would be used to represent the per cent protein of a certain cow's production on a given day, where X is the per cent butterfat of that particular cow's milk on that particular day.

By normal regression theory we know that the conditional distribution of y given x has a mean of $\beta_0 + \beta_1 X$ and variance

$$\sigma_2^2\left(1 - \frac{\sigma_{12}^2}{\sigma_1^2\sigma_2^2}\right) = \sigma^2$$

So it follows that, if $\beta_0 + \beta_1 X$ is used to represent the per cent protein of a quantity of milk when X is the per cent butterfat of the same quantity, then, about 95 per cent of the time, the true value y will be within 2σ units of $\beta_0 + \beta_1 X$. Then, since $\sigma^2 \leqslant \sigma_2^2$, the regression estimate is generally better than using μ_2.

Since x is never known, the quantity $\beta_0 + \beta_1 X$ cannot be used even if β_0 and β_1 are given. What is needed is a regression function based on u, the *observed* butterfat content. The regression equation $Y_u = \alpha_0 + \alpha_1 U$, based on the distribution of u and y, can be used. However, α_0 and α_1 are unknowns. So, as explained in the preceding article, the regression of u on v can be used to obtain an unbiased estimate of α_0 and α_1, and then $\hat{Y}_u = \hat{\alpha}_0 + \hat{\alpha}_1 U$ can be used as the prediction equation.

Problems

10.1 If (x_i, y_i), where $i = 1, 2, \ldots, n$, is an n-sample from the bivariate normal distribution defined in Art. 10.2.2, show:

(*a*) $\hat{\mu}_1 = \bar{x}$ is an unbiased estimate of μ_1.

(*b*) $\hat{\mu}_2 = \bar{y}$ is an unbiased estimate of μ_2.

(*c*) $\dfrac{1}{n-1}\Sigma(x_i - \bar{x})^2$ is an unbiased estimate of σ_{11}.

(*d*) $\dfrac{1}{n-1}\Sigma(y_i - \bar{y})^2$ is an unbiased estimate of σ_{22}.

(*e*) $\dfrac{1}{n-1}\Sigma(x_i - \bar{x})(y_i - \bar{y})$ is an unbiased estimate of σ_{12}.

10.2 The following data are assumed to be a random sample of n vectors from a bivariate normal:

x_i	6.8	17.3	15.1	26.7	13.2	5.2	17.1	8.3
y_i	27.0	35.1	36.2	45.1	32.8	27.1	39.5	30.1

(*a*) Use Theorem 10.9 to test $\boldsymbol{\mu}' = (15, 30)$.
(*b*) Find $\hat{\beta}_0$, $\hat{\beta}_1$ and test $\beta_1 = 0$.
(*c*) Set a 95 per cent confidence interval on β_1.
(*d*) Set a 95 per cent confidence interval on β_0.
10.3 Evaluate K in the integral in Eq. (10.4).
10.4 Prove Theorem 10.5.
10.5 Prove Theorem 10.12 by using Theorem 10.11.
10.6 Prove Theorem 10.13.
10.7 Prove Theorem 10.14.
10.8 Suppose the following estimates of correlation coefficients were computed from bivariate normal distributions (n_i are the respective sample sizes):

$\hat{\rho}_i$	.04	.05	.81	.72
n_i	100	120	90	95

(*a*) Use Fisher's transformation to test $\rho_1 = 0$ against the alternative $\rho_1 > 0$.
(*b*) Use Student's t to test $\rho_1 = 0$ against the alternative $\rho_1 > 0$.
10.9 In Prob. 10.8, use Fisher's transformation to test $\rho_1 = \rho_2$ against the alternative $\rho_1 \neq \rho_2$.
10.10 (*a*) In Prob. 10.8, use Fisher's transformation to test the hypothesis $\rho_4 = .80$ against the alternative $\rho_4 \neq .80$.
(*b*) Compute the power of the test for $\rho = .70$, $\rho = .75$, $\rho = .85$.
10.11 In Prob. 10.8, use Fisher's transformation to test $\rho_3 = \rho_4 = .75$.
10.12 In Prob. 10.8, use Fisher's transformation to set a 95 per cent confidence interval on ρ_3.
10.13 In Prob. 10.8, use Fisher's transformation to set a 95 per cent confidence interval on ρ_4.
10.14 If $\rho_3 = \rho_4$, use the data in Prob. 10.8 to find a combined estimate.
10.15 The 3×1 vector $\mathbf{Y}$ is distributed $N(\boldsymbol{\mu}, \mathbf{V})$. On the basis of 100 observations the following matrix was computed:

$$\mathbf{S} = \begin{pmatrix} 96 & 67 & -10 \\ 67 & 48 & -8 \\ -10 & -8 & 20 \end{pmatrix}$$

Find $\hat{\rho}_{12\cdot3}$ and $\hat{\rho}_{13\cdot2}$.
10.16 In Prob. 10.15, test the hypothesis $\rho_{12\cdot3} = 0$ against the alternative $\rho_{12\cdot3} \neq 0$ at the 5 per cent level.
10.17 In Prob. 10.15, set a 95 per cent confidence interval on $\rho_{13\cdot2}$.

Further Reading

1 A. M. Mood: "Introduction to the Theory of Statistics," McGraw-Hill Book Company, Inc., New York, 1950.

2 P. G. Hoel: "Introduction to Mathematical Statistics," John Wiley & Sons, Inc., New York, 1956.

3 M. G. Kendall: "The Advanced Theory of Statistics," vols. I, II, Charles Griffin & Co. Ltd., London, 1946.

4 G. W. Snedecor: "Statistical Methods," Iowa State College Press, Ames, Iowa, 1958.

5 R. L. Anderson and T. A. Bancroft: "Statistical Theory in Research," McGraw-Hill Book Company, Inc., New York, 1952.

6 T. W. Anderson: "An Introduction to Multivariate Statistical Analysis," John Wiley & Sons, Inc., New York, 1957.

7 S. S. Wilks: "Mathematical Statistics," Princeton University Press, Princeton, N.J., 1943.

8 C. R. Rao: "Advanced Statistical Methods in Biometric Research," John Wiley & Sons, Inc., New York, 1952.

9 F. N. David: "Tables of the Correlation Coefficient," Biometrika Office, University College, London, 1938.

10 R. A. Fisher: Frequency Distribution of the Values of the Correlation Coefficient in Samples from an Indefinitely Large Population, *Biometrika*, vol. 10, pp. 507–521, 1915.

11 R. A. Fisher: On the Probable Error of a Coefficient of Correlation Deduced from a Small Sample, *Metron*, vol. 1, part 4, pp. 3–32, 1921.

12 R. A. Fisher: The Distribution of the Partial Correlation Coefficient, *Metron*, vol. 3, pp. 329–332, 1921.

13 R. A. Fisher: The General Sampling Distribution of the Multiple Correlation Coefficient, *Proc. Roy. Soc. London*, ser. A, vol. 121, pp. 654–673, 1928.

11

Model 4: Experimental Design Models

11.1 Introduction

In Art. 5.3.4 it was stated that model 4 is defined as

$$y = \beta_1 X_1 + \beta_2 X_2 + \cdots + \beta_k X_k + e$$

where y and e are random variables, β_i are unknown parameters, and the X_i take only the values 0 and 1. To explain this in more detail, let us consider two examples.

Suppose a manufacturer of light bulbs, looking for ways to increase the life of his product, has developed two chemical coatings to put on the filaments of the bulbs. If no coating is put on the filament, the bulbs will last an average of μ (unknown) hr. The manufacturer assumes that, if he puts coating 1 on the filament, it will increase the filament life by τ_1 (unknown) hr and that, if he uses coating 2, it will increase the life by τ_2 hr. Now, he would like to find τ_1 and τ_2, the number of hours that coatings 1 and 2, respectively, lengthen the life of the bulbs (τ_1 or τ_2 may be zero or negative), and $\tau_1 - \tau_2$, the difference between the effects of the two coatings. To evaluate these constants the manufacturer could make a bulb with coating 1 and test it to see how many hours it will last, then make a bulb with coating 2 and see how many hours it will last. The model can be written

$$y_1 = \mu + \tau_1 + e_1$$
$$y_2 = \mu + \tau_2 + e_2$$

where y_i is the number of hours that the bulb with coating i lasts, and where e_i is a random error due to all the uncontrolled factors, such as nonconstant voltage, imperfections in the manufacture of the various components, temperature, and humidity. Now, the manufacturer will probably not be content to draw conclusions based on observation of only one light bulb for each coating. Let us say that he decides to make six bulbs and coat three with chemical 1 and three with chemical 2. The model can be written

$$\begin{aligned} y_{11} &= \mu + \tau_1 + e_{11} \\ y_{12} &= \mu + \tau_1 + e_{12} \\ y_{13} &= \mu + \tau_1 + e_{13} \\ y_{21} &= \mu + \tau_2 + e_{21} \\ y_{22} &= \mu + \tau_2 + e_{22} \\ y_{23} &= \mu + \tau_2 + e_{23} \end{aligned} \tag{11.1}$$

or, more compactly,

$$y_{ij} = \mu + \tau_i + e_{ij} \qquad i = 1, 2; \quad j = 1, 2, 3$$

where y_{ij} is the observed hours of life of the jth bulb (by some system of identification) that has received the ith chemical coating, and where e_{ij} is the associated random error.

If we write the above system in matrix form similar to that used for model 1, we get $\mathbf{Y} = \mathbf{X}\boldsymbol{\beta} + \mathbf{e}$, or

$$\begin{pmatrix} y_{11} \\ y_{12} \\ y_{13} \\ y_{21} \\ y_{22} \\ y_{23} \end{pmatrix} = \begin{pmatrix} 1 & 1 & 0 \\ 1 & 1 & 0 \\ 1 & 1 & 0 \\ 1 & 0 & 1 \\ 1 & 0 & 1 \\ 1 & 0 & 1 \end{pmatrix} \begin{pmatrix} \mu \\ \tau_1 \\ \tau_2 \end{pmatrix} + \begin{pmatrix} e_{11} \\ e_{12} \\ e_{13} \\ e_{21} \\ e_{22} \\ e_{23} \end{pmatrix} \tag{11.2}$$

In this case $\mathbf{X}$ is a 6×3 matrix, and its rank is 2. Hence, this model is not of full rank, and the theorems of Chap. 6 are not immediately applicable for estimating and testing hypotheses about the components of $\boldsymbol{\beta}$, that is, μ, τ_1, τ_2.

Suppose the manufacturer is interested only in $\mu + \tau_1$ and $\mu + \tau_2$ rather than in τ_1 and τ_2. If we write $\mu + \tau_1 = \alpha_1$ and $\mu + \tau_2 = \alpha_2$, we can write the model as

$$\begin{aligned} y_{11} &= \alpha_1 + e_{11} \\ y_{12} &= \alpha_1 + e_{12} \\ y_{13} &= \alpha_1 + e_{13} \\ y_{21} &= \alpha_2 + e_{21} \\ y_{22} &= \alpha_2 + e_{22} \\ y_{23} &= \alpha_2 + e_{23} \end{aligned} \tag{11.3}$$

or in matrix notation as $\mathbf{Y} = \mathbf{U}\boldsymbol{\alpha} + \mathbf{e}$, or

$$\begin{pmatrix} y_{11} \\ y_{12} \\ y_{13} \\ y_{21} \\ y_{22} \\ y_{23} \end{pmatrix} = \begin{pmatrix} 1 & 0 \\ 1 & 0 \\ 1 & 0 \\ 0 & 1 \\ 0 & 1 \\ 0 & 1 \end{pmatrix} \begin{pmatrix} \alpha_1 \\ \alpha_2 \end{pmatrix} + \begin{pmatrix} e_{11} \\ e_{12} \\ e_{13} \\ e_{21} \\ e_{22} \\ e_{23} \end{pmatrix} \tag{11.4}$$

The $\mathbf{U}$ matrix is 6×2, and its rank is 2. Hence, this matrix is of full rank, and all the theorems of Chap. 6 are applicable for estimating and testing hypotheses about the elements α_1 and α_2. So, when the matrix is not of full rank, we use this alternative approach. The model (11.4) is called a *reparametrization* of the model (11.2). We shall define this transformation later.

As another example, suppose a researcher wants to measure the effect of two different chemical compounds and two different methods of applying these compounds on the yield of a certain variety of corn. Suppose τ_i is the effect of the ith chemical ($i = 1, 2$) and β_j is the effect of the jth method of applying the compounds ($j = 1, 2$). If the researcher assumes that the effects are additive, he might assume the following model:

$$\begin{aligned} y_{11} &= \mu + \tau_1 + \beta_1 + e_{11} \\ y_{12} &= \mu + \tau_1 + \beta_2 + e_{12} \\ y_{21} &= \mu + \tau_2 + \beta_1 + e_{21} \\ y_{22} &= \mu + \tau_2 + \beta_2 + e_{22} \end{aligned} \tag{11.5}$$

or

$$y_{ij} = \mu + \tau_i + \beta_j + e_{ij} \qquad i = 1, 2; \quad j = 1, 2$$

where y_{ij} is the observed yield of corn on a certain plot of ground. The researcher is assuming that the observation y_{ij} is equal to a constant μ (the average yield when no chemical is applied) plus the effect due to the ith chemical, plus the effect due to the jth method of application, plus a random error due to all the uncontrolled factors, such as differences of fertility among the plots. The experimenter may desire to test or estimate the parameters μ, τ_i, β_j. In matrix notation this model is

$$\begin{pmatrix} y_{11} \\ y_{12} \\ y_{21} \\ y_{22} \end{pmatrix} = \begin{pmatrix} 1 & 1 & 0 & 1 & 0 \\ 1 & 1 & 0 & 0 & 1 \\ 1 & 0 & 1 & 1 & 0 \\ 1 & 0 & 1 & 0 & 1 \end{pmatrix} \begin{pmatrix} \mu \\ \tau_1 \\ \tau_2 \\ \beta_1 \\ \beta_2 \end{pmatrix} + \begin{pmatrix} e_{11} \\ e_{12} \\ e_{21} \\ e_{22} \end{pmatrix}$$

The **X** matrix is 4×5 with rank 3, and does not fit into the framework of model 1. In defining model 4, we have stressed the fact that the X_i can take only the values 0 and 1. However, if this were the only distinction between model 4 and model 1, all the theorems in Chap. 6 would be applicable to model 4 also. The main distinction to be noticed between model 1 and model 4 is that in model 4 the rank of the $n \times p$ ($p \leqslant n$) matrix **X** is equal to $k < p$. Thus the matrix $\mathbf{X}'\mathbf{X}$, which occurs so frequently, does not have an inverse. Hence the theorems in Chap. 6 are not directly applicable to model 4.

We shall now formulate a definition of model 4.

◆ **Definition 11.1** If the linear matrix model $\mathbf{Y} = \mathbf{X}\boldsymbol{\beta} + \mathbf{e}$ is as defined in Definition 6.1 except that the matrix **X** consists of 0's and 1's and that the rank of **X** is equal to $k < p$, then $\mathbf{Y} = \mathbf{X}\boldsymbol{\beta} + \mathbf{e}$ will be called model 4, sometimes referred to as a general-linear-hypothesis model of less than full rank or an experimental design model.

Two cases will be discussed:

Case A: The vector **e** is distributed $N(\mathbf{0},\sigma^2\mathbf{I})$; that is, the errors are independent normal variables with means 0 and variances σ^2.

Case B: The vector **e** is such that $E(\mathbf{e}) = \mathbf{0}$ and $\text{cov}(\mathbf{e}) = \sigma^2\mathbf{I}$; that is, the errors are uncorrelated, with means 0 and variances σ^2.

We shall be interested in the following:

1. Point estimates of σ^2 and certain linear functions of the β_i
2. Interval estimates of certain linear functions of the β_i
3. Tests of the hypothesis that certain linear functions of the parameters are equal to known constants
4. Tests of the hypothesis $\beta_1 = \beta_2 = \cdots = \beta_q$ ($q \leqslant p$)

We shall examine item 1 for both case A and case B, but 2, 3, and 4 for case A only.

11.2 Point Estimation

We shall first discuss point estimates of linear functions of the β_i under case B. We shall assume model 4, $\mathbf{Y} = \mathbf{X}\boldsymbol{\beta} + \mathbf{e}$, where **X** is of order $n \times p$ and rank $k < p \leqslant n$, and where the errors are such that $E(\mathbf{e}) = \mathbf{0}$, $E(\mathbf{ee}') = \sigma^2\mathbf{I}$.

11.2.1 Point Estimate of a Linear Combination of β_i under Case B. If we use least squares, we must minimize $\mathbf{e}'\mathbf{e} = (\mathbf{Y} - \mathbf{X}\boldsymbol{\beta})'(\mathbf{Y} - \mathbf{X}\boldsymbol{\beta})$. The partial derivative of $\mathbf{e}'\mathbf{e}$ with respect to $\boldsymbol{\beta}$ gives

$$\frac{\partial(\mathbf{e}'\mathbf{e})}{\partial\boldsymbol{\beta}} = 2\mathbf{X}'\mathbf{X}\boldsymbol{\beta} - 2\mathbf{X}'\mathbf{Y}$$

If this is set equal to $\mathbf{0}$, the resulting normal equations are $\mathbf{X}'\mathbf{X}\hat{\boldsymbol{\beta}} = \mathbf{X}'\mathbf{Y}$.

If $\mathbf{X}$ is of full rank, $\mathbf{X}'\mathbf{X}$ has an inverse and the solution for $\hat{\boldsymbol{\beta}}$ is unique, and this is an unbiased estimate of $\boldsymbol{\beta}$. However, if $\mathbf{X}$ is of less than full rank, then $\mathbf{X}'\mathbf{X}$ has no inverse, and we must examine the system to see whether a solution exists. Using Theorem 1.51, we see that the coefficient matrix is $\mathbf{X}'\mathbf{X} = \mathbf{A}$ and that the augmented matrix is $(\mathbf{X}'\mathbf{X} \mid \mathbf{X}'\mathbf{Y}) = \mathbf{B}$. We can write $\mathbf{B}$ as $\mathbf{X}'(\mathbf{X} \mid \mathbf{Y})$; thus, the rank of $\mathbf{B}$ is less than or equal to the rank of $\mathbf{X}'$ (which is precisely the rank of $\mathbf{A}$). But the rank of the augmented matrix must be greater than or equal to the rank of the coefficient matrix (since augmenting a matrix by another matrix or vector cannot decrease the dimension of the largest non-vanishing determinant in the original matrix). Hence the rank of $(\mathbf{X}'\mathbf{X} \mid \mathbf{X}'\mathbf{Y})$ must be equal to the rank of $\mathbf{X}'\mathbf{X}$, and the system is consistent. However, since $\mathbf{X}'\mathbf{X}$ is of dimension $p \times p$ and rank $k < p$, Theorem 1.52 states that there are an infinite number of different vectors $\hat{\boldsymbol{\beta}}$ that satisfy $\mathbf{X}'\mathbf{X}\hat{\boldsymbol{\beta}} = \mathbf{X}'\mathbf{Y}$. This does not seem to be a very happy state of affairs, for two researchers with the same data, both using the same method of estimation, can draw different conclusions.

Since any solution to the normal equations $\mathbf{X}'\mathbf{X}\hat{\boldsymbol{\beta}} = \mathbf{X}'\mathbf{Y}$ for $\hat{\boldsymbol{\beta}}$ will be linear functions of the observation vector $\mathbf{Y}$, it might be interesting to see whether there are *any* linear functions of $\mathbf{Y}$ whatsoever that give rise to *unbiased* estimates of $\boldsymbol{\beta}$. Let $\mathbf{C}$ be a $p \times n$ matrix whose elements are constants independent of $\boldsymbol{\beta}$. Does there exist a $\mathbf{C}$ such that $E(\mathbf{CY}) = \boldsymbol{\beta}$? If so, we get $E[\mathbf{C}(\mathbf{X}\boldsymbol{\beta} + \mathbf{e})] = \mathbf{CX}\boldsymbol{\beta}$ and, if $\mathbf{CY}$ is unbiased, this must equal $\boldsymbol{\beta}$, or $\mathbf{CX} = \mathbf{I}$. But this result is impossible, since $\mathbf{CX}$ is at most of rank k and $\mathbf{I}$ is of rank $p > k$. So there exist no linear functions of the observations y_i that yield unbiased estimates of $\boldsymbol{\beta}$.

It would seem natural to investigate next whether there exists an unbiased estimate of any linear combination of the β_i. That is to say, if $\boldsymbol{\lambda}$ is a known $p \times 1$ vector of constants, does there exist an unbiased estimate of $\boldsymbol{\lambda}'\boldsymbol{\beta}$, where $\boldsymbol{\lambda}'\boldsymbol{\beta} = \Sigma\lambda_i\beta_i$? This is certainly not true for every $\boldsymbol{\lambda}$; for, if $\lambda_t = 0$ for all t except $t = i$ and if $\lambda_i = 1$, then $\boldsymbol{\lambda}'\boldsymbol{\beta} = \beta_i$, and if there were an unbiased estimate of $\boldsymbol{\lambda}'\boldsymbol{\beta} = \beta_i$ there would then be an unbiased estimate of $\boldsymbol{\beta}$, and we have shown above that there is none. Before proceeding further we shall formulate two useful definitions.

◆ **Definition 11.2** A parameter (a function of parameters) is said to be estimable if there exists an unbiased estimate of the parameter (of the function of the parameters).

◆ **Definition 11.3** A parameter (a function of parameters) is said to be *linearly* estimable if there exists a linear combination of the observations whose expected value is equal to the parameter (the function of parameters), i.e., if there exists an unbiased estimate.

Throughout this chapter, *unless specifically stated otherwise*, we shall mean *linearly estimable* when we say *estimable*.

We shall now prove a theorem that will aid in determining whether a given linear function is estimable.

◆ **Theorem 11.1** In model 4, case B, the linear combination $\boldsymbol{\lambda}'\boldsymbol{\beta}$ ($\boldsymbol{\lambda}$ is a vector of known constants) is estimable if and only if there exists a solution for $\mathbf{r}$ in the equations $\mathbf{X}'\mathbf{X}\mathbf{r} = \boldsymbol{\lambda}$.

Proof: We shall let $\mathbf{S} = \mathbf{X}'\mathbf{X}$. We need to show that there is a vector $\mathbf{b}$ such that $E(\mathbf{b}'\mathbf{Y}) = \boldsymbol{\lambda}'\boldsymbol{\beta}$ if and only if there exists a vector $\mathbf{r}$ such that $\mathbf{S}\mathbf{r} = \boldsymbol{\lambda}$. If $\boldsymbol{\lambda}'\boldsymbol{\beta}$ is estimable, there is a vector $\mathbf{b}$ such that

$$E(\mathbf{b}'\mathbf{Y}) = \mathbf{b}'\mathbf{X}\boldsymbol{\beta} = \boldsymbol{\lambda}'\boldsymbol{\beta}$$

which implies that $\mathbf{b}'\mathbf{X} = \boldsymbol{\lambda}'$, or $\mathbf{X}'\mathbf{b} = \boldsymbol{\lambda}$. This implies that the rank of the coefficient matrix $\mathbf{X}'$ equals the rank of the augmented matrix $(\mathbf{X}' \mid \boldsymbol{\lambda})$. If this is true, then the rank of $\mathbf{X}'\mathbf{X}$ equals the rank of $(\mathbf{X}'\mathbf{X} \mid \boldsymbol{\lambda})$ and, hence, $\mathbf{S}\mathbf{r} = \boldsymbol{\lambda}$ has a solution for $\mathbf{r}$. On the other hand, if $\mathbf{S}\mathbf{r} = \boldsymbol{\lambda}$ has a solution for $\mathbf{r}$, then we can write $\mathbf{X}'(\mathbf{X}\mathbf{r}) = \boldsymbol{\lambda}$ and let $\mathbf{b} = \mathbf{X}\mathbf{r}$.

We shall now prove a theorem for model 4 analogous to the Gauss-Markoff theorem for model 1.

◆ **Theorem 11.2** In model 4, case B, the best linear unbiased estimate for any estimable function $\boldsymbol{\lambda}'\boldsymbol{\beta}$ is $\mathbf{r}'\mathbf{X}'\mathbf{Y}$, where $\mathbf{r}$ satisfies the equation $\mathbf{S}\mathbf{r} = \boldsymbol{\lambda}$.

Proof: Let us assume that the best linear unbiased estimator of $\boldsymbol{\lambda}'\boldsymbol{\beta}$ is $\mathbf{b}'\mathbf{Y}$, where $\mathbf{b}' = \mathbf{r}'\mathbf{X}' + \mathbf{a}'$. Thus $\mathbf{b}$ is completely general, since $\mathbf{a}$ is general. We must determine the vector $\mathbf{a}$ such that

(1) $E(\mathbf{b}'\mathbf{Y}) = \boldsymbol{\lambda}'\boldsymbol{\beta}$

(2) $\text{var}(\mathbf{b}'\mathbf{Y})$ is less than any other linear function of $\mathbf{Y}$ that satisfies (1)

For unbiasedness, we get $E(\mathbf{b}'\mathbf{Y}) = \mathbf{b}'\mathbf{X}\boldsymbol{\beta} = (\mathbf{r}'\mathbf{X}'\mathbf{X} + \mathbf{a}'\mathbf{X})\boldsymbol{\beta} = \boldsymbol{\lambda}'\boldsymbol{\beta}$. Since our hypothesis states that $\mathbf{S}\mathbf{r} = \boldsymbol{\lambda}$, we must have $\mathbf{a}'\mathbf{X} = \mathbf{0}$. Also,

$$\begin{aligned}
\text{var}(\mathbf{b}'\mathbf{Y}) &= E(\mathbf{b}'\mathbf{Y} - \boldsymbol{\lambda}'\boldsymbol{\beta})^2 \\
&= E[(\mathbf{b}'\mathbf{Y} - \boldsymbol{\lambda}'\boldsymbol{\beta})(\mathbf{b}'\mathbf{Y} - \boldsymbol{\lambda}'\boldsymbol{\beta})'] \\
&= E[(\mathbf{b}'\mathbf{X}\boldsymbol{\beta} + \mathbf{b}'\mathbf{e} - \boldsymbol{\lambda}'\boldsymbol{\beta})(\mathbf{b}'\mathbf{X}\boldsymbol{\beta} + \mathbf{b}'\mathbf{e} - \boldsymbol{\lambda}'\boldsymbol{\beta})'] \\
&= E[(\mathbf{r}'\mathbf{X}'\mathbf{X}\boldsymbol{\beta} + \mathbf{a}'\mathbf{X}\boldsymbol{\beta} + \mathbf{b}'\mathbf{e} - \boldsymbol{\lambda}'\boldsymbol{\beta}) \\
&\qquad \times (\mathbf{r}'\mathbf{X}'\mathbf{X}\boldsymbol{\beta} + \mathbf{a}'\mathbf{X}\boldsymbol{\beta} + \mathbf{b}'\mathbf{e} - \boldsymbol{\lambda}'\boldsymbol{\beta})'] \\
&= E(\mathbf{b}'\mathbf{e}\mathbf{e}'\mathbf{b}) = \sigma^2\mathbf{b}'\mathbf{b} = \sigma^2(\mathbf{r}'\mathbf{X}' + \mathbf{a}')(\mathbf{X}\mathbf{r} + \mathbf{a}) \\
&= \sigma^2\mathbf{r}'\mathbf{X}'\mathbf{X}\mathbf{r} + \sigma^2\mathbf{a}'\mathbf{a}
\end{aligned}$$

(using the fact that $\mathbf{a}'\mathbf{X} = \mathbf{0}$).

Therefore, to minimize $\text{var}(\mathbf{b}'\mathbf{Y})$ we must minimize $\mathbf{a}'\mathbf{a} = \Sigma a_i^2$. This is a minimum when $\mathbf{a} = \mathbf{0}$ and this is consistent with $\mathbf{a}'\mathbf{X} = \mathbf{0}$; hence, $\mathbf{b}' = \mathbf{r}'\mathbf{X}'$, and the best linear unbiased estimate of the estimable function $\boldsymbol{\lambda}'\boldsymbol{\beta}$ is $\mathbf{r}'\mathbf{X}'\mathbf{Y}$.

Since $\mathbf{S}$ is not of full rank, there are infinitely many solutions to the system $\mathbf{Sr} = \boldsymbol{\lambda}$ if $\boldsymbol{\lambda}'\boldsymbol{\beta}$ is an estimable function. However, we shall now prove a theorem on uniqueness.

◆ **Theorem 11.3** If $\boldsymbol{\lambda}'\boldsymbol{\beta}$ is an estimable function, then any vector $\mathbf{r}$ that satisfies $\mathbf{Sr} = \boldsymbol{\lambda}$ gives the same estimate of $\boldsymbol{\lambda}'\boldsymbol{\beta}$; that is, if $\mathbf{r}_1$ and $\mathbf{r}_2$ both satisfy $\mathbf{Sr} = \boldsymbol{\lambda}$, then $\mathbf{r}_1'\mathbf{X}'\mathbf{Y} = \mathbf{r}_2'\mathbf{X}'\mathbf{Y} = \widehat{\boldsymbol{\lambda}'\boldsymbol{\beta}} = \boldsymbol{\lambda}'\hat{\boldsymbol{\beta}}$.

Proof: Let $\hat{\boldsymbol{\beta}}$ be any solution to the system $\mathbf{X}'\mathbf{X}\hat{\boldsymbol{\beta}} = \mathbf{X}'\mathbf{Y}$. Using $\mathbf{r}_1$ and $\mathbf{r}_2$, we have $\mathbf{r}_1'\mathbf{X}'\mathbf{X}\hat{\boldsymbol{\beta}} = \mathbf{r}_1'\mathbf{X}'\mathbf{Y}$ and $\mathbf{r}_2'\mathbf{X}'\mathbf{X}\hat{\boldsymbol{\beta}} = \mathbf{r}_2'\mathbf{X}'\mathbf{Y}$. But $\mathbf{r}_1'\mathbf{X}'\mathbf{X} = \mathbf{r}_2'\mathbf{X}'\mathbf{X} = \boldsymbol{\lambda}'$ by hypothesis; hence, $\widehat{\boldsymbol{\lambda}'\boldsymbol{\beta}} = \mathbf{r}_1'\mathbf{X}'\mathbf{Y} = \boldsymbol{\lambda}'\hat{\boldsymbol{\beta}}$ and $\widehat{\boldsymbol{\lambda}'\boldsymbol{\beta}} = \mathbf{r}_2'\mathbf{X}'\mathbf{Y} = \boldsymbol{\lambda}'\hat{\boldsymbol{\beta}}$.

That $\widehat{\boldsymbol{\lambda}'\boldsymbol{\beta}} = \boldsymbol{\lambda}'\hat{\boldsymbol{\beta}}$ means that, if $\boldsymbol{\lambda}'\boldsymbol{\beta}$ is estimable, $\boldsymbol{\lambda}'\hat{\boldsymbol{\beta}}$ is the estimate, where $\hat{\boldsymbol{\beta}}$ is any solution to the normal equations.

It will be important to know how many linear functions are estimable, but first we shall formulate the following.

◆ **Definition 11.4** The estimable functions $\boldsymbol{\lambda}_1'\boldsymbol{\beta}, \boldsymbol{\lambda}_2'\boldsymbol{\beta}, \ldots, \boldsymbol{\lambda}_q'\boldsymbol{\beta}$ are said to be *linearly independent estimable functions* if there exist vectors $\mathbf{r}_1, \mathbf{r}_2, \ldots, \mathbf{r}_q$ such that $\mathbf{Sr}_1 = \boldsymbol{\lambda}_1, \mathbf{Sr}_2 = \boldsymbol{\lambda}_2, \ldots, \mathbf{Sr}_q = \boldsymbol{\lambda}_q$ and if the vectors $\boldsymbol{\lambda}_1, \boldsymbol{\lambda}_2, \ldots, \boldsymbol{\lambda}_q$ are linearly independent. If the $\boldsymbol{\lambda}_i$ are not linearly independent but if the matrix $\boldsymbol{\Lambda} = (\boldsymbol{\lambda}_1, \boldsymbol{\lambda}_2, \ldots, \boldsymbol{\lambda}_q)$ has rank t, then the set contains t linearly independent estimable functions.

◆ **Theorem 11.4** There are exactly k linearly independent estimable functions, where k is the rank of $\mathbf{X}$.

Proof: We must show that there are exactly k linearly independent vectors $\boldsymbol{\lambda}$ that lead to a solution for $\mathbf{Sr} = \boldsymbol{\lambda}$, where k is the rank of $\mathbf{X}$. If there are q vectors satisfying $\mathbf{Sr} = \boldsymbol{\lambda}$, let them be $\boldsymbol{\lambda}_1, \boldsymbol{\lambda}_2, \ldots, \boldsymbol{\lambda}_q$ $(q > k)$. Then, by Theorem 11.1, there exist q vectors $\mathbf{r}_1, \mathbf{r}_2, \ldots, \mathbf{r}_q$ such that $\mathbf{Sr}_1 = \boldsymbol{\lambda}_1, \mathbf{Sr}_2 = \boldsymbol{\lambda}_2, \ldots, \mathbf{Sr}_q = \boldsymbol{\lambda}_q$, or $\mathbf{S}(\mathbf{r}_1, \mathbf{r}_2, \ldots, \mathbf{r}_q) = (\boldsymbol{\lambda}_1, \boldsymbol{\lambda}_2, \ldots, \boldsymbol{\lambda}_q)$, or $\mathbf{SR} = \boldsymbol{\Lambda}$, where $\mathbf{R}$ is the $p \times q$ matrix $(\mathbf{r}_1, \mathbf{r}_2, \ldots, \mathbf{r}_q)$ and $\boldsymbol{\Lambda} = (\boldsymbol{\lambda}_1, \boldsymbol{\lambda}_2, \ldots, \boldsymbol{\lambda}_q)$.

But $\mathbf{S}$ is of rank k; so $\boldsymbol{\Lambda}$ must be at most of rank k; hence there can be at most k linearly independent estimable functions.

Let $\mathbf{X}_i'$ be the ith row of $\mathbf{X}$. We shall show that $\mathbf{X}_i'\boldsymbol{\beta}$ is estimable for every i. To show that $\mathbf{X}_i'\boldsymbol{\beta}$ is estimable we must show that $\mathbf{X}'\mathbf{Xr} = \mathbf{X}_i$ admits a solution $\mathbf{r}$. The rank of $\mathbf{X}'\mathbf{X}$ equals the rank of $(\mathbf{X}'\mathbf{X} \mid \mathbf{X}_i)$. Hence, $\mathbf{X}_i'\boldsymbol{\beta}$ is estimable for $i = 1, 2, \ldots, n$. But these $\mathbf{X}_i$ form a set of k linearly independent vectors, and the theorem is proved.

Next we shall extend Definition 11.3.

◆ **Definition 11.5** Let $\mathbf{A}$ be a matrix such that $\mathbf{A} = (\mathbf{A}_1, \mathbf{A}_2, \ldots, \mathbf{A}_m)$ where $\mathbf{A}_i$ is $p \times 1$. When we say that the matrix function $\mathbf{A}'\boldsymbol{\beta}$ is estimable, we shall mean that each $\mathbf{A}_i'\boldsymbol{\beta}$ is estimable ($i = 1, 2, \ldots, m$).

◆ **Theorem 11.5** $\mathbf{X}\boldsymbol{\beta}$ and $\mathbf{X}'\mathbf{X}\boldsymbol{\beta}$ are estimable.

The proof will be left for the reader. As a consequence of this theorem, $E(y_i)$ is estimable for every i, since $E(\mathbf{Y}) = \mathbf{X}\boldsymbol{\beta}$.

◆ **Theorem 11.6** If $\boldsymbol{\lambda}_1'\boldsymbol{\beta}, \boldsymbol{\lambda}_2'\boldsymbol{\beta}, \ldots, \boldsymbol{\lambda}_q'\boldsymbol{\beta}$ are estimable, any linear combination of these is estimable.

Proof: Let $a_1, a_2, \ldots, a_q$ be constant scalars. We must show that $\boldsymbol{\lambda}'\boldsymbol{\beta}$ is an estimable function if $\boldsymbol{\lambda} = \Sigma a_i \boldsymbol{\lambda}_i$. Let $\mathbf{r}_i$ be such that $\mathbf{S}\mathbf{r}_i = \boldsymbol{\lambda}_i$ and $\mathbf{r} = \Sigma a_i \mathbf{r}_i$. Then $\mathbf{r}$ satisfies $\mathbf{S}\mathbf{r} = \boldsymbol{\lambda}$, and the proof is complete.

◆ **Corollary 11.6.1** If $\boldsymbol{\lambda}'\boldsymbol{\beta}$ is estimable, then $\boldsymbol{\lambda}' = \mathbf{a}'\mathbf{X}$, where $\mathbf{a}'$ is a $1 \times n$ vector; that is to say, if $\boldsymbol{\lambda}'\boldsymbol{\beta}$ is estimable, $\boldsymbol{\lambda}'$ must be a linear combination of the rows of the matrix $\mathbf{X}$.

Proof: $\mathbf{X}\boldsymbol{\beta}$ contains k linearly independent estimable functions; since there are exactly k such functions, $\boldsymbol{\lambda}'$ must be a linear combination of the rows of $\mathbf{X}$.

This corollary is valuable in determining what functions are estimable by inspection of the model $\mathbf{Y} = \mathbf{X}\boldsymbol{\beta} + \mathbf{e}$. This will be illustrated in detail later.

◆ **Corollary 11.6.2** If $\boldsymbol{\lambda}'\boldsymbol{\beta}$ is estimable, there exists a vector $\mathbf{b}$ such that $\boldsymbol{\lambda}' = \mathbf{b}'\mathbf{X}'\mathbf{X}$, and the best linear unbiased estimate of $\boldsymbol{\lambda}'\boldsymbol{\beta}$ is $\mathbf{b}'\mathbf{X}'\mathbf{Y}$.

Thus we can determine from the normal equations what is estimable and what the best linear unbiased estimate is.

◆ **Corollary 11.6.3** Let $\mathbf{C}$ be any $m \times n$ matrix; then $E(\mathbf{CY})$ is estimable.

Proof: Since $E(\mathbf{CY}) = \mathbf{CX}\boldsymbol{\beta}$, the result follows from Theorems 11.5 and 11.6.

◆ **Theorem 11.7** The best linear unbiased estimate of a linear combination of estimable functions is given by the same linear combination of the best linear unbiased estimates of the estimable functions; i.e., if $\mathbf{r}_i'\mathbf{X}'\mathbf{Y}$ $(i = 1, 2, \ldots, q)$ are the best linear unbiased estimates of the estimable functions $\boldsymbol{\lambda}_i'\boldsymbol{\beta}$ $(i = 1, 2, \ldots, q)$, then $\Sigma a_i \mathbf{r}_i'\mathbf{X}'\mathbf{Y}$ is the best linear unbiased estimate of $\Sigma a_i \boldsymbol{\lambda}_i'\boldsymbol{\beta}$.

The proof will be left for the reader.

11.2.2 Examples. To illustrate the preceding theorems, let us consider the model given in Eqs. (11.1) and (11.2).

$$y_{ij} = \mu + \tau_i + e_{ij} \qquad i = 1, 2; \quad j = 1, 2, 3$$

We have already shown that $\boldsymbol{\beta}$ is not estimable, since $\mathbf{X}$ is of less than full rank. The first question we shall answer is: Is $\tau_1 - \tau_2$ estimable? It is clear that

$$\tau_1 - \tau_2 = (0, 1, -1)\begin{pmatrix} \mu \\ \tau_1 \\ \tau_2 \end{pmatrix} = \boldsymbol{\lambda}'\boldsymbol{\beta}$$

We must see whether there exists a vector $\mathbf{r}$ that is a solution to the equations $\mathbf{X}'\mathbf{X}\mathbf{r} = \boldsymbol{\lambda}$. Now

$$\mathbf{X}'\mathbf{X} = \begin{pmatrix} 6 & 3 & 3 \\ 3 & 3 & 0 \\ 3 & 0 & 3 \end{pmatrix}$$

and $\mathbf{S}\mathbf{r} = \boldsymbol{\lambda}$ gives three equations and three unknowns.

$$\begin{pmatrix} 6 & 3 & 3 \\ 3 & 3 & 0 \\ 3 & 0 & 3 \end{pmatrix}\begin{pmatrix} r_1 \\ r_2 \\ r_3 \end{pmatrix} = \begin{pmatrix} 0 \\ 1 \\ -1 \end{pmatrix}$$

To investigate the existence of a solution, we see that the rank of $\mathbf{X}'\mathbf{X}$ is 2. The augmented matrix is

$$\left(\begin{array}{ccc|c} 6 & 3 & 3 & 0 \\ 3 & 3 & 0 & 1 \\ 3 & 0 & 3 & -1 \end{array}\right)$$

and this matrix also has rank 2. Writing out the system and solving in terms of r_1

$$r_2 = \tfrac{1}{3} - r_1$$
$$r_3 = -\tfrac{1}{3} - r_1$$

By fixing r_1 at any value we please, we get a solution to the equations $\mathbf{X'Xr} = \boldsymbol{\lambda}$. For example, if we let $r_1 = 0$, we get

$$\mathbf{r} = \begin{pmatrix} 0 \\ \frac{1}{3} \\ -\frac{1}{3} \end{pmatrix}$$

and
$$\mathbf{r'X'Y} = (0, \tfrac{1}{3}, -\tfrac{1}{3}) \begin{pmatrix} Y_{..} \\ Y_{1.} \\ Y_{2.} \end{pmatrix} = \frac{Y_{1.} - Y_{2.}}{3}$$

since
$$\mathbf{X'Y} = \begin{pmatrix} Y_{..} \\ Y_{1.} \\ Y_{2.} \end{pmatrix}$$

where $Y_{i.} = \sum_j y_{ij}$; and $Y_{..} = \sum_{ij} y_{ij}$. So $\tau_1 - \tau_2$ is estimable, and the best linear unbiased estimate is

$$\frac{Y_{1.} - Y_{2.}}{3}$$

In further demonstration of Theorem 11.1, we see that

$$E\left(\frac{Y_{1.} - Y_{2.}}{3}\right) = E\left(\mu + \tau_1 + \frac{\sum_j e_{1j}}{3} - \mu - \tau_2 - \frac{\sum_j e_{2j}}{3}\right) = \tau_1 - \tau_2$$

i.e., there does exist an unbiased estimate of $\tau_1 - \tau_2$.

To demonstrate Theorem 11.3, suppose we take another solution of $\mathbf{X'Xr} = \boldsymbol{\lambda}$; that is, let $r_1 = \frac{1}{3}$. We get $\mathbf{r}' = (\frac{1}{3}, 0, -\frac{2}{3})$, and

$$\mathbf{r'X'Y} = (\tfrac{1}{3}, 0, -\tfrac{2}{3}) \begin{pmatrix} Y_{..} \\ Y_{1.} \\ Y_{2.} \end{pmatrix} = \frac{Y_{..} - 2Y_{2.}}{3} = \frac{Y_{1.} - Y_{2.}}{3}$$

the same estimate as before.

Now we shall investigate whether $\tau_1 + \tau_2$ is estimable. In this case $\boldsymbol{\lambda} = (0, 1, 1)$, and $\mathbf{X'Xr} = \boldsymbol{\lambda}$ becomes

$$\begin{pmatrix} 6 & 3 & 3 \\ 3 & 3 & 0 \\ 3 & 0 & 3 \end{pmatrix} \begin{pmatrix} r_1 \\ r_2 \\ r_3 \end{pmatrix} = \begin{pmatrix} 0 \\ 1 \\ 1 \end{pmatrix}$$

By investigating the ranks we see that the rank of the coefficient matrix is 2 and the rank of the augmented matrix is 3. Hence $\tau_1 + \tau_2$ is *not* estimable.

This is quite a difficult procedure (and becomes increasingly laborious if the dimension of $\mathbf{X}'\mathbf{X}$ is large) for examining the existence of estimable functions. By using the model and Theorem 11.5 (instead of Theorem 11.1) we can sometimes tell immediately whether a certain function of the parameters is estimable. For example, $\mathbf{X}\boldsymbol{\beta}$ is estimable and it equals

$$\mathbf{X}\boldsymbol{\beta} = \begin{pmatrix} \mu + \tau_1 \\ \mu + \tau_1 \\ \mu + \tau_1 \\ \mu + \tau_2 \\ \mu + \tau_2 \\ \mu + \tau_2 \end{pmatrix} \tag{11.6}$$

or the general term of this vector can be written

$$E(y_{ij}) = \mu + \tau_i \qquad i = 1, 2; \quad j = 1, 2, 3$$

By Theorem 11.6, any linear combination of $\mu + \tau_i$ is estimable.

Also, since $\mathbf{X}\boldsymbol{\beta}$ contains $k = 2$ linearly independent estimable functions, every estimable function must be a linear combination of the rows of $\mathbf{X}\boldsymbol{\beta}$. Since μ occurs in every row of $\mathbf{X}\boldsymbol{\beta}$, any estimable function that *does not* contain μ must be such that the coefficients of the τ_i add to zero. For example, $3\mu + 2\tau_1 + \tau_2$ is estimable (first row plus second row plus fourth row), but there is no linear combination of the rows of $\mathbf{X}\boldsymbol{\beta}$ that gives $\tau_1 + \tau_2$; hence $\tau_1 + \tau_2$ is not estimable. Therefore, by displaying $\mathbf{X}\boldsymbol{\beta} = E(y_{ij})$ we can generally see what is and what is not estimable.

After we decide that a certain function is estimable, there still remains the problem of finding the best estimate. This can be accomplished by using the normal equations and Corollary 11.6.2. For example, here the normal equations are $\mathbf{X}'\mathbf{X}\hat{\boldsymbol{\beta}} = \mathbf{X}'\mathbf{Y}$, which are equal to

$$\begin{aligned}
&(a) \qquad & 6\hat{\mu} + 3\hat{\tau}_1 + 3\hat{\tau}_2 &= Y_{..} \\
&(b) \qquad & 3\hat{\mu} + 3\hat{\tau}_1 \qquad\quad &= Y_{1.} \\
&(c) \qquad & 3\hat{\mu} \qquad\quad + 3\hat{\tau}_2 &= Y_{2.}
\end{aligned} \tag{11.7}$$

By Theorems 11.5 and 11.6, any linear combination of the rows on the left side (with hats removed) is estimable, and, by Corollary 11.6.2, the best linear unbiased estimate is given by the same linear combination of rows on the right-hand side of the equations. Suppose we take Eq. (11.7*b*) minus Eq. (11.7*c*); we get

$$3\hat{\tau}_1 - 3\hat{\tau}_2 = Y_{1.} - Y_{2.}$$

or

$$\hat{\tau}_1 - \hat{\tau}_2 = \frac{Y_{1.} - Y_{2.}}{3}$$

But, by Theorem 11.3,

$$\widehat{\tau_1 - \tau_2} = \hat{\tau}_1 - \hat{\tau}_2 = \frac{Y_{1.} - Y_{2.}}{3}$$

so the best linear unbiased estimate of $\tau_1 - \tau_2$ is

$$\frac{Y_{1.} - Y_{2.}}{3}$$

the same estimate that we obtained by solving $\mathbf{X'Xr} = \boldsymbol{\lambda}$ and using $\mathbf{r'X'Y}$. By examining the normal equations we see that $3\mu + 3\tau_1$ and $3\mu + 3\tau_2$ are each estimable (since they are rows of $\mathbf{X'X}$) and that they are linearly independent. That is to say, $\boldsymbol{\lambda}_1' = (3, 3, 0)$; $\boldsymbol{\lambda}_2' = (3, 0, 3)$ are linearly independent. Since the rank of $\mathbf{X'X}$ is 2, we know from Theorem 11.4 that every estimable function $\boldsymbol{\lambda}'\boldsymbol{\beta}$ must be a linear combination of $\boldsymbol{\lambda}_1'\boldsymbol{\beta}$ and $\boldsymbol{\lambda}_2'\boldsymbol{\beta}$.

For another example, let us examine Eq. (11.5). From

$$E(y_{ij}) = \mu + \tau_i + \beta_j$$

we see that any linear combination of $\mu + \tau_i + \beta_j$ is estimable. If we examine the normal equations, we get for $\mathbf{X'X}\hat{\boldsymbol{\beta}}$ and $\mathbf{X'Y}$

$$\mathbf{X'X} = \begin{pmatrix} 4 & 2 & 2 & 2 & 2 \\ 2 & 2 & 0 & 1 & 1 \\ 2 & 0 & 2 & 1 & 1 \\ 2 & 1 & 1 & 2 & 0 \\ 2 & 1 & 1 & 0 & 2 \end{pmatrix} \qquad \mathbf{X'Y} = \begin{pmatrix} Y_{..} \\ Y_{1.} \\ Y_{2.} \\ Y_{.1} \\ Y_{.2} \end{pmatrix}$$

A dot in the subscript of Y_{ij} means that the y_{ij} are to be summed over the subscript that is replaced by the dot. Therefore, we get

$$\begin{aligned}
&(a) \quad & 4\hat{\mu} + 2\hat{\tau}_1 + 2\hat{\tau}_2 + 2\hat{\beta}_1 + 2\hat{\beta}_2 &= Y_{..} \\
&(b) \quad & 2\hat{\mu} + 2\hat{\tau}_1 \qquad\quad + \hat{\beta}_1 + \hat{\beta}_2 &= Y_{1.} \\
&(c) \quad & 2\hat{\mu} \qquad\quad + 2\hat{\tau}_2 + \hat{\beta}_1 + \hat{\beta}_2 &= Y_{2.} \qquad (11.8) \\
&(d) \quad & 2\hat{\mu} + \hat{\tau}_1 + \hat{\tau}_2 + 2\hat{\beta}_1 \qquad\quad &= Y_{.1} \\
&(e) \quad & 2\hat{\mu} + \hat{\tau}_1 + \hat{\tau}_2 \qquad\quad + 2\hat{\beta}_2 &= Y_{.2}
\end{aligned}$$

As in the previous illustration and by Theorem 11.5, each row of $\mathbf{X'X}$ leads to an estimable function, and the estimate is equal to the right-hand side of the equation; e.g., from row (c) we know that $2\mu + 2\tau_2 + \beta_1 + \beta_2$ is estimable and that the best linear unbiased estimate (Corollary 11.6.2) of this function is equal to $Y_{2.}$. We see that the following are estimable:

$$\begin{aligned} &\tau_1 - \tau_2 \\ &\beta_1 - \beta_2 \\ &\tau_1 - \tau_2 + \beta_2 - \beta_1 \\ &\tau_1 - \tau_2 + \beta_1 - \beta_2 \end{aligned}$$

etc., and that the best unbiased estimates are immediately available from Eq. (11.8). For example, the best linear unbiased estimate of $\tau_1 - \tau_2$ is obtained by taking $\frac{1}{2}$ times Eq. (11.8b) minus $\frac{1}{2}$ times Eq. (11.8c). This gives

$$\hat{\tau}_1 - \hat{\tau}_2 = \tfrac{1}{2}(Y_{1.} - Y_{2.}) = y_{1.} - y_{2.}$$

Notice that the only linear combination of the τ_i that is estimable is $\Sigma c_i \tau_i$, where $\Sigma c_i = 0$. This prompts the following definition.

◆ **Definition 11.6** In the model $\mathbf{Y} = \mathbf{X}\boldsymbol{\beta} + \mathbf{e}$ the linear combination of parameters $\Sigma c_i \beta_i$ is called a *contrast* if $\Sigma c_i = 0$.

11.2.3 Reparametrization. In order to utilize the theorems in Chap. 6 on model 1 where $\mathbf{X'X}$ is of full rank, we shall reparametrize the model of *less* than full rank to a model of *full* rank.

◆ **Definition 11.7** By a reparametrization of the model $\mathbf{Y} = \mathbf{X}\boldsymbol{\beta} + \mathbf{e}$ we shall mean a transformation from the vector $\boldsymbol{\beta}$ to the vector $\boldsymbol{\alpha}$ by $\boldsymbol{\alpha} = \mathbf{U}\boldsymbol{\beta}$, where each element of $\boldsymbol{\alpha} = \mathbf{U}\boldsymbol{\beta}$ is an estimable function.

Since $\mathbf{X'X}$ is positive semidefinite of rank k, we know that there exists a nonsingular matrix $\mathbf{W}^*$ ($p \times p$) such that

$$(\mathbf{W}^*)'\mathbf{X'XW}^* = \begin{pmatrix} \mathbf{B} & \mathbf{0} \\ \mathbf{0} & \mathbf{0} \end{pmatrix}$$

where $\mathbf{B}$ is $k \times k$ of rank k. If we partition $\mathbf{W}^*$ into $\mathbf{W}^* = (\mathbf{W}, \mathbf{W}_1)$, where $\mathbf{W}$ is $p \times k$, we get

$$\begin{pmatrix} \mathbf{W}' \\ \mathbf{W}_1' \end{pmatrix} \mathbf{X'X}(\mathbf{W}, \mathbf{W}_1) = \begin{pmatrix} \mathbf{B} & \mathbf{0} \\ \mathbf{0} & \mathbf{0} \end{pmatrix}$$

which gives $\mathbf{W'X'XW} = \mathbf{B}$ and $\mathbf{W}_1'\mathbf{X'XW}_1 = \mathbf{0}$. This implies that

$\mathbf{W}'\mathbf{X}'$ is of rank k and that $\mathbf{W}_1'\mathbf{X}' = \mathbf{0}$. We can write the model $\mathbf{Y} = \mathbf{X}\boldsymbol{\beta} + \mathbf{e}$ as $\mathbf{Y} = \mathbf{X}\mathbf{W}^*(\mathbf{W}^*)^{-1}\boldsymbol{\beta} + \mathbf{e}$, and, if we let $(\mathbf{W}^*)^{-1} = \mathbf{U}^* = \begin{pmatrix} \mathbf{U} \\ \mathbf{U}_1 \end{pmatrix}$, we get

$$\mathbf{Y} = \mathbf{X}(\mathbf{W}, \mathbf{W}_1)\begin{pmatrix} \mathbf{U} \\ \mathbf{U}_1 \end{pmatrix}\boldsymbol{\beta} + \mathbf{e}$$

or

$$\mathbf{Y} = (\mathbf{X}\mathbf{W})(\mathbf{U}\boldsymbol{\beta}) + (\mathbf{X}\mathbf{W}_1)(\mathbf{U}_1\boldsymbol{\beta}) + \mathbf{e}$$

which reduces to $\mathbf{Y} = (\mathbf{X}\mathbf{W})(\mathbf{U}\boldsymbol{\beta}) + \mathbf{e}$, since $\mathbf{X}\mathbf{W}_1 = \mathbf{0}$. Letting $\mathbf{X}\mathbf{W} = \mathbf{Z}$ and $\mathbf{U}\boldsymbol{\beta} = \boldsymbol{\alpha}$, we can write $\mathbf{Y} = \mathbf{X}\boldsymbol{\beta} + \mathbf{e}$ as $\mathbf{Y} = \mathbf{Z}\boldsymbol{\alpha} + \mathbf{e}$, where $\mathbf{Z}$ is $n \times k$ and has rank k; hence we have a full-rank reparametrization, i.e., from a *less*-than-full-rank model to a *full*-rank model. Thus we have the following useful theorem.

◆ **Theorem 11.8** Let $\boldsymbol{\alpha}$ be a $k \times 1$ vector of k linearly independent estimable functions of the parameters $\boldsymbol{\beta}$ in the model $\mathbf{Y} = \mathbf{X}\boldsymbol{\beta} + \mathbf{e}$, where $\mathbf{X}$ is an $n \times p$ matrix of rank k. Then there exists a reparametrization to the full-rank model $\mathbf{Y} = \mathbf{Z}\boldsymbol{\alpha} + \mathbf{e}$.

To estimate $\boldsymbol{\alpha}$ or any linear combination of $\boldsymbol{\alpha}$ we use the normal equations $(\mathbf{Z}'\mathbf{Z})\hat{\boldsymbol{\alpha}} = \mathbf{Z}'\mathbf{Y}$, and all the theorems of Chap. 6 apply.

Another method of proving that $\boldsymbol{\alpha} = \mathbf{U}\boldsymbol{\beta}$ forms a set of k linearly independent estimable functions is to show that $\alpha_i = \mathbf{u}_i\boldsymbol{\beta}$ is estimable for each $i = 1, 2, \ldots, k$, where $\mathbf{u}_i$ is the ith row of $\mathbf{U}$. By Theorem 11.1, this means that there must exist vectors $\mathbf{r}_1, \mathbf{r}_2, \ldots, \mathbf{r}_k$ such that $\mathbf{X}'\mathbf{X}\mathbf{r}_1 = \mathbf{u}_1', \ldots, \mathbf{X}'\mathbf{X}\mathbf{r}_k = \mathbf{u}_k'$. Putting these into a single matrix equation, we get $\mathbf{X}'\mathbf{X}\mathbf{R} = \mathbf{U}'$, where $\mathbf{R} = (\mathbf{r}_1, \mathbf{r}_2, \ldots, \mathbf{r}_k)$. So we must find a matrix $\mathbf{R}$ satisfying $\mathbf{X}'\mathbf{X}\mathbf{R} = \mathbf{U}'$. Using $(\mathbf{W}^*)'\mathbf{X}'\mathbf{X}(\mathbf{W}^*) = \begin{pmatrix} \mathbf{B} & \mathbf{0} \\ \mathbf{0} & \mathbf{0} \end{pmatrix}$, we can obtain $\mathbf{X}'\mathbf{X}(\mathbf{W}, \mathbf{W}_1) = (\mathbf{U}'\mathbf{B}, \mathbf{0})$ by multiplying on the left by $(\mathbf{W}'^*)^{-1}$. From this it follows that $\mathbf{X}'\mathbf{X}\mathbf{W}\mathbf{B}^{-1} = \mathbf{U}'$ and $\mathbf{R} = \mathbf{W}\mathbf{B}^{-1}$; so $\boldsymbol{\alpha} = \mathbf{U}\boldsymbol{\beta}$ is estimable.

There may be different matrices $\mathbf{W}^*$ that will diagonalize $\mathbf{X}'\mathbf{X}$ and, hence, many different full-rank reparametrizations. For example, in the model $\mathbf{Y} = \mathbf{X}\boldsymbol{\beta} + \mathbf{e}$, suppose two full-rank reparametrizations are represented by $\mathbf{Y} = \mathbf{Z}\boldsymbol{\alpha} + \mathbf{e}$ and $\mathbf{Y} = \mathbf{T}\boldsymbol{\delta} + \mathbf{e}$. By the definition of reparametrization, this means that there exist matrices $\mathbf{V}$ and $\mathbf{U}$ of rank k and of dimension $k \times p$ such that $\boldsymbol{\alpha} = \mathbf{U}\boldsymbol{\beta}$ and $\boldsymbol{\delta} = \mathbf{V}\boldsymbol{\beta}$. But there exists a nonsingular $k \times k$ matrix $\mathbf{A}$ such that $\mathbf{U} = \mathbf{A}\mathbf{V}$. This shows that there is a linear relationship between full-rank reparametrizations. That is to say, if the full-rank reparametrized model $\mathbf{Y} = \mathbf{Z}\boldsymbol{\alpha} + \mathbf{e}$ is used to estimate the estimable function $\boldsymbol{\lambda}'\boldsymbol{\beta}$ and if another full-rank reparametrized model $\mathbf{Y} = \mathbf{T\delta} + \mathbf{e}$ is used to estimate $\boldsymbol{\lambda}'\boldsymbol{\beta}$, then the two estimates of $\boldsymbol{\lambda}'\boldsymbol{\beta}$ are identical. Therefore, we shall state an important corollary to Theorem 11.8 on the uniqueness of estimates.

◆ **Corollary 11.8.1** Any full-rank reparametrization gives the same estimate of the estimable function $\boldsymbol{\lambda}'\boldsymbol{\beta}$.

In the full-rank model in Chap. 6 we found that, if the matrix $\mathbf{X}'\mathbf{X}$ is diagonal, this leads to some simplifying results. We shall now show that a model can always be so reparametrized that the resulting matrix of normal equations is diagonal. We shall first state the following definition.

◆ **Definition 11.8** By an orthogonal reparametrization of the model $\mathbf{Y} = \mathbf{X\beta} + \mathbf{e}$ we shall mean a full-rank reparametrization to the model $\mathbf{Y} = \mathbf{Z\alpha} + \mathbf{e}$, where $\mathbf{Z}'\mathbf{Z}$ is a diagonal matrix.

We shall now show that there always exists an orthogonal reparametrization.

◆ **Theorem 11.9** If $\mathbf{W}^*$ is an orthogonal matrix such that

$$(\mathbf{W}^*)'\mathbf{X}'\mathbf{X}\mathbf{W}^* = \mathbf{D}$$

($\mathbf{D}$ is a diagonal matrix), there exists a partition of $\mathbf{W}^*$, say, $\mathbf{W}^* = (\mathbf{W}, \mathbf{W}_1)$ (where $\mathbf{W}$ is $p \times k$), such that $\mathbf{X}\mathbf{W}^*\mathbf{W}^{*\prime} = \mathbf{X}\mathbf{W}\mathbf{W}'$ and $\boldsymbol{\alpha} = \mathbf{W}'\boldsymbol{\beta}$ is estimable and $\mathbf{W}'\mathbf{X}'\mathbf{X}\mathbf{W}$ is diagonal of full rank.

Proof: This follows from the fact that there exists an orthogonal matrix $\mathbf{W}^*$ such that $(\mathbf{W}^*)'\mathbf{X}'\mathbf{X}\mathbf{W}^* = \mathbf{D}$, where the characteristic roots of $\mathbf{X}'\mathbf{X}$ are displayed down the diagonal of $\mathbf{D}$. Exactly k of these roots are nonzero. We can write $\mathbf{W}^* = (\mathbf{W}, \mathbf{W}_1)$, and

$$(\mathbf{W}^*)'\mathbf{X}'\mathbf{X}\mathbf{W}^* = \begin{pmatrix} \mathbf{W}' \\ \mathbf{W}_1' \end{pmatrix} \mathbf{X}'\mathbf{X}(\mathbf{W}, \mathbf{W}_1) = \begin{pmatrix} \mathbf{D}_1 & \mathbf{0} \\ \mathbf{0} & \mathbf{0} \end{pmatrix}$$

where $\mathbf{D}_1$ is $k \times k$ of rank k (nonzero characteristic roots of $\mathbf{X}'\mathbf{X}$ are on the diagonal of $\mathbf{D}_1$). Thus $\mathbf{W}'\mathbf{X}'\mathbf{X}\mathbf{W} = \mathbf{D}_1$ and $\mathbf{W}_1'\mathbf{X}'\mathbf{X}\mathbf{W}_1 = \mathbf{0}$, which implies that $\mathbf{W}_1'\mathbf{X}' = \mathbf{0}$. Now we can write $\mathbf{Y} = \mathbf{X\beta} + \mathbf{e}$ as $\mathbf{Y} = \mathbf{X}\mathbf{W}^*(\mathbf{W}^*)'\boldsymbol{\beta} + \mathbf{e}$,

$$\text{or} \qquad \mathbf{Y} = \mathbf{X}(\mathbf{W}, \mathbf{W}_1)\begin{pmatrix} \mathbf{W}' \\ \mathbf{W}_1' \end{pmatrix} \boldsymbol{\beta} + \mathbf{e} = \mathbf{X}\mathbf{W}\mathbf{W}'\boldsymbol{\beta} + \mathbf{X}\mathbf{W}_1\mathbf{W}_1'\boldsymbol{\beta} + \mathbf{e}$$

or $\mathbf{Y} = (\mathbf{X}\mathbf{W})(\mathbf{W}'\boldsymbol{\beta}) + \mathbf{e}$, since $\mathbf{X}\mathbf{W}_1 = \mathbf{0}$. If we let $\mathbf{X}\mathbf{W} = \mathbf{Z}$ and $\mathbf{W}'\boldsymbol{\beta} = \boldsymbol{\alpha}$, we get $\mathbf{Y} = \mathbf{Z\alpha} + \mathbf{e}$, and, since $\mathbf{Z}'\mathbf{Z} = \mathbf{W}'\mathbf{X}'\mathbf{X}\mathbf{W} = \mathbf{D}_1$ (diagonal), the theorem is proved.

◆ **Corollary 11.9.1** If $\mathbf{W}$ is as defined in Theorem 11.9 and if we denote $\mathbf{XW}$ by $\mathbf{Z}$ and $\mathbf{W}'\boldsymbol{\beta}$ by $\boldsymbol{\alpha}$, then the model $\mathbf{Y} = \mathbf{Z}\boldsymbol{\alpha} + \mathbf{e}$ is an orthogonal reparametrization of the model $\mathbf{Y} = \mathbf{X}\boldsymbol{\beta} + \mathbf{e}$.

◆ **Corollary 11.9.2** If $\mathbf{W}'\boldsymbol{\beta}$ is an orthogonal reparametrization of the model $\mathbf{Y} = \mathbf{X}\boldsymbol{\beta} + \mathbf{e}$, then $\mathbf{X} = \mathbf{XWW}'$.

This corollary will be quite useful in reparametrizing.

11.2.4 Examples. We shall return to Eq. (11.2) to illustrate reparametrization. From Eq. (11.2) we see that $\mu + \tau_1$ and $\mu + \tau_2$ are two linearly independent estimable functions. They are estimable, since $E(y_{11}) = \mu + \tau_1$ and $E(y_{21}) = \mu + \tau_2$; they are linearly independent, since

$$\mu + \tau_1 = (1, 1, 0)\begin{pmatrix}\mu \\ \tau_1 \\ \tau_2\end{pmatrix} = \boldsymbol{\lambda}_1'\boldsymbol{\beta}$$

and

$$\mu + \tau_2 = (1, 0, 1)\begin{pmatrix}\mu \\ \tau_1 \\ \tau_2\end{pmatrix} = \boldsymbol{\lambda}_2'\boldsymbol{\beta}$$

and since $\boldsymbol{\lambda}_1$ and $\boldsymbol{\lambda}_2$ are linearly independent. To illustrate Theorem 11.8, let $\alpha_1 = \mu + \tau_1$, $\alpha_2 = \mu + \tau_2$; we get

$$\boldsymbol{\alpha} = \begin{pmatrix}\alpha_1 \\ \alpha_2\end{pmatrix} = \begin{pmatrix}1 & 1 & 0 \\ 1 & 0 & 1\end{pmatrix}\begin{pmatrix}\mu_1 \\ \tau_1 \\ \tau_2\end{pmatrix} = \mathbf{U}\boldsymbol{\beta}$$

To find $\mathbf{Z}$ in the new model $\mathbf{Y} = \mathbf{Z}\boldsymbol{\alpha} + \mathbf{e}$, we need to find $\mathbf{W}^* = (\mathbf{W}, \mathbf{W}_1) = (\mathbf{U}^*)^{-1}$, where

$$\mathbf{U}^* = \begin{pmatrix}\mathbf{U} \\ \mathbf{U}_1\end{pmatrix}$$

Since

$$\mathbf{U} = \begin{pmatrix}1 & 1 & 0 \\ 1 & 0 & 1\end{pmatrix}$$

we need to find $\mathbf{U}_1$ such that $\mathbf{U}^*$ is nonsingular. It is easy to see that $\mathbf{U}_1 = (0, 1, 1)$ will satisfy this condition. So

$$\mathbf{U}^* = \begin{pmatrix}1 & 1 & 0 \\ 1 & 0 & 1 \\ 0 & 1 & 1\end{pmatrix}$$

and
$$(\mathbf{U}^*)^{-1} = \mathbf{W}^* = \begin{pmatrix} \frac{1}{2} & +\frac{1}{2} & -\frac{1}{2} \\ \frac{1}{2} & -\frac{1}{2} & +\frac{1}{2} \\ -\frac{1}{2} & +\frac{1}{2} & +\frac{1}{2} \end{pmatrix}$$

so
$$\mathbf{W} = \begin{pmatrix} \frac{1}{2} & \frac{1}{2} \\ \frac{1}{2} & -\frac{1}{2} \\ -\frac{1}{2} & \frac{1}{2} \end{pmatrix}$$

and
$$\mathbf{Z} = \mathbf{XW} = \begin{pmatrix} 1 & 0 \\ 1 & 0 \\ 1 & 0 \\ 0 & 1 \\ 0 & 1 \\ 0 & 1 \end{pmatrix}$$

The normal equations are $\mathbf{Z}'\mathbf{Z}\hat{\boldsymbol{\alpha}} = \mathbf{Z}'\mathbf{Y}$; so
$$\begin{pmatrix} 3 & 0 \\ 0 & 3 \end{pmatrix} \begin{pmatrix} \hat{\alpha}_1 \\ \hat{\alpha}_2 \end{pmatrix} = \begin{pmatrix} Y_{1.} \\ Y_{2.} \end{pmatrix}$$

which implies that $\hat{\alpha}_1 = \frac{1}{3}Y_{1.}$, and $\hat{\alpha}_2 = \frac{1}{3}Y_{2.}$. Now $\alpha_1 - \alpha_2$ is estimable, and $\alpha_1 - \alpha_2 = \tau_1 - \tau_2$; so $\hat{\tau}_1 - \hat{\tau}_2 = \frac{1}{3}Y_{1.} - \frac{1}{3}Y_{2.}$, just as in the model before it was reparametrized.

11.2.5 Variances and Covariances of Estimable Functions. In Theorem 11.2 we proved that, if $\boldsymbol{\lambda}'\boldsymbol{\beta}$ is estimable, $\mathbf{r}'\mathbf{X}'\mathbf{Y}$ is the best (minimum-variance) linear unbiased estimate of $\boldsymbol{\lambda}'\boldsymbol{\beta}$, where $\mathbf{r}$ satisfies $\mathbf{Sr} = \boldsymbol{\lambda}$. We also found the variance of $\boldsymbol{\lambda}'\hat{\boldsymbol{\beta}}$, and now we state the following:

◆ **Theorem 11.10** If $\boldsymbol{\lambda}_1'\boldsymbol{\beta}$ and $\boldsymbol{\lambda}_2'\boldsymbol{\beta}$ are two estimable functions, the respective variances of the best linear unbiased estimates are $\sigma^2\mathbf{r}_1'\mathbf{X}'\mathbf{X}\mathbf{r}_1$ and $\sigma^2\mathbf{r}_2'\mathbf{X}'\mathbf{X}\mathbf{r}_2$, where $\mathbf{r}_1$ and $\mathbf{r}_2$, respectively, satisfy $\mathbf{X}'\mathbf{X}\mathbf{r}_1 = \boldsymbol{\lambda}_1$ and $\mathbf{X}'\mathbf{X}\mathbf{r}_2 = \boldsymbol{\lambda}_2$. The covariance of the estimates of $\boldsymbol{\lambda}_1'\boldsymbol{\beta}$ and $\boldsymbol{\lambda}_2'\boldsymbol{\beta}$ is equal to $\sigma^2\mathbf{r}_1'\boldsymbol{\lambda}_2 = \sigma^2\boldsymbol{\lambda}_1'\mathbf{r}_2$.

Proof: We get
$$\begin{aligned} \text{cov}(\boldsymbol{\lambda}_1'\hat{\boldsymbol{\beta}}, \boldsymbol{\lambda}_2'\hat{\boldsymbol{\beta}}) &= E[(\boldsymbol{\lambda}_1'\hat{\boldsymbol{\beta}} - \boldsymbol{\lambda}_1'\boldsymbol{\beta})(\boldsymbol{\lambda}_2'\hat{\boldsymbol{\beta}} - \boldsymbol{\lambda}_2'\boldsymbol{\beta})'] \\ &= E[(\mathbf{r}_1'\mathbf{X}'\mathbf{Y} - \boldsymbol{\lambda}_1'\boldsymbol{\beta})(\mathbf{r}_2'\mathbf{X}'\mathbf{Y} - \boldsymbol{\lambda}_2'\boldsymbol{\beta})'] \\ &= E[(\mathbf{r}_1'\mathbf{X}'\mathbf{e})(\mathbf{e}'\mathbf{X}\mathbf{r}_2)] \\ &= \sigma^2\mathbf{r}_1'\mathbf{X}'\mathbf{X}\mathbf{r}_2 = \sigma^2\boldsymbol{\lambda}_1'\mathbf{r}_2 = \sigma^2\mathbf{r}_1'\boldsymbol{\lambda}_2 \end{aligned}$$

To find the variances, $\boldsymbol{\lambda}_1$ can be set equal to $\boldsymbol{\lambda}_2$, and then $\mathbf{r}_1$ becomes $\mathbf{r}_2$.

To show why an orthogonal reparametrization is sometimes useful, we shall prove the following theorem.

◆ **Theorem 11.11** If $\mathbf{Y} = \mathbf{Z}\boldsymbol{\alpha} + \mathbf{e}$ is an orthogonal reparametrization of the model $\mathbf{Y} = \mathbf{X}\boldsymbol{\beta} + \mathbf{e}$, the elements of $\hat{\boldsymbol{\alpha}}$ are uncorrelated.

Proof: $\hat{\boldsymbol{\alpha}} = (\mathbf{Z}'\mathbf{Z})^{-1}\mathbf{Z}'\mathbf{Y} = \mathbf{D}_1'\mathbf{Z}'\mathbf{Y}$, where $\mathbf{D}_1$ is diagonal.

$$\begin{aligned}\text{cov}(\hat{\boldsymbol{\alpha}}) = E[(\hat{\boldsymbol{\alpha}} - \boldsymbol{\alpha})(\hat{\boldsymbol{\alpha}} - \boldsymbol{\alpha})'] &= E\{[(\mathbf{Z}'\mathbf{Z})^{-1}\mathbf{Z}'\mathbf{Y} - \boldsymbol{\alpha}][(\mathbf{Z}'\mathbf{Z})^{-1}\mathbf{Z}'\mathbf{Y} - \boldsymbol{\alpha}]'\} \\ &= E\{[(\mathbf{Z}'\mathbf{Z})^{-1}\mathbf{Z}'(\mathbf{Z}\boldsymbol{\alpha} + \mathbf{e}) - \boldsymbol{\alpha}] \\ &\quad \times [(\mathbf{Z}'\mathbf{Z})^{-1}\mathbf{Z}'(\mathbf{Z}\boldsymbol{\alpha} + \mathbf{e}) - \boldsymbol{\alpha}]'\} \\ &= E\{[(\mathbf{Z}'\mathbf{Z})^{-1}\mathbf{Z}'\mathbf{e}][(\mathbf{Z}'\mathbf{Z})^{-1}\mathbf{Z}'\mathbf{e}]'\} \\ &= \sigma^2(\mathbf{Z}'\mathbf{Z})^{-1}\mathbf{Z}'\mathbf{Z}(\mathbf{Z}'\mathbf{Z})^{-1} = \sigma^2\mathbf{D}_1^{-1}\end{aligned}$$

So $\text{cov}(\hat{\alpha}_i,\hat{\alpha}_j) = 0$ if $i \neq j$, and the theorem is proved.

11.2.6 Point Estimates of a Linear Combination of β_i under Case A. If $\mathbf{e}$ is distributed $N(\mathbf{0},\sigma^2\mathbf{I})$, maximizing the likelihood equation leads to the same normal equations as the least-squares method under case B.

Let us reparametrize the model $\mathbf{Y} = \mathbf{X}\boldsymbol{\beta} + \mathbf{e}$ to a model of full rank such as $\mathbf{Y} = \mathbf{Z}\boldsymbol{\alpha} + \mathbf{e}$. Then, by the invariance property of maximum likelihood, the maximum-likelihood estimate of $\boldsymbol{\alpha}$ is $\hat{\boldsymbol{\alpha}} = (\mathbf{Z}'\mathbf{Z})^{-1}\mathbf{Z}'\mathbf{Y}$, and all the properties listed in Theorem 6.1 apply to $\hat{\boldsymbol{\alpha}}$ (remember $\boldsymbol{\alpha}$ is a linearly independent set of estimable functions). This result is given in the following theorem.

◆ **Theorem 11.12** In model 4, the model of less than full rank, under case A, the least-squares maximum-likelihood estimate of any estimable function of $\boldsymbol{\beta}$ has all the properties listed in Theorem 6.1, where

$$\hat{\sigma}^2 = \frac{1}{n-k}(\mathbf{Y} - \mathbf{Z}\hat{\boldsymbol{\alpha}})'(\mathbf{Y} - \mathbf{Z}\hat{\boldsymbol{\alpha}})$$

where $\mathbf{Y} = \mathbf{Z}\boldsymbol{\alpha} + \mathbf{e}$ is the reparametrized model, and k is the rank of $\mathbf{X}$.

The proof of this theorem will be left for the reader.

By the theorems in this chapter we know that, if $\boldsymbol{\lambda}'\boldsymbol{\beta}$ is estimable, the best linear unbiased estimate is given by $\boldsymbol{\lambda}'\hat{\boldsymbol{\beta}}$, where $\hat{\boldsymbol{\beta}}$ is any solution to the normal equations. We shall state a similar theorem for the estimation of σ^2.

◆ **Theorem 11.13** In model 4, $\mathbf{Y} = \mathbf{X}\boldsymbol{\beta} + \mathbf{e}$, where $\mathbf{X}$ is of rank k, the quantity

$$\hat{\sigma}^2 = \frac{1}{n-k}(\mathbf{Y} - \mathbf{X}\hat{\boldsymbol{\beta}})'(\mathbf{Y} - \mathbf{X}\hat{\boldsymbol{\beta}})$$

is invariant for any $\hat{\boldsymbol{\beta}}$ that is a solution to the normal equations $\mathbf{X}'\mathbf{X}\hat{\boldsymbol{\beta}} = \mathbf{X}'\mathbf{Y}$.

When we say that $\hat{\sigma}^2$ is *invariant* we mean that $\hat{\sigma}^2$ is the same regardless of which solution to the normal equations is used. The proof will be left for the reader.

◆ **Theorem 11.14** In Theorem 11.13 the quantity

$$\frac{(n-k)\hat{\sigma}^2}{\sigma^2}$$

is distributed as $\chi^2(n-k)$, and $\hat{\sigma}^2$ is an unbiased estimate of σ^2.

The proof of this theorem will be left for the reader.

11.3 Interval Estimation

11.3.1 Interval Estimate of Estimable Functions. To set a confidence interval on an estimable function $\boldsymbol{\lambda}'\boldsymbol{\beta}$, the procedure is to reparametrize the model $\mathbf{Y} = \mathbf{X}\boldsymbol{\beta} + \mathbf{e}$, where $\mathbf{e}$ is distributed $N(\mathbf{0},\sigma^2\mathbf{I})$, into a full-rank model $\mathbf{Y} = \mathbf{Z}\boldsymbol{\alpha} + \mathbf{e}$, where, say, $\alpha_1 = \boldsymbol{\lambda}'\boldsymbol{\beta}$. Then, from the normal equations $(\mathbf{Z}'\mathbf{Z})\hat{\boldsymbol{\alpha}} = \mathbf{Z}'\mathbf{Y}$, the estimate $\hat{\alpha}_1$ can be obtained. By Theorems 11.12 and 11.14 it follows that $\hat{\alpha}_1$ is distributed $N(\alpha_1,c_{11}\sigma^2)$, where c_{11} is the first diagonal element of $(\mathbf{Z}'\mathbf{Z})^{-1}$, that

$$(n-k)\frac{\hat{\sigma}^2}{\sigma^2} = \frac{(\mathbf{Y}-\mathbf{X}\hat{\boldsymbol{\beta}})'(\mathbf{Y}-\mathbf{X}\hat{\boldsymbol{\beta}})}{\sigma^2}$$

is distributed $\chi^2(n-k)$, and that $\hat{\sigma}^2$ and the $\hat{\alpha}_i$ are independent. Therefore

$$v = \frac{\hat{\alpha}_1 - \alpha_1}{\sqrt{c_{11}\hat{\sigma}^2}}$$

is distributed as Student's t with $n-k$ degrees of freedom. This can be used to set a confidence interval on any estimable function $\alpha_1 = \boldsymbol{\lambda}'\boldsymbol{\beta}$. Later we shall show that we need not find $\mathbf{Z}$ in order to find v.

11.4 Testing Hypotheses

11.4.1 Test of the Hypothesis that an Estimable Function of β Is Equal to a Known Constant. To test the hypothesis $H_0: \boldsymbol{\lambda}'\boldsymbol{\beta} = \alpha_0^*$ (where α_0^* is a known constant), where $\boldsymbol{\lambda}'\boldsymbol{\beta}$ is an estimable function, we can reparametrize $\mathbf{Y} = \mathbf{X}\boldsymbol{\beta} + \mathbf{e}$ to a full-rank model

$\mathbf{Y} = \mathbf{Z}\boldsymbol{\alpha} + \mathbf{e}$, where $\alpha_1 = \boldsymbol{\lambda}'\boldsymbol{\beta}$, and use the test function in Art. 11.3.1 with α_1 replaced by α_0^*; hence, if H_0 is true, the quantity

$$v^2 = \frac{(\hat{\alpha}_1 - \alpha_0^*)^2}{c_{11}\hat{\sigma}^2}$$

is distributed as $F'(1, n - k, \lambda)$, where c_{11} is the first diagonal element of $(\mathbf{Z}'\mathbf{Z})^{-1}$ and where the noncentrality $\lambda = \dfrac{1}{2c_{11}\sigma^2}(\boldsymbol{\lambda}'\boldsymbol{\beta} - \alpha_0^*)^2$.

11.4.2 Test of the Hypothesis H_0: $\beta_1 = \beta_2 = \cdots = \beta_q$ $(q \leqslant k)$. This is one of the most useful tests in the general linear hypothesis of less than full rank. For example, in Prob. 11.2, it may be desirable to test $\tau_1 = \tau_2 = \tau_3$.

We shall discuss the testing of estimable hypotheses only. That is, we shall consider the testing of the hypothesis $\beta_1 = \beta_2 = \cdots = \beta_q$ if and only if there exists a set of linearly independent estimable functions $\boldsymbol{\lambda}_1'\boldsymbol{\beta}, \boldsymbol{\lambda}_2'\boldsymbol{\beta}, \ldots, \boldsymbol{\lambda}_s'\boldsymbol{\beta}$ such that H_0 is true if and only if $\boldsymbol{\lambda}_1'\boldsymbol{\beta} = \boldsymbol{\lambda}_2'\boldsymbol{\beta} = \cdots = \boldsymbol{\lambda}_s'\boldsymbol{\beta} = 0$. For example, consider the model given by

$$y_{ij} = \mu + \tau_i + \delta_j + e_{ij} \qquad i = 1, 2, 3; \quad j = 1, 2, 3$$

Suppose we wish to test H_0: $\tau_1 = \tau_2 = \tau_3$. By examining the model we see that $\tau_1 - \tau_2$ and $\tau_1 + \tau_2 - 2\tau_3$ are two linearly independent estimable functions. It is also evident that H_0 is true if and only if $\tau_1 - \tau_2$ and $\tau_1 + \tau_2 - 2\tau_3$ are simultaneously equal to zero. Thus, H_0 is an estimable hypothesis, and we shall test $\tau_1 = \tau_2 = \tau_3$ by testing

$$\begin{pmatrix} \tau_1 - \tau_2 \\ \tau_1 + \tau_2 - 2\tau_3 \end{pmatrix} = \mathbf{0}$$

◆ **Definition 11.9** A hypothesis H_0 will be called estimable if there exists a set of linearly independent estimable functions $\boldsymbol{\lambda}_1'\boldsymbol{\beta}, \boldsymbol{\lambda}_2'\boldsymbol{\beta}, \ldots, \boldsymbol{\lambda}_s'\boldsymbol{\beta}$ such that H_0 is true if and only if $\boldsymbol{\lambda}_1'\boldsymbol{\beta} = \boldsymbol{\lambda}_2'\boldsymbol{\beta} = \cdots = \boldsymbol{\lambda}_s'\boldsymbol{\beta} = 0$.

In general, suppose we wish to test the hypothesis H_0: $\beta_1 = \beta_2 = \cdots = \beta_q$ in model 4, the less-than-full-rank model $\mathbf{Y} = \mathbf{X}\boldsymbol{\beta} + \mathbf{e}$. Suppose there exists a set of linearly independent estimable functions $\boldsymbol{\lambda}_1'\boldsymbol{\beta}, \boldsymbol{\lambda}_2'\boldsymbol{\beta}, \ldots, \boldsymbol{\lambda}_s'\boldsymbol{\beta}$ such that H_0 is true if and only if

$$\begin{pmatrix} \boldsymbol{\lambda}_1'\boldsymbol{\beta} \\ \boldsymbol{\lambda}_2'\boldsymbol{\beta} \\ \cdots \\ \boldsymbol{\lambda}_s'\boldsymbol{\beta} \end{pmatrix} = \mathbf{0}$$

Let the $s \times 1$ vector $\boldsymbol{\alpha}$ be

$$\boldsymbol{\alpha} = \begin{pmatrix} \boldsymbol{\lambda}_1'\boldsymbol{\beta} \\ \boldsymbol{\lambda}_2'\boldsymbol{\beta} \\ \cdots \\ \boldsymbol{\lambda}_s'\boldsymbol{\beta} \end{pmatrix}$$

Further, let $\boldsymbol{\lambda}_{s+1}'\boldsymbol{\beta}, \ldots, \boldsymbol{\lambda}_k'\boldsymbol{\beta}$ be $k - s$ estimable functions such that $\boldsymbol{\lambda}_1'\boldsymbol{\beta}, \boldsymbol{\lambda}_2'\boldsymbol{\beta}, \ldots, \boldsymbol{\lambda}_k'\boldsymbol{\beta}$ form a set of k linearly independent estimable functions. Also, let

$$\boldsymbol{\gamma} = \begin{pmatrix} \boldsymbol{\lambda}_{s+1}'\boldsymbol{\beta} \\ \boldsymbol{\lambda}_{s+2}'\boldsymbol{\beta} \\ \cdots \\ \boldsymbol{\lambda}_k'\boldsymbol{\beta} \end{pmatrix}$$

If we let $\boldsymbol{\delta} = \begin{pmatrix} \boldsymbol{\alpha} \\ \boldsymbol{\gamma} \end{pmatrix}$, then, by Theorem 11.8, it follows that we can reparametrize from the less-than-full-rank model $\mathbf{Y} = \mathbf{X}\boldsymbol{\beta} + \mathbf{e}$ to the full-rank model $\mathbf{Y} = \mathbf{Z}\boldsymbol{\delta} + \mathbf{e}$. This can be written

$$\mathbf{Y} = \mathbf{Z}_1\boldsymbol{\alpha} + \mathbf{Z}_2\boldsymbol{\gamma} + \mathbf{e}$$

Now H_0 is true if and only if $\boldsymbol{\alpha} = \mathbf{0}$; therefore, to test H_0 we use Theorem 6.6 and test $\boldsymbol{\alpha} = \mathbf{0}$. This is demonstrated in Table 11.1. The important quantities are:

1. Sum of squares due to error, which equals $(\mathbf{Y} - \mathbf{Z}\hat{\boldsymbol{\delta}})'(\mathbf{Y} - \mathbf{Z}\hat{\boldsymbol{\delta}})$, where $\hat{\boldsymbol{\delta}}$ is the solution to the normal equations $\mathbf{Z}'\mathbf{Z}\hat{\boldsymbol{\delta}} = \mathbf{Z}'\mathbf{Y}$.

2. Sum of squares due to $\boldsymbol{\alpha}$ adjusted for $\boldsymbol{\gamma}$, which is equal to $R(\boldsymbol{\alpha} \mid \boldsymbol{\gamma}) = \hat{\boldsymbol{\delta}}'\mathbf{Z}'\mathbf{Y} - \tilde{\boldsymbol{\gamma}}'\mathbf{Z}_2'\mathbf{Y}$, where $\tilde{\boldsymbol{\gamma}}$ is the solution to the normal equations $\mathbf{Z}_2'\mathbf{Z}_2\tilde{\boldsymbol{\gamma}} = \mathbf{Z}_2'\mathbf{Y}$.

These normal equations are derived from the model $\mathbf{Y} = \mathbf{Z}_2\boldsymbol{\gamma} + \mathbf{e}$. If we set

$$v = \frac{\hat{\boldsymbol{\delta}}\mathbf{Z}'\mathbf{Y} - \tilde{\boldsymbol{\gamma}}'\mathbf{Z}_2'\mathbf{Y}}{\mathbf{Y}'\mathbf{Y} - \hat{\boldsymbol{\delta}}'\mathbf{Z}'\mathbf{Y}} \frac{n - k}{s}$$

v is distributed as $F'(s, n - k, \lambda)$. The value of λ will be discussed later.

It would be quite a tedious job to find the matrix $\mathbf{Z}$ if this were necessary in order to find v, but fortunately it is not necessary. We can use Theorem 11.13 to show that $(\mathbf{Y} - \mathbf{Z}\hat{\boldsymbol{\delta}})'(\mathbf{Y} - \mathbf{Z}\hat{\boldsymbol{\delta}}) = (\mathbf{Y} - \mathbf{X}\hat{\boldsymbol{\beta}})'(\mathbf{Y} - \mathbf{X}\hat{\boldsymbol{\beta}})$. So the denominator of v can be obtained from the normal equations $\mathbf{X}'\mathbf{X}\hat{\boldsymbol{\beta}} = \mathbf{X}'\mathbf{Y}$. By the same token the term $(\mathbf{Y} - \mathbf{Z}_2\tilde{\boldsymbol{\gamma}})'(\mathbf{Y} - \mathbf{Z}_2\tilde{\boldsymbol{\gamma}})$ can be obtained from the normal equations that are derived from the model $\mathbf{Y} = \mathbf{X}\boldsymbol{\beta} + \mathbf{e}$ with the condition $\beta_1 = \beta_2 = \cdots = \beta_q$. We have therefore proved the following theorem.

◆ **Theorem 11.15** In model 4, $\mathbf{Y} = \mathbf{X}\boldsymbol{\beta} + \mathbf{e}$, to test the hypothesis $\beta_1 = \beta_2 = \cdots = \beta_q$ ($q \leqslant k$), which we assume is equivalent to testing the linearly estimable functions $\boldsymbol{\lambda}_1'\boldsymbol{\beta} = \boldsymbol{\lambda}_2'\boldsymbol{\beta} = \cdots = \boldsymbol{\lambda}_s'\boldsymbol{\beta} = 0$,

(1) Obtain any solution to the normal equations $\mathbf{X}'\mathbf{X}\hat{\boldsymbol{\beta}} = \mathbf{X}'\mathbf{Y}$, and form $Q_0 = (\mathbf{Y} - \mathbf{X}\hat{\boldsymbol{\beta}})'(\mathbf{Y} - \mathbf{X}\hat{\boldsymbol{\beta}})$.

(2) Obtain any solution to the normal equations $\mathbf{Z}_2'\mathbf{Z}_2\tilde{\boldsymbol{\gamma}} = \mathbf{Z}_2'\mathbf{Y}$, which are derived from the model $\mathbf{Y} = \mathbf{X}\boldsymbol{\beta} + \mathbf{e}$ under the conditions $\beta_1 = \beta_2 = \cdots = \beta_q$ (denote the reduced model by $\mathbf{Y} = \mathbf{Z}_2\boldsymbol{\gamma} + \mathbf{e}$), and form $Q_0 + Q_1 = (\mathbf{Y} - \mathbf{Z}_2\tilde{\boldsymbol{\gamma}})'(\mathbf{Y} - \mathbf{Z}_2\tilde{\boldsymbol{\gamma}})$.

Then
$$v = \frac{n-k}{s}\frac{Q_1}{Q_0} = \frac{\hat{\boldsymbol{\beta}}'\mathbf{X}'\mathbf{Y} - \tilde{\boldsymbol{\gamma}}'\mathbf{Z}_2'\mathbf{Y}}{\mathbf{Y}'\mathbf{Y} - \hat{\boldsymbol{\beta}}'\mathbf{X}'\mathbf{Y}}\frac{n-k}{s}$$

is distributed as $F'(s, n-k, \lambda)$.

This can be put in the form of an AOV table such as Table 11.1.

TABLE 11.1 ANALYSIS OF VARIANCE FOR TESTING $\boldsymbol{\lambda}_1'\boldsymbol{\beta} = \cdots = \boldsymbol{\lambda}_s'\boldsymbol{\beta} = 0$

SV	DF	SS	F
Total	n	$\mathbf{Y}'\mathbf{Y}$	
Due to $\boldsymbol{\beta}$	k	$\hat{\boldsymbol{\beta}}'\mathbf{X}'\mathbf{Y}$	
Due to $\boldsymbol{\gamma}$	$k-s$	$\tilde{\boldsymbol{\gamma}}'\mathbf{Z}_2'\mathbf{Y}$	
$\boldsymbol{\alpha}$ (adj)	s	$\hat{\boldsymbol{\beta}}'\mathbf{X}'\mathbf{Y} - \tilde{\boldsymbol{\gamma}}'\mathbf{Z}_2'\mathbf{Y} = \alpha_{ss}$	$\dfrac{\alpha_{ms}}{E_{ms}}$
Error	$n-k$	$\mathbf{Y}'\mathbf{Y} - \hat{\boldsymbol{\beta}}'\mathbf{X}'\mathbf{Y} = E_{ss}$	

Note the use of the notation α_{ss} and E_{ss} for the sum of squares, and α_{ms} and E_{ms} for the respective mean squares, which is the sum of squares divided by the degrees of freedom. This notation will be used throughout.

The quantity α_{ss} is the adjusted sum of squares for testing the hypothesis. By using the fact that α_{ss}/σ^2 is distributed as $\chi'^2(s,\lambda)$, and Prob. 4.1, we see that $E(\alpha_{ss}/\sigma^2) = s + 2\lambda$. Hence a theorem to aid in computing the noncentrality is as follows.

◆ **Theorem 11.16** Under the conditions of Theorem 11.15 the noncentrality parameter is

$$\lambda = \frac{s[E(\alpha_{ms})]}{2\sigma^2} - \frac{s}{2}$$

Next we shall prove a theorem that will be helpful in determining whether a specified hypothesis is estimable.

◆ **Theorem 11.17** Let $\beta_1, \beta_2, \ldots, \beta_q$ be a subset of the elements of $\boldsymbol{\beta}$ in model 4. Suppose that $\sum_{i=1}^{q} c_i\beta_i$ is estimable for *every* set of constants c_i such that $\sum_{i=1}^{q} c_i = 0$. Then the hypothesis $H_0: \beta_1 = \beta_2 = \cdots = \beta_q$ is an estimable hypothesis.

Proof: Since $\Sigma\, c_i\beta_i$ is estimable for every set of c_i such that $\Sigma c_i = 0$, we know in particular that the following are estimable:

$$\begin{aligned}
\boldsymbol{\lambda}_1'\boldsymbol{\beta} &= \beta_1 - \beta_2\\
\boldsymbol{\lambda}_2'\boldsymbol{\beta} &= \beta_1 + \beta_2 - 2\beta_3\\
\boldsymbol{\lambda}_3'\boldsymbol{\beta} &= \beta_1 + \beta_2 + \beta_3 - 3\beta_4\\
&\cdots\cdots\cdots\cdots\cdots\cdots\\
\boldsymbol{\lambda}_{q-1}'\boldsymbol{\beta} &= \beta_1 + \beta_2 + \beta_3 + \cdots + \beta_{q-1} - (q-1)\beta_q
\end{aligned}$$

It is easy, then, to show that $\boldsymbol{\lambda}_1'\boldsymbol{\beta}, \boldsymbol{\lambda}_2'\boldsymbol{\beta}, \ldots, \boldsymbol{\lambda}_{q-1}'\boldsymbol{\beta}$ constitute a set of linearly independent estimable functions that are all zero if and only if $\beta_1 = \beta_2 = \cdots = \beta_q$, and the theorem is established.

11.5 Normal Equations and Computing

11.5.1 Obtaining the Normal Equations. Since the normal equations are so important, a method for finding them will be discussed. In model 1, the normal equations were found by getting the sums of squares and cross products of the x_i and y_i. This method can also be used for the normal equations in model 4. However, since the $\mathbf{X}$ matrix for model 4 consists entirely of 0's and 1's, a method will be presented that may save some computational labor. In model 4 the model is not usually written out so completely as it is in model 1; i.e., the $\mathbf{X}$ matrix is not given explicitly. Instead, the model is usually written something like $y_{ijh} = \mu + \tau_i + \alpha_j + (\tau\alpha)_{ij} + e_{ijh}$, etc.; that is to say, only the parameters are given, indexed on certain subscripts. From this the $\mathbf{X}$ matrix can be obtained, but the work may be tedious if there are many observations. The normal equations could also be found by equating to zero the various derivatives of the quantity $\mathbf{e}'\mathbf{e}$, but this too can become laborious.

An alternative easy method for finding the normal equations will be given. The normal equations are $\mathbf{X}'\mathbf{X}\hat{\boldsymbol{\beta}} = \mathbf{X}'\mathbf{Y}$. There are p equations and p parameters. Therefore, we shall associate one equation with each parameter. The procedure will be to find the right-hand side of the normal equations $\mathbf{X}'\mathbf{Y}$ and to take the expected value to get the left-hand side; that is,

$$E(\mathbf{X}'\mathbf{Y}) = E[\mathbf{X}'(\mathbf{X}\boldsymbol{\beta} + e)] = \mathbf{X}'\mathbf{X}\boldsymbol{\beta}$$

Let us write $\mathbf{X} = (\mathbf{X}_1, \mathbf{X}_2, \ldots, \mathbf{X}_p)$, where $\mathbf{X}_i$ is the ith column of $\mathbf{X}$. The model $\mathbf{Y} = \mathbf{X\beta} + \mathbf{e}$ can be written as $\mathbf{Y} = \Sigma\mathbf{X}_i\beta_i + \mathbf{e}$, where β_i is the ith element of $\boldsymbol{\beta}$. Let us examine $\mathbf{X}_1\beta_1$. The elements of the $n \times 1$ vector $\mathbf{X}_1$ are only 0's or 1's. Therefore, the parameter β_1 occurs in the nth observation of the model if and only if the nth element of $\mathbf{X}_1$ is equal to 1. Now examine $\mathbf{X}_1'\mathbf{Y}$. This is the sum of the elements of $\mathbf{Y}$ for those elements of $\mathbf{X}_1$ that are equal to 1. Therefore, the term $\mathbf{X}_1'\mathbf{Y}$ can be found by summing the elements of $\mathbf{Y}$ over those elements in the model that contain β_1. This same procedure applies to all the β_i. Now the normal equations can be obtained by finding $\mathbf{X}_i'\mathbf{Y}$ for each β_i, taking the expectation of each $\mathbf{X}_i'\mathbf{Y}$ to obtain $\mathbf{X}_i'\mathbf{X\beta}$, then putting the sign $\frown$ over each parameter. This gives

$$
\begin{array}{ll}
\beta_1: & \mathbf{X}_1'\mathbf{Y} = \mathbf{X}_1'\mathbf{X}\hat{\boldsymbol{\beta}} \\
\beta_2: & \mathbf{X}_2'\mathbf{Y} = \mathbf{X}_2'\mathbf{X}\hat{\boldsymbol{\beta}} \\
 & \cdots\cdots\cdots\cdots \\
\beta_p: & \mathbf{X}_p'\mathbf{Y} = \mathbf{X}_p'\mathbf{X}\hat{\boldsymbol{\beta}}
\end{array}
$$

These ideas will be demonstrated further by considering the following model:

$$y_{ij} = \mu + \tau_i + \alpha_j + e_{ij} \qquad i = 1, 2, 3; \quad j = 1, 2$$

The set of normal equations consists of six equations, one corresponding to each of the parameters $\mu, \tau_1, \tau_2, \tau_3, \alpha_1, \alpha_2$. First find the normal equation corresponding to μ. The element of $\mathbf{X}'\mathbf{Y}$ corresponding to μ is equal to $\mathbf{X}_1'\mathbf{Y}$ and is the sum of the elements y_{ij} over all the equations in the model in which μ occurs. By examining the model we see that μ occurs in every one of the equations of the model; hence,

$$\mathbf{X}_1'\mathbf{Y} = \sum_{ij} y_{ij} = Y_{..}$$

The next quantity to find is the expected value of $Y_{..}$. This is

$$
\begin{aligned}
E(Y_{..}) &= E\left[\sum_{ij}(\mu + \tau_i + \alpha_j + e_{ij})\right] \\
&= \sum_{ij}(\mu + \tau_i + \alpha_j) \\
&= 6\mu + 2\tau_1 + 2\tau_2 + 2\tau_3 + 3\alpha_1 + 3\alpha_2
\end{aligned}
$$

Putting the symbol $\frown$ on the parameters, we get for the first equation in the set of normal equations

$$\mu: \qquad 6\hat{\mu} + 2\hat{\tau}_1 + 2\hat{\tau}_2 + 2\hat{\tau}_3 + 3\hat{\alpha}_1 + 3\hat{\alpha}_2 = Y_{..}$$

Next we shall look for the equation corresponding to τ_1. τ_1 occurs in those equations in the model which involve y_{1j}. So the sum of y_{1j} over those which contain τ_1 is

$$\sum_j y_{1j} = Y_{1.}$$

Next we evaluate

$$\begin{aligned} E(Y_{1.}) &= E\left[\sum_j (\mu + \tau_1 + \alpha_j + e_{1j})\right] \\ &= \sum_j (\mu + \tau_1 + \alpha_j) \\ &= 2\mu + 2\tau_1 + \alpha_1 + \alpha_2 \end{aligned}$$

This gives the equation corresponding to τ_1, which is

$$\tau_1: \qquad 2\hat{\mu} + 2\hat{\tau}_1 + \hat{\alpha}_1 + \hat{\alpha}_2 = Y_{1.}$$

The equations for τ_2 and τ_3 are obtained in a similar fashion. The equation corresponding to α_2, for example, gives

$$\sum_i y_{i2} = Y_{.2}$$

since α_2 occurs only in those equations in the model which involve y_{i2}. Also,

$$E(Y_{.2}) = 3\mu + \tau_1 + \tau_2 + \tau_3 + 3\alpha_2$$

The complete set of normal equations can now be written, as follows:

$$\begin{array}{ll} \mu: & 6\hat{\mu} + 2\hat{\tau}_1 + 2\hat{\tau}_2 + 2\hat{\tau}_3 + 3\hat{\alpha}_1 + 3\hat{\alpha}_2 = Y_{..} \\ \tau_1: & 2\hat{\mu} + 2\hat{\tau}_1 + \hat{\alpha}_1 + \hat{\alpha}_2 = Y_{1.} \\ \tau_2: & 2\hat{\mu} + 2\hat{\tau}_2 + \hat{\alpha}_1 + \hat{\alpha}_2 = Y_{2.} \\ \tau_3: & 2\hat{\mu} + 2\hat{\tau}_3 + \hat{\alpha}_1 + \hat{\alpha}_2 = Y_{3.} \\ \alpha_1: & 3\hat{\mu} + \hat{\tau}_1 + \hat{\tau}_2 + \hat{\tau}_3 + 3\hat{\alpha}_1 = Y_{.1} \\ \alpha_2: & 3\hat{\mu} + \hat{\tau}_1 + \hat{\tau}_2 + \hat{\tau}_3 + 3\hat{\alpha}_2 = Y_{.2} \end{array} \qquad (11.9)$$

The work of evaluating the normal equations can be shortened many times by finding only a representative equation in each set of the parameters. For instance, in the example above, $y_{ij} = \mu + \tau_i + \alpha_j + e_{ij}$, there are three sets of parameters, μ, τ, and α. Instead of finding all the normal equations, suppose we find the normal equation corresponding to μ, that corresponding to τ_q (the qth τ), and that for α_s (the sth α). This gives $Y_{q.}$ for the right-hand side of τ_q and $Y_{.s}$ for the right-hand side of α_s. Hence we obtain

$$\begin{array}{lll} \mu: & 6\hat{\mu} + 2\hat{\tau}_. + 3\hat{\alpha}_. = Y_{..} & \\ \tau_q: & 2\hat{\mu} + 2\hat{\tau}_q + \hat{\alpha}_. = Y_{q.} & q = 1, 2, 3 \\ \alpha_s: & 3\hat{\mu} + \hat{\tau}_. + 3\hat{\alpha}_s = Y_{.s} & s = 1, 2 \end{array}$$

where $\hat{\alpha}_. = \Sigma\hat{\alpha}_j$ and $\hat{\tau}_. = \Sigma\hat{\tau}_{i.}$. This is a very compact representation of the normal equations and easy to obtain from the model.

11.5.2 Solving the Normal Equations. Since the $\mathbf{X}'\mathbf{X}$ matrix is $p \times p$ of rank $k < p$, there are infinitely many vectors $\hat{\boldsymbol{\beta}}$ that satisfy the normal equations $\mathbf{X}'\mathbf{X}\hat{\boldsymbol{\beta}} = \mathbf{X}'\mathbf{Y}$. We need to find only one such vector. Now, by Theorem 11.5, $\mathbf{X}'\mathbf{X}\hat{\boldsymbol{\beta}}$ represents a set of k linearly independent estimable functions, and every estimable function is a linear combination of the rows of $\mathbf{X}'\mathbf{X}\hat{\boldsymbol{\beta}}$. Let $\mathbf{V}$ be a $(p - k) \times p$ matrix of rank $p - k$ such that $\mathbf{V}\boldsymbol{\beta}$ is a set of $p - k$ nonestimable functions and such that no linear combination of $\mathbf{V}\boldsymbol{\beta}$ is an estimable function. It can be shown that there exists a vector $\hat{\boldsymbol{\beta}}$ that satisfies the $2p - k$ equations $\mathbf{X}'\mathbf{X}\hat{\boldsymbol{\beta}} = \mathbf{X}'\mathbf{Y}$ and $\mathbf{V}\hat{\boldsymbol{\beta}} = \mathbf{0}$. This follows from the fact that the matrix $\begin{pmatrix}\mathbf{X}'\mathbf{X} \\ \mathbf{V}\end{pmatrix}$ has rank p, since $\mathbf{V}$ is of rank $p - k$ (by the way it is chosen) and since the rows of $\mathbf{V}$ are linearly independent of the rows of $\mathbf{X}'\mathbf{X}$ or else some linear combination of $\mathbf{V}\boldsymbol{\beta}$ would be a set of estimable functions. Now $\mathbf{V}\boldsymbol{\beta}$ can often be so chosen as to make the solution of the normal equations quite easy. Also, nonestimable functions $\mathbf{V}\boldsymbol{\beta}$ can often be found by examining the model.

11.5.3 Example. Let us find the solution to the normal equations given in Eq. (11.8), which come from the model

$$y_{ij} = \mu + \tau_i + \alpha_j + e_{ij} \qquad i = 1, 2; \quad j = 1, 2$$

Examination of these normal equations reveals the fact that Eq. (11.8*d*) plus Eq. (11.8*e*) minus Eq. (11.8*a*) is equal to zero; also, Eq. (11.8*b*) plus Eq. (11.8*c*) minus Eq. (11.8*a*) is equal to zero. Hence, there are at most three linearly independent equations. Therefore, we must find two linearly independent equations that are not estimable. By examining the model it can be seen that we can use

$$\mathbf{V}\boldsymbol{\beta} = \begin{pmatrix}\alpha_1 + \alpha_2 \\ \tau_1 + \tau_2\end{pmatrix}$$

We could have found many different functions that could be taken for $\mathbf{V}\boldsymbol{\beta}$, but we chose these because they make the solution of the normal equations quite easy. Setting $\mathbf{V}\hat{\boldsymbol{\beta}} = \mathbf{0}$ gives $\hat{\alpha}_1 + \hat{\alpha}_2 = 0$ and $\hat{\tau}_1 + \hat{\tau}_2 = 0$. Then, from Eq. (11.8*a*) we get $\hat{\mu} = \frac{1}{4}Y_{..} = y_{..}$; from Eq. (11.8*b*), $\hat{\tau}_1 = \frac{1}{2}Y_{1.} - y_{..} = y_{1.} - y_{..}$. Continuing, we get $\hat{\tau}_2 = y_{2.} - y_{..}$; $\hat{\alpha}_1 = y_{.1} - y_{..}$; $\hat{\alpha}_2 = y_{.2} - y_{..}$.

11.5.4 Testing Hypotheses. Suppose we desire to test the hypothesis $\tau_1 = \tau_2 = \tau_3$ in the model

$$y_{ij} = \mu + \tau_i + \alpha_j + e_{ij} \qquad i = 1, 2, 3; j = 1, 2$$

The normal equations are given in (11.9). From these we see that $\tau_1 - \tau_2$ and $\tau_1 - \tau_3$ are estimable. Hence H_0: $\tau_1 = \tau_2 = \tau_3$ is an estimable hypothesis. Using the nonestimable conditions $\hat{\tau}_1 + \hat{\tau}_2 + \hat{\tau}_3 = 0$ and $\hat{\alpha}_1 + \hat{\alpha}_2 = 0$ makes the normal equations easy to solve. The solution is

$$\begin{aligned}\hat{\mu} &= y_{..}\\ \hat{\tau}_1 &= y_{1.} - y_{..}\\ \hat{\tau}_2 &= y_{2.} - y_{..}\\ \hat{\tau}_3 &= y_{3.} - y_{..}\\ \hat{\alpha}_1 &= y_{.1} - y_{..}\\ \hat{\alpha}_2 &= y_{.2} - y_{..}\end{aligned}$$

Therefore, the quantity $\hat{\boldsymbol{\beta}}'\mathbf{X}'\mathbf{Y}$ in Theorem 11.15 is

$$R(\mu,\tau,\alpha) = \hat{\mu}Y_{..} + \hat{\tau}_1 Y_{1.} + \hat{\tau}_2 Y_{2.} + \hat{\tau}_3 Y_{3.} + \hat{\alpha}_1 Y_{.1} + \hat{\alpha}_2 Y_{.2}$$

The error sum of squares is

$$\Sigma\, y_{ij}^2 - \hat{\boldsymbol{\beta}}'\mathbf{X}'\mathbf{Y} = \Sigma\, y_{ij}^2 - \frac{Y_{..}^2}{6} - \left(\sum_{i=1}^{3} \frac{Y_{i.}^2}{2} - \frac{Y_{..}^2}{6}\right) - \left(\sum_{j=1}^{2} \frac{Y_{.j}^2}{3} - \frac{Y_{..}^2}{6}\right)$$

To get $\tilde{\boldsymbol{\gamma}}'\mathbf{Z}_2'\mathbf{Y}$, we use the model $y_{ij} = \mu + \tau + \alpha_j + e_{ij}$; we have put $\tau_1 = \tau_2 = \tau_3 = \tau$. Now $\mu + \tau$ will be replaced with μ^*. The normal equations are

$$\begin{aligned}&\mu^*: && 6\tilde{\mu}^* + 3\tilde{\alpha}_1 + 3\tilde{\alpha}_2 = Y_{..}\\ &\alpha_1: && 3\tilde{\mu}^* + 3\tilde{\alpha}_1 \phantom{+ 3\tilde{\alpha}_2} = Y_{.1}\\ &\alpha_2: && 3\tilde{\mu}^* \phantom{+ 3\tilde{\alpha}_1} + 3\tilde{\alpha}_2 = Y_{.2}\end{aligned}$$

These can be obtained from (11.9) by letting $\tau_1 = \tau_2 = \tau_3 = 0$ and using only the equations for μ, α_1, and α_2. The solution is easy to obtain if we use the nonestimable condition $\tilde{\alpha}_1 + \tilde{\alpha}_2 = 0$. The solution is

$$\begin{aligned}\tilde{\mu}^* &= y_{..}\\ \tilde{\alpha}_1 &= y_{.1} - y_{..}\\ \tilde{\alpha}_2 &= y_{.2} - y_{..}\end{aligned}$$

Also,

$$R(\mu^*,\alpha) = \tilde{\boldsymbol{\gamma}}'\mathbf{Z}_2'\mathbf{Y} = \tilde{\mu}^* Y_{..} + \tilde{\alpha}_1 Y_{.1} + \tilde{\alpha}_2 Y_{.2} = \frac{Y_{..}^2}{6} + \left(\sum_{j=1}^{2} \frac{Y_{.j}^2}{3} - \frac{Y_{..}^2}{6}\right)$$

$$R(\tau \mid \mu,\alpha) = R(\mu,\tau,\alpha) - R(\mu^*,\alpha) = \sum_i \frac{Y_{i.}^2}{2} - \frac{Y_{..}^2}{6}$$

This is generally written in a table such as Table 11.2.

TABLE 11.2 ANALYSIS OF VARIANCE FOR TESTING $\tau_1 = \tau_2 = \tau_3$

SV	DF	SS	F	EMS
Total	6	$\sum_{ij} y_{ij}^2$		
Due to μ, τ, α	4	$S_1 = R(\mu,\tau,\alpha)$		
Due to μ, α	2	$S_2 = R(\mu,\alpha)$		
Due to τ (adj)	2	$S_1 - S_2 = \tau_{ss}$	$\dfrac{\tau_{ms}}{E_{ms}}$	$\sigma^2 + \sum_i (\tau_i - \bar{\tau}_.)^2$
Error	2	$\sum_{ij} y_{ij}^2 - S_1 = E_{ss}$		

The EMS is the *expected mean square* and is obtained only for the quantity $R(\tau \mid \mu,\alpha)$. By Theorem 11.15, the quantity τ_{ss}/σ^2 is distributed as $\chi'^2(2,\lambda)$. To find λ we use Theorem 11.16. This gives

$$\frac{E(\tau_{ss})}{\sigma^2} = p + 2\lambda = 2 + 2\lambda$$

But, by the term in the EMS column, we get

$$\frac{E(\tau_{ss})}{\sigma^2} = 2 + \frac{2}{\sigma^2} \sum_i (\tau_i - \bar{\tau}_.)^2$$

and

$$\lambda = \frac{1}{\sigma^2} \sum_i (\tau_i - \bar{\tau}_.)^2$$

This can be used to evaluate the power function by using Tang's tables.

11.6 Optimum Properties of the Tests of Hypotheses

The test procedure outlined in Theorem 11.15 is equivalent to the likelihood-ratio test given in Chap. 6. This test has certain desirable properties, which will be stated without proof in the following theorem.

◆ **Theorem 11.18** To test the hypothesis $\boldsymbol{\lambda}_1'\boldsymbol{\beta} = \boldsymbol{\lambda}_2'\boldsymbol{\beta} = \cdots = \boldsymbol{\lambda}_s'\boldsymbol{\beta} = 0$ in model 4, with the alternative that at least one $\boldsymbol{\lambda}_i'\boldsymbol{\beta} \neq 0$, the test procedure given in Theorem 11.15 is a uniformly most powerful test in the class of tests whose power depends only on the single parameter λ (the noncentrality).

Problems

11.1 For the model

$$y_{ij} = \mu + \tau_i + e_{ij} \qquad i = 1, 2, 3;\ j = 1, 2, 3$$

(a) Find the normal equations.
(b) Find two linearly independent estimable functions.
(c) Find the variances and covariances of the estimates of the functions in (b).

11.2 For the model

$$y_{ij} = \mu + \tau_i + \delta_j + e_{ij} \qquad i = 1, 2, 3;\ j = 1, 2, 3$$

(a) Find the normal equations.
(b) Find two linearly independent estimable functions.
(c) Find the variances and covariances of the estimates of the functions in (b).

11.3 For the model

$$y_{ijk} = \mu + \tau_i + \delta_j + \pi_{ij} + e_{ijk} \qquad \begin{cases} i = 1, 2 \\ j = 1, 2 \\ k = 1, 2, 3 \end{cases}$$

(a) Find the normal equations.
(b) Find two linearly independent estimable functions.
(c) Find the variances and covariances of the estimates of the functions in (b).

11.4 Prove Theorem 11.5.

11.5 Prove Theorem 11.7.

Problems 6 to 15 refer to the model $\mathbf{Y} = \mathbf{X}\boldsymbol{\beta} + e$, which is explicitly written

$$\begin{aligned} y_1 &= \tau_1 + \tau_2 + \tau_3 + e_1 \\ y_2 &= \tau_1 + \tau_3 + e_2 \\ y_3 &= + \tau_2 + e_3 \end{aligned}$$

11.6 Find $\mathbf{X}$.

11.7 Find $\mathbf{X}'\mathbf{X}$.

11.8 Find the rank of $\mathbf{X}$.

11.9 Find the normal equations.

11.10 From the model, show that $\tau_1 + \tau_3 - 2\tau_2$ is estimable (use Theorem 11.5).

11.11 From the normal equations, find the best linear unbiased estimate of $\tau_1 + \tau_3 - 2\tau_2$ (use Theorem 11.5).

11.12 Find two full-rank reparametrizations, and show that the estimate of $\tau_1 - \tau_2 + \tau_3$ is the same for both.

11.13 Find an orthogonal reparametrization of the model.

11.14 Find the variance of the estimates of $\tau_1 - 2\tau_2 + \tau_3$, $\tau_1 + 2\tau_2 + \tau_3$.

11.15 Find the covariance of the estimates in Prob. 11.14.

11.16 Prove Corollary 11.9.1.

11.17 Prove Corollary 11.9.2.

11.18 Prove Corollary 11.6.2.

11.19 Prove Corollary 11.8.1 in detail.

In Probs. 11.20 to 11.36, assume that the vector of errors $\mathbf{e}$ is distributed $N(\mathbf{0},\sigma^2\mathbf{I})$.

11.20 Find the normal equations for the model

$$y_{ijm} = \mu + \alpha_i + \beta_j + \gamma_m + e_{ijm} \qquad \begin{cases} i = 1, 2 \\ j = 1, 2 \\ m = 1, 2 \end{cases}$$

11.21 In Prob. 11.20, find a set of linearly independent estimable functions.

11.22 In Prob. 11.20, find $\hat{\boldsymbol{\beta}}'\mathbf{X}'\mathbf{Y}$.

11.23 Find a solution to the normal equations given in Eq. (11.9) by using two linearly independent nonestimable functions other than $\Sigma\hat{\tau}_i = 0$ and $\Sigma\hat{\alpha}_j = 0$.

11.24 In Prob. 11.20, find the estimate of σ^2.

11.25 In Prob. 11.20, construct an AOV table to test the hypothesis $\alpha_1 = \alpha_2$.

11.26 In Prob. 11.25, find the noncentrality parameter λ.

11.27 Prove Theorem 11.13.

11.28 Prove Theorem 11.14.

11.29 In the model

$$y_{ij} = \mu + \tau_i + \beta_j + e_{ij} \qquad i = 1, 2, \ldots, n; \; j = 1, 2, \ldots, m$$

show that $\sum_{i=1}^{n} a_i\tau_i$ is estimable if and only if $\sum_{i=1}^{n} a_i = 0$.

11.30 In Table 11.2, show that $R(\alpha \mid \mu,\tau) = R(\alpha \mid \mu)$.

11.31 In Prob. 11.29 with $n = 3$ and $m = 4$, the following values are given:

y_{11}	y_{12}	y_{13}	y_{14}	y_{21}	y_{22}	y_{23}	y_{24}	y_{31}	y_{32}	y_{33}	y_{34}
6.2	2.0	5.8	6.3	5.1	5.3	5.8	5.4	6.1	6.2	5.7	5.9

Compute the quantities in Table 11.2 to test the hypothesis $\tau_1 = \tau_2 = \tau_3$, the alternative being that at least one inequality holds. Use a probability of type I error equal to .05.

11.32 In Prob. 11.31, find the power of the test for the following values

τ_1	τ_2	τ_3	σ^2
1	1	2	.25
1	2	2	.25
1	2	2	.10

11.33 In Prob. 11.31, set a 95 per cent confidence interval on $\tau_1 + 3\tau_2 - 4\tau_3$.

11.34 Find the width and the expected width of the interval in Prob. 11.33.

11.35 Prove Theorem 11.12.

11.36 Let $\mathbf{b}$ be a vector of constants such that $E(\mathbf{b}'\mathbf{Y}) = 0$; then $\mathbf{b}'\mathbf{Y}$ will be called an *unbiased estimate of zero*. Prove that an unbiased estimate of zero is uncorrelated with the best linear unbiased estimate of any estimable function.

Further Reading

1 G. W. Snedecor: "Statistical Methods," Iowa State College Press, Ames, Iowa, 1956.

2 W. G. Cochran and G. M. Cox: "Experimental Designs," John Wiley & Sons, Inc., New York, 1957.

3 O. Kempthorne: "Design and Analysis of Experiments," John Wiley & Sons, Inc., New York, 1952.

4 C. H. Goulden: "Methods of Statistical Analysis," John Wiley & Sons, Inc., New York, 1939.

5 O. L. Davies: "Design and Analysis of Industrial Experiments," Oliver and Boyd, Ltd., London, 1954.

6 R. L. Anderson and T. A. Bancroft: "Statistical Theory in Research," McGraw-Hill Book Company, Inc., New York, 1952.

7 C. R. Rao: "Advanced Statistical Methods in Biometric Research," John Wiley & Sons, Inc., New York, 1952.

8 J. Wolfowitz: The Power of the Classical Tests Associated with the Normal Distribution, *Ann. Math. Statist.*, vol. 20, pp. 540–551, 1949.

9 S. W. Nash: Note on Power of the *F* Test, *Ann. Math. Statist.*, vol. 19, p. 434, 1948.

10 A. Wald: On the Power Function of the Analysis of Variance Test, *Ann. Math. Statist.*, vol. 13, pp. 434–439, 1942.

11 P. C. Tang: "The Power Function of AOV Tests," *Statist. Research Mem.*, vol. 2, 1938.

12 C. R. Rao: Generalisation of Markoff's Theorem and Tests of Linear Hypotheses, *Sankhya*, vol. 7, pp. 9–19, 1945–1946.

13 C. R. Rao: On the Linear Combination of Observations and the General Theory of Least Squares, *Sankhya*, vol. 7, pp. 237–256, 1945–1946.

14 S. Kolodziejczyk: On an Important Class of Statistical Hypotheses, *Biometrika*, vol. 27, 1935.

15 H. Scheffé: "The Analysis of Variance," John Wiley & Sons, Inc., New York, 1959.

12

The Cross-classification or Factorial Model

12.1 Introduction

In this chapter and some of the following chapters various special cases of model 4, the model of less than full rank, will be discussed. In the general case of model 4 discussed in Chap. 11, the **X** matrix was defined to consist only of 0's and 1's. Many of the models underlying controlled scientific experiments can be written in this fashion; that is to say, many scientific experiments are so designed that the observed quantities are assumed to have a structure of the form of the model in Chap. 11. Sometimes, also, the observations from uncontrolled or semicontrolled experiments are assumed to follow this model. Therefore, it seems worthwhile to study various special cases of model 4. The study of these special cases may help the experimenter to design "better" experiments or to interpret better those which have been designed.

Throughout this chapter we shall assume that there exists a set of observable random variables $y_{ijmpq\ldots t}$ whose structure is such that $y_{ijmpq\ldots t} = \mu_{ijmpq\ldots t} + e_{ijmpq\ldots t}$, where $\mu_{ijmpq\ldots t}$ are a set of constants and $e_{ijmpq\ldots t}$ are a set of uncorrelated random variables with means 0 and constant variance σ^2. For confidence intervals and tests of hypotheses we shall make the further assumption that the $e_{ijmpq\ldots t}$ are normally distributed.

We shall start with the simplest model in terms of the number of elements present—the one-way classification model, which is sometimes referred to as the *completely randomized model.* We shall then discuss the two-way classification model and later extend our discussion to an N-way classification model.

12.2 The One-way Classification Model

12.2.1 Definition and Notation. Suppose the observed random variables $y_{11}, y_{12}, \ldots, y_{rm_r}$ (the subscripts are for identification purposes) are assumed to have the following structure:

$$y_{ij} = \mu + \alpha_i + e_{ij} \qquad j = 1, 2, \ldots, m_i; \quad i = 1, 2, \ldots, r$$
$$\Sigma m_i = M \qquad \text{each } m_i > 0 \tag{12.1}$$

where μ and α_i ($i = 1, 2, \ldots, r$) are unknown constants, which will be referred to as the *mean* and *additive treatment constants*, respectively, and where the e_{ij} are random variables. If $m_i = m$ for all i, the model is said to be *balanced*; if $m_i \neq m$ for some i, the model is said to be *unbalanced*. For what follows we assume $m_i \neq m$ for some i.

The model can also be written as $y_{ij} = (\mu + \bar{\alpha}_.) + (\alpha_i - \bar{\alpha}_.) + e_{ij}$, which we rewrite as

$$y_{ij} = \mu^* + \alpha_i^* + e_{ij} \qquad \text{where } \bar{\alpha}_. = \frac{1}{M}\sum \alpha_i m_i$$

We note that $\sum_i \alpha_i^* m_i = 0$ and that $\mu^* = \sum_{ij} E(y_{ij})/M$. This is a reparametrization of the original model. Although it seems that the model $y_{ij} = \mu^* + \alpha_i^* + e_{ij}$ may have advantages of interpretation in experimental work, we shall in general work with the model $y_{ij} = \mu + \alpha_i + e_{ij}$, where the α_i do not sum to zero. Since one of these models is a reparametrization of the other, the estimates of estimable functions will be the same regardless of which is used.

12.2.2 Point Estimation. Using the procedure in Sec. 11.5, we can obtain the following normal equations:

$$\begin{array}{lllll}
\mu: & M\hat{\mu} + m_1\hat{\alpha}_1 + m_2\hat{\alpha}_2 + \cdots + m_r\hat{\alpha}_r = Y_{..} \\
\alpha_1: & m_1\hat{\mu} + m_1\hat{\alpha}_1 = Y_{1.} \\
\alpha_2: & m_2\hat{\mu} + m_2\hat{\alpha}_2 = Y_{2.} \\
\cdots & \cdots\cdots\cdots\cdots\cdots\cdots \\
\alpha_r: & m_r\hat{\mu} + m_r\hat{\alpha}_r = Y_{r.}
\end{array} \tag{12.2}$$

or, more compactly,

$$\begin{array}{ll}
\mu: & M\hat{\mu} + \Sigma m_i\hat{\alpha}_i = Y_{..} \\
\alpha_q: & m_q\hat{\mu} + m_q\hat{\alpha}_q = Y_{q.} \qquad q = 1, 2, \ldots, r
\end{array} \tag{12.3}$$

Since the sum of the last r rows of Eq. (12.2) is equal to the first row, there is at least one linear dependence. It can be seen also that the last r rows are linearly independent. Thus, the rank of $\mathbf{X}'\mathbf{X}$ is r, and there are r linearly independent estimable functions (by Theorem 11.4). Also, by Theorem 11.6, the quantities $E(y_{ij})$ and any linear function of $E(y_{ij})$ are estimable. Thus, $\mu + \alpha_i$ and any linear function of the $\mu + \alpha_i$ are estimable. We can use the $(i + 1)$st equation of the normal equations (12.2) and Corollary 11.6.2, which states that the best linear unbiased estimate is the right-hand side of the corresponding equation; that is, $\widehat{\mu + \alpha_i} = Y_{i.}/m_i = y_{i.}$. If $\Sigma c_i = 0$, then $\Sigma c_i\alpha_i$ is estimable and the estimate is $\Sigma c_i y_{i.}$. Suppose $\Sigma d_i\alpha_i$ and $\Sigma f_i\alpha_i$ are two estimable functions of the α_i. Since they are both estimable, some linear combination of the left-hand side of the normal equations (12.2) must yield $\Sigma d_i\hat{\alpha}_i$. Then the same linear combination of the right-hand side of Eq. (12.2) will give the best linear unbiased estimate. The linear combination is obtained by multiplying the $(i + 1)$st equation by d_i/m_i and adding. For $\Sigma d_i\alpha_i$ to be estimable it must be true that $\Sigma d_i = 0$. The estimate is $\sum_i Y_{i.}d_i/m_{i.}$. Similarly, the estimate of $\Sigma f_i\alpha_i$ is $\sum_i Y_{i.}f_i/m_{i.}$. The covariance of the estimates is

$$\text{cov}(\widehat{\Sigma d_i\alpha_i}, \widehat{\Sigma f_j\alpha_j}) = E[(\Sigma d_i y_{i.} - \Sigma d_i\alpha_i)(\Sigma f_j y_{j.} - \Sigma f_j\alpha_j)] = \sigma^2\Sigma \frac{d_i f_i}{m_i}$$

The two estimates are uncorrelated if and only if

$$\Sigma \frac{d_i f_i}{m_i} = 0$$

If $m_i = m$, the two estimates are uncorrelated if and only if $\Sigma d_i f_i = 0$. The variance of the estimate of $\Sigma d_i\alpha_i$ is

$$\text{var}(\widehat{\Sigma d_i\alpha_i}) = \text{var}\left(\Sigma \frac{d_i Y_{i.}}{m_i}\right) = \Sigma \frac{d_i^2}{m_i}\sigma^2$$

If $m_i = m$, this variance of a contrast is $\sigma^2\Sigma d_i^2/m$.

12.2.3 Tests of Hypotheses. Suppose it is desired to test the hypothesis $\alpha_1 = \alpha_2 = \cdots = \alpha_r$. By Theorem 11.17, this is an estimable hypothesis, since $\Sigma c_i\alpha_i$ is estimable if $\Sigma c_i = 0$. The quantities needed are $R(\mu,\alpha)$, $R(\mu)$, and finally $R(\alpha \mid \mu) = R(\mu,\alpha) - R(\mu)$. To obtain a solution to the normal equations (12.2), we use the nonestimable condition $\Sigma m_i\hat{\alpha}_i = 0$. This gives $\hat{\mu} = y_{..}$; $\hat{\alpha}_q = y_{q.} - y_{..}$ for $q = 1, 2, \ldots, r$. Therefore,

$$R(\mu,\alpha) = \hat{\boldsymbol{\beta}}'\mathbf{X}'\mathbf{Y} = Y_{..}\hat{\mu} + \Sigma Y_{i.}\hat{\alpha}_i = \frac{Y_{..}^2}{M} + \left(\Sigma \frac{Y_{i.}^2}{m_i} - \frac{Y_{..}^2}{M}\right)$$

To find $R(\mu)$, we can write the model $y_{ij} = \mu^* + e_{ij}$, where $\mu^* =$

$\mu + \alpha$, where $\alpha_1 = \alpha_2 = \cdots = \alpha_r = \alpha$. There is only one normal equation, and, by the procedure of Sec. 11.5, this is

$$\mu^*: \qquad M\tilde{\mu}^* = Y_{..}$$

Hence, $R(\mu) = Y_{..}\tilde{\mu}^* = Y_{..}^2/M$. The AOV is presented in Table 12.1.

TABLE 12.1 ANALYSIS OF VARIANCE FOR TESTING $\alpha_1 = \alpha_2 = \cdots = \alpha_r$

SV	DF	SS	MS	F	EMS
Total	M	$\sum_{ij} y_{ij}^2$			
Due to mean	1	$\frac{Y_{..}^2}{M}$			
Due to α (adj)	$r - 1$	$\alpha_{SS} = \sum_i \frac{Y_{i.}^2}{m_i} - \frac{Y_{..}^2}{M}$	$\frac{\alpha_{SS}}{r-1}$	$\frac{\alpha_{ms}}{W_{ms}}$	$\sigma^2 + \frac{1}{r-1}\sum_i m_i(\alpha_i - \bar{\alpha}.)^2$
Error	$M - r$	$W_{SS} = \sum_{ij} y_{ij}^2 - \sum_i \frac{Y_{i.}^2}{m_i}$	$\frac{W_{SS}}{M-r}$		σ^2

To find the expected mean square of treatments, consider that

$$\alpha_{SS} = \sum_i \frac{Y_{i.}^2}{m_i} - \frac{Y_{..}^2}{M} = \sum_{ij} (y_{i.} - y_{..})^2$$

But
$$y_{i.} = \frac{1}{m_i}\sum_j y_{ij} = \frac{1}{m_i}\sum_j (\mu + \alpha_i + e_{ij}) = \mu + \alpha_i + \frac{1}{m_i}\sum_j e_{ij}$$

and
$$\begin{aligned} y_{..} &= \frac{1}{M}\sum_{ij} y_{ij} \\ &= \frac{1}{M}\left[\sum_{ij} (\mu + \alpha_i + e_{ij})\right] \\ &= \frac{1}{M}\left[\sum_i m_i(\mu + \alpha_i) + \sum_{ij} e_{ij}\right] \\ &= \mu + \frac{1}{M}\sum_i m_i\alpha_i + \frac{1}{M}\sum_{ij} e_{ij} \end{aligned}$$

Substituting gives

$$\begin{aligned} E\left[\sum_{ij} (y_{i.} - y_{..})^2\right] &= E\left[\sum_{ij}\left(\mu + \alpha_i + \frac{1}{m_i}\sum_j e_{ij} - \mu - \frac{1}{M}\sum_p m_p\alpha_p - \frac{1}{M}\sum_{pq} e_{pq}\right)^2\right] \\ &= \sum_i m_i(\alpha_i - \bar{\alpha}_.)^2 + (r-1)\sigma^2 \end{aligned}$$

where $\bar{\alpha}_. = \frac{1}{M}\sum_i m_i\alpha_i$. Therefore, by Theorem 11.16, the noncentrality λ is

$$\lambda = \frac{(r-1)\left[\frac{\sum m_i(\alpha_i - \bar{\alpha}_.)^2}{(r-1)} + \sigma^2\right]}{2\sigma^2} - \frac{r-1}{2}$$

$$= \frac{1}{2\sigma^2}\sum_i m_i(\alpha_i - \bar{\alpha}_.)^2$$

If $m_i = m$, this is

$$\lambda = \frac{m}{2\sigma^2}\sum_i (\alpha_i - \bar{\alpha}_.)^2$$

where

$$\bar{\alpha}_. = \frac{1}{r}\sum_i \alpha_i$$

Since the power of the test is an increasing function of λ, this formula for λ gives an experimenter some insight into how he can improve his experiment, i.e., by either increasing m or decreasing σ^2, or both.

To set a confidence interval on any treatment contrast $\Sigma c_i\alpha_i$, we can use the following facts:

1. $\Sigma c_i y_{i.}$ is distributed $N\left(\Sigma c_i\alpha_i,\ \sigma^2\Sigma \frac{c_i^2}{m_i}\right)$.
2. W_{ss}/σ^2 is distributed as $\chi^2(M - r)$.
3. W_{ss} and $\Sigma c_i y_{i.}$ are independent.

Hence,

$$\frac{(\Sigma c_i y_{i.} - \Sigma c_i\alpha_i)(M - r)^{\frac{1}{2}}}{(W_{ss}\Sigma c_i^2/m_i)^{\frac{1}{2}}}$$

is distributed as Student's t with $M - r$ degrees of freedom. The $1 - \alpha$ confidence interval is

$$\Sigma c_i y_{i.} - t_{\alpha/2}\left(\frac{W_{ss}}{M-r}\Sigma\frac{c_i^2}{m_i}\right)^{\frac{1}{2}} < \Sigma c_i\alpha_i < \Sigma c_i y_{i.} + t_{\alpha/2}\left(\frac{W_{ss}}{M-r}\Sigma\frac{c_i^2}{m_i}\right)^{\frac{1}{2}}$$

The width is

$$2t_{\alpha/2}\left(\frac{W_{ss}}{M-r}\Sigma\frac{c_i^2}{m_i}\right)^{\frac{1}{2}}$$

12.3 Two-way Cross Classification: No Interaction

12.3.1 Definition and Introduction. The two-way balanced cross classification without interaction is defined by the model

$$y_{ij} = \mu + \tau_i + \beta_j + e_{ij} \qquad i = 1, 2, \ldots, t; j = 1, 2, \ldots, b \qquad (12.4)$$

where y_{ij} is an observation, μ is an unknown parameter, the τ_i are one set of parameters, and the β_j are a second set of parameters. The set τ_i will be associated with a certain factor, say, factor A, which might be a set of machines or a set of chemical treatments; the set β_j will be associated with a certain factor, say, factor B, which might be a set of temperatures.

For example, suppose an experimenter wants to measure the effect of different ovens and different temperatures on the strength of a certain metal. If he desires to use four temperatures, let these be factor A, and let the set of parameters be $\tau_1, \tau_2, \tau_3, \tau_4$, the respective additive treatment constants due to temperatures. If, say, three ovens are available, let these be factor B, and let the set of parameters $\beta_1, \beta_2, \beta_3$ be the respective additive treatment constants due to ovens. If μ is a general constant, the model might be written as Eq. (12.4), where y_{ij} is the observed strength of the material when the ith temperature is applied in the jth oven.

Some might prefer to write the model as

$$y_{ij} = (\mu + \bar{\tau}_. + \bar{\beta}_.) + (\tau_i - \bar{\tau}_.) + (\beta_j - \bar{\beta}_.) + e_{ij}$$

which we can write as

$$y_{ij} = \mu^* + \tau_i^* + \beta_j^* + e_{ij} \tag{12.5}$$

This is a reparametrization of (12.4) and, by Corollary 11.8.1, will lead to the same estimates of estimable functions. Notice that $\sum_i \tau_i^* = \sum_j \beta_j^* = 0$ and that $E\left(\sum_{ij} y_{ij}\right) = bt\mu^*$. Model (12.5) seems to be superior for interpretation in many experimental situations, but we shall sometimes use model (12.4) to develop the theory. Model (12.5) can of course be derived from (12.4). The e_{ij} are random errors, which are assumed to be uncorrelated with mean 0 and variance σ^2. For tests of hypotheses and interval estimation, the errors will be assumed to be normally distributed.

The object in this model is generally to estimate the contrasts $\Sigma c_i \tau_i$ and $\Sigma d_j \beta_j$ and to test the hypotheses $\tau_1 = \tau_2 = \cdots = \tau_t$ and $\beta_1 = \beta_2 = \cdots = \beta_b$.

12.3.2 Point Estimation. The normal equations are

$$\begin{aligned}
\mu:&\quad bt\hat{\mu} + b\hat{\tau}_1 + \cdots + b\hat{\tau}_t + t\hat{\beta}_1 + \cdots + t\hat{\beta}_b = Y_{..} \\
\tau_p:&\quad b\hat{\mu} + b\hat{\tau}_p + \hat{\beta}_1 + \cdots + \hat{\beta}_b = Y_{p.} \qquad p = 1, 2, \ldots, t \\
\beta_q:&\quad t\hat{\mu} + \hat{\tau}_1 + \cdots + \hat{\tau}_t + t\hat{\beta}_q = Y_{.q} \qquad q = 1, 2, \ldots, b
\end{aligned} \tag{12.6}$$

The only linear combinations of the τ_i or of the β_j that are estimable are contrasts. The contrast $\Sigma c_i\tau_i$ is estimated by $\Sigma c_i y_{i.}$, and $\Sigma d_j\beta_j$ by $\Sigma d_j y_{.j}$. The variances of these estimates are $\Sigma c_i^2\sigma^2/b$ and $\Sigma d_j^2\sigma^2/t$, respectively.

12.3.3 Tests of Hypotheses. To test $\tau_1 = \tau_2 = \cdots = \tau_t$, we must find $R(\mu,\tau,\beta)$. To solve the normal equations, we can use the two nonestimable conditions $\Sigma\hat{\tau}_i = 0$ and $\Sigma\hat{\beta}_j = 0$. This gives $\hat{\mu} = y_{..}$, $\hat{\tau}_p = y_{p.} - y_{..}$, and $\hat{\beta}_q = y_{.q} - y_{..}$. Therefore,

$$R(\mu,\tau,\beta) = \hat{\mu}Y_{..} + \Sigma\hat{\tau}_p Y_{p.} + \Sigma\hat{\beta}_q Y_{.q}$$

$$= \frac{Y_{..}^2}{bt} + \left(\frac{\Sigma Y_{p.}^2}{b} - \frac{Y_{..}^2}{bt}\right) + \left(\frac{\Sigma Y_{.q}^2}{t} - \frac{Y_{..}^2}{bt}\right)$$

The model under the hypothesis leads to the normal equations:

$$\mu: \qquad bt\tilde{\mu} + t\tilde{\beta}_1 + \cdots + t\tilde{\beta}_b = Y_{..}$$

$$\beta_q: \qquad t\tilde{\mu} + t\tilde{\beta}_q = Y_{.q} \qquad q = 1, 2, \ldots, b$$

To solve these we can use the nonestimable condition $\Sigma\tilde{\beta}_j = 0$. This gives $\tilde{\mu} = y_{..}$ and $\tilde{\beta}_q = y_{.q} - y_{..}$. Therefore,

$$R(\mu,\beta) = \tilde{\mu}Y_{..} + \sum_j \tilde{\beta}_j Y_{.j} = \frac{Y_{..}^2}{bt} + \left(\Sigma\frac{Y_{.q}^2}{t} - \frac{Y_{..}^2}{bt}\right)$$

We get

$$R(\tau \mid \mu,\beta) = R(\mu,\tau,\beta) - R(\mu,\beta) = \sum_i \frac{Y_{i.}^2}{b} - \frac{Y_{..}^2}{bt} = \sum_{ij}(y_{i.} - y_{..})^2$$

By a similar procedure it can be shown that

$$R(\beta \mid \mu,\tau) = \Sigma\frac{Y_{.j}^2}{t} - \frac{Y_{..}^2}{bt} = \sum_{ij}(y_{.j} - y_{..})^2$$

Also, the error sum of squares, which is $\sum_{ij} y_{ij}^2 - R(\mu,\beta,\tau)$, is

$$\sum_{ij} y_{ij}^2 - \frac{\sum_i Y_{i.}^2}{b} - \frac{\sum_j Y_{.j}^2}{t} + \frac{Y_{..}^2}{bt} = \sum_{ij}(y_{ij} - y_{i.} - y_{.j} + y_{..})^2$$

We see that

$$\sum_{ij} y_{ij}^2 = \frac{Y_{..}^2}{bt} + R(\beta \mid \mu,\tau) + R(\tau \mid \mu.\beta) + E_{\text{ss}}$$

The AOV is shown in Table 12.2.

The noncentrality for testing $\beta_1 = \beta_2 = \cdots = \beta_b$ is

$$\frac{t}{2\sigma^2}\sum_j(\beta_j - \bar{\beta}_.)^2$$

TABLE 12.2 ANALYSIS OF VARIANCE FOR TWO-WAY CLASSIFICATION MODEL

SV	DF	SS	MS	F	EMS
Total	bt	$\sum_{ij} y_{ij}^2$			
Mean	1	$\frac{Y_{..}^2}{bt}$			
β (adj)	$b-1$	$B_{ss} = \frac{\sum_j Y_{.j}^2}{t} - \frac{Y_{..}^2}{bt}$	$\frac{B_{ss}}{b-1}$	$\frac{B_{ms}}{E_{ms}}$	$\sigma^2 + \frac{t}{b-1}\sum_j (\beta_j - \bar{\beta}_.)^2$
τ (adj)	$t-1$	$T_{ss} = \frac{\sum_i Y_{i.}^2}{b} - \frac{Y_{..}^2}{bt}$	$\frac{T_{ss}}{t-1}$	$\frac{T_{ms}}{E_{ms}}$	$\sigma^2 + \frac{b}{t-1}\sum_i (\tau_i - \bar{\tau}_.)^2$
Error	$(b-1)(t-1)$	E_{ss}	$\frac{E_{ss}}{(t-1)(b-1)}$		σ^2

and that for testing $\tau_1 = \tau_2 = \cdots = \tau_t$ is

$$\frac{b}{2\sigma^2}\sum_i (\tau_i - \bar{\tau}_.)^2$$

12.3.4 Interval Estimation. A confidence interval on $\Sigma c_i \tau_i$ of size $1 - \alpha$ is

$$\Sigma c_i y_{i.} - t_{\alpha/2}\left(\frac{E_{ms}\Sigma c_i^2}{b}\right)^{\frac{1}{2}} < \Sigma c_i \tau_i < \Sigma c_i y_{i.} + t_{\alpha/2}\left(\frac{E_{ms}\Sigma c_i^2}{b}\right)^{\frac{1}{2}}$$

The width is
$$2t_{\alpha/2}\left(\frac{E_{ms}\Sigma c_i^2}{b}\right)^{\frac{1}{2}}$$

Similarly, for $\Sigma d_j \beta_j$,

$$\Sigma d_j y_{.j} - t_{\alpha/2}\left(\frac{E_{ms}\Sigma d_j^2}{t}\right)^{\frac{1}{2}} < \Sigma c_j \beta_j < \Sigma d_j y_{.j} + t_{\alpha/2}\left(\frac{E_{ms}\Sigma d_j^2}{t}\right)^{\frac{1}{2}}$$

The width is
$$2t_{\alpha/2}\left(\frac{E_{ms}\Sigma d_j^2}{t}\right)^{\frac{1}{2}}$$

$t_{\alpha/2}$ is the appropriate value of Student's t with $(b-1)(t-1)$ degrees of freedom.

12.4 N-way Cross Classification: No Interaction

The N-way cross classification is an extension of the two-way cross classification to include N sets of parameters (N factors). We shall demonstrate by discussing a four-way model. The reader should have no trouble extending this.

Let α_i ($i = 1, 2, \ldots, p$), β_j ($j = 1, 2, \ldots, q$), γ_k ($k = 1, 2, \ldots, r$), and τ_m ($m = 1, 2, \ldots, s$) be four sets of parameters (treatment constants). Let the observation y_{ijkm} be defined by

$$y_{ijkm} = \mu + \alpha_i + \beta_j + \gamma_k + \tau_m + e_{ijkm} \tag{12.7}$$

or

$$y_{ijkm} = (\mu + \bar{\alpha}_. + \bar{\beta}_. + \bar{\gamma}_. + \bar{\tau}_.) + (\alpha_i - \bar{\alpha}_.) + (\beta_j - \bar{\beta}_.) + (\gamma_k - \bar{\gamma}_.) + (\tau_m - \bar{\tau}_.) + e_{ijkm}$$

which we write as

$$y_{ijkm} = \mu^* + \alpha_i^* + \beta_j^* + \gamma_k^* + \tau_m^* + e_{ijkm}$$

where the e_{ijkm} are uncorrelated random errors with mean 0 and variance σ^2.

By the procedure of Sec. 11.5, the normal equations are as follows:

$$\begin{aligned}
\mu:&\quad pqrs\hat{\mu} + qrs\Sigma\hat{\alpha}_i + prs\Sigma\hat{\beta}_j + pqs\Sigma\hat{\gamma}_k + pqr\Sigma\hat{\tau}_m = Y_{....} \\
\alpha_i:&\quad qrs\hat{\mu} + qrs\hat{\alpha}_i + rs\Sigma\hat{\beta}_j + qs\Sigma\hat{\gamma}_k + qr\Sigma\hat{\tau}_m = Y_{i...} \qquad i = 1, 2, \ldots p \\
\beta_j:&\quad prs\hat{\mu} + rs\Sigma\hat{\alpha}_i + prs\hat{\beta}_j + ps\Sigma\hat{\gamma}_k + pr\Sigma\hat{\tau}_m = Y_{.j..} \qquad j = 1, 2, \ldots q \\
\gamma_k:&\quad pqs\hat{\mu} + qs\Sigma\hat{\alpha}_i + ps\Sigma\hat{\beta}_j + pqs\hat{\gamma}_k + pq\Sigma\hat{\tau}_m = Y_{..k.} \qquad k = 1, 2, \ldots r \\
\tau_m:&\quad pqr\hat{\mu} + qr\Sigma\hat{\alpha}_i + pr\Sigma\hat{\beta}_j + pq\Sigma\hat{\gamma}_k + pqr\hat{\tau}_m = Y_{...m} \qquad m = 1, 2, \ldots s
\end{aligned} \tag{12.8}$$

It follows that:

1. $\Sigma d_i\alpha_i$ is estimable if $\Sigma d_i = 0$, and the estimate is $\Sigma d_i y_{i...}$; the variance is $\sigma^2 \Sigma d_i^2/qrs$.
2. $\Sigma f_j\beta_j$ is estimable if $\Sigma f_j = 0$, and the estimate is $\Sigma f_j y_{.j..}$; the variance is $\sigma^2\Sigma f_j^2/prs$.
3. $\Sigma g_k\gamma_k$ is estimable if $\Sigma g_k = 0$, and the estimate is $\Sigma g_k y_{..k.}$; the variance is $\sigma^2\Sigma g_k^2/pqs$.
4. $\Sigma h_m\tau_m$ is estimable if $\Sigma h_m = 0$, and the estimate is $\Sigma h_m y_{...m}$; the variance is $\sigma^2\Sigma h_m^2/pqr$.

Use of the nonestimable conditions $\Sigma\hat{\alpha}_i = 0, \Sigma\hat{\beta}_j = 0, \Sigma\hat{\gamma}_k = 0, \Sigma\hat{\tau}_m = 0$ allows for easy solution of the normal equations. The solution is

$$\begin{aligned}
\hat{\mu} &= y_{....} \\
\hat{\alpha}_i &= y_{i...} - y_{....} \\
\hat{\beta}_j &= y_{.j..} - y_{....} \\
\hat{\gamma}_k &= y_{..k.} - y_{....} \\
\hat{\tau}_m &= y_{...m} - y_{....}
\end{aligned}$$

This gives

$$R(\mu,\alpha,\beta,\gamma,\tau) = \hat{\mu}Y_{....} + \sum_i \hat{\alpha}_i Y_{i...} + \sum_j \hat{\beta}_j Y_{.j..} + \sum_k \hat{\gamma}_k Y_{..k.} + \sum_m \hat{\tau}_m Y_{...m}$$

$$= \frac{Y^2_{....}}{pqrs} + \left(\frac{\sum_i Y^2_{i...}}{qrs} - \frac{Y^2_{....}}{pqrs}\right) + \left(\frac{\sum_j Y^2_{.j..}}{prs} - \frac{Y^2_{....}}{pqrs}\right)$$

$$+ \left(\frac{\sum_k Y^2_{..k.}}{pqs} - \frac{Y^2_{....}}{pqrs}\right) + \left(\frac{\sum_m Y^2_{...m}}{pqr} - \frac{Y^2_{....}}{pqrs}\right)$$

Hence, the error sum of squares is

$$\sum_i\sum_j\sum_k\sum_m y^2_{ijkm} - R(\mu,\alpha,\beta,\gamma,\tau)$$

$$= \sum_i\sum_j\sum_k\sum_m (y_{ijkm} - y_{i...} - y_{.j..} - y_{..k.} - y_{...m} + 3y_{....})^2$$

Suppose we desire to test the hypothesis $\alpha_1 = \alpha_2 = \cdots = \alpha_p$. This is an estimable hypothesis, since $\Sigma c_i\alpha_i$ is estimable if $\Sigma c_i = 0$. The new model is

$$y_{ijkm} = \bar{\mu} + \beta_j + \gamma_k + \tau_m + e_{ijkm}$$

where we have substituted $\bar{\mu}$ for $\alpha + \mu$ in (12.7), where we let $\alpha_1 = \alpha_2 = \cdots = \alpha_p = \alpha$. The normal equations for this model can be obtained from the normal equations (12.8) by letting $\hat{\alpha}_1 = \hat{\alpha}_2 = \cdots = \hat{\alpha}_p = 0$ and by eliminating the equations for α_i. The normal equations are

$$\begin{array}{ll}
\bar{\mu}: & pqrs\tilde{\bar{\mu}} + prs\Sigma\tilde{\beta}_j + pqs\Sigma\tilde{\gamma}_k + pqr\Sigma\tilde{\tau}_m = Y_{....} \\
\beta_j: & prs\tilde{\bar{\mu}} + prs\tilde{\beta}_j + ps\Sigma\tilde{\gamma}_k + pr\Sigma\tilde{\tau}_m = Y_{.j..} \\
\gamma_k: & pqs\tilde{\bar{\mu}} + ps\Sigma\tilde{\beta}_j + pqs\tilde{\gamma}_k + pq\Sigma\tilde{\tau}_m = Y_{..k.} \\
\tau_m: & pqr\tilde{\bar{\mu}} + pr\Sigma\tilde{\beta}_j + pq\Sigma\tilde{\gamma}_k + pqr\tilde{\tau}_m = Y_{...m}
\end{array}$$

Using the nonestimable conditions $\Sigma\tilde{\beta}_j = 0$, $\Sigma\tilde{\gamma}_k = 0$, $\Sigma\tilde{\tau}_m = 0$, we solve these normal equations and get

$$\tilde{\bar{\mu}} = y_{....}$$
$$\tilde{\beta}_j = y_{.j..} - y_{....}$$
$$\tilde{\gamma}_k = y_{..k.} - y_{....}$$
$$\tilde{\tau}_m = y_{...m} - y_{....}$$

This yields

$$R(\mu,\beta,\gamma,\tau) = \tilde{\bar{\mu}}Y_{....} + \sum_j \tilde{\beta}_j Y_{.j..} + \sum_k \tilde{\gamma}_k Y_{..k.} + \sum_m \tilde{\tau}_m Y_{...m}$$

$$= \frac{Y^2_{....}}{pqrs} + \left(\frac{\sum_j Y^2_{.j..}}{prs} - \frac{Y^2_{....}}{pqrs}\right) + \left(\frac{\sum_k Y^2_{..k.}}{pqs} - \frac{Y^2_{....}}{pqrs}\right)$$

$$+ \left(\frac{\sum_m Y^2_{...m}}{pqr} - \frac{Y^2_{....}}{pqrs}\right)$$

We have
$$R(\alpha \mid \mu,\tau,\beta,\gamma) = R(\mu,\alpha,\beta,\gamma,\tau) - R(\mu,\beta,\gamma,\tau)$$
$$= \frac{\sum_i Y_{i...}^2}{qrs} - \frac{Y_{....}^2}{pqrs}$$
$$= \sum_i\sum_j\sum_k\sum_m (y_{i...} - y_{....})^2$$

denoted by α_{ss}. Similarly, to test $\beta_1 = \beta_2 = \cdots = \beta_q$, we could get
$$\sum_i\sum_j\sum_k\sum_m (y_{.j..} - y_{....})^2$$

denoted by β_{ss}, and we could get similar results if we wanted to test $\gamma_1 = \gamma_2 = \cdots = \gamma_r$ or $\tau_1 = \tau_2 = \cdots = \tau_s$.

It is interesting to observe that
$$\frac{Y_{....}^2}{pqrs} + R(\alpha \mid \mu,\beta,\gamma,\tau) + R(\beta \mid \mu,\alpha,\gamma,\tau) + R(\gamma \mid \mu,\alpha,\beta,\tau) + R(\tau \mid \mu,\alpha,\beta,\gamma)$$
$$+ E_{ss} = \sum_i\sum_j\sum_k\sum_m y_{ijkm}^2$$

The quantity $Y_{....}^2/pqrs$ is sometimes called the *sum of squares due to the mean*, and $R(\alpha \mid \mu,\beta,\tau,\gamma)$ is called the *sum of squares due to* α (adjusted), etc. Thus we can get the AOV shown in Table 12.3.

TABLE 12.3 ANALYSIS OF VARIANCE FOR FOUR-WAY MODEL

SV	DF	SS	MS	EMS
Total	$pqrs$	$\sum_{ijkm} y_{ijkm}^2$		
Mean	1	$\frac{Y_{....}^2}{pqrs}$		
α (adj)	$p - 1$	α_{ss}	α_{ms}	$\sigma^2 + \frac{qrs}{p-1}\sum_i (\alpha_i - \bar{\alpha}_.)^2$
β (adj)	$q - 1$	β_{ss}	β_{ms}	$\sigma^2 + \frac{prs}{q-1}\sum_j (\beta_j - \bar{\beta}_.)^2$
γ (adj)	$r - 1$	γ_{ss}	γ_{ms}	$\sigma^2 + \frac{pqs}{r-1}\sum_k (\gamma_k - \bar{\gamma}_.)^2$
τ (adj)	$s - 1$	τ_{ss}	τ_{ms}	$\sigma^2 + \frac{pqr}{s-1}\sum_m (\tau_m - \bar{\tau}_.)^2$
Error	$pqrs - p - q - r - s + 3$	E_{ss}	E_{ms}	σ^2

By Theorem 11.15, the quantities α_{ss}/σ^2, β_{ss}/σ^2, γ_{ss}/σ^2, and τ_{ss}/σ^2 are each distributed as a noncentral chi-square; E_{ss}/σ^2 is a central chi-square; and all are independent. Therefore, the AOV Table 12.3 can be used to test the four separate hypotheses:

(1) $\alpha_1 = \alpha_2 = \cdots = \alpha_p$ Noncentrality $= \dfrac{qrs}{2\sigma^2} \sum_i (\alpha_i - \bar{\alpha}_.)^2$

(2) $\beta_1 = \beta_2 = \cdots = \beta_q$ Noncentrality $= \dfrac{prs}{2\sigma^2} \sum_j (\beta_j - \bar{\beta}_.)^2$

(3) $\gamma_1 = \gamma_2 = \cdots = \gamma_r$ Noncentrality $= \dfrac{pqs}{2\sigma^2} \sum_k (\gamma_k - \bar{\gamma}_.)^2$

(4) $\tau_1 = \tau_2 = \cdots = \tau_s$ Noncentrality $= \dfrac{pqr}{2\sigma^2} \sum_m (\tau_m - \bar{\tau}_.)^2$

The noncentrality parameters are obtained from the EMS column and Theorem 11.16.

Confidence intervals can be set on any contrast by using E_{ms} as the estimate of σ^2.

12.5 Two-way Classification with Interaction

Before discussing the two-way classification with interaction, we shall define what we mean by the term *interaction*.

12.5.1 Definition of Interaction. To define the concept of interaction, consider the function of two variables $f(x,z)$. We formulate a definition as follows.

◆ **Definition 12.1** $f(x,z)$ will be defined to be a function with no interaction if and only if there exist functions $g(x)$ and $h(z)$ such that $f(x,z) = g(x) + h(z)$.

For example, the functions $x^2 + xz$, $x^2 + \log z + xz^2$, e^{zx}, and e^{z+x} have interaction, but the functions $x + z$, $\log xz$, and $x^2 + 2x + z^2 + 2z$ have no interaction. The above definition can be extended to any number of variables; i.e., the function $f(x,u,v, \ldots ,z)$ has no interaction if $f(x,u,v, \ldots ,z) = h_1(x) + h_2(u) + h_3(v) + \cdots + h_t(z)$, and has interaction otherwise.

The thing to notice in the model with no interaction, $f(x,z) = g(x) + h(z)$, is that, for two x values $x = a$ and $x = b$, the quantity $f(a,z) - f(b,z)$ is independent of z. Similarly for two z values.

The two-way classification model can be written as $y_{ij} = \mu_{ij} + e_{ij}$, where μ_{ij} is the "true" total effect of the combinations of element i from factor A and element j from factor B. If this "true" total effect is just the sum of the ith effect of A, which is τ_i, plus the jth effect of B, which is β_j, plus μ, then $\mu_{ij} = \mu + \tau_i + \beta_j$. To consider this more carefully, we see that

$$\mu_{1j} - \mu_{2j} = \tau_1 - \tau_2 \qquad \text{and} \qquad \mu_{1j'} - \mu_{2j'} = \tau_1 - \tau_2$$

which implies that $(\mu_{1j} - \mu_{2j}) - (\mu_{1j'} - \mu_{2j'}) = 0$

or, more generally,

$$(\mu_{ij} - \mu_{i'j}) - (\mu_{ij'} - \mu_{i'j'}) = 0 \qquad \text{for all } i, i', j, j'$$

Notice that the terms can be rearranged as follows:

$$(\mu_{ij} - \mu_{ij'}) - (\mu_{i'j} - \mu_{i'j'}) = (\beta_j - \beta_{j'}) - (\beta_j - \beta_{j'}) = 0$$

We now formulate the following definition.

◆ **Definition 12.2** The two-way classification model $y_{ij} = \mu_{ij} + e_{ij}$ will be said to be an additive model, or, in other words, a model with no interaction, if μ_{ij} is such that $(\mu_{ij} - \mu_{i'j}) - (\mu_{ij'} - \mu_{i'j'}) = 0$ for all i, i', j, j'. In the contrary case it will be said to be a nonadditive model, or a model with interaction.

Suppose that $\mu_{ij} = \mu + \tau_i + \beta_j + (\tau\beta)_{ij}$; then $(\mu_{ij} - \mu_{i'j}) - (\mu_{ij'} - \mu_{i'j'}) = (\tau\beta)_{ij} - (\tau\beta)_{i'j} - (\tau\beta)_{ij'} + (\tau\beta)_{i'j'}$. If this is not zero, the implication is that the "true" effect of the difference of two elements of A depends on which element of B is used. The quantity

$$\bar{\mu}_{i.} = \frac{1}{b}\sum_j \mu_{ij} = \mu + \tau_i + \bar{\beta}_. + \overline{(\tau\beta)}_{i.}$$

will be called the "true" mean effect of the ith element of A in the presence of the elements of B. A similar definition applies to $\bar{\mu}_{.j}$.

12.5.2 The Two-way Classification Model with Interaction. The two-way cross-classification model with interaction will be defined by

$$y_{ij} = \mu + \tau_i + \beta_j + (\tau\beta)_{ij} + e_{ij} \qquad i = 1, 2, \ldots, t; \quad j = 1, 2, \ldots, b \tag{12.9}$$

The normal equations can be found by the method of Sec. 11.5. They are as follows:

$$\begin{aligned}
&\mu: && bt\hat{\mu} + b\hat{\tau}_. + t\hat{\beta}_. + \widehat{(\tau\beta)}_{..} = Y_{..} \\
&\tau_p: && b\hat{\mu} + b\hat{\tau}_p + \hat{\beta}_. + \widehat{(\tau\beta)}_{p.} = Y_{p.} && p = 1, 2, \ldots, t \\
&\beta_q: && t\hat{\mu} + \hat{\tau}_. + t\hat{\beta}_q + \widehat{(\tau\beta)}_{.q} = Y_{.q} && q = 1, 2, \ldots, b \\
&(\tau\beta)_{pq}: && \hat{\mu} + \hat{\tau}_p + \hat{\beta}_q + \widehat{(\tau\beta)}_{pq} = Y_{pq} && \begin{cases} p = 1, 2, \ldots, t \\ q = 1, 2, \ldots, b \end{cases}
\end{aligned} \tag{12.10}$$

In the two-way classification model without interaction, the contrasts of the τ_i and of the β_j were estimable. By examining the normal equations for model (12.9) it can be seen that here the contrasts of the τ_i or the β_j are *not* estimable. In other words, the parameter $(\tau\beta)_{p.}$ cannot be separated from τ_p. The model can be written as

$$y_{ij} = [\mu + \bar{\tau}_. + \bar{\beta}_. + \overline{(\tau\beta)}_{..}] + [\tau_i - \bar{\tau}_. + \overline{(\tau\beta)}_{i.} - \overline{(\tau\beta)}_{..}] + [\beta_j - \bar{\beta}_. + \overline{(\tau\beta)}_{.j} - \overline{(\tau\beta)}_{..}] + [(\tau\beta)_{ij} - \overline{(\tau\beta)}_{i.} - \overline{(\tau\beta)}_{.j} + \overline{(\tau\beta)}_{..}] + e_{ij}$$

or

$$y_{ij} = \mu^* + \tau_i^* + \beta_j^* + (\tau\beta)_{ij}^* + e_{ij}$$

The "true" mean effect of the ith element of factor A could be defined as τ_i^*, and this is estimable, since $\tau_i^* = \tau_i - \bar{\tau}_. + \overline{(\tau\beta)}_{i.} - \overline{(\tau\beta)}_{..}$. We get $\hat{\tau}_i^* = y_{i.} - y_{..}$ from (12.10). If we use the nonestimable conditions $\hat{\tau}_. = 0$; $\hat{\beta}_. = 0$; $\widehat{(\tau\beta)}_{1.} = 0, \ldots, \widehat{(\tau\beta)}_{t.} = 0$; and $\widehat{(\tau\beta)}_{.1} = 0, \ldots, \widehat{(\tau\beta)}_{.b} = 0$, the normal equations (12.10) are easily solved. The solution is

$$\begin{aligned} \hat{\mu} &= y_{..} \\ \hat{\tau}_p &= y_{p.} - y_{..} \\ \hat{\beta}_q &= y_{.q} - y_{..} \\ \widehat{(\tau\beta)}_{pq} &= y_{pq} - y_{p.} - y_{.q} + y_{..} \end{aligned}$$

The quantity

$$\begin{aligned} R[\mu,\tau,\beta,(\tau\beta)] &= \hat{\mu} Y_{..} + \sum_i \hat{\tau}_i Y_{i.} + \sum_j \hat{\beta}_j Y_{.j} + \sum_{ij} \widehat{(\tau\beta)}_{ij} y_{ij} \\ &= \frac{Y_{..}^2}{bt} + \left(\frac{\sum_i Y_{i.}^2}{b} - \frac{Y_{..}^2}{bt}\right) + \left(\frac{\sum_j Y_{.j}^2}{t} - \frac{Y_{..}^2}{bt}\right) \\ &\quad + \sum_{ij} y_{ij}(y_{ij} - y_{i.} - y_{.j} + y_{..}) = \sum_{ij} y_{ij}^2 \end{aligned}$$

So the error sum of squares, which is $\sum_{ij} y_{ij}^2 - R[(\mu,\tau,\beta,(\tau\beta)]$, is equal to zero. Hence, in this model there is no estimate of the variance. So we cannot make the conventional tests of hypotheses or set confidence intervals.

To get an estimate of σ^2 there must be more than one observation for at least one ij combination. For example, suppose there are r observations per ij combination. Then the model becomes

$$y_{ijm} = \mu + \tau_i + \beta_j + (\tau\beta)_{ij} + e_{ijm} \qquad \begin{cases} i = 1, 2, \ldots, t \\ j = 1, 2, \ldots, b \\ m = 1, 2, \ldots, r \end{cases} \qquad (12.11)$$

The normal equations are

$$
\begin{aligned}
&\mu: && rbt\hat{\mu} + rb\hat{\tau}_. + rt\hat{\beta}_. + r(\widehat{\tau\beta})_{..} = Y_{...} \\
&\tau_p: && rb\hat{\mu} + rb\hat{\tau}_p + r\hat{\beta}_. + r(\widehat{\tau\beta})_{p.} = Y_{p..} && p = 1, 2, \ldots, t \\
&\beta_p: && rt\hat{\mu} + r\hat{\tau}_. + rt\hat{\beta}_q + r(\widehat{\tau\beta})_{.q} = Y_{.q.} && q = 1, 2, \ldots, b \qquad (12.12) \\
&(\tau\beta)_{pq}: && r\hat{\mu} + r\hat{\tau}_p + r\hat{\beta}_q + r(\widehat{\tau\beta})_{pq} = Y_{pq.} && \begin{cases} p = 1, 2, \ldots, t \\ q = 1, 2, \ldots, b \end{cases}
\end{aligned}
$$

The rank of this system (the number of linearly independent estimable functions) is bt. Using the nonestimable conditions $\hat{\tau}_. = 0$; $\hat{\beta}_. = 0$; and $(\widehat{\tau\beta})_{1.} = (\widehat{\tau\beta})_{2.} = \cdots = (\widehat{\tau\beta})_{t.} = (\widehat{\tau\beta})_{.1} = (\widehat{\tau\beta})_{.2} = \cdots = (\widehat{\tau\beta})_{.b} = 0$ gives the solution

$$
\begin{aligned}
\hat{\mu} &= y_{...} \\
\hat{\tau}_p &= y_{p..} - y_{...} \\
\hat{\beta}_q &= y_{.q.} - y_{...} \\
(\widehat{\tau\beta})_{pq} &= y_{pq.} - y_{p..} - y_{.q.} + y_{...}
\end{aligned}
$$

This gives
$$R[\mu,\tau,\beta,(\tau\beta)] = \frac{\sum_{ij} Y_{ij.}^2}{r}$$

So the error sum of squares is

$$\sum_{ijm} y_{ijm}^2 - \frac{\sum_{ij} Y_{ij.}^2}{r} = \sum_{ijm} (y_{ijm} - y_{ij.})^2$$

We shall next devise a test to see whether interaction is present in (12.11). We shall rewrite (12.11) in the following way:

$$
\begin{aligned}
y_{ijm} = [\mu + \bar{\tau}_. + \bar{\beta}_. + \overline{(\tau\beta)}_{..}] &+ [\tau_i - \bar{\tau}_. + \overline{(\tau\beta)}_{i.} - \overline{(\tau\beta)}_{..}] \\
&+ [\beta_j - \bar{\beta}_. + \overline{(\tau\beta)}_{.j} - \overline{(\tau\beta)}_{..}] \\
&+ [(\tau\beta)_{ij} - \overline{(\tau\beta)}_{i.} - \overline{(\tau\beta)}_{.j} + \overline{(\tau\beta)}_{..}] + e_{ijm}
\end{aligned}
$$

or
$$y_{ijm} = \mu^* + \tau_i^* + \beta_j^* + (\tau\beta)_{ij}^* + e_{ijm} \qquad (12.13)$$

It will be shown that there is no interaction in the model if and only if $(\tau\beta)_{ij} - \overline{(\tau\beta)}_{i.} - \overline{(\tau\beta)}_{.j} + \overline{(\tau\beta)}_{..}$ is equal to zero for all i and j. By Definition 12.2, we must show that $(\tau\beta)_{ij}^* = (\tau\beta)_{ij} - \overline{(\tau\beta)}_{i.} - \overline{(\tau\beta)}_{.j} + \overline{(\tau\beta)}_{..} = 0$ for all i and j if and only if $\gamma^* = (\mu_{ij} - \mu_{i'j}) - (\mu_{ij'} - \mu_{i'j'}) = 0$ for all i, i', j, j', where $\mu_{ij} = \mu + \tau_i + \beta_j + (\tau\beta)_{ij}$. Now, if $\gamma^* = 0$,

$$\frac{1}{bt} \sum_{i'j'} \gamma^* = \mu_{ij} - \bar{\mu}_{.j} - \bar{\mu}_{i.} + \bar{\mu}_{..} = 0 \qquad \text{for all } i \text{ and } j$$

But this equals $(\overline{\tau\beta})_{ij} - (\overline{\tau\beta})_{.j} - (\overline{\tau\beta})_{i.} + (\overline{\tau\beta})_{..}$. Next assume that $(\tau\beta)^*_{ij} = 0$ for all i and j. Then it is also true that $(\tau\beta)^*_{ij} - (\tau\beta)^*_{i'j} - (\tau\beta)^*_{ij'} + (\tau\beta)^*_{i'j'}$ must be equal to zero for all i, i', j, j'. By substituting $\mu_{ij} - \mu - \tau_i - \beta_j$ for $(\tau\beta)_{ij}$, we get $0 = (\tau\beta)^*_{ij} - (\tau\beta)^*_{i'j} - (\tau\beta)^*_{ij'} + (\tau\beta)^*_{i'j'} = (\mu_{ij} - \mu_{i'j}) - (\mu_{ij'} - \mu_{i'j'})$. We have, therefore, shown that there is no interaction if and only if $(\tau\beta)^*_{ij} = 0$ for all i and j. So we shall test the hypothesis that $(\tau\beta)^*_{ij} = 0$ for all i and j. By (12.10) it follows that $(\tau\beta)^*_{ij}$ is estimable, and, hence, that $(\tau\beta)^*_{ij} = 0$ (all i and j) is an estimable hypothesis.

To test this hypothesis we shall work with model (12.13), realizing that $\tau^*_. = \beta^*_. = (\tau\beta)^*_{i.} = (\tau\beta)^*_{.j} = 0$. The normal equations are

$$\begin{aligned}
\mu^*: &\quad rbt\hat{\mu}^* &&= Y_{...}\\
\tau^*_p: &\quad rb\hat{\mu}^* + rb\hat{\tau}^*_p &&= Y_{p..}\\
\beta^*_q: &\quad rt\hat{\mu}^* \qquad\qquad + rt\hat{\beta}^*_q &&= Y_{.q.}\\
(\tau\beta)^*_{pq}: &\quad r\hat{\mu}^* + r\hat{\tau}^*_p + r\hat{\beta}^*_q + r\widehat{(\tau\beta)}^*_{pq} &&= Y_{pq.}
\end{aligned}$$

This gives

$$R[\mu^*,\tau^*,\beta^*,(\tau\beta)^*] = \frac{Y^2_{...}}{rbt} + \left(\sum_i \frac{Y^2_{i..}}{rb} - \frac{Y^2_{...}}{rbt}\right) + \left(\sum_j \frac{Y^2_{.j.}}{rt} - \frac{Y^2_{...}}{rbt}\right)$$
$$+ \left(\sum_{ij} \frac{Y^2_{ij.}}{r} - \sum_i \frac{Y^2_{i..}}{rb} - \sum_j \frac{Y^2_{.j.}}{rt} + \frac{Y^2_{...}}{rbt}\right)$$

To find $R(\mu^*,\tau^*,\beta^*)$, we put $(\tau\beta)^*_{pq} = 0$ for all p and q. This gives the following normal equations:

$$\begin{aligned}
\mu^*: &\quad rbt\tilde{\mu}^* &&= Y_{...}\\
\tau^*_p: &\quad rb\tilde{\mu}^* + rb\tilde{\tau}^*_p &&= Y_{p..}\\
\beta^*_q: &\quad rt\tilde{\mu}^* \qquad\qquad + rt\tilde{\beta}^*_q &&= Y_{.q.}
\end{aligned}$$

Solving (using $\tilde{\tau}^*_. = \tilde{\beta}^*_. = 0$) gives

$$R(\mu^*,\tau^*,\beta^*) = \frac{Y^2_{...}}{rbt} + \left(\sum_i \frac{Y^2_{i..}}{rb} - \frac{Y^2_{...}}{rbt}\right) + \left(\sum_j \frac{Y^2_{.j.}}{rt} - \frac{Y^2_{...}}{rbt}\right)$$

and

$$\begin{aligned}
R[(\tau\beta)^* \mid \mu^*,\tau^*,\beta^*] &= R[\mu^*,\tau^*,\beta^*,(\tau\beta)^*] - R(\mu^*,\tau^*,\beta^*)\\
&= \sum_{ij} \frac{Y^2_{ij.}}{r} - \sum_i \frac{Y^2_{i..}}{rb} - \sum_j \frac{Y^2_{.j.}}{rt} + \frac{Y^2_{...}}{rbt}\\
&= \sum_{ijm} (y_{ij.} - y_{i..} - y_{.j.} + y_{...})^2
\end{aligned}$$

The error sum of squares is

$$\sum_{ijm} y^2_{ijm} - R[\mu^*,\tau^*,\beta^*,(\tau\beta)^*] = \sum_{ijm} y^2_{ijm} - \sum_{ij} \frac{Y^2_{ij.}}{r} = \sum_{ijm} (y_{ijm} - y_{ij.})^2$$

This leads to the AOV presented in Table 12.4.

TABLE 12.4 ANALYSIS OF VARIANCE FOR TESTING INTERACTION IN A TWO-WAY CLASSIFICATION

SV	DF	SS	EMS
Total	rbt	$\sum_{ijm} y_{ijm}^2$	
Due to $(\tau\beta)^*$ (adj)	$(t-1)(b-1)$	$\sum_{ijm} (y_{ij.} - y_{i..} - y_{.j.} + y_{...})^2$	$\sigma^2 + \frac{r}{(b-1)(t-1)} \sum_{ij} [(\tau\beta)_{ij}^*]^2$
Error	$bt(r-1)$	$\sum_{ijm} (y_{ijm} - y_{ij.})^2$	σ^2

The noncentrality parameter is

$$\lambda = \frac{r}{2\sigma^2} \sum_{ij} [(\tau\beta)_{ij}^*]^2 = \frac{r}{2\sigma^2} \sum_{ij} [(\tau\beta)_{ij} - (\overline{\tau\beta})_{i.} - (\overline{\tau\beta})_{.j} + (\overline{\tau\beta})_{..}]^2$$

If it is concluded that there is no interaction, model (12.4) is appropriate and Sec. 12.3 applies; otherwise model (12.11) is appropriate.

If it is decided that the interaction model (12.11)

$$y_{ijm} = \mu + \tau_i + \beta_j + (\tau\beta)_{ij} + e_{ijm}$$

is appropriate, then by using the normal equations (12.12) it can be determined what quantities are estimable. Contrasts of the τ_i and β_j are not estimable, as they are in the model with no interaction (Art. 12.3.2). Also, the hypotheses $\tau_1 = \tau_2 = \cdots = \tau_t$ and $\beta_1 = \beta_2 = \cdots = \beta_b$ are not estimable hypotheses. The "true" difference between the ith and jth effect of factor A can be defined as

$$\tau_i + (\overline{\tau\beta})_{i.} - \tau_j - (\overline{\tau\beta})_{j.}$$

This is estimable, and its best linear unbiased estimate is $y_{i..} - y_{j..}$.

12.5.3 Example. Assume that an experiment was run on three ovens, each at four temperatures, to ascertain the strength of the final product. Suppose that three runs were made in each oven at each temperature. The observations are given in Table 12.5*a*.

We shall assume the model

$$y_{ijm} = \mu + \tau_i + \beta_j + (\tau\beta)_{ij} + e_{ijm} \qquad \begin{cases} i = 1, 2, 3 \\ j = 1, 2, 3, 4 \\ m = 1, 2, 3 \end{cases}$$

where τ refers to ovens and β to temperature.

TABLE 12.5

β \ τ	1	2	3
1	3 3 2	4 3 2	3 4 6
2	6 3 4	6 2 7	8 5 9
3	3 4 4	3 6 7	5 6 9
4	4 5 4	3 7 9	7 8 10

(a) Entries are observations y_{ijk}.

β \ τ	1	2	3	$Y_{.j.}$
1	8	9	13	30
2	13	15	22	50
3	11	16	20	47
4	13	19	25	57
$Y_{i..}$	45	59	80	184

(b) Entries are cell totals $Y_{ij.}$

The cell totals $Y_{ij.}$ are given in Table 12.5*b*. The following values are easily obtained:

$$\tfrac{1}{3}\sum_{ij} Y_{ij.}^2 = 1{,}041.333$$

$$\tfrac{1}{36} Y_{...}^2 = 940.444$$

$$\tfrac{1}{12}\sum_{i} Y_{i..}^2 = 992.167$$

$$\tfrac{1}{9}\sum_{j} Y_{.j.}^2 = 984.222$$

$$\sum_{ij}\sum_{m} y_{ijm}^2 = 1{,}118$$

Therefore, we have

$$\begin{aligned}\text{SS due to } (\tau\beta)^* \text{ adj} &= \sum_{ij}\sum_{m} (y_{ij.} - y_{i..} - y_{.j.} + y_{...})^2 \\ &= \tfrac{1}{3}\sum_{ij} Y_{ij.}^2 - \tfrac{1}{12}\sum_{i} Y_{i..}^2 - \tfrac{1}{9}\sum_{j} Y_{.j.}^2 + \tfrac{1}{36} Y_{...}^2 \\ &= 5.388\end{aligned}$$

Also, the error sum of squares is

$$\sum_{ijm} (y_{ijm} - y_{ij.})^2 = \sum_{ijm} y_{ijm}^2 - \tfrac{1}{3}\sum_{ij} Y_{ij.}^2 = 76.67$$

The AOV is given in Table 12.6. There is no indication of interaction.

TABLE 12.6 ANALYSIS OF VARIANCE FOR TESTING INTERACTION IN THE EXAMPLE OF ART. 12.5.3

SV	DF	SS	MS	F
Total	36	1,118		
Due to $(\tau\beta)^*$ (adj)	6	5.39	.90	<1
Error	24	76.67	3.19	

12.6 N-way Classification with Interaction

In an N-way classification with interaction the model could have two-factor interaction, three-factor interaction, . . . , N-factor interaction. We shall demonstrate for $N = 3$; the reader should have no trouble extending the demonstration. The model can be written

$$y_{ijkm} = \mu_{ijk} + e_{ijkm}$$

or, more explicitly,

$$y_{ijkm} = \mu + \tau_i + \beta_j + (\tau\beta)_{ij} + \gamma_k + (\tau\gamma)_{ik} + (\beta\gamma)_{jk} + (\tau\beta\gamma)_{ijk} + e_{ijkm} \tag{12.14}$$

where $i = 1, 2, \ldots, I$; $j = 1, 2, \ldots, J$; $k = 1, 2, \ldots, K$; $m = 1, 2, \ldots, M$. The quantities $(\tau\beta)_{ij}$, $(\tau\gamma)_{ik}$, $(\beta\gamma)_{jk}$ are called *two-factor interaction terms.* $(\tau\beta\gamma)_{ijk}$ is called a *three-factor interaction term.* The normal equations are given in the group labeled (12.15) on page 273.

Suppose we rewrite model (12.14) as shown in (12.16) on page 274. Model (12.16) will be written as (12.17).

$$y_{ijkm} = \mu^* + \tau_i^* + \beta_j^* + (\tau\beta)_{ij}^* + \gamma_k^* + (\tau\gamma)_{ik}^* + (\beta\gamma)_{jk}^* + (\tau\beta\gamma)_{ijk}^* + e_{ijkm} \tag{12.17}$$

where τ_i^*, etc., can be identified with the corresponding quantities in brackets in (12.16). The following restrictions on (12.17) should be noticed:

$$\tau_.^* = \beta_.^* = \gamma_.^* = (\tau\beta)_{..}^* = (\tau\gamma)_{..}^* = (\beta\gamma)_{..}^* = (\tau\beta\gamma)_{...}^* = 0$$

Also,

$$(\tau\beta)_{i.}^* = 0 \qquad (\tau\beta)_{.j}^* = 0 \qquad (\beta\gamma)_{j.}^* = 0 \qquad (\beta\gamma)_{.k}^* = 0$$

$$(\tau\gamma)_{i.}^* = 0 \qquad (\tau\gamma)_{.k}^* = 0$$

$$(\tau\beta\gamma)_{ij.}^* = 0 \qquad (\tau\beta\gamma)_{i.k}^* = 0 \qquad (\tau\beta\gamma)_{.jk}^* = 0$$

for all i, j, k.

$$
\begin{array}{ll}
\mu\colon & IJKM\hat{\mu} + JKM\hat{\tau}_{.} + IKM\hat{\beta}_{.} + KM\widehat{(\tau\beta)}_{..} + IJM\hat{\gamma}_{.} + JM\widehat{(\tau\gamma)}_{..} + IM\widehat{(\beta\gamma)}_{..} + M\widehat{(\tau\beta\gamma)}_{...} = Y_{....} \\
\tau_i\colon & JKM\hat{\mu} + JKM\hat{\tau}_i + KM\hat{\beta}_{.} + KM\widehat{(\tau\beta)}_{i.} + JM\hat{\gamma}_{.} + JM\widehat{(\tau\gamma)}_{i.} + M\widehat{(\beta\gamma)}_{..} + M\widehat{(\tau\beta\gamma)}_{i..} = Y_{i...} \\
\beta_j\colon & IKM\hat{\mu} + KM\hat{\tau}_{.} + IKM\hat{\beta}_j + KM\widehat{(\tau\beta)}_{.j} + IM\hat{\gamma}_{.} + M\widehat{(\tau\gamma)}_{..} + IM\widehat{(\beta\gamma)}_{j.} + M\widehat{(\tau\beta\gamma)}_{.j.} = Y_{.j..} \\
(\tau\beta)_{ij}\colon & KM\hat{\mu} + KM\hat{\tau}_i + KM\hat{\beta}_j + KM\widehat{(\tau\beta)}_{ij} + M\hat{\gamma}_{.} + M\widehat{(\tau\gamma)}_{i.} + M\widehat{(\beta\gamma)}_{j.} + M\widehat{(\tau\beta\gamma)}_{ij.} = Y_{ij..} \\
\gamma_k\colon & IJM\hat{\mu} + JM\hat{\tau}_{.} + IM\hat{\beta}_{.} + M\widehat{(\tau\beta)}_{..} + IJM\hat{\gamma}_k + JM\widehat{(\tau\gamma)}_{.k} + IM\widehat{(\beta\gamma)}_{.k} + M\widehat{(\tau\beta\gamma)}_{..k} = Y_{..k.} \\
(\tau\gamma)_{ik}\colon & JM\hat{\mu} + JM\hat{\tau}_i + M\hat{\beta}_{.} + M\widehat{(\tau\beta)}_{i.} + JM\hat{\gamma}_k + JM\widehat{(\tau\gamma)}_{ik} + M\widehat{(\beta\gamma)}_{.k} + M\widehat{(\tau\beta\gamma)}_{i.k} = Y_{i.k.} \\
(\beta\gamma)_{jk}\colon & IM\hat{\mu} + M\hat{\tau}_{.} + IM\hat{\beta}_j + M\widehat{(\tau\beta)}_{.j} + IM\hat{\gamma}_k + M\widehat{(\tau\gamma)}_{.k} + IM\widehat{(\beta\gamma)}_{jk} + M\widehat{(\tau\beta\gamma)}_{.jk} = Y_{.jk.} \\
(\tau\beta\gamma)_{ijk}\colon & M\hat{\mu} + M\hat{\tau}_i + M\hat{\beta}_j + M\widehat{(\tau\beta)}_{ij} + M\hat{\gamma}_k + M\widehat{(\tau\gamma)}_{ik} + M\widehat{(\beta\gamma)}_{jk} + M\widehat{(\tau\beta\gamma)}_{ijk} = Y_{ijk.}
\end{array}
$$

Equation (12.15)

$$
\begin{aligned}
Y_{ijkm} = {} & [\mu + \bar{\tau}_{.} + \bar{\beta}_{.} + \overline{(\tau\beta)}_{..} + \bar{\gamma}_{.} + \overline{(\tau\gamma)}_{..} + \overline{(\beta\gamma)}_{..} + \overline{(\tau\beta\gamma)}_{...}] \\
& + [\tau_i - \bar{\tau}_{.} + \overline{(\tau\beta)}_{i.} - \overline{(\tau\beta)}_{..} + \overline{(\tau\gamma)}_{i.} - \overline{(\tau\gamma)}_{..} + \overline{(\tau\beta\gamma)}_{i..} - \overline{(\tau\beta\gamma)}_{...}] \\
& + [\beta_j - \bar{\beta}_{.} + \overline{(\tau\beta)}_{.j} - \overline{(\tau\beta)}_{..} + \overline{(\beta\gamma)}_{j.} - \overline{(\beta\gamma)}_{..} + \overline{(\tau\beta\gamma)}_{.j.} - \overline{(\tau\beta\gamma)}_{...}] \\
& + [(\tau\beta)_{ij} - \overline{(\tau\beta)}_{i.} - \overline{(\tau\beta)}_{.j} + \overline{(\tau\beta)}_{..} + \overline{(\tau\beta\gamma)}_{ij.} - \overline{(\tau\beta\gamma)}_{i..} - \overline{(\tau\beta\gamma)}_{.j.} + \overline{(\tau\beta\gamma)}_{...}] \\
& + [\gamma_k - \bar{\gamma}_{.} + \overline{(\tau\gamma)}_{.k} - \overline{(\tau\gamma)}_{..} + \overline{(\beta\gamma)}_{.k} - \overline{(\beta\gamma)}_{..} + \overline{(\tau\beta\gamma)}_{..k} - \overline{(\tau\beta\gamma)}_{...}] \\
& + [(\tau\gamma)_{ik} - \overline{(\tau\gamma)}_{i.} - \overline{(\tau\gamma)}_{.k} + \overline{(\tau\gamma)}_{..} + \overline{(\tau\beta\gamma)}_{i.k} - \overline{(\tau\beta\gamma)}_{i..} - \overline{(\tau\beta\gamma)}_{..k} + \overline{(\tau\beta\gamma)}_{...}] \\
& + [(\beta\gamma)_{jk} - \overline{(\beta\gamma)}_{j.} - \overline{(\beta\gamma)}_{.k} + \overline{(\beta\gamma)}_{..} + \overline{(\tau\beta\gamma)}_{.jk} - \overline{(\tau\beta\gamma)}_{.j.} - \overline{(\tau\beta\gamma)}_{..k} + \overline{(\tau\beta\gamma)}_{...}] \\
& + [(\tau\beta\gamma)_{ijk} - \overline{(\tau\beta\gamma)}_{ij.} - \overline{(\tau\beta\gamma)}_{i.k} - \overline{(\tau\beta\gamma)}_{.jk} + \overline{(\tau\beta\gamma)}_{i..} + \overline{(\tau\beta\gamma)}_{.j.} + \overline{(\tau\beta\gamma)}_{..k} - \overline{(\tau\beta\gamma)}_{...}] + e_{ijkm}
\end{aligned}
$$

Equation (12.16)

The normal equations for the model written as (12.17) can be found from (12.15). This gives the group of equations labeled (12.18) on page 276.

The rank of this system is IJK. The estimate of the parameters using the normal equations (12.18) is the same as the estimate of the corresponding elements in (12.16) by using the normal equations (12.15). For example, from (12.18) we have

$$\hat{\tau}_i^* = y_{i...} - y_{....}$$

But, by (12.16),

$$\tau_i^* = \tau_i - \bar{\tau}_. + (\overline{\tau\beta})_{i.} - (\overline{\tau\beta})_{..} + (\overline{\tau\gamma})_{i.} - (\overline{\tau\gamma})_{..} + (\overline{\tau\beta\gamma})_{i..} - (\overline{\tau\beta\gamma})_{...}$$

which can be written

$$\tau_i^* = [\mu + \tau_i + \bar{\beta}_. + (\overline{\tau\beta})_{i.} + \bar{\gamma}_. + (\overline{\tau\gamma})_{i.} + (\overline{\tau\beta\gamma})_{i..}]$$
$$- [\mu + \bar{\tau}_. + \bar{\beta}_. + (\overline{\tau\beta})_{..} + \bar{\gamma}_. + (\overline{\tau\gamma})_{..} + (\overline{\tau\beta\gamma})_{...}]$$

Using the equations (12.15), the estimate of the first quantity in brackets is $y_{i...}$ and the second is $y_{....}$; hence, using (12.15), we have $\hat{\tau}_i^* = y_{i...} - y_{....}$, the same expression that we obtained using model (12.17) and the normal equations (12.18). A similar situation holds for the other parameters of model (12.17). Of course, model (12.17) is a reparametrization of model (12.14). Hence, by Corollary 11.8.1, the estimate of any estimable function (say, τ_i^*) is the same regardless of which of the two models is used. So far as point estimation of the parameters is concerned, it seems to be more straightforward to use the model as written in (12.14) and the normal equations (12.15). It is clear from the normal equations in (12.15) that contrasts of the τ_i or the β_j or the γ_k are not estimable (that is to say, there exists no unbiased estimate of $\Sigma c_i \tau_i$, where $\Sigma c_i = 0$) unless there is no interaction. Therefore, it may be desirable to test the following hypotheses concerning model (12.14):

1. μ_{ijk} is free of three-factor interaction; that is, $(\tau\beta\gamma)_{ijk}^* = 0$ for all i, j, k

If μ_{ijk} is free of three-factor interaction, it might be desirable to test for two-factor interaction. In this case we might continue, and test

2. $(\tau\beta)_{ij}^* = 0$ for all i and j
3. $(\tau\gamma)_{ik}^* = 0$ for all i and k
4. $(\beta\gamma)_{jk}^* = 0$ for all j and k

We shall demonstrate the method of testing $(\tau\beta\gamma)_{ijk}^* = 0$ for all i, j, k. The quantities needed will be obtained from the normal equations (12.18) although they could just as well be obtained from (12.15).

$$
\begin{aligned}
&\mu^*: && IJKM\hat{\mu}^* && && && && && && && = Y_{....}\\
&\tau_i^*: && JKM\hat{\mu}^* + JKM\hat{\tau}_i^* && && && && && && && = Y_{i...}\\
&\beta_j^*: && IKM\hat{\mu}^* && + IKM\hat{\beta}_j^* && && && && && && = Y_{.j..}\\
&(\tau\beta)_{ij}^*: && KM\hat{\mu}^* + KM\hat{\tau}_i^* && + KM\hat{\beta}_j^* && + KM(\widehat{\tau\beta})_{ij}^* && && && && && = Y_{ij..}\\
&\gamma_k^*: && IJM\hat{\mu}^* && && && + IJM\hat{\gamma}_k^* && && && && = Y_{..k.}\\
&(\tau\gamma)_{ik}^*: && JM\hat{\mu}^* + JM\hat{\tau}_i^* && && && + JM\hat{\gamma}_k^* && + JM(\widehat{\tau\gamma})_{ik}^* && && && = Y_{i.k.}\\
&(\beta\gamma)_{jk}^*: && IM\hat{\mu}^* && + IM\hat{\beta}_j^* && && + IM\hat{\gamma}_k^* && && + IM(\widehat{\beta\gamma})_{jk}^* && && = Y_{.jk.}\\
&(\tau\beta\gamma)_{ijk}^*: && M\hat{\mu}^* + M\hat{\tau}_i^* && + M\hat{\beta}_j^* && + M(\widehat{\tau\beta})_{ij}^* && + M\hat{\gamma}_k^* && + M(\widehat{\tau\gamma})_{ik}^* && + M(\widehat{\beta\gamma})_{jk}^* && + M(\widehat{\tau\beta\gamma})_{ijk}^* && = Y_{ijk.}
\end{aligned}
$$

Equation (12.18)

We get the following estimates:

$$\hat{\mu}^* = y_{....}$$

$$\hat{\tau}_i^* = y_{i...} - y_{....}$$

$$\hat{\beta}_j^* = y_{.j..} - y_{....}$$

$$\hat{\gamma}_k^* = y_{..k.} - y_{....}$$

$$\widehat{(\tau\beta)}_{ij}^* = y_{ij..} - y_{i...} - y_{.j..} + y_{....}$$

$$\widehat{(\tau\gamma)}_{ik}^* = y_{i.k.} - y_{i...} - y_{..k.} + y_{....}$$

$$\widehat{(\beta\gamma)}_{jk}^* = y_{.jk.} - y_{.j..} - y_{..k.} + y_{....}$$

$$\widehat{(\tau\beta\gamma)}_{ijk}^* = y_{ijk.} - y_{ij..} - y_{i.k.} - y_{.jk.} + y_{i...} + y_{.j..} + y_{..k.} - y_{....}$$

$$R[\mu^*,\tau^*,\beta^*,\gamma^*,(\tau\beta)^*,(\tau\gamma)^*,(\beta\gamma)^*,(\tau\beta\gamma)^*]$$

$$= \hat{\mu}^* Y_{....} + \sum_i \hat{\tau}_i^* Y_{i...} + \sum_j \hat{\beta}_j^* Y_{.j..} + \sum_k \hat{\gamma}_k^* Y_{..k.} + \sum_{ij} \widehat{(\tau\beta)}_{ij}^* Y_{ij..} + \sum_{ik} \widehat{(\tau\gamma)}_{ik}^* Y_{i.k.} + \sum_{jk} \widehat{(\beta\gamma)}_{jk}^* Y_{.jk.} + \sum_{ijk} \widehat{(\tau\beta\gamma)}_{ijk}^* Y_{ijk.}$$

which simplifies to $$\sum_{ijk} \frac{Y_{ijk.}^2}{M}$$

So the error sum of squares is

$$\sum_{ijkm} y_{ijkm}^2 - \frac{\sum_{ijk} Y_{ijk.}^2}{M} = \sum_{ijkm} (y_{ijkm} - y_{ijk.})^2 = E_{\text{ss}}$$

If we set $(\tau\beta\gamma)_{ijk}^* = 0$ in model (12.17), the normal equations are exactly the same as (12.18) with the equations for $(\tau\beta\gamma)_{ijk}^*$ omitted. The rank of the resulting system is $IJK - (I - 1)(J - 1)(K - 1)$. Also, from this reduced system we have

$$\tilde{\mu}^* = y_{....}$$

$$\tilde{\tau}_i^* = y_{i...} - y_{....}$$

$$\tilde{\beta}_j^* = y_{.j..} - y_{....}$$

$$\tilde{\gamma}_k^* = y_{..k.} - y_{....}$$

$$\widetilde{(\tau\beta)}_{ij}^* = y_{ij..} - y_{i...} - y_{.j..} + y_{....}$$

$$\widetilde{(\tau\gamma)}_{ik}^* = y_{i.k.} - y_{i...} - y_{..k.} + y_{....}$$

$$\widetilde{(\beta\gamma)}_{jk}^* = y_{.jk.} - y_{.j..} - y_{..k.} + y_{....}$$

This gives

$$R[\mu^*,\tau^*,\beta^*,\gamma^*,(\tau\beta)^*,(\beta\gamma)^*] = \tilde{\mu}^* Y_{....} + \sum_i \tilde{\tau}_i^* Y_{i...} + \sum_j \tilde{\beta}_j^* Y_{.j..}$$
$$+ \sum_k \tilde{\gamma}_k^* Y_{..k.} + \sum_{ij} (\widetilde{\tau\beta})_{ij}^* Y_{ij..}$$
$$+ \sum_{ik} (\widetilde{\tau\gamma})_{ik}^* Y_{i.k.} + \sum_{jk} (\widetilde{\beta\gamma})_{jk}^* Y_{.jk.}$$
$$= \frac{\sum_{ij} Y_{ij..}^2}{KM} + \frac{\sum_{ik} Y_{i.k.}^2}{JM} + \frac{\sum_{jk} Y_{.jk.}^2}{IM} - \frac{\sum_i Y_{i...}^2}{JKM}$$
$$- \frac{\sum_j Y_{.j..}^2}{IKM} - \frac{\sum_k Y_{..k.}^2}{IJM} + \frac{Y_{....}^2}{IJKM}$$

Then the adjusted sum of squares for three-factor interaction is

$$R[(\tau\beta\gamma)^* \mid \mu^*,\tau^*,\beta^*,\gamma^*,(\tau\beta)^*,(\tau\gamma)^*,(\beta\gamma)^*]$$
$$= \frac{\sum_{ijk} Y_{ijk.}^2}{M} - \frac{\sum_{ij} Y_{ij..}^2}{KM} - \frac{\sum_{ik} Y_{i.k.}^2}{JM} - \frac{\sum_{jk} Y_{.jk.}^2}{IM} + \frac{\sum_i Y_{i...}^2}{JKM}$$
$$+ \frac{\sum_j Y_{.j..}^2}{IKM} + \frac{\sum_k Y_{..k.}^2}{IJM} - \frac{Y_{....}^2}{IJKM}$$
$$= \sum_{ijkm} (y_{ijk.} - y_{ij..} - y_{i.k.} - y_{.jk.} + y_{i...} + y_{.j..} + y_{..k.} - y_{....})^2$$
$$= (\tau\beta\gamma)_{\text{ss}}^*$$

and the rank is

$$IJK - [IJK - (I-1)(J-1)(K-1)] = (I-1)(J-1)(K-1)$$

By a similar procedure it can be shown that

$$\tau_{\text{ss}}^* = \frac{\sum_i Y_{i...}^2}{JKM} - \frac{Y_{....}^2}{IJKM} = \sum_{ijkm} (y_{i...} - y_{....})^2 \qquad \text{and} \qquad \text{DF} = I - 1$$

$$\beta_{\text{ss}}^* = \frac{\sum_j Y_{.j..}^2}{IKM} - \frac{Y_{....}^2}{IJKM} = \sum_{ijkm} (y_{.j..} - y_{....})^2 \qquad \text{and} \qquad \text{DF} = J - 1$$

$$\gamma_{\text{ss}}^* = \frac{\sum_k Y_{..k.}^2}{IJM} - \frac{Y_{....}^2}{IJKM} = \sum_{ijkm} (y_{..k.} - y_{....})^2 \qquad \text{and} \qquad \text{DF} = K - 1$$

$$(\tau\beta)^*_{\text{ss}} = \frac{\sum_{ij} Y^2_{ij..}}{KM} - \frac{\sum_i Y^2_{i...}}{JKM} - \frac{\sum_j Y^2_{.j..}}{IKM} + \frac{Y^2_{....}}{IJKM}$$

$$= \sum_{ijkm} (y_{ij..} - y_{i...} - y_{.j..} + y_{....})^2 \qquad \text{and} \qquad \text{DF} = (I-1)(J-1)$$

$$(\tau\gamma)^*_{\text{ss}} = \frac{\sum_{ik} Y^2_{i.k.}}{JM} - \frac{\sum_i Y^2_{i...}}{JKM} - \frac{\sum_k Y^2_{..k.}}{IJM} + \frac{Y^2_{....}}{IJKM}$$

$$= \sum_{ijkm} (y_{i.k.} - y_{i...} - y_{..k.} + y_{....})^2 \qquad \text{and} \qquad \text{DF} = (I-1)(K-1)$$

$$(\beta\gamma)^*_{\text{ss}} = \frac{\sum_{jk} Y^2_{.jk.}}{IM} - \frac{\sum_j Y^2_{.j..}}{IKM} - \frac{\sum_k Y^2_{..k.}}{IJM} + \frac{Y^2_{....}}{IJKM}$$

$$= \sum_{ijkm} (y_{.jk.} - y_{.j..} - y_{..k.} + y_{....})^2 \qquad \text{and} \qquad \text{DF} = (J-1)(K-1)$$

It is interesting to note that

$$\sum_{ijkm} y^2_{ijkm} = \frac{Y^2_{....}}{IJKM} + \tau^*_{\text{ss}} + \beta^*_{\text{ss}} + \gamma^*_{\text{ss}} + (\tau\beta)^*_{\text{ss}}$$
$$+ (\tau\gamma)^*_{\text{ss}} + (\beta\gamma)^*_{\text{ss}} + (\tau\beta\gamma)^*_{\text{ss}} + E_{\text{ss}}$$

The AOV is given in Table 12.7. Confidence intervals on any estimable functions can be found by the usual method.

We have gone to great lengths to exhibit methods of estimation and tests of hypotheses for the three-way cross classification with interaction. With this background, the reader should have no difficulty extending these methods to four-way and higher classifications.

12.7 2^n Factorial Models

The 2^n factorial model is the n-way cross-classification model where each factor contains a set of only two elements. For example, the 2^3 model contains three sets of factors A, B, C, and each set contains two elements; that is, a_1, a_2 are in A; b_1, b_2 in B; and c_1, c_2 in C. If every element in A is combined with every element of B and C, the treatment combinations are

$a_1b_1c_1 \quad a_2b_1c_1 \quad a_1b_2c_1 \quad a_2b_2c_1 \quad a_1b_1c_2 \quad a_2b_1c_2 \quad a_1b_2c_2 \quad a_2b_2c_2$

For example, suppose A represents concentrations of a certain chemical; a_1 = concentration 1, a_2 = concentration 2. Suppose B represents pressures of a certain manufacturing process; b_1 = pressure 1, b_2 = pressure 2. Suppose C represents temperatures of the process; c_1 = temperature 1, c_2 = temperature 2. Then $a_1b_2c_1$ represents the

TABLE 12.7 ANALYSIS OF VARIANCE FOR THREE-WAY CLASSIFICATION WITH INTERACTION

SV	DF	SS	MS	F	λ
Total	$IJKM$	$\sum_{ijkm} y_{ijkm}^2$			
Mean	1	$\frac{Y_{....}^2}{IJKM}$			
τ^* (adj)	$I - 1$	τ_{ss}^*	τ_{ms}^*	$\frac{\tau_{ms}^*}{E_{ms}}$	$\frac{JKM}{2\sigma^2} \sum_i (\tau_i^*)^2$
β^* (adj)	$J - 1$	β_{ss}^*	β_{ms}^*	$\frac{\beta_{ms}^*}{E_{ms}}$	$\frac{IKM}{2\sigma^2} \sum_j (\beta_j^*)^2$
γ^* (adj)	$K - 1$	γ_{ss}^*	γ_{ms}^*	$\frac{\gamma_{ms}^*}{E_{ms}}$	$\frac{IJM}{2\sigma^2} \sum_k (\gamma_k^*)^2$
$(\tau\beta)^*$ (adj)	$(I - 1)(J - 1)$	$(\tau\beta)_{ss}^*$	$(\tau\beta)_{ms}^*$	$\frac{(\tau\beta)_{ms}^*}{E_{ms}}$	$\frac{KM}{2\sigma^2} \sum_{ij} [(\tau\beta)_{ij}^*]^2$
$(\tau\gamma)^*$ (adj)	$(I - 1)(K - 1)$	$(\tau\gamma)_{ss}^*$	$(\tau\gamma)_{ms}^*$	$\frac{(\tau\gamma)_{ms}^*}{E_{ms}}$	$\frac{JM}{2\sigma^2} \sum_{ik} [(\tau\gamma)_{ik}^*]^2$
$(\beta\gamma)^*$ (adj)	$(J - 1)(K - 1)$	$(\beta\gamma)_{ss}^*$	$(\beta\gamma)_{ms}^*$	$\frac{(\beta\gamma)_{ms}^*}{E_{ms}}$	$\frac{IM}{2\sigma^2} \sum_{jk} [(\beta\gamma)_{jk}^*]^2$
$(\tau\beta\gamma)^*$ (adj)	$(I - 1)(J - 1)(K - 1)$	$(\tau\beta\gamma)_{ss}^*$	$(\tau\beta\gamma)_{ms}^*$	$\frac{(\tau\beta\gamma)_{ms}^*}{E_{ms}}$	$\frac{M}{2\sigma^2} \sum_{ijk} [(\tau\beta\gamma)_{ijk}^*]^2$
Error	$IJK(M - 1)$	E_{ss}	E_{ms}		

combination of concentration 1, pressure 2, and temperature 1. Similarly for the other $a_ib_jc_k$.

The model will be written

$$y_{ijkm} = \mu_{ijk} + e_{ijkm}$$

where y_{ijkm} is the mth observation of treatment $a_ib_jc_k$ and where μ_{ijk} is the "true" effect of this particular treatment. We shall write

$$\begin{aligned} \mu_{ijk} &= \mu + \tau_i + \beta_j + (\tau\beta)_{ij} + \gamma_k + (\tau\gamma)_{ik} + (\beta\gamma)_{jk} + (\tau\beta\gamma)_{ijk} \\ &= \mu^* + \tau_i^* + \beta_j^* + (\tau\beta)_{ij}^* + \gamma_k^* + (\tau\gamma)_{ik}^* + (\beta\gamma)_{jk}^* + (\tau\beta\gamma)_{ijk}^* \end{aligned}$$

where these quantities are defined by Eqs. (12.16) and (12.17). The normal equations are given in (12.15) with $I = J = K = 2$. We shall use the model

$$y_{ijkm} = \mu^* + \tau_i^* + \beta_j^* + (\tau\beta)_{ij}^* + \gamma_k^* + (\tau\gamma)_{ik}^* + (\beta\gamma)_{jk}^* + (\tau\beta\gamma)_{ijk}^* + e_{ijkm}$$

The normal equations are given in Eq. (12.18), where $I = J = K = 2$. Table 12.8 will be useful in examining the normal equations.

TABLE 12.8 DEFINITION OF THE ESTIMATES IN 2^3 FACTORIAL

	$Y_{111\cdot}$	$Y_{211\cdot}$	$Y_{121\cdot}$	$Y_{221\cdot}$	$Y_{112\cdot}$	$Y_{212\cdot}$	$Y_{122\cdot}$	$Y_{222\cdot}$	z
$8M(\hat{\mu})$*	+1	+1	+1	+1	+1	+1	+1	+1	z_1
$4M(\hat{A})$	−1	+1	−1	+1	−1	+1	−1	+1	z_2
$4M(\hat{B})$	−1	−1	+1	+1	−1	−1	+1	+1	z_3
$4M(\widehat{AB})$	+1	−1	−1	+1	+1	−1	−1	+1	z_4
$4M(\hat{C})$	−1	−1	−1	−1	+1	+1	+1	+1	z_5
$4M(\widehat{AC})$	+1	−1	+1	−1	−1	+1	−1	+1	z_6
$4M(\widehat{BC})$	+1	+1	−1	−1	−1	−1	+1	+1	z_7
$4M(\widehat{ABC})$	−1	+1	+1	−1	+1	−1	−1	+1	z_8

Note: An example of the use of the table is

$$\begin{aligned} 4M(\hat{B}) = &(-1)Y_{111.} + (-1)Y_{211.} + (+1)Y_{121.} + (+1)Y_{221.} \\ &+ (-1)Y_{112.} + (-1)Y_{212.} + (+1)Y_{122.} + (+1)Y_{222.} = z_3 \end{aligned}$$

There are four independent unbiased estimates of $\tau_2 - \tau_1$ if all the interactions are zero. They are $y_{211.} - y_{111.}$, $y_{221.} - y_{121.}$, $y_{212.} - y_{112.}$, and $y_{222.} - y_{122.}$. Similarly for β and γ. If the interactions are not zero, then $\hat{A}$ in Table 12.8 can be taken as the estimate of the A effect. This is equal to

$$\begin{aligned} E(y_{2\ldots} - y_{1\ldots}) &= \tau_2^* - \tau_1^* \\ &= \tau_2 - \tau_1 + (\overline{\tau\beta})_{2.} - (\overline{\tau\beta})_{1.} + (\overline{\tau\gamma})_{2.} \\ &\quad - (\overline{\tau\gamma})_{1.} + (\overline{\tau\beta\gamma})_{2..} - (\overline{\tau\beta\gamma})_{1..} \end{aligned}$$

If the three-factor interaction is zero, there are two independent unbiased estimates of $4(\tau\beta)^*_{11}$. They are

$$y_{111.} - y_{211.} - y_{121.} + y_{221.} \qquad \text{and} \qquad y_{112.} - y_{212.} - y_{122.} + y_{222.}$$

For example,

$$E[y_{111.} - y_{211.} - y_{121.} + y_{221.}] = 4(\tau\beta)^*_{11}$$

since $$(\tau\beta)^*_{11} + (\tau\beta)^*_{21} = (\tau\beta)^*_{21} + (\tau\beta)^*_{22} = 0$$

and $$(\tau\beta)^*_{11} + (\tau\beta)^*_{12} = (\tau\beta)^*_{12} + (\tau\beta)^*_{22} = 0$$

If the three-factor interaction is not zero, then $\widehat{AB}$ in Table 12.8 can be taken as the estimate of the AB interaction averaged over the C factor. The square array in Table 12.8 is an orthogonal matrix if each element is divided by $\sqrt{8}$. Let the z_i be so defined that they are equal to the indicated linear combination in the ith row. Then the z_i are independent normal variables with variance $8M\sigma^2$. The adjusted sum of squares due to the mean μ^* is $z_1^2/8M$, and similarly for the other effects. This leads to the AOV shown in Table 12.9. Similar methods are applicable for $n > 3$.

TABLE 12.9 ANALYSIS OF VARIANCE OF 2^3 FACTORIAL

SV	DF	SS	λ
Total	$8M$	$\sum_{ijkm} y^2_{ijkm}$	
μ^*	1	$\frac{z_1^2}{8M}$	
A	1	$\frac{z_2^2}{8M}$	$\frac{2M}{\sigma^2}\Sigma(\tau_i^*)^2$
B	1	$\frac{z_3^2}{8M}$	$\frac{2M}{\sigma^2}\Sigma(\beta_j^*)^2$
AB	1	$\frac{z_4^2}{8M}$	$\frac{M}{\sigma^2}\Sigma[(\tau\beta)^*_{ij}]^2$
C	1	$\frac{z_5^2}{8M}$	$\frac{2M}{\sigma^2}\Sigma(\gamma_k^*)^2$
AC	1	$\frac{z_6^2}{8M}$	$\frac{M}{\sigma^2}\Sigma[(\tau\gamma)^*_{ik}]^2$
BC	1	$\frac{z_7^2}{8M}$	$\frac{M}{\sigma^2}\Sigma[(\beta\gamma)^*_{jk}]^2$
ABC	1	$\frac{z_8^2}{8M}$	$\frac{M}{2\sigma^2}\Sigma[(\tau\beta\gamma)^*_{ijk}]^2$
Error	$8(M-1)$	$\sum_{ijkm}(y_{ijkm} - y_{ijk.})^2$	

12.8 Using Interaction Sum of Squares for Error Sum of Squares

In a three-way factorial model as explained in Sec. 12.6, the total number of observations is $IJKM$. Often it is impossible to repeat the treatment combinations, and then $M = 1$. If this is the case there is no estimate of σ^2 and no method for testing hypotheses or setting confidence intervals. However, in many experiments the researcher is willing to assume that some of the interactions are zero. Then the corresponding sums of squares can be used as an estimate of σ^2. Suppose, for example, there is no three-factor interaction; that is, suppose $(\tau\beta\gamma)^*_{ijk} = 0$ for all i, j, k. Then the noncentrality for this term in Table 12.7 is zero. Therefore, $(\tau\beta\gamma)^*_{ss}/\sigma^2$ is distributed as $\chi^2(p)$, where $p = (I - 1)(J - 1)(K - 1)$. Then $(\tau\beta\gamma)^*_{ss}$ can be used as the error term.

If $M > 1$ and $(\tau\beta\gamma)^*_{ijk} = 0$, then $(\tau\beta\gamma)^*_{ss}$ can be pooled with E_{ss} to give an error term with $IJK(M - 1) + (I - 1)(J - 1)(K - 1)$ degrees of freedom. It must be remembered that this is under the assumption that $(\tau\beta\gamma)^*_{ijk}$ is zero for all i, j, k.

Suppose that an experimenter does not know whether the three-factor interaction is zero, but he decides to assume that it is zero if the appropriate F test from Table 12.7 is not significant at a chosen probability level. Then, if he decides on the basis of the F test that it is zero, he will pool $(\tau\beta\gamma)^*_{ss}$ with E_{ss} and obtain a new error term for testing main effects and two-factor interaction. If this rule is followed, the tests for main effects and two-factor interaction are not exact. These tests are of the nature of conditional tests, since the decision to pool $(\tau\beta\gamma)^*_{ss}$ with E_{ss} depends on a preliminary test of significance. The reader is referred to the bibliography at the end of this chapter for further reading on this subject.

Next we shall show that, in a two-way model with only one observation per ij combination, if the interaction is *assumed* to be zero but is *really* not zero, it vitiates the test of significance. Let us go back to the model given by Eq. (12.4). Under this model the error sum of squares divided by σ^2, that is, E_{ss}/σ^2, is distributed as $\chi^2[(b - 1)(t - 1)]$. If, however, the model contains a two-factor interaction term and is given by Eq. (12.9), then E_{ss}/σ^2 is distributed as $\chi'^2[(t - 1)(b - 1), \lambda]$, where

$$\lambda = \frac{1}{2\sigma^2} \sum_{ij} [(\tau\beta)^*_{ij}]^2$$

Hence the quantity

$$u = \frac{\tau^*_{ms}}{E_{ms}}$$

used to test the hypothesis $\tau_1^* = \tau_2^* = \cdots = \tau_t^*$ is not distributed as a central F even if the hypothesis is true. If no interaction is present, $E(E_{\text{ms}}) = \sigma^2$. If interaction is present,

$$E(E_{\text{ms}}) = \sigma^2 + \frac{1}{(t-1)(b-1)} \sum_{ij} [(\tau\beta)_{ij}^*]^2$$

Hence the error term is increased on the average, and the test will not reject the hypothesis as often as the level of significance indicates. Similar situations can be shown for other models.

Problems

12.1 Find the normal equations for the reparametrized model of the one-way classification

$$y_{ij} = \mu^* + \alpha_i^* + e_{ij}$$

where $\mu^* = \mu + \bar{\alpha}.$ and $\alpha_i^* = \alpha_i - \bar{\alpha}$ $(i = 1, 2, \ldots t; j = 1, 2, \ldots r)$.

12.2 In the one-way classification model 12.1, find the sum of squares due to α (adjusted) for testing the hypothesis H_0: $\alpha_1 = \alpha_2 = \cdots = \alpha_k$, with $k < t$.

12.3 Assume that the data of Table 12.10 fit a one-way classification model:

TABLE 12.10

Treatment				
1	2	3	4	5
12.3	11.4	15.2	14.3	15.3
14.1	10.9	16.8		16.2
13.6	12.0	18.6		14.8
12.5				

(*a*) Test the hypothesis $\alpha_1 = \alpha_2 = \alpha_3 = \alpha_4 = \alpha_5$. Use a type I error probability of .05.
(*b*) Set a 95 per cent confidence interval on $\alpha_1 - \alpha_2$.
(*c*) Find $\hat{\mu} + \hat{\alpha}_i$ $(i = 1, 2, \ldots, 5)$ and the variance of each estimate.
(*d*) Find the power of the test in (*a*) if

$$\frac{\alpha_i}{\sigma} = \begin{cases} 2 & \text{for } i = 1, 2 \\ 4 & \text{for } i = 3, 4, 5 \end{cases}$$

(*e*) Test the hypothesis H_0: $\alpha_1 = \alpha_2 = \alpha_3$.

12.4 In the two-way classification model given in Art. 12.3.1, find the treatment sum of squares for testing H_0: $\tau_1 = \tau_2 = \cdots = \tau_s$, where $s < t$.

12.5 Find the noncentrality for Prob. 12.4.

12.6 In a two-way classification model as defined in Art. 12.3.1, work out in detail the expected mean squares that are given in Table 12.2.

12.7 Assume that the following data of Table 12.11 satisfy the two-way classification model

$$y_{ij} = \mu + \tau_i + \beta_j + e_{ij}$$

TABLE 12.11

β \ τ	1	2	3	4
1	13	16	20	20
2	22	25	29	29
3	25	30	34	32

(*a*) Test the hypothesis $\tau_1 = \tau_2 = \tau_3 = \tau_4$ with a type I error probability of 5 per cent.
(*b*) Set a 95 per cent confidence interval on $\tau_3 - \tau_4$.
(*c*) Find the estimates of $\tau_i - \bar{\tau}.$ $(i = 1, 2, 3, 4)$.
(*d*) Find the variance of each of the estimates in (*c*).
(*e*) Find the width and expected width in (*b*).

12.8 Find the noncentrality parameter for Prob. 12.7*a*.

12.9 Work out the four noncentrality parameters for Table 12.3.

12.10 By using Definition 12.1, show that, if $f(x,z)$ has first partial derivatives, then it is a function with no interaction if $\partial f/\partial x$ is independent of z and $\partial f/\partial z$ is independent of x.

12.11 Go through the algebra to find a solution to the normal equations (12.10) using the nonestimable conditions

$$\hat{\tau}_1 = \hat{\beta}_1 = \widehat{(\tau\beta)}_{1\cdot} = \widehat{(\tau\beta)}_{2\cdot} = \cdots = \widehat{(\tau\beta)}_{t\cdot} = \widehat{(\tau\beta)}_{\cdot 1} = \cdots = \widehat{(\tau\beta)}_{\cdot b} = 0$$

12.12 Work out the noncentrality parameter for Table 12.4.

12.13 Assume that the data of Table 12.12 satisfy the two-way classification model

$$y_{ijk} = \mu + \tau_i + \beta_j + (\tau\beta)_{ij} + e_{ijk}$$

TABLE 12.12

β \ τ	1	2	3	4
1	6 6	8 10	11 10	11 9
2	11 12	11 12	14 16	13 16
3	13 12	16 14	18 16	16 17

Test the hypothesis that there is no interaction at the 5 per cent level of significance.

12.14 Show that model (12.14) is the same as (12.16).

12.15 Solve the normal equations (12.18).

12.16 Check the noncentrality parameters in Table 12.7.

12.17 In a 2^3 factorial, suppose that the three-factor interaction is zero. Show that $E(\widehat{ABC})$ is zero in Table 12.8.

12.18 In Table 12.8, show that

(*a*) $\hat{A}$ and $\hat{B}$ are uncorrelated.

(*b*) $\hat{A}$ and $\widehat{AB}$ are uncorrelated.

12.19 In Table 12.9, work out the fact that the adjusted sum of squares for testing the A effect equal to zero is $z_2^2/8M$.

Further Reading

1 G. W. Snedecor: "Statistical Methods," Iowa State College Press, Ames, Iowa, 1956.

2 W. G. Cochran and G. M. Cox: "Experimental Designs," John Wiley & Sons, Inc., New York, 1957.

3 O. Kempthorne: "Design and Analysis of Experiments," John Wiley & Sons, Inc., New York, 1952.

4 C. H. Goulden: "Methods of Statistical Analysis," John Wiley & Sons, Inc., New York, 1939.

5 O. L. Davies: "Design and Analysis of Industrial Experiments," Oliver & Boyd, Ltd., London, 1954.

6 R. L. Anderson and T. A. Bancroft: "Statistical Theory in Research," McGraw-Hill Book Company, Inc., New York, 1952.

7 C. R. Rao: "Advanced Statistical Methods in Biometric Research," John Wiley & Sons, Inc., New York, 1952.

8 H. B. Mann: "Analysis and Design of Experiments," Dover Publications, New York, 1949.

9 A. E. Paull: On a Preliminary Test for Pooling Mean Squares in the Analysis of Variance, *Ann. Math. Statist.*, vol. 21, pp. 539–556, 1950.

10 H. Scheffé: "The Analysis of Variance," John Wiley & Sons, Inc., New York, 1959.

13

Two-way Classification with Unequal Numbers in Subclasses

Chapter 12 was devoted to a study of various cross-classification models with the common feature that in each cell there were an equal number of observations. In many cases it is not possible for an experimenter to control his experiment so as to ensure this type of balance. Also, a researcher may purposely design his experiment to have unequal numbers in the subclasses. This is the case in incomplete block designs. The two-way classification model with unequal numbers in the cells and with no interaction will be discussed in some generality in this chapter. Certain special cases, such as incomplete blocks, will be discussed in later chapters.

13.1 Two-way Classification Model with No Interaction

13.1.1 Definition and Notation. Let the model be given by

$$y_{ijk} = \mu + \tau_i + \beta_j + e_{ijk} \qquad \begin{cases} k = 0, 1, 2, \ldots, n_{ij} \\ j = 1, 2, \ldots, b \\ i = 1, 2, \ldots, t \end{cases} \tag{13.1}$$

where y_{ijk} is the kth observation in the ijth cell; μ, τ_i, β_j are unknown parameters; and e_{ijk} are random variables with the conventional distributional properties. The ijth cell contains n_{ij} observations, and, if

$n_{ij} = 0$, the cell contains no observations; hence, the observation y_{ij0} does not exist. The notation

$$N_{i.} = \sum_{j=1}^{b} n_{ij}$$
$$N_{.j} = \sum_{i=1}^{t} n_{ij}$$
$$N_{..} = \sum_{ij} n_{ij}$$

will be used. Notice that

$$Y_{ij.} = \sum_{k=0}^{n_{ij}} y_{ijk} = \sum_{k=0}^{n_{ij}} (\mu + \tau_i + \beta_j + e_{ijk}) = n_{ij}\mu + n_{ij}\tau_i + n_{ij}\beta_j + e_{ij.}$$

where

$$\sum_{k=0}^{n_{ij}} e_{ijk} = e_{ij.}$$

Also, we shall use the notation

$$\frac{1}{n_{ij}} \sum_{k=0}^{n_{ij}} e_{ijk} = \bar{e}_{ij.}$$

Similar notation will be used for $e_{i..}, \bar{e}_{i..}, e_{.j.}, \bar{e}_{.j.}$, etc. Also notice that

$$Y_{i..} = \sum_{j=1}^{b} \sum_{k=0}^{n_{ij}} y_{ijk} = \sum_{j=1}^{b} n_{ij}(\mu + \tau_i + \beta_j) + \sum_{jk} e_{ijk}$$
$$= N_{i.}\mu + N_{i.}\tau_i + \sum_{j} n_{ij}\beta_j + e_{i..}$$

Further, we have

$$Y_{.j.} = \sum_{i=1}^{t} \sum_{k=0}^{n_{ij}} y_{ijk} = \sum_{i=1}^{t} n_{ij}(\mu + \tau_i + \beta_j) + \sum_{ik} e_{ijk}$$
$$= N_{.j}\mu + \sum_{i=1}^{t} n_{ij}\tau_i + N_{.j}\beta_j + e_{.j.}$$

and

$$Y_{...} = \sum_{i=1}^{t} \sum_{j=1}^{b} \sum_{k=0}^{n_{ij}} y_{ijk} = \sum_{i} \sum_{j} n_{ij}(\mu + \tau_i + \beta_j) + e_{...}$$
$$= N_{..}\mu + \sum_{i} N_{i.}\tau_i + \sum_{j} N_{.j}\beta_j + e_{...}$$

Also,

$$y_{ij.} = \frac{1}{n_{ij}} \sum_{k} y_{ijk}$$
$$y_{i..} = \frac{1}{N_{i.}} \sum_{j} \sum_{k} y_{ijk}$$
$$y_{.j.} = \frac{1}{N_{.j}} \sum_{i} \sum_{k} y_{ijk}$$
$$y_{...} = \frac{1}{N_{..}} \sum_{i} \sum_{j} \sum_{k} y_{ijk}$$

A study of this notation will help in the theory that follows.

13.1.2 Point Estimation. In this section the normal equations will be derived, estimable functions will be examined, and the variance of certain estimable functions will be found.

By the method of Sec. 11.5, the normal equations are

$$\begin{aligned}
\mu: &\quad N_{..}\hat{\mu} + \sum_i N_{i.}\hat{\tau}_i + \sum_j N_{.j}\hat{\beta}_j = Y_{...} \\
\tau_r: &\quad N_{r.}\hat{\mu} + N_{r.}\hat{\tau}_r + \sum_j n_{rj}\hat{\beta}_j = Y_{r..} \qquad r = 1, 2, \ldots, t \\
\beta_s: &\quad N_{.s}\hat{\mu} + \sum_i n_{is}\hat{\tau}_i + N_{.s}\hat{\beta}_s = Y_{.s.} \qquad s = 1, 2, \ldots, b
\end{aligned} \tag{13.2}$$

To examine these equations in detail we shall make the following assumption.

◆ **Assumption 13.1** The n_{ij} are values such that $\tau_i - \tau_j$ is estimable for every $i \neq j = 1, 2, \ldots, t$ and $\beta_i - \beta_j$ is estimable for every $i \neq j = 1, 2, \ldots, b$.

Thus a design such as that shown in Table 13.1*a* is admissible, but that of Table 13.1*b* is not. The symbol $*$ in a cell means that there are observations in that particular cell. The symbol 0 means that there is no observation. In general, experiments will satisfy the conditions set forth in this assumption. If the assumption is not satisfied, all parameter differences are not estimable; in that case the experiment might be broken up into parts such that the parameter difference in each part is estimable.

TABLE 13.1 DESIGNS FOR EXPERIMENTS, TWO-WAY CLASSIFICATION MODEL

β \ τ	1	2	3
1	*	*	0
2	0	*	*
3	0	0	*

(*a*)

β \ τ	1	2	3
1	*	0	0
2	0	*	0
3	0	0	*

(*b*)

◆ **Theorem 13.1** If the n_{ij} in model (13.1) are such that $\tau_i - \tau_j$, $\beta_{i'} - \beta_{j'}$ are estimable for all $i \neq j$ and all $i' \neq j'$, then

(1) There are exactly $b + t - 1$ linearly independent estimable functions.

(2) $\Sigma c_i \tau_i$ and $\Sigma d_j \beta_j$ are estimable if $\Sigma c_i = \Sigma d_j = 0$.

Proof: It is immediately clear that the $b + t - 1$ estimable functions $\tau_1 - \tau_2, \tau_1 - \tau_3, \ldots, \tau_1 - \tau_t, \beta_1 - \beta_2, \beta_1 - \beta_3, \ldots, \beta_1 - \beta_b,$

and $N_{..}\mu + \sum_i N_{i.}\tau_i + \sum_j N_{.j}\beta_j$ are linearly independent. There are $b + t + 1$ parameters and, therefore, $b + t + 1$ equations in the normal equations (13.2). The sum of the t equations represented by τ_r is equal to the equation represented by μ. Also, the sum of the b equations represented by β_s is equal to the equation represented by μ. Hence, there are at least two linearly dependent equations among the $b + t + 1$ equations. This, coupled with the fact that there are at *least* $b + t - 1$ linearly independent estimable functions, implies that the rank of the normal equations is exactly $b + t - 1$. Since the $\tau_i - \tau_j$ are estimable, every linear combination of these is estimable. Consider

$$\sum_{j=1}^{t} (\tau_i - \tau_j)\frac{1}{t} \qquad j \neq i$$

This gives

$$\frac{t-1}{t}\tau_i - \frac{1}{t}\sum_{\substack{j=1\\ j\neq i}}^{t} \tau_j = \frac{t-1}{t}\tau_i - \frac{1}{t}(\tau_. - \tau_i) = \tau_i - \bar{\tau}_.$$

This shows that $\tau_i - \bar{\tau}_.$ is estimable for all i; so $\Sigma c_i(\tau_i - \bar{\tau}_.)$ is estimable. But this is $\Sigma c_i\tau_i$ if $\Sigma c_i = 0$. A similar argument holds for $\Sigma d_j\beta_j$.

We shall solve the equations for contrasts of the τ_i. We can use the β_j equations and obtain

$$\hat{\mu} + \hat{\beta}_j = y_{.j.} - \frac{1}{N_{.j}}\sum_i n_{ij}\hat{\tau}_i \qquad j = 1, 2, \ldots, b \tag{13.3}$$

The τ_r equation is

$$\sum_j n_{rj}(\hat{\mu} + \hat{\beta}_j) + N_{r.}\hat{\tau}_r = Y_{r..} \qquad r = 1, 2, \ldots, t \tag{13.4}$$

Substituting $\hat{\mu} + \hat{\beta}_j$ from Eq. (13.3) into Eq. (13.4) gives

$$\sum_j n_{rj}\left(y_{.j.} - \frac{1}{N_{.j}}\sum_i n_{ij}\hat{\tau}_i\right) + N_{r.}\hat{\tau}_r = Y_{r..} \qquad r = 1, 2, \ldots, t \tag{13.5}$$

This gives

$$N_{r.}\hat{\tau}_r - \sum_{i=1}^{t}\sum_{j=1}^{b}\frac{n_{rj}n_{ij}}{N_{.j}}\hat{\tau}_i = Y_{r..} - \sum_{j=1}^{b} n_{rj}y_{.j.} \qquad r = 1, 2, \ldots, t \tag{13.6}$$

Let

$$q_r = Y_{r..} - \sum_{j=1}^{b} n_{rj}y_{.j.} \tag{13.7}$$

The left-hand side of Eq. (13.6) is

$$N_{r.}\hat{\tau}_r - \sum_{i=1}^{t}\sum_{j=1}^{b}\frac{n_{rj}n_{ij}}{N_{.j}}\hat{\tau}_i$$

If we isolate the quantity involving $\hat{\tau}_r$ from the second term, we get

$$N_{r.}\hat{\tau}_r - \hat{\tau}_r \sum_{j=1}^{b} \frac{n_{rj}^2}{N_{.j}} - \sum_{\substack{i=1\\i\neq r}}^{t} \sum_{j=1}^{b} \frac{n_{rj}n_{ij}}{N_{.j}} \hat{\tau}_i$$

Therefore, Eq. (13.6) becomes

$$\left(N_{r.} - \sum_{j=1}^{b} \frac{n_{rj}^2}{N_{.j}}\right)\hat{\tau}_r - \sum_{\substack{i=1\\i\neq r}}^{t} \sum_{j=1}^{b} \frac{n_{rj}n_{ij}}{N_{.j}} \hat{\tau}_i = q_r \qquad r = 1, 2, \ldots, t \qquad (13.8)$$

The system (13.8) represents t equations in the t unknowns $\hat{\tau}_i$. They can be represented in matrix form as

$$\mathbf{A}\hat{\boldsymbol{\tau}} = \mathbf{q} \qquad (13.9)$$

where the $t \times t$ matrix $\mathbf{A} = (a_{rs})$ has elements as follows:

$$\text{Diagonal elements:} \quad a_{rr} = N_{r.} - \sum_{j=1}^{b} \frac{n_{rj}^2}{N_{.j}} \qquad r = 1, 2, \ldots, t$$

$$\text{Off-diagonal elements:} \quad a_{rs} = -\sum_{j=1}^{b} \frac{n_{rj}n_{sj}}{N_{.j}} \qquad r \neq s = 1, 2, \ldots, t \qquad (13.10)$$

The $t \times 1$ vector $\hat{\boldsymbol{\tau}}$ has elements $\hat{\tau}_r$, and $\mathbf{q}$ has elements q_r, as defined by Eq. (13.7). If there are equal numbers in the cells (say, m), then $n_{ij} = m$, $N_{.j} = tm$, $N_{i.} = bm$, and $N_{..} = btm$. In this case $a_{rr} = bm - bm/t$ and $a_{rs} = -bm/t$, and the system of normal equations is easily solved. This case, in fact, was discussed in the preceding chapter. There are other situations in which the n_{ij} values are such that the normal equations are fairly easy to solve, for example, in incomplete block models. However, in the general two-way classification model with unequal numbers in the subcells, the matrix $\mathbf{A}$ has no special feature to make the normal equations easy to solve.

The system $\mathbf{A}\hat{\boldsymbol{\tau}} = \mathbf{q}$ will be studied in some detail, since from this we must obtain estimates of estimable functions of the τ_i and the variance of these estimates. In Art. 13.1.4 it will be shown that a solution to this system will be needed for the AOV table used to test the hypothesis $\tau_1 = \tau_2 = \cdots = \tau_t$.

◆ **Theorem 13.2** The rank of the matrix $\mathbf{A}$ in Eq. (13.9) is $t - 1$.

Proof: By Theorem 13.1 and Assumption 13.1 concerning the estimability of $\tau_i - \tau_j$, there are $t - 1$ linearly independent estimable functions of the τ_i. These must come from Eq. (13.9); hence, the rank of $\mathbf{A}$ must be at *least* $t - 1$.

If the equations of Eq. (13.8) are added together (summed over $r = 1$ to $r = t$), the result is zero, which shows that there is at least one dependent equation in the system. Therefore, $\mathbf{A}$ is $t \times t$ with one linearly dependent row; hence its rank must be at *most* $t - 1$.

Therefore, its rank is exactly $t - 1$.

Thus there are an infinite number of vectors $\hat{\boldsymbol{\tau}}$ that will satisfy $\mathbf{A}\hat{\boldsymbol{\tau}} = \mathbf{q}$. To solve the system any nonestimable condition can be used. We shall discuss first the use of the nonestimable condition $\Sigma\hat{\tau}_i = 0$. This means that we want to find a vector $\hat{\boldsymbol{\tau}}$ that will satisfy the equation $\mathbf{A}\hat{\boldsymbol{\tau}} = \mathbf{q}$ and the equation $\mathbf{1}'\hat{\boldsymbol{\tau}} = 0$, where $\mathbf{1}$ is a vector with each element equal to unity (note that $\mathbf{1}'\hat{\boldsymbol{\tau}} = \Sigma\hat{\tau}_i$). These two matrix equations can be written as a single matrix equation, as follows:

$$\begin{pmatrix} \mathbf{A} & \mathbf{1} \\ \mathbf{1}' & 0 \end{pmatrix} \begin{pmatrix} \hat{\boldsymbol{\tau}} \\ 0 \end{pmatrix} = \begin{pmatrix} \mathbf{q} \\ 0 \end{pmatrix} \tag{13.11}$$

The τ_i are clearly not estimable; therefore, we need to find what the solution $\hat{\tau}_i$ to the system (13.11) represents. That is to say, what does $E(\hat{\boldsymbol{\tau}})$ equal in Eq. (13.11)? We shall first state a theorem on the rank of the matrix

$$\mathbf{A}^* = \begin{pmatrix} \mathbf{A} & \mathbf{1} \\ \mathbf{1}' & 0 \end{pmatrix}$$

◆ **Theorem 13.3** Under the conditions of Assumption 13.1, the rank of $\mathbf{A}^*$ is $t + 1$; that is, $\mathbf{A}^*$ is nonsingular.

The proof will be left for the reader.

Let
$$\mathbf{A}^{*-1} = \begin{pmatrix} \mathbf{B}_{11} & \mathbf{B}_{12} \\ \mathbf{B}_{21} & B_{22} \end{pmatrix}$$

where $\mathbf{B}_{11}$ is a $t \times t$ matrix, $\mathbf{B}_{12}$ is $t \times 1$, $\mathbf{B}_{21} = \mathbf{B}'_{12}$ is $1 \times t$, and B_{22} is 1×1. Then, from (13.11),

$$\begin{pmatrix} \hat{\boldsymbol{\tau}} \\ 0 \end{pmatrix} = \begin{pmatrix} \mathbf{B}_{11} & \mathbf{B}_{12} \\ \mathbf{B}_{21} & B_{22} \end{pmatrix} \begin{pmatrix} \mathbf{q} \\ 0 \end{pmatrix}$$

and
$$\hat{\boldsymbol{\tau}} = \mathbf{B}_{11}\mathbf{q} \tag{13.12}$$

Next we shall prove a theorem concerning the elements of $\mathbf{A}^{*-1}$.

◆ **Theorem 13.4** The elements of $\mathbf{A}^{*-1}$ are such that

(1) $\mathbf{B}_{21}$ (and $\mathbf{B}_{12}$) has all its elements equal to $1/t$.

(2) $B_{22} = 0$.

(3) $\mathbf{B}_{11}\mathbf{A}\mathbf{B}_{11} = \mathbf{B}_{11}$.

(4) $\mathbf{A}\mathbf{B}_{11}$ is idempotent of rank $t - 1$, with diagonal elements each equal to $(t - 1)/t$ and with off-diagonal elements each equal to $-1/t$.

(5) The rows of $\mathbf{B}_{11}$ add to zero.

Proof: By the equation $\mathbf{A}^*\mathbf{A}^{*-1} = \mathbf{I}$, we get

$$\begin{pmatrix} \mathbf{A} & \mathbf{1} \\ \mathbf{1}' & 0 \end{pmatrix} \begin{pmatrix} \mathbf{B}_{11} & \mathbf{B}_{12} \\ \mathbf{B}_{21} & B_{22} \end{pmatrix} = \begin{pmatrix} \mathbf{I} & \mathbf{0} \\ \mathbf{0} & 1 \end{pmatrix}$$

and the equations

$$\begin{aligned} &(a) \qquad \mathbf{A}\mathbf{B}_{11} + \mathbf{1}\mathbf{B}_{21} = \mathbf{I} \\ &(b) \qquad \mathbf{1}'\mathbf{B}_{11} = \mathbf{0} \\ &(c) \qquad \mathbf{1}'\mathbf{B}_{12} = 1 \\ &(d) \qquad \mathbf{A}\mathbf{B}_{12} + \mathbf{1}B_{22} = \mathbf{0} \end{aligned} \tag{13.13}$$

To prove the theorem we shall also use the matrix equations

$$\begin{aligned} &(a) \qquad \mathbf{1}'\mathbf{A} = \mathbf{0} \\ &(b) \qquad \mathbf{1}'\mathbf{1} = t \end{aligned} \tag{13.14}$$

Equation (13.14*a*) expresses that fact that the rows of $\mathbf{A}$ sum to zero; (13.14*b*) shows the sum of the t elements of $\mathbf{1}$.

1. If we multiply Eq. (13.13*a*) by $\mathbf{1}'$ we get

$$\mathbf{1}'\mathbf{A}\mathbf{B}_{11} + \mathbf{1}'\mathbf{1}\mathbf{B}_{21} = \mathbf{1}'$$

and using (13.14) gives $t\mathbf{B}_{21} = \mathbf{1}'$, or $\mathbf{B}_{21} = (1/t)\mathbf{1}'$, which means that each element in $\mathbf{B}_{21}$ is equal to $1/t$.

2. To show that $B_{22} = 0$, multiply Eq. (13.13*d*) by $\mathbf{1}'$. This gives $\mathbf{1}'\mathbf{A}\mathbf{B}_{12} + \mathbf{1}'\mathbf{1}B_{22} = 0$, or $tB_{22} = 0$ by (13.14*a*).

3. Multiply (13.13*a*) by $\mathbf{B}_{11}$; this yields $\mathbf{B}_{11}\mathbf{A}\mathbf{B}_{11} + \mathbf{B}_{11}\mathbf{1}\mathbf{B}_{21} = \mathbf{B}_{11}$, or $\mathbf{B}_{11}\mathbf{A}\mathbf{B}_{11} = \mathbf{B}_{11}$, since $\mathbf{B}_{11}\mathbf{1} = \mathbf{0}$ by (13.13*b*).

4. By Eq. (13.13*a*) we get $\mathbf{A}\mathbf{B}_{11} = \mathbf{I} - (1/t)\mathbf{1}\mathbf{1}'$. Thus, the diagonal elements are equal to $(t - 1)/t$, and the off-diagonal elements are equal to $- 1/t$.

5. Eq. (13.13*b*) implies that the rows of $\mathbf{B}_{11}$ add to zero.

This concludes the proof of the theorem.

◆ **Theorem 13.5** In Eq. (13.12), $E(\hat{\tau}_i) = \tau_i - \bar{\tau}_{.}$.

Proof: Since $\mathbf{A}\hat{\boldsymbol{\tau}} = \mathbf{q}$ is derived from linear combinations of the normal equations, it follows that $E(\mathbf{q}) = \mathbf{A}\boldsymbol{\tau}$. Hence,

$$E(\hat{\boldsymbol{\tau}}) = E(\mathbf{B}_{11}\mathbf{q}) = \mathbf{B}_{11}E(\mathbf{q}) = \mathbf{B}_{11}\mathbf{A}\boldsymbol{\tau} = \left(\mathbf{I} - \frac{1}{t}\mathbf{1}\mathbf{1}'\right)\boldsymbol{\tau} = \boldsymbol{\tau} - \frac{1}{t}\mathbf{1}\bar{\tau}.$$

so the ith element of $E(\hat{\boldsymbol{\tau}})$ gives $E(\hat{\tau}_i) = \tau_i - \bar{\tau}$.

◆ **Theorem 13.6** $\text{cov}(\hat{\boldsymbol{\tau}}) = \sigma^2\mathbf{B}_{11}$, where $\hat{\boldsymbol{\tau}} = \mathbf{B}_{11}\mathbf{q}$.

Proof: By definition,

$$\begin{aligned}\text{cov}(\hat{\boldsymbol{\tau}}) = \text{cov}(\mathbf{B}_{11}\mathbf{q}) &= E\{[\hat{\boldsymbol{\tau}} - E(\hat{\boldsymbol{\tau}})][\hat{\boldsymbol{\tau}} - E(\hat{\boldsymbol{\tau}})]'\}\\ &= E[(\mathbf{B}_{11}\mathbf{q} - \mathbf{B}_{11}\mathbf{A}\boldsymbol{\tau})(\mathbf{B}_{11}\mathbf{q} - \mathbf{B}_{11}\mathbf{A}\boldsymbol{\tau})']\\ &= \mathbf{B}_{11}E\{[\mathbf{q} - E(\mathbf{q})][\mathbf{q} - E(\mathbf{q})]'\}\mathbf{B}_{11}\end{aligned}$$

Let $\mathbf{C} = \text{cov}(\mathbf{q}) = E\{[\mathbf{q} - E(\mathbf{q})][\mathbf{q} - E(\mathbf{q})]'\}$; then

$$\begin{aligned}c_{ii} = \text{var}(q_i) &= \text{var}\left(Y_{i..} - \sum_{j=1}^{b} n_{ij}y_{.j.}\right)\\ &= \text{var}(Y_{i..}) + \text{var}\left(\sum_{j=1}^{b} n_{ij}y_{.j.}\right) - 2\,\text{cov}\left(\sum_{j=1}^{b} n_{ij}y_{.j.}, Y_{i..}\right)\\ &= N_{i.}\sigma^2 + \sigma^2\sum_{j=1}^{b}\frac{n_{ij}^2}{N_{.j}} - 2\sigma^2\sum_{j=1}^{b}\frac{n_{ij}^2}{N_{.j}}\\ &= \left(N_{i.} - \sum_{j=1}^{b}\frac{n_{ij}^2}{N_{.j}}\right)\sigma^2\end{aligned}$$

Also, for $r \neq s$,

$$\begin{aligned}c_{rs} = \text{cov}(q_r, q_s) &= \text{cov}\left[\left(Y_{r..} - \sum_{j=1}^{b} n_{rj}y_{.j.}\right), \left(Y_{s..} - \sum_{k=1}^{b} n_{sk}y_{.k.}\right)\right]\\ &= \text{cov}(Y_{r..}, Y_{s..}) - \text{cov}\left(Y_{s..}, \sum_{j=1}^{b} n_{rj}y_{.j.}\right) - \text{cov}\left(Y_{r..}, \sum_{k=1}^{b} n_{sk}y_{.k.}\right)\\ &\qquad + \text{cov}\left(\sum_{j=1}^{b} n_{rj}y_{.j.}, \sum_{k=1}^{b} n_{sk}y_{.k.}\right)\\ &= 0 - \sigma^2\sum_{j=1}^{b}\frac{n_{sj}n_{rj}}{N_{.j}} - \sigma^2\sum_{j=1}^{b}\frac{n_{sj}n_{rj}}{N_{.j}} + \sigma^2\sum_{j=1}^{b}\frac{n_{sj}n_{rj}}{N_{.j}}\\ &= -\sigma^2\sum_{j=1}^{b}\frac{n_{sj}n_{rj}}{N_{.j}}\end{aligned}$$

We have, therefore, shown that $\mathbf{C} = \mathbf{A}\sigma^2$. So $\text{cov}(\hat{\boldsymbol{\tau}}) = \mathbf{B}_{11}\mathbf{A}\mathbf{B}_{11}\sigma^2 = \sigma^2\mathbf{B}_{11}$, by Theorem 13.4.

The procedure for solving the normal equations is as follows:

1. Eliminate $\hat{\mu}$ and $\hat{\beta}_j$ and obtain $\mathbf{A}\hat{\boldsymbol{\tau}} = \mathbf{q}$.
2. Augment this by $\Sigma\hat{\tau}_i = 0$; this gives $\mathbf{A}^*$.
3. Find $\mathbf{A}^{*-1}$ and $\hat{\boldsymbol{\tau}} = \mathbf{B}_{11}\mathbf{q}$. Then $E(\hat{\tau}_i) = \tau_i - \bar{\tau}_.$ and $\text{cov}(\hat{\boldsymbol{\tau}}) = \sigma^2\mathbf{B}_{11}$.

Using the condition $\Sigma\hat{\tau}_i = 0$ means that we must solve a system of $t + 1$ equations for $t + 1$ unknowns. The work can be reduced by using other nonestimable conditions. For example, suppose we use the nonestimable condition $\hat{\tau}_t = 0$. This immediately reduces the system to $t - 1$ equations by crossing out the tth equation in (13.9) and setting $\hat{\tau}_t = 0$.

◆ **Theorem 13.7** If the nonestimable condition $\hat{\tau}_t = 0$ is used with the system of equations $\mathbf{A}\hat{\boldsymbol{\tau}} = \mathbf{q}$, the solution is such that $E(\hat{\tau}_i) = \tau_i - \tau_t$ and $\text{cov}(\hat{\boldsymbol{\tau}}) = \sigma^2\mathbf{B}^*_{11}$, where $\mathbf{B}^*_{11} = (\mathbf{A}^{**})^{-1}$, where $\mathbf{A}^{**}$ is the matrix $\mathbf{A}$ with the tth row and column deleted. If $\hat{\tau}_s = 0$ is used, a similar situation holds, with t replaced by s.

The proof will be left for the reader.

13.1.3 Example. Suppose the normal equations have been reduced so that the system $\mathbf{A}\hat{\boldsymbol{\tau}} = \mathbf{q}$ is

$$\begin{aligned} 2\hat{\tau}_1 - \hat{\tau}_2 - \hat{\tau}_3 &= 12 \\ -\hat{\tau}_1 + \hat{\tau}_2 \qquad &= -5 \\ -\hat{\tau}_1 \qquad + \hat{\tau}_3 &= -7 \end{aligned}$$

Then

$$\mathbf{A} = \begin{pmatrix} 2 & -1 & -1 \\ -1 & 1 & 0 \\ -1 & 0 & 1 \end{pmatrix}$$

If we use the nonestimable condition $\Sigma\hat{\tau}_i = 0$, we get

$$\mathbf{A}^* = \begin{pmatrix} 2 & -1 & -1 & 1 \\ -1 & 1 & 0 & 1 \\ -1 & 0 & 1 & 1 \\ 1 & 1 & 1 & 0 \end{pmatrix}$$

and

$$\mathbf{A}^{*^{-1}} = \tfrac{1}{9}\begin{pmatrix} 2 & -1 & -1 & 3 \\ -1 & 5 & -4 & 3 \\ -1 & -4 & 5 & 3 \\ 3 & 3 & 3 & 0 \end{pmatrix}$$

So

$$\begin{aligned} \hat{\tau}_1 &= 4 & E(\hat{\tau}_1) &= \tau_1 - \bar{\tau}_. \\ \hat{\tau}_2 &= -1 & E(\hat{\tau}_2) &= \tau_2 - \bar{\tau}_. \\ \hat{\tau}_3 &= -3 & E(\hat{\tau}_3) &= \tau_3 - \bar{\tau}_. \end{aligned} \tag{13.15}$$

$$\text{cov}(\hat{\boldsymbol{\tau}}) = \frac{\sigma^2}{9}\begin{pmatrix} 2 & -1 & -1 \\ -1 & 5 & -4 \\ -1 & -4 & 5 \end{pmatrix} \tag{13.16}$$

For example,

$$\begin{aligned}\operatorname{var}(\hat{\tau}_1 - \hat{\tau}_2) &= \operatorname{var}[(\hat{\tau}_1 - \hat{\bar{\tau}}_.) - (\hat{\tau}_2 - \hat{\bar{\tau}}_.)] \\ &= \operatorname{var}(\hat{\tau}_1 - \hat{\bar{\tau}}_.) + \operatorname{var}(\hat{\tau}_2 - \hat{\bar{\tau}}_.) - 2\operatorname{cov}(\hat{\tau}_1 - \hat{\bar{\tau}}_., \hat{\tau}_2 - \hat{\bar{\tau}}_.) \\ &= \tfrac{2}{9}\sigma^2 + \tfrac{5}{9}\sigma^2 - \tfrac{2}{9}(-\sigma^2) = \sigma^2\end{aligned}$$

This solution required inversion of a 4×4 matrix.

Next let us use the nonestimable equation $\hat{\tau}_3 = 0$. Thus, $\mathbf{A}\hat{\boldsymbol{\tau}} = \mathbf{q}$ is immediately reduced (eliminate the third equation and set $\hat{\tau}_3 = 0$) to

$$\begin{aligned} 2\hat{\tau}_1 - \hat{\tau}_2 &= 12 \\ -\hat{\tau}_1 + \hat{\tau}_2 &= -5 \end{aligned}$$

We shall denote this system of equations as $\bar{\mathbf{A}}\tilde{\boldsymbol{\tau}} = \bar{\mathbf{q}}$. The matrix is

$$\bar{\mathbf{A}} = \begin{pmatrix} 2 & -1 \\ -1 & 1 \end{pmatrix}$$

and its inverse is

$$\begin{pmatrix} 1 & 1 \\ 1 & 2 \end{pmatrix}$$

So $\tilde{\tau}_1 = 7$; $\tilde{\tau}_2 = 2$ (of course, $\tilde{\tau}_3 = 0$). By Theorem 13.7, $E(\tilde{\tau}_1) = \tau_1 - \tau_3$; $E(\tilde{\tau}_2) = \tau_2 - \tau_3$. Now consider $\hat{\alpha}_3$, where

$$\hat{\alpha}_3 = -\tfrac{1}{3}(\tilde{\tau}_1 + \tilde{\tau}_2 + \tilde{\tau}_3) = -\tfrac{1}{3}(7 + 2) = -3$$

But $$E(\hat{\alpha}_3) = -\tfrac{1}{3}[(\tau_1 - \tau_3) + (\tau_2 - \tau_3) + (\tau_3 - \tau_3)] = \tau_3 - \bar{\tau}_.$$

So -3 is an estimate of $\tau_3 - \bar{\tau}_.$. Also, let $\hat{\alpha}_2 = (\tilde{\tau}_1 + \hat{\alpha}_3) = 7 - 3 = 4$. But $E(\hat{\alpha}_2) = \tau_1 - \bar{\tau}_.$. So 4 is an estimate of $\tau_1 - \bar{\tau}_.$. By the same process -1 is an estimate of $\tau_2 - \bar{\tau}_.$. These are the same estimates as those of Eq. (13.15). This verifies the fact that the estimates of these estimable functions are the same whether the nonestimable condition $\tilde{\tau}_3 = 0$ or $\Sigma\hat{\tau}_i = 0$ is used. Also, if we use $\tilde{\tau}_3 = 0$, we get

$$\operatorname{cov}(\tilde{\boldsymbol{\tau}}) = \sigma^2\begin{pmatrix} 1 & 1 \\ 1 & 2 \end{pmatrix}$$

so $\operatorname{cov}(\tilde{\tau}_1, \tilde{\tau}_2) = \sigma^2$; $\operatorname{var}(\tilde{\tau}_1) = \sigma^2$. But it can also be shown, by using the covariance matrix in Eq. (13.16), that

$$\begin{aligned}\operatorname{var}(\tilde{\tau}_1) &= \operatorname{var}(\hat{\tau}_1 - \hat{\tau}_3) = \operatorname{var}(\hat{\tau}_1) + \operatorname{var}(\hat{\tau}_3) - 2\operatorname{cov}(\hat{\tau}_1, \hat{\tau}_3) \\ &= \tfrac{2}{9}\sigma^2 + \tfrac{5}{9}\sigma^2 - (2)(-\tfrac{1}{9})\sigma^2 = \sigma^2\end{aligned}$$

where $\hat{\tau}_i$ is the solution to $\mathbf{A}\hat{\tau} = \mathbf{q}$ with $\Sigma\hat{\tau}_i = 0$.

After finding a solution to $\mathbf{A}\hat{\boldsymbol{\tau}} = \mathbf{q}$, we can substitute these values into Eq. (13.3) and obtain $\hat{\mu} + \hat{\beta}_j$, from which contrasts of the β_j can be estimated.

13.1.4 Test of the Hypothesis $\tau_1 = \tau_2 = \cdots = \tau_t$. To test this hypothesis we need to evaluate $R(\mu,\tau,\beta)$ from Eq. (13.2) and then to evaluate $R(\mu,\beta)$, the unadjusted sum of squares due to μ and β, from Eq. (13.2) with $\tau_1 = \tau_2 = \cdots = \tau_t$. Now $R(\mu,\tau,\beta) = \hat{\mu}Y_{...} + \sum_i \hat{\tau}_i Y_{i..} + \sum_j \hat{\beta}_j Y_{.j.}$, where $\hat{\mu}$, $\hat{\tau}_i$, $\hat{\beta}_j$ are *any* solution to Eq. (13.2). This gives us $R(\mu,\tau,\beta) = \sum_j (\hat{\mu} + \hat{\beta}_j)Y_{.j.} + \Sigma\hat{\tau}_i Y_{i..}$, and, using the value of $\hat{\mu} + \hat{\beta}_j$ from Eq. (13.3), we have

$$R(\mu,\tau,\beta) = \sum_{j=1}^{b}\left(y_{.j.} - \frac{1}{N_{.j}}\sum_{i=1}^{t} n_{ij}\hat{\tau}_i\right)Y_{.j.} + \sum_{i=1}^{t}\hat{\tau}_i Y_{i..}$$

$$= \sum_j \left(\frac{Y_{.j.}^2}{N_{.j}}\right) + \sum_{i=1}^{t}\hat{\tau}_i\left(Y_{i..} - \sum_{j=1}^{b}\frac{n_{ij}}{N_{.j}}Y_{.j.}\right) = \sum_{j=1}^{b}\frac{Y_{.j.}^2}{N_{.j}} + \sum_{i=1}^{t}\hat{\tau}_i q_i \tag{13.17}$$

We shall return to this equation after we obtain $R(\mu,\beta)$.

To get $R(\mu,\beta)$ we use the normal equations (13.2) with $\hat{\tau}_1 = \hat{\tau}_2 = \cdots = \hat{\tau}_t = \hat{\tau}$ (replace $\hat{\mu} + \hat{\tau}$ with $\tilde{\mu}^*$) and eliminate the t equations represented by τ_r. This gives the reduced normal equations:

$$\mu^*: \qquad N_{..}\tilde{\mu}^* + \sum_{j=1}^{b} N_{.j}\tilde{\beta}_j = Y_{...}$$

$$\beta_s: \qquad N_{.s}\tilde{\mu}^* + \quad N_{.s}\tilde{\beta}_s = Y_{.s.} \qquad s = 1, 2, \ldots, b$$

Using the nonestimable equation $\Sigma N_{.j}\tilde{\beta}_j = 0$ gives the solution

$$\tilde{\mu}^* = y_{...}$$

$$\tilde{\beta}_s = y_{.s.} - y_{...}$$

and

$$R(\mu,\beta) = Y_{...}\tilde{\mu}^* + \sum_{j=1}^{b} Y_{.j.}\tilde{\beta}_j = \sum_j \frac{Y_{.j.}^2}{N_{.j}}$$

Therefore, the adjusted sum of squares due to τ is

$$R(\tau \mid \mu,\beta) = R(\mu,\tau,\beta) - R(\mu,\beta) = \sum_{i=1}^{t}\hat{\tau}_i\left(Y_{i..} - \sum_{j=1}^{b} n_{ij}y_{.j.}\right)$$

$$= \sum_{i=1}^{t}\hat{\tau}_i q_i = \hat{\boldsymbol{\tau}}'\mathbf{q}$$

By the theorems of preceding articles, we can write $\hat{\boldsymbol{\tau}}'\mathbf{q} = \hat{\boldsymbol{\tau}}'\mathbf{A}\hat{\boldsymbol{\tau}} = \mathbf{q}'\mathbf{B}_{11}\mathbf{q}$. We have proved the following theorem.

◆ **Theorem 13.8** The adjusted sum of squares for testing the hypothesis $\tau_1 = \tau_2 = \cdots = \tau_t$ is $\Sigma \hat{\tau}_i q_i$, where $\hat{\tau}_i$ is any solution to $\mathbf{A}\hat{\boldsymbol{\tau}} = \mathbf{q}$.

The AOV is given in Table 13.2.

TABLE 13.2 ANALYSIS OF VARIANCE FOR TWO-WAY CLASSIFICATION WITH UNEQUAL NUMBERS IN THE SUBCLASSES

SV	DF	SS	MS	F	λ
Total	$N_{..}$	$\sum_{ijk} y_{ijk}^2$			
Due to μ,β (unadj)	b	$\sum_j \frac{Y_{.j.}^2}{N_{.j}}$			
Due to τ (adj)	$t-1$	$\sum_{i=1}^{t} \hat{\tau}_i q_i = \tau_{ss}$	τ_{ms}	$\frac{\tau_{ms}}{E_{ms}}$	$\frac{1}{2\sigma^2}\left[\sum_{ij} n_{ij}\left(\tau_i - \sum_p \frac{n_{pj}\tau_p}{N_{.j}}\right)^2\right]$
Error	$N_{..} - b - t + 1$	Subtraction $= E_{ss}$	E_{ms}		

The abbreviated Doolittle technique described in Art. 7.2.2 can be used to good advantage to obtain $\hat{\boldsymbol{\tau}}'\mathbf{q}$. This is seen from the fact that from the equation $\mathbf{X}'\mathbf{X}\hat{\boldsymbol{\beta}} = \mathbf{X}'\mathbf{Y}$ the Doolittle technique gives $\hat{\boldsymbol{\beta}}'\mathbf{X}'\mathbf{Y}$. The procedure is to let $\hat{\tau}_t = 0$ and eliminate the last equation in $\mathbf{A}\hat{\boldsymbol{\tau}} = \mathbf{q}$. This leaves (say)

$$\mathbf{A}_{11}\hat{\boldsymbol{\tau}}^* = \mathbf{q}^*$$

where $\mathbf{A}_{11}$ is obtained by crossing out the last row and column of $\mathbf{A}$. $\hat{\boldsymbol{\tau}}^*$ is $\hat{\boldsymbol{\tau}}$ with $\hat{\tau}_t$ eliminated, and $\mathbf{q}^*$ is $\mathbf{q}$ with q_t eliminated. Theorem 7.1 is applied to $\mathbf{A}_{11}\hat{\boldsymbol{\tau}}^* = \mathbf{q}^*$, and the result is $(\hat{\boldsymbol{\tau}}^*)'\mathbf{q}^*$, which is equal to $\hat{\boldsymbol{\tau}}'\mathbf{q}$, since $\hat{\tau}_t = 0$.

To find the noncentrality parameter λ in Table 13.2, the expected value of the quantity τ_{ss} must be found. By Theorem 11.14, the error mean square is an unbiased estimate of σ^2; hence, $E(E_{ss}) = (N_{..} - b - t + 1)\sigma^2$. The identity

$$\sum_{ijk} y_{ijk}^2 = \sum_j \frac{Y_{.j.}^2}{N_{.j}} + \tau_{ss} + E_{ss}$$

or

$$\tau_{ss} = \sum_{ijk} y_{ijk}^2 - \sum_j \frac{Y_{.j.}^2}{N_{.j}} - E_{ss}$$

can be used to find $E(\tau_{ss})$. This gives

$$E(\tau_{ss}) = E\left(\sum_{ijk} y_{ijk}^2\right) - E\left(\sum_j \frac{Y_{.j.}^2}{N_{.j}}\right) - (N_{..} - b - t + 1)\sigma^2$$

$$= \left[\sum_{ij} n_{ij}(\mu + \tau_i + \beta_j)^2 + N_{..}\sigma^2\right]$$

$$- \left[\sum_j \left(\mu + \sum_i \frac{n_{ij}\tau_i}{N_{.j}} + \frac{N_{.j}\beta_j}{N_{.j}}\right)^2 + b\sigma^2\right] - (N_{..} - b - t + 1)\sigma^2$$

$$= \sum_i N_{i.}\tau_i^2 - \sum_j \frac{\left(\sum_i n_{ij}\tau_i\right)^2}{N_{.j}} + (t - 1)\sigma^2$$

Using Theorem 11.16 gives us λ.

The best method for solving $\mathbf{A}\hat{\boldsymbol{\tau}} = \mathbf{q}$ seems to be to set one of the $\hat{\tau}_i$'s equal to zero (say, we set $\hat{\tau}_t = 0$). Then the solution $\hat{\tau}_i$ is an unbiased estimate of $\tau_i - \tau_t$, and σ^2 times the ijth element of the inverse matrix is the covariance of $\hat{\tau}_i - \hat{\tau}_t$ and $\hat{\tau}_j - \hat{\tau}_t$. Therefore, we shall discuss setting confidence limits on a linear combination of $\tau_i - \tau_t$, say, on $\Sigma c_i(\tau_i - \tau_t)$. We can use the fact that $\Sigma c_i \hat{\tau}_i$ is distributed

$$N[\Sigma c_i(\tau_i - \tau_t), \sigma^2 \Sigma c_i c_j b_{ij}]$$

where b_{ij} is the ijth element of the matrix $\mathbf{B}_{11}^*$ in Theorem 13.7 and $\hat{\tau}_i$ is the solution to the equations $\mathbf{A}\hat{\boldsymbol{\tau}} = \mathbf{q}$ augmented by $\hat{\tau}_t = 0$. This gives

$$u = \frac{\Sigma c_i \hat{\tau}_i - \Sigma c_i(\tau_i - \tau_t)}{\sigma\sqrt{\Sigma c_i c_j b_{ij}}}$$

which is distributed $N(0,1)$. Further,

$$v = \frac{E_{ss}}{\sigma^2}$$

is distributed as $\chi^2(N_{..} - b - t + 1)$ and is independent of u. Hence, a $1 - \alpha$ confidence interval is given by

$$\Sigma c_i \hat{\tau}_i - t_{\alpha/2}\sqrt{E_{ms}}\sqrt{\Sigma c_i c_j b_{ij}} \leqslant \Sigma c_i(\tau_i - \tau_t) \leqslant \Sigma c_i \hat{\tau}_i + t_{\alpha/2}\sqrt{E_{ms}}\sqrt{\Sigma c_i c_j b_{ij}} \tag{13.18}$$

where $t_{\alpha/2}$ is the appropriate value of Student's t with $N_{..} - b - t + 1$ degrees of freedom.

13.1.5 Example. First, suppose the reduced equations $\mathbf{A}\hat{\boldsymbol{\tau}} = \mathbf{q}$ are as given in Art. 13.1.3; that is,

$$\begin{aligned} 2\hat{\tau}_1 - \hat{\tau}_2 - \hat{\tau}_3 &= 12 \\ -\hat{\tau}_1 + \hat{\tau}_2 \qquad &= -5 \\ -\hat{\tau}_1 \qquad + \hat{\tau}_3 &= -7 \end{aligned}$$

If we put $\hat{\tau}_3 = 0$ and omit the third equation, we are left with

$$\begin{aligned} 2\hat{\tau}_1 - \hat{\tau}_2 &= 12 \\ -\hat{\tau}_1 + \hat{\tau}_2 &= -5 \end{aligned}$$

The solution is $\hat{\tau}_1 = 7$, $\hat{\tau}_2 = 2$; and $R(\mu,\tau,\beta) = \hat{\boldsymbol{\tau}}'\mathbf{q} = (7)(12) + (2)(-5) = 74$. If we use the nonestimable function $\Sigma\hat{\tau}_i = 0$, we can use Eq. (13.15) and get $R(\mu,\tau,\beta) = \hat{\boldsymbol{\tau}}'\mathbf{q} = (4)(12) + (-1)(-5) + (-3)(-7) = 74$.

13.1.6 Example. In a two-way classification with no interaction, the data in Table 13.3 are to be analyzed.

TABLE 13.3 DATA FOR EXAMPLE OF ART. 13.1.6

$\tau \backslash \beta$	1	2	3	$Y_{i..}$
1	3		1	4
2	6 2	4 1	1 2	16
3	1	3	2	6
$Y_{.j.}$	12	8	6	26

(*a*) Entries are y_{ijk}

$\tau \backslash \beta$	1	2	3	$N_{i.}$
1	1	0	1	2
2	2	2	2	6
3	1	1	1	3
$N_{.j}$	4	3	4	11

(*b*) Entries are n_{ij}

The normal equations are:

$$\begin{array}{llllllll} 11\hat{\mu} & + 2\hat{\tau}_1 & + 6\hat{\tau}_2 & + 3\hat{\tau}_3 & + 4\hat{\beta}_1 & + 3\hat{\beta}_2 & + 4\hat{\beta}_3 & = 26 \\ 2\hat{\mu} & + 2\hat{\tau}_1 & & & \hat{\beta}_1 & & \hat{\beta}_3 & = 4 \\ 6\hat{\mu} & & + 6\hat{\tau}_2 & & + 2\hat{\beta}_1 & + 2\hat{\beta}_2 & + 2\hat{\beta}_3 & = 16 \\ 3\hat{\mu} & & & + 3\hat{\tau}_3 & + \hat{\beta}_1 & + \hat{\beta}_2 & + \hat{\beta}_3 & = 6 \\ 4\hat{\mu} & + \hat{\tau}_1 & + 2\hat{\tau}_2 & + \hat{\tau}_3 & + 4\hat{\beta}_1 & & & = 12 \\ 3\hat{\mu} & & + 2\hat{\tau}_2 & + \hat{\tau}_3 & & + 3\hat{\beta}_2 & & = 8 \\ 4\hat{\mu} & + \hat{\tau}_1 & + 2\hat{\tau}_2 & + \hat{\tau}_3 & & & + 4\hat{\beta}_3 & = 6 \end{array}$$

The quantities $\hat{\mu} + \hat{\beta}_j$ can be eliminated from these equations, or else the data of Table 13.3*b* and Eq. (13.10) can be used to obtain **A**. The system $\mathbf{A}\hat{\boldsymbol{\tau}} = \mathbf{q}$ is

$$\frac{1}{6}\begin{pmatrix} 9 & -6 & -3 \\ -6 & 16 & -10 \\ -3 & -10 & 13 \end{pmatrix}\begin{pmatrix} \hat{\tau}_1 \\ \hat{\tau}_2 \\ \hat{\tau}_3 \end{pmatrix} = \begin{pmatrix} -\frac{3}{6} \\ +\frac{10}{6} \\ -\frac{7}{6} \end{pmatrix}$$

If we let $\hat{\tau}_3 = 0$, we get

$$\begin{pmatrix} \frac{9}{6} & -\frac{6}{6} \\ -\frac{6}{6} & \frac{16}{6} \end{pmatrix}\begin{pmatrix} \hat{\tau}_1 \\ \hat{\tau}_2 \end{pmatrix} = \begin{pmatrix} -\frac{3}{6} \\ +\frac{10}{6} \end{pmatrix}$$

This yields $\hat{\tau}_1 = \frac{1}{9}$, which is an estimate of $\tau_1 - \tau_3$, and $\hat{\tau}_2 = \frac{2}{3}$, which is an estimate of $\tau_2 - \tau_3$. The adjusted sum of the squares due to $\boldsymbol{\tau}$, which is $R(\tau \mid \mu,\beta)$, is

$$(\tfrac{1}{9})(-\tfrac{3}{6}) + (\tfrac{2}{3})(\tfrac{10}{6}) = \tfrac{19}{18} = 1.06$$

The inverse of the matrix is

$$\frac{1}{18}\begin{pmatrix} 16 & 6 \\ 6 & 9 \end{pmatrix}$$

$$\text{var}(\hat{\tau}_1) = \tfrac{16}{18}\sigma^2 \qquad \text{var}\hat{\tau}_2 = \tfrac{9}{18}\sigma^2 \qquad \text{cov}(\hat{\tau}_1,\hat{\tau}_2) = \tfrac{6}{18}\sigma^2$$

From Table 13.3*a*, the sum of squares for μ and β unadjusted is

$$\sum_j \frac{Y_{.j.}^2}{N_{.j}} = \frac{(12)^2}{4} + \frac{8^2}{3} + \frac{6^2}{4} = 66.33$$

The total sum of squares is 86. The AOV is given in Table 13.4. To

TABLE 13.4 ANALYSIS OF VARIANCE OF DATA IN EXAMPLE OF ART. 13.1.6

SV	DF	SS	MS
Total	11	86.00	
Due to μ, β (unadj)	3	66.33	
Due to τ (adj)	2	1.06	.53
Error	6	18.61	3.10

set a 97.5 per cent confidence interval on $\tau_1 - \tau_3$, we can use Eq. (13.18) with $c_1 = 1$, $c_2 = 0$. We get

$$\frac{1}{9} - 2.97\sqrt{\frac{(3.10)(16)}{18}} \leqslant \tau_1 - \tau_3 \leqslant \frac{1}{9} + 2.97\sqrt{\frac{(3.10)(16)}{18}}$$

which reduces to

$$-4.82 \leqslant \tau_1 - \tau_3 \leqslant 5.04$$

13.2 Computing Instructions

In the preceding sections the theory of the two-way classification with unequal numbers in the subclasses has been developed. Now our attention will be turned to the problem of computing. The procedure will be to impose conditions on the normal equations to make them full rank, and to use the Doolittle technique. From these we shall obtain estimates of estimable functions, the covariance matrix of these estimates, and an AOV table.

TABLE 13.5 NUMBER OF OBSERVATIONS IN THE *ij*th CELL

$\tau \backslash \beta$	1	2	3	$\cdots$	b	
1	n_{11}	n_{12}	n_{13}	$\cdots$	n_{1b}	$N_{1.}$
2	n_{21}	n_{22}	n_{23}	$\cdots$	n_{2b}	$N_{2.}$
3	n_{31}	n_{32}	n_{33}	$\cdots$	n_{3b}	$N_{3.}$
.	.			.		.
.	.			.		.
.	.			.		.
$t-1$						$N_{t-1.}$
	$N_{.1}$	$N_{.2}$	$N_{.3}$	$\cdots$	$N_{.b}$	

Suppose that the τ_i are the parameters that are of interest in model (13.1). The normal equations are given in Eq. (13.2). They will be reproduced here, except that the τ_r and β_s equations will be interchanged. They are

$$\mu: \qquad N_{..}\hat{\mu} + \sum_j N_{.j}\hat{\beta}_j + \sum_i N_{i.}\hat{\tau}_i = Y_{...}$$

$$\beta_s: \qquad N_{.s}\hat{\mu} + N_{.s}\hat{\beta}_s + \sum_i n_{is}\hat{\tau}_i = Y_{.s.} \qquad s = 1, 2, \ldots, b$$

$$\tau_r: \qquad N_{r.}\hat{\mu} + \sum_j n_{rj}\hat{\beta}_j + N_{r.}\hat{\tau}_r = Y_{r..} \qquad r = 1, 2, \ldots, t$$

The nonestimable conditions $\hat{\mu} = 0$ and $\hat{\tau}_t = 0$ will be imposed. The first equation for (μ) and the last equation for (τ_t) will be omitted; the resulting system is composed of $b + t - 1$ equations with $b + t - 1$ unknowns and is of full rank. The system can be written

$$\begin{pmatrix} \mathbf{C}_{11} & \mathbf{C}_{12} \\ \mathbf{C}_{21} & \mathbf{C}_{22} \end{pmatrix} \begin{pmatrix} \hat{\boldsymbol{\beta}} \\ \hat{\boldsymbol{\tau}} \end{pmatrix} = \begin{pmatrix} \mathbf{g}_1 \\ \mathbf{g}_2 \end{pmatrix}$$

where $\hat{\boldsymbol{\beta}} = (\hat{\beta}_s)$ has dimension $b \times 1$; $\mathbf{g}_1 = (Y_{.s.})$ has dimension $b \times 1$; $\hat{\boldsymbol{\tau}} = (\hat{\tau}_r)$ has dimension $(t-1) \times 1$ ($\hat{\tau}_t$ is omitted); and $\mathbf{g}_2 = (Y_{r..})$ has dimension $(t-1) \times 1$. The $\mathbf{C}_{ij}$ have the n_{rs} for elements. The elements of $\mathbf{C}_{ij}$ can be found from Table 13.5. $\mathbf{C}_{11}$ is a diagonal matrix with the column totals $N_{.s}$ as elements; $\mathbf{C}_{22}$ is a diagonal matrix with the row totals $N_{r.}$ as elements. $\mathbf{C}_{21}$ is a $t - 1 \times b$ matrix equal to the body of Table 13.5. Also, $\mathbf{C}'_{21} = \mathbf{C}_{12}$. The sum of squares for error for testing the hypothesis $\tau_1 = \cdots = \tau_t$ can be obtained by reducing the system

$$\begin{pmatrix} \mathbf{C}_{11} & \mathbf{C}_{12} & \acute{\mathbf{g}}_1 \\ \mathbf{C}_{21} & \mathbf{C}_{22} & \acute{\mathbf{g}}_2 \end{pmatrix}$$

by the abbreviated Doolittle technique and using Theorem 7.1. We shall illustrate with the example of Art. 13.1.6. From data in Table 13.3b the following are computed

$$\mathbf{C}_{11} = \begin{pmatrix} 4 & 0 & 0 \\ 0 & 3 & 0 \\ 0 & 0 & 4 \end{pmatrix} \qquad \mathbf{C}_{12} = \begin{pmatrix} 1 & 2 \\ 0 & 2 \\ 1 & 2 \end{pmatrix} \qquad \mathbf{C}_{22} = \begin{pmatrix} 2 & 0 \\ 0 & 6 \end{pmatrix}$$

From the border totals of Table 13.3a we get

$$\acute{\mathbf{g}}_1 = \begin{pmatrix} 12 \\ 8 \\ 6 \end{pmatrix} \qquad \acute{\mathbf{g}}_2 = \begin{pmatrix} 4 \\ 16 \end{pmatrix}$$

The format for the Doolittle technique is given in Table 13.6. This is

TABLE 13.6 DOOLITTLE TECHNIQUE FOR TESTING $\tau_1 = \tau_2 = \cdots = \tau_t$

Row	C_1	C_2	C_3	C_4	C_5	C_0	Check
R'_1	4	0	0	1	2	12	19
R'_2		3	0	0	2	8	13
R'_3			4	1	2	6	13
R'_4				2	0	4	8
R'_5					6	16	28
R_1	4	0	0	1	2	12	19
r_1	1	0	0	$\frac{1}{4}$	$\frac{1}{2}$	3	$\frac{19}{4}$
R_2		3	0	0	2	8	13
r_2		1	0	0	$\frac{2}{3}$	$\frac{8}{3}$	$\frac{13}{3}$
R_3			4	1	2	6	13
r_3			1	$\frac{1}{4}$	$\frac{1}{2}$	$\frac{3}{2}$	$\frac{13}{4}$
R_4				$\frac{3}{2}$	-1	$-\frac{1}{2}$	0
r_4				1	$-\frac{2}{3}$	$-\frac{1}{3}$	0
R_5					2	$\frac{4}{3}$	$\frac{10}{3}$
r_5					1	$\frac{2}{3}$	$\frac{5}{3}$

full rank, and the quantities $R(\mu,\tau,\beta)$ and $R(\tau \mid \mu,\beta)$ are easily computed from the C_0 column.

$$R(\mu,\tau,\beta) = (\tfrac{4}{3})(\tfrac{2}{3}) + (-\tfrac{1}{2})(-\tfrac{1}{3}) + (6)(\tfrac{3}{2}) + (\tfrac{8}{3})(8) + (3)(12) = 67.39$$

and $R(\tau \mid \mu,\beta) = (\tfrac{4}{3})(\tfrac{2}{3}) + (-\tfrac{1}{3})(-\tfrac{1}{2}) = \tfrac{19}{18} = 1.06$

These values are, of course, the same as those given in Art. 13.1.6.

Problems

13.1 The data given in Table 13.7 are assumed to follow a two-way classification model with unequal numbers in the subclasses.

TABLE 13.7

τ \ β	1	2	3	4
1	6 8	7	3 1 4	7 10
2	5		6	4 8 7
3	3 6	5		

Find n_{11}, n_{12}, n_{13}, n_{14}, n_{21}, n_{22}, n_{23}, n_{24}, n_{31}, n_{32}, n_{33}, and n_{34}.

13.2 In Prob. 13.1, find $N_{1.}$, $N_{2.}$, $N_{3.}$, $N_{.1}$, $N_{.2}$, $N_{.3}$, and $N_{.4}$.

13.3 In Prob. 13.1, find $Y_{1..}$, $Y_{2..}$, $Y_{3..}$, $Y_{.1.}$, $Y_{.2.}$, $Y_{.3.}$, and $Y_{.4.}$.

13.4 In Prob. 13.1, find the normal equations.

13.5 In Prob. 13.1, find the equations $\mathbf{A}\hat{\boldsymbol{\tau}} = \mathbf{q}$.

13.6 In a two-way classification model with unequal numbers in the subclasses, suppose that the set of equations $\mathbf{A}\hat{\boldsymbol{\tau}} = \mathbf{q}$ is given by

$$\begin{aligned} 10\hat{\tau}_1 - 2\hat{\tau}_2 - 8\tau_3 &= 12 \\ -2\hat{\tau}_1 + 5\hat{\tau}_2 - 3\hat{\tau}_3 &= 16 \\ -8\hat{\tau}_1 - 3\hat{\tau}_2 + 11\hat{\tau}_3 &= -28 \end{aligned}$$

(*a*) Find the rank of this system of equations.

(*b*) Find a solution, using the condition $\Sigma\hat{\tau}_i = 0$.

13.7 In Prob. 13.6, construct the $\mathbf{A}^*$ matrix given in Theorem 13.3.

13.8 In Prob. 13.7, find the rank of $\mathbf{A}^*$.

13.9 In Prob. 13.7, find $(\mathbf{A}^*)^{-1}$.

13.10 Using the equations in Prob. 13.6, verify all the statements made in Theorem 13.4.

13.11 Prove Theorem 13.3.

13.12 Using the data in Prob. 13.6, find a solution to the equations by using the condition $\hat{\tau}_3 = 0$.

13.13 In Prob. 13.6, find the value of the unbiased estimate of $\tau_1 - \tau_3$, using the condition $\Sigma\hat{\tau}_i = 0$.

13.14 In Prob. 13.6, find the value of the unbiased estimate of $\tau_1 - \tau_3$, using the condition $\hat{\tau}_3 = 0$. Compare this value with the value obtained in Prob. 13.13.

13.15 In Prob. 13.13, find the covariance of the estimate of $\tau_1 - \tau_3$.

13.16 In Prob. 13.14, find the covariance of the estimate of $\tau_1 - \tau_3$.

13.17 Using the equations in Prob. 13.6, find the sum of squares for τ (adj) by using

(*a*) The condition $\Sigma\hat{\tau}_i = 0$.

(*b*) The condition $\hat{\tau}_3 = 0$.

13.18 Using the data in Prob. 13.1, test the hypothesis $\tau_1 = \tau_2 = \tau_3$ with a 5 per cent type I error probability.

13.19 Using the data in Prob. 13.1, set a 95 per cent confidence interval on $\tau_1 - \tau_3$.

13.20 Show that $\mathbf{A} + (1/t)\mathbf{J}$ has an inverse, where $\mathbf{A}$ is defined in Eq. (13.10).

13.21 In Prob. 13.20, show that $\mathbf{B}_{11} + (1/t)\mathbf{J}$ is the inverse, where $\mathbf{B}_{11}$ is defined immediately following Theorem 13.3.

Further Reading

1 M. G. Kendall: "The Advanced Theory of Statistics," vols. I, II, Charles Griffin & Co., Ltd., London, 1946.

2 G. W. Snedecor: "Statistical Methods," Iowa State College Press, Ames, Iowa, 1958.

3 O. Kempthorne: "Design and Analysis of Experiments," John Wiley & Sons, Inc., New York, 1952.

4 R. L. Anderson and T. A. Bancroft: "Statistical Theory in Research," McGraw-Hill Book Company, Inc., New York, 1952.

5 C. R. Rao: "Advanced Statistical Methods in Biometric Research," John Wiley & Sons, Inc., New York, 1952.

6 S. Kolodziejczyk: On an Important Class of Statistical Hypotheses, *Biometrika*, vol. 27, 1935.

7 H. Scheffé: "The Analysis of Variance," John Wiley & Sons, Inc., New York, 1959.

14

Incomplete Block Models

14.1 Introduction

Suppose a researcher, in conducting an experiment on various methods of mixing cookie dough, has seven treatments (methods of mixing), and he wishes to investigate the texture of the resulting mixture after the dough is baked under uniform conditions. In order that all treatments may be handled uniformly, it is desirable to bake the dough from all seven treatments in one oven. As many ovens and repetitions of the seven treatments will be used as the experimenter thinks are necessary. However, suppose only material representing three treatments can be baked in one oven. In general, the ovens are not the important factor being studied; it is desirable to remove the effect of the ovens so that the treatments may be more effectively studied. The model can be written

$$y_{ijm} = \mu + \tau_i + \beta_j + e_{ijm}$$

where τ_i is the ith treatment constant, β_j is the jth oven constant, $m = 0$ or 1, and y_{ij1} is the observation when the ith treatment occurs in the jth oven. The observation y_{ij0} does not exist, since, if $m = 0$ for a certain ij combination, this means that the ith treatment does not occur in the jth oven.

Another example is an experiment in which a researcher is studying the effect of laundering on different fabrics. It is desired to study, say, 25 different fabrics, but a washing machine can launder only ten fabrics at one time. Washing machines are not the important factors under study. If many machines are available, the model could be a two-way classification where each treatment does not occur with each machine.

Instead of giving other examples, we shall talk in terms of two general factors: blocks and treatments. The *blocks* will be similar to ovens, washing machines, etc., that is, factors that are not under study by the experimenter. Only cases will be considered in which each block contains k experimental units (sometimes referred to as k *plots*). The number of treatments will be denoted by t; therefore, $k < t$. Also, only those cases will be considered in which each treatment is *replicated* (appears) in exactly r blocks, and the number of blocks is b.

There are various types of incomplete blocks, but we shall discuss only what are termed *balanced incomplete blocks*.

◆ **Definition 14.1** An arrangement of blocks and treatments will be called a balanced incomplete block if the following conditions are satisfied:

(1) Every block contains k experimental units.

(2) There are more treatments than there are plots in a block ($k < t$).

(3) Every treatment appears in r blocks.

(4) Every pair of treatments occurs together in the same number of blocks (let λ denote this number).

For example, consider the arrangement of Table 14.1*a*. In this arrangement $b = 7, t = 7, k = 3, r = 3$, and $\lambda = 1$. By Definition 14.1,

TABLE 14.1

Block no.	Treatment nos.		
1	1	2	4
2	2	3	5
3	3	4	6
4	4	5	7
5	5	6	1
6	6	7	2
7	7	1	3

(*a*)

Block no.	Treatment nos.		
1	1	2	3
2	4	5	6
3	7	8	9
4	1	4	7
5	2	5	8
6	3	6	9

(*b*)

this is a balanced incomplete block arrangement. The arrangement of Table 14.1*b* is not a balanced incomplete block, even though each block contains the same number of treatments ($k = 3$) and even though each treatment is repeated the same number ($r = 2$) of times. It fails to satisfy the definition, since every pair of treatments does not occur together in the same number of blocks (condition 4). For example, treatments 4 and 5 occur together in one block (block 2), but treatments 6 and 7 occur together in no blocks.

There may be various reasons why an experimenter would want to consider a balanced incomplete block arrangement. It may be that, by grouping the experimental units into homogeneous sets, the error variance can be reduced; this may lead to blocks with less than t units in each block and, therefore, require a balanced incomplete block arrangement. The problems connected with the construction of balanced incomplete blocks for a given t and k will not be discussed. The reader can consult papers in the bibliography for a discussion of these.

14.1.1 Definitions and Notation. The balanced incomplete block model will be considered a special case of the two-way classification model with unequal numbers in the subclasses. There will be either one or no observations in each subclass; there is one observation in the ijth subclass if the ith treatment occurs in the jth block, and there is no observation in the ijth subclass if the ith treatment does not occur in the jth block. Therefore, the model will be written

$$y_{ijm} = \mu + \tau_i + \beta_j + e_{ijm} \qquad \begin{cases} i = 1, 2, \ldots, t \\ j = 1, 2, \ldots, b \\ m = n_{ij} \end{cases} \tag{14.1}$$

where $n_{ij} = 1$ if the ith treatment appears in the jth block, and $n_{ij} = 0$ if the ith treatment does not appear in the jth block. y_{ij1} will be used to represent the observation in the ijth cell, and y_{ij0} will mean that there is no observation in that particular cell. τ_i will represent the ith treatment constant, and β_j will represent the jth block constant. The error terms e_{ijm} will be assumed to be uncorrelated, with means 0 and variances σ^2, for point estimation. For interval estimation and testing hypotheses, the e_{ijm} will be assumed to be normally distributed. The following equations are seen to hold:

$$N_{i.} = \sum_{j=1}^{b} n_{ij} = r \qquad \text{(number of blocks in which the } i\text{th treatment appears)}$$

$$N_{.j} = \sum_{i=1}^{t} n_{ij} = k \qquad \text{(number of treatments that appear in the } j\text{th block)} \tag{14.2}$$

$$N_{..} = rt = bk \qquad \text{(total number of observations)}$$

$$\sum_{j=1}^{b} n_{ij}n_{i'j} = \lambda \qquad \text{for all } i \neq i'$$

(this is the number of blocks in which the ith treatment and the i'th treatment occur together. This should not be confused with the symbol for noncentrality).

A relationship among the quantities b, t, r, k, and λ is

$$\lambda = \frac{r(k-1)}{t-1}$$

To show this, consider any treatment, say, treatment 1. This treatment appears in exactly r blocks. The number of plots in these r blocks that is occupied by the remaining treatments is $r(k-1)$. This number is also equal to $\lambda(t-1)$. Equating these gives the result. Also

$Y_{i..}$ = total of observations of the ith treatment
$Y_{.j.}$ = total of the observations in the jth block
$Y_{...}$ = total of all the observations

One more quantity will be needed; this is $\sum_{j=1}^{b} n_{ij} Y_{.j.}$, which is the total of the observations in all the blocks in which the ith treatment appears. The notation

$$B_i = \sum_{j=1}^{b} n_{ij} Y_{.j.}$$

will be used.

14.2 Point Estimation

Since the balanced incomplete block model has been defined as a special case of the two-way classification with unequal numbers in the subclasses, the normal equations can be derived from those in Eq. (13.2) by using the notation of Sec. 14.1. The normal equations are:

$$\begin{aligned}
\mu:&\quad bk\hat{\mu} + r\sum_i \hat{\tau}_i + k\sum_j \hat{\beta}_j = Y_{...} \\
\tau_p:&\quad r\hat{\mu} + r\hat{\tau}_p + \sum_j n_{pj}\hat{\beta}_j = Y_{p..} \qquad p = 1, 2, \ldots, t \\
\beta_s:&\quad k\hat{\mu} + \sum_i n_{is}\hat{\tau}_i + k\hat{\beta}_s = Y_{.s.} \qquad s = 1, 2, \ldots, b
\end{aligned} \tag{14.3}$$

The equations represented by β_s can be solved for $\hat{\mu} + \hat{\beta}_s$. This gives

$$\hat{\mu} + \hat{\beta}_s = \frac{1}{k}\left(Y_{.s.} - \sum_i n_{is}\hat{\tau}_i\right) \qquad s = 1, 2, \ldots, b \tag{14.4}$$

The equations represented by τ_p in (14.3) can be written

$$r\hat{\tau}_p + \sum_j n_{pj}(\hat{\mu} + \hat{\beta}_j) = Y_{p..} \qquad p = 1, 2, \ldots, t \tag{14.5}$$

If (14.4) is substituted into (14.5), we get

$$r\hat{\tau}_p + \sum_{j=1}^{b} n_{pj} \frac{1}{k}\left(Y_{.j.} - \sum_{i=1}^{t} n_{ij}\hat{\tau}_i\right) = Y_{p..} \qquad p = 1, 2, \ldots, t$$

This becomes

$$r\hat{\tau}_p - \frac{1}{k}\sum_{i=1}^{t}\sum_{j=1}^{b} n_{pj}n_{ij}\hat{\tau}_i = Y_{p..} - \frac{1}{k}\sum_{j=1}^{b} n_{pj}Y_{.j.}$$

The quantity $\hat{\tau}_p$ will be factored out of the second term on the left of the equals sign. This gives

$$r\hat{\tau}_p - \frac{1}{k}\sum_{j=1}^{b} n_{pj}n_{pj}\hat{\tau}_p - \frac{1}{k}\sum_{\substack{i=1\\ i\neq p}}^{t}\sum_{j=1}^{b} n_{pj}n_{ij}\hat{\tau}_i = Y_{p..} - \frac{1}{k}\sum_{j=1}^{b} n_{pj}Y_{.j.} \qquad p = 1, 2, \ldots, t$$

From Eq. (14.2), the quantities

$$\lambda = \sum_{j=1}^{b} n_{pj}n_{ij} \qquad \text{if } p \neq i$$

and $n_{pj}^2 = n_{pj}$ (since $n_{pj} = 0$ or 1) can be used to give

$$\left(r - \frac{r}{k}\right)\hat{\tau}_p - \frac{\lambda}{k}\sum_{\substack{i=1\\ i\neq p}}^{t} \hat{\tau}_i = Y_{p..} - \frac{1}{k}\sum_{j=1}^{b} n_{pj}Y_{.j.} \qquad p = 1, 2, \ldots, t \tag{14.6}$$

The nonestimable condition $\sum_{i=1}^{t} \hat{\tau}_i = 0$ can be used to obtain

$$\sum_{\substack{i=1\\ i\neq p}}^{t} \hat{\tau}_i = -\hat{\tau}_p$$

If this is substituted into (14.6), it gives

$$\left(r - \frac{r}{k} + \frac{\lambda}{k}\right)\hat{\tau}_p = Y_{p..} - \frac{1}{k}\sum_{j=1}^{b} n_{pj}Y_{.j.} \qquad p = 1, 2, \ldots, t \tag{14.7}$$

The coefficient of $\hat{\tau}_p$ is $(rk - r + \lambda)/k$. If the value of λ in terms of r, k, and t is used, this expression becomes $\lambda t/k$. The quantity

$$Y_{p..} - \frac{1}{k}\sum_{j=1}^{b} n_{pj}Y_{.j.} = Y_{p..} - \frac{1}{k}B_p$$

will be denoted by q_p. This is the total of the observations of the pth treatment minus $1/k$ times the total of the observations of all blocks in which the pth treatment occurs. The quantity q_p does not contain μ or the block constants β_i. This can be seen if Eq. (14.1) is substituted into the definition of q_p. The result is

$$q_p = \sum_{jm} (\mu + \tau_p + \beta_j + e_{pjm}) - \frac{1}{k} \sum_{j=1}^{b} n_{pj} \sum_{im} (\mu + \tau_i + \beta_j + e_{ijm})$$

$$= r\mu + r\tau_p + \sum_j n_{pj}\beta_j + \sum_{jm} e_{pjm}$$

$$- \frac{1}{k} \sum_{j=1}^{b} n_{pj} \left(k\mu + \sum_i n_{ij}\tau_i + k\beta_j + \sum_{im} e_{ijm} \right)$$

$$= r\mu + r\tau_p + \sum_j n_{pj}\beta_j + \sum_{jm} e_{pjm} - r\mu - \frac{1}{k} \sum_{j=1}^{b} \sum_{i=1}^{t} n_{pj}n_{ij}\tau_i - \sum_j n_{pj}\beta_j$$

$$- \frac{1}{k} \sum_{ijm} n_{pj}e_{ijm}$$

$$= r\tau_p - \frac{1}{k} \sum_{j=1}^{b} \sum_{i=1}^{t} n_{pj}n_{ij}\tau_i + \sum_{jm} e_{pjm} - \frac{1}{k} \sum_{ijm} n_{pj}e_{ijm}$$

$$= r\tau_p - \frac{1}{k} \sum_{j=1}^{b} n_{pj}n_{pj}\tau_p - \frac{1}{k} \sum_{\substack{i=1 \\ i \neq p}}^{t} \lambda\tau_i + \sum_{jm} e_{pjm} - \frac{1}{k} \sum_{ijm} n_{pj}e_{ijm}$$

$$= \frac{\lambda t}{k} (\tau_p - \bar{\tau}_.) + \left(\sum_{jm} e_{pjm} - \frac{1}{k} \sum_{ijm} n_{pj}e_{ijm} \right) \tag{14.8}$$

From this we get

$$E(q_p) = \frac{\lambda t}{k} (\tau_p - \bar{\tau}_.) \tag{14.9}$$

and

$$\text{var}(q_p) = E\left[\sum_{jm} e_{pjm} - \frac{1}{k} \sum_{ijm} n_{pj}e_{ijm} \right]^2 = \frac{r(k-1)}{k} \sigma^2 \tag{14.10}$$

$$\text{cov}(q_p,q_{p'}) = E\left[\left(\sum_{jm} e_{pjm} - \frac{1}{k} \sum_{ijm} n_{pj}e_{ijm} \right) \left(\sum_{j'm'} e_{p'j'm'} - \frac{1}{k} \sum_{i'j'm'} n_{p'j'}e_{i'j'm'} \right) \right]$$

$$= -\frac{\lambda\sigma^2}{k} \qquad \text{for } p \neq p'$$

From these facts we get the following theorem.

◆ **Theorem 14.1** In the balanced incomplete block model given by Eq. (14.1), the linear minimum-variance unbiased estimate of any contrast of the τ_p is given by $(k/\lambda t)\Sigma c_p q_p$. The variance of the estimate is $(k\sigma^2/\lambda t)\Sigma c_p^2$. If the e_{ijm} are normally distributed, the estimate is maximum likelihood and is the best unbiased estimate.

14.3 Interval Estimation and Tests of Hypotheses

We have considered the balanced incomplete block model as a special case of the two-way classification model with unequal numbers in the subclasses. Therefore, the AOV is given by Table 13.2. The quantity $\Sigma\hat{\tau}_i q_i$ is $(k/\lambda t)\Sigma q_i^2$, since $\hat{\tau}_i = (k/\lambda t)q_i$ is a solution to the system $\mathbf{A}\hat{\boldsymbol{\tau}} = \mathbf{q}$. The labeling in the AOV table will be altered slightly. It is found in Table 14.2.

TABLE 14.2 ANALYSIS OF VARIANCE OF BALANCED INCOMPLETE BLOCK

SV	DF	SS	MS	Noncentrality
Total	bk	$\sum_{ijm} y_{ijm}^2$		
Mean	1	$\frac{Y_{...}^2}{bk}$		
Blocks (unadj)	$b - 1$	$\frac{\sum_j Y_{.j.}^2}{k} - \frac{Y_{...}^2}{bk}$		
Treatments (adj)	$t - 1$	$\frac{k}{\lambda t}\Sigma q_i^2 = \tau_{\text{ss}}$	τ_{ms}	$\frac{\lambda t}{k}\frac{\sum_i (\tau_i - \bar{\tau}_.)^2}{2\sigma^2}$
Error	$bk - b - t + 1$	Subtraction $= E_{\text{ss}}$	E_{ms}	

The quantity $\tau_{\text{ms}}/E_{\text{ms}}$ is distributed as the noncentral F and reduces to the central F distribution if and only if $\tau_1 = \tau_2 = \cdots = \tau_t$. Hence, this quantity can be used to test the hypothesis that the treatment constants are equal.

To set a confidence interval on a contrast of the τ_i, we note that

$$u = \frac{\Sigma c_i\hat{\tau}_i - \Sigma c_i\tau_i}{\sigma\sqrt{\Sigma c_i^2(k/\lambda t)}}$$

is distributed $N(0,1)$ and is independent of E_{ss}/σ^2, which is distributed as $\chi^2(bk - t - b + 1)$. Therefore,

$$v = \frac{\Sigma c_i\hat{\tau}_i - \Sigma c_i\tau_i}{\sqrt{E_{\text{ms}}\Sigma c_i^2(k/\lambda t)}}$$

is distributed as Student's t with $bk - t - b + 1$ degrees of freedom. A $1 - \alpha$ confidence interval is

$$\Sigma c_i\hat{\tau}_i - t_{\alpha/2}\sqrt{E_{\text{ms}}\Sigma c_i^2\frac{k}{\lambda t}} \leqslant \Sigma c_i\tau_i \leqslant \Sigma c_i\hat{\tau}_i + t_{\alpha/2}\sqrt{E_{\text{ms}}\Sigma c_i^2\frac{k}{\lambda t}}$$

The width of the interval is $2t_{\alpha/2}\sqrt{E_{\text{ms}}\Sigma c_i^2(k/\lambda t)}$.

14.4 Computing

The computing for the balanced incomplete block model is quite easy. Construct a format such as Table 14.3, and for each treatment compute the total $Y_{i..}$ and the total of all blocks that contain the ith treatment.

TABLE 14.3 FORM FOR COMPUTATIONS

Treatment number	Total	Total of blocks containing treatment i	q_i	$q_i\left(\frac{k}{\lambda t}\right) = \hat{\tau}_i - \hat{\bar{\tau}}.$
1	$Y_{1..}$	B_1	$Y_{1..} - \frac{1}{k} B_1$	$\left(Y_{1..} - \frac{1}{k} B_1\right)\frac{k}{\lambda t}$
2	$Y_{2..}$	B_2	$Y_{2..} - \frac{1}{k} B_2$	$\left(Y_{2..} - \frac{1}{k} B_2\right)\frac{k}{\lambda t}$
.	.	.	.	.
.	.	.	.	.
.	.	.	.	.
t	$Y_{t..}$	B_t	$Y_{t..} - \frac{1}{k} B_t$	$\left(Y_{t..} - \frac{1}{k} B_t\right)\frac{k}{\lambda t}$

From this the remaining quantities for the AOV can be computed, simply by squaring and cumulating. The column $q_i\left(\frac{k}{\lambda t}\right)$ gives unbiased estimates of $\tau_i - \bar{\tau}$. $\frac{\lambda t}{k}$ times the sum of squares of the elements in the fourth column gives the sum of squares for treatments adjusted.

14.4.1 Example. A greenhouse experiment was run to test the effects of various fertilizer treatments on the nitrogen content of alfalfa forage. Because of the effect of location on the greenhouse bench, it was decided that block sizes of three plots would be desirable. Since there were seven treatments, a balanced incomplete block was used. The yields are given in Table 14.4. The numbers in parentheses are treatment numbers.

The treatment totals and the total of blocks containing each treatment are needed and are given in Table 14.5.

Since $\sum_i Y_{i..} = Y_{...}$, the second column in Table 14.5 sums to the grand total 58.30. Also, $\sum_i(\sum_j n_{ij} Y_{.j.}) = \sum_j(\sum_i n_{ij} Y_{.j.}) = \sum_j k Y_{.j.} = k Y_{...} =$ number of plots per block times the grand total $= (3)(58.30) = 174.90$. Since $\Sigma q_i = 0$, the last two columns will sum to zero except for rounding

Table 14.4 Total Weights of Nitrogen in Grams of Six Cuttings of Alfalfa Forage Grown in Greenhouse Pots

Block j	Treatments			Block total $Y_{.j.}$
1	2.10 (1)	2.67 (2)	2.91 (4)	7.68
2	1.14 (2)	3.00 (3)	3.10 (5)	7.24
3	2.92 (3)	3.14 (4)	2.99 (6)	9.05
4	3.13 (4)	2.63 (5)	2.75 (7)	8.51
5	2.84 (5)	3.13 (6)	1.85 (1)	7.82
6	3.01 (6)	2.99 (7)	1.82 (2)	7.82
7	2.86 (7)	3.54 (1)	3.78 (3)	10.18

$k = 3,\ t = 7,\ b = 7,\ r = 3,\ \lambda = 1$

Grand total $Y_{...} = 58.30$

Table 14.5 Computations for the Balanced Incomplete Block

Treatment number	Total	Total of the blocks containing treatment i	q_i	Treatment estimator
i	$Y_{i..}$	$\sum_j n_{ij} Y_{.j.}$	$Y_{i..} - \frac{1}{k} \sum_j n_{ij} Y_{.j.}$	$\hat{\tau}_i - \hat{\bar{\tau}}_. = \frac{k}{\lambda t} q_i$
1	7.49	25.68	−1.0700	−.459
2	5.63	22.74	−1.9500	−.836
3	9.70	26.47	.8767	.376
4	9.18	25.24	.7667	.329
5	8.57	23.57	.7133	.306
6	9.13	24.69	.9000	.386
7	8.60	26.51	−.2367	−.101
Check	58.30	174.90	.0000	.001

errors. These checks should be performed at this stage. The total sum of squares is equal to

$$\Sigma y_{ijm}^2 = 169.1162$$

The sum of squares due to mean is

$$\frac{Y_{\cdots}^2}{bk} = 161.8519$$

The sum of squares due to blocks (unadjusted) is

$$\sum_j \frac{Y_{.j.}^2}{3} - \frac{Y_{\cdots}^2}{21} = 2.0347$$

The sum of squares due to treatments (adjusted) is

$$\frac{k}{\lambda t} \Sigma q_i^2 = \frac{3}{(1)(7)} \Sigma q_i^2 = 3.2908$$

The AOV is given in Table 14.6.

TABLE 14.6 ANALYSIS OF VARIANCE FOR EXAMPLE OF ART. 14.4.1

SV	DF	SS	MS	F	Noncentrality
Total	21	169.1162			
Mean	1	161.8519			
Block (unadj)	6	2.0347			
Treatments (adj)	6	3.2908	.5485	2.263	$\frac{7}{6\sigma^2} \Sigma(\tau_i - \bar{\tau}_.)^2$
Error	8	1.9388	.2424		

The tabulated F at the 5 per cent level for 6 and 8 degrees of freedom is 3.58. Hence these data do not suggest rejecting the hypothesis $\tau_1 = \tau_2 = \cdots = \tau_t$ at the 5 per cent level of significance. The estimated variance of a treatment contrast $\Sigma c_i \tau_i$ is equal to

$$\frac{k}{\lambda t} \Sigma c_i^2 \hat{\sigma}^2 = E_{\text{ms}} \frac{k}{\lambda t} \Sigma c_i^2 = \left(\frac{3}{7}\right)(.2424)\Sigma c_i^2 = .1039 \Sigma c_i^2$$

Problems

14.1 Which of the designs shown in Table 14.7 are balanced incomplete block designs? The entries are treatment numbers.

14.2 For the balanced incomplete blocks in Prob. 14.1, find the values of r, b, t, k, and λ.

TABLE 14.7

Block no.	Treatment nos.	
1	1	2
2	1	3
3	2	3

(*a*)

Block no.	Treatment nos.	
1	1	2
2	1	3
3	1	4
4	2	3
5	2	4
6	3	4

(*b*)

Block no.	Treatment nos.		
1	1	2	3
2	1	2	4
3	2	3	4
4	1	3	4

(*c*)

Block no.	Treatment nos.			
1	1	2	3	4
2	1	2	3	5
3	2	3	4	5
4	1	3	4	5

(*d*)

14.3 The data of Table 14.8 are in a balanced incomplete block.
(*a*) Find the values of r, b, t, k, and λ.
(*b*) Construct the AOV table.
(*c*) Estimate $\tau_i - \bar{\tau}.$ for each i.
(*d*) Find the standard error of the estimates in (*c*).

TABLE 14.8

Block	Treatments	
1	(1) 16.1	(2) 18.2
2	(1) 13.4	(3) 16.0
3	(2) 18.5	(3) 20.1

The numbers in parentheses are treatment numbers.
14.4 In Prob. 14.3, write out the normal equations.
14.5 In Prob. 14.4, solve the normal equations, using the condition $\hat{\tau}_3 = 0$.

14.6 In Prob. 14.4, solve the normal equations, using the condition $\Sigma\hat{\tau}_i = 0$.

14.7 Work out in detail the noncentrality for testing $\tau_1 + \tau_2 - 2\tau_3 = 0$ in Prob. 14.3.

14.8 In Prob. 14.3, set a 95 per cent confidence interval on $\tau_1 - \tau_2$.

14.9 From the equations $\mathbf{A}\hat{\boldsymbol{\tau}} = \mathbf{q}$ for the balanced incomplete block model show that there exists a constant c such that $c\mathbf{A}$ is idempotent.

Further Reading

1 W. G. Cochran and G. M. Cox: "Experimental Designs," John Wiley & Sons, Inc., New York, 1957.

2 O. Kempthorne: "Design and Analysis of Experiments," John Wiley & Sons, Inc., New York, 1952.

3 R. L. Anderson and T. A. Bancroft: "Statistical Theory in Research," McGraw-Hill Book Company, Inc., New York, 1952.

4 H. B. Mann: "Analysis and Design of Experiments," Dover Publications, New York, 1949.

5 F. Yates: Incomplete Randomized Blocks, *Ann. Eugenics*, vol. 7, 1936.

6 R. C. Bose: On the Construction of Balanced Incomplete Block Designs, *Ann. Eugenics*, vol. 9, 1939.

7 C. R. Rao: General Methods of Analysis for Incomplete Block Designs, *J. Am. Statist. Assoc.*, vol. 42, 1947.

15

Some Additional Topics about Model 4

15.1 Introduction

In this chapter the assumptions underlying model 4 will be investigated further. The work that has been done in this area is vast, and no attempt will be made to be complete. Some of the consequences that follow when the assumptions do not hold will be investigated.

Only linear models will be considered. These have been defined as observable random variables y that are equal to linear functions of parameters and unobservable random variables. For example, the model

$$y_{ij} = \mu_{ij}e_{ij}$$

is not a linear model according to Definition 5.2. On the other hand, taking the logarithm of y_{ij} gives

$$\log y_{ij} = \log \mu_{ij} + \log e_{ij}$$

which *is* a linear model in the logarithms. There are some models that cannot be made linear by transformation, for instance,

$$y_{ij} = \mu_{ij}^* + \mu_{ij}^{e_{ij}}$$

These are important, but they will not be discussed here. Only models will be considered that either are linear or can be made linear by transformation.

15.2 Assumptions for Model 4

Most of the ideas in this section will be discussed for a two-way classification model. However, almost all the results will generalize to more complex situations.

Let the observable random variable y_{ij} have the structure

$$y_{ij} = \mu_{ij} + e_{ij} \tag{15.1}$$

where μ_{ij} is an unknown parameter and e_{ij} is an unobservable random variable with mean equal to zero. Stating that $E(e_{ij}) = 0$ is no restriction, since, if $E(e_{ij}) = \mu \neq 0$, then μ can become part of μ_{ij}. We should note that this is the only assumption that is necessary for μ_{ij} to be estimable for every i and j; that is, $E(y_{ij}) = \mu_{ij}$. So μ_{ij} and every linear function of μ_{ij} is estimable if $E(e_{ij}) = 0$. However, an experimenter generally wants to do more than obtain an unbiased estimate of the unknown parameters; he may want to be able to say something about the variance of the estimate and test hypotheses about the parameters. Therefore, it would seem desirable to put conditions on model (15.1) such that

1. Unbiased estimates ($\hat{\mu}_{ij}$) of μ_{ij} and certain functions of μ_{ij} will be obtainable.
2. Estimates of the variance of $\hat{\mu}_{ij}$ will be obtainable.
3. Tests of certain hypotheses about μ_{ij} will be obtainable.
4. Confidence intervals on μ_{ij} will be obtainable.

The conditions often used are as follows:

A. $E(e_{ij}) = 0$.

B. μ_{ij} can be written $\mu_{ij} = \mu + \tau_i + \beta_j$; that is, there is no interaction.

C. The e_{ij} are uncorrelated.

D. The e_{ij} have the same variance σ^2 for all i and j.

E. The e_{ij} are normally distributed.

As stated before, condition A, that $E(e_{ij}) = 0$, is the only assumption that is necessary to obtain unbiased estimates of μ_{ij} and linear combinations of μ_{ij} for all i and j. Condition B is a condition on the parameters. If we write

$$\begin{aligned} \mu_{ij} &= \bar{\mu}_{..} + (\bar{\mu}_{i.} - \bar{\mu}_{..}) + (\bar{\mu}_{.j} - \bar{\mu}_{..}) + (\mu_{ij} - \bar{\mu}_{i.} - \bar{\mu}_{.j} + \bar{\mu}_{..}) \\ &= \mu^* + \tau_i^* + \beta_j^* + (\tau\beta)_{ij}^* \end{aligned}$$

then $E(y_{ij} - y_{i'j}) = \tau_i^* - \tau_{i'}^*$ for all j and $i \neq i'$ if and only if $(\tau\beta)_{ij}^* = 0$. Although it may be desirable from some points of view for $(\tau\beta)_{ij}^*$ to be zero, it should be noticed that $\tau_i^* - \tau_k^*$ is estimable whether $(\tau\beta)_{ij}^*$ is zero or not; the estimate is $y_{i.} - y_{k.}$. Condition B is also necessary for the analysis of variance, to give an unbiased estimate of the variance of e_{ij} and to give a test of the hypothesis that the τ_i are equal. This has been shown in Art. 12.5.2. Suppose that $(\tau\beta)_{ij}^* \neq 0$ for some i and j and that there is more than one observation per cell. Then the model can be written as

$$y_{ijm} = \mu_{ij} + e_{ijm}$$

If the variance of e_{ijm} equals σ^2 for all i, j, and m, σ^2 can be estimated by the usual analysis of variance; the variance of $\hat{\mu}_{ij}$ can also be estimated. This problem has been discussed in Chap. 12.

Next we shall turn our attention to the conditions made on the random variables e_{ij}. The normality condition E is not essential for point estimation, but is used for interval estimation and tests of hypotheses. However, for the conventional estimates of μ_{ij} to have the optimum properties stated in Chap. 6, it is essential that conditions C and D be met. We have shown, however, that, if C and D do not hold but the variances and covariances are *known*, then estimates of μ_{ij} can be found such that the optimum properties in Chap. 6 hold. This is stated in Theorem 6.4.

15.3 Tests of Hypotheses

Next we shall turn our attention to the testing of hypotheses. The normality condition E is necessary if the conventional tests are to be strictly valid. If the normality condition does not hold, then some nonparametric test can be used. If the normality condition E is satisfied, but if either C or D is not, then a nonparametric test could be used. However, there are other methods available for this case. These will be discussed next.

For the remainder of this section we shall assume the model

$$y_{ij} = \mu + \tau_i + \beta_j + e_{ij} \qquad i = 1, 2, \ldots, t; \quad j = 1, 2, \ldots, b \tag{15.2}$$

where $\Sigma\tau_i = \Sigma\beta_j = 0$, and we shall assume that the random variables e_{ij} are normally distributed with mean zero. We shall call the τ_i treatments and the β_j blocks. We shall further assume that the errors satisfy the following:

$$\begin{aligned} E(e_{ij}^2) &= \sigma_{ii} \\ E(e_{ij}e_{i'j}) &= \sigma_{ii'} \\ E(e_{ij}e_{i'j'}) &= 0 \qquad \text{if } j \neq j' \end{aligned} \tag{15.3}$$

We shall also assume that the $t \times t$ matrix (σ_{rs}) is positive definite. These assumptions on e_{ij} state that the observations are uncorrelated if they are in different blocks, that the variance of the ith treatment observation is σ_{ii}, and that the covariance of the ith treatment observation and i'th treatment observation in the same block is $\sigma_{ii'}$. In some instances these assumptions seem to be more realistic than those generally made, i.e., that the e_{ij} are distributed independently $N(0,\sigma^2)$.

We desire a test of the hypothesis $\tau_1 = \tau_2 = \cdots = \tau_t$. Let the $(t - 1) \times 1$ vector $\mathbf{Y}_j$ be

$$\mathbf{Y}_j = \begin{pmatrix} y_{2j} - y_{1j} \\ y_{3j} - y_{1j} \\ \cdots \\ y_{tj} - y_{1j} \end{pmatrix} \qquad j = 1, 2, \ldots, b$$

We get
$$E(\mathbf{Y}_j) = \begin{pmatrix} \tau_2 - \tau_1 \\ \tau_3 - \tau_1 \\ \cdots \\ \tau_t - \tau_1 \end{pmatrix} = \boldsymbol{\gamma}$$

Let the covariance matrix of $\mathbf{Y}_j$ be $\mathbf{V}$. The pqth element of $\mathbf{V}$ is

$$v_{pq} = E\{[(y_{pj} - y_{1j}) - E(y_{pj} - y_{1j})][(y_{qj} - y_{1j}) - E(y_{qj} - y_{1j})]\}$$

Using Eq. (15.2) and the distributional properties in (15.3), we get

$$v_{pq} = E[(e_{pj} - e_{1j})(e_{qj} - e_{1j})] = \sigma_{pq} - \sigma_{p1} - \sigma_{1q} + \sigma_{11} \qquad p, q = 2, 3, \ldots, t \tag{15.4}$$

It can be shown that $\mathbf{V}$ is positive definite, since we have assumed that (σ_{pq}) is positive definite. Also, by (15.3), $\mathbf{Y}_j$ and $\mathbf{Y}_{j'}$ are independent if $j \neq j'$. Therefore, we have $\mathbf{Y}_j$ as a random sample from the density $N(\boldsymbol{\gamma}, \mathbf{V})$ for $j = 1, 2, \ldots, b$ and $\tau_1 = \tau_2 = \cdots = \tau_t$ if and only if $\boldsymbol{\gamma} = 0$. Therefore, we can let

$$\frac{(b - t + 1)b}{(b - 1)(t - 1)} \bar{\mathbf{Y}}'\mathbf{S}^{-1}\bar{\mathbf{Y}} = u$$

where
$$\bar{\mathbf{Y}} = \frac{1}{b} \Sigma \mathbf{Y}_i \qquad \mathbf{S} = \frac{1}{b - 1} \sum_{i=1}^{b} (\mathbf{Y}_i - \bar{\mathbf{Y}})(\mathbf{Y}_i - \bar{\mathbf{Y}})'$$

and we can use Theorem 10.9 to state: u is distributed as $F(t - 1, b - t + 1)$ if $\tau_1 = \tau_2 = \cdots = \tau_t$. Thus we have the following theorem.

◆ **Theorem 15.1** Let the two-way classification model $y_{ij} = \mu + \tau_i + \beta_j + e_{ij}$ be such that $E(e_{ij}^2) = \sigma_{ii}$, $E(e_{ij}e_{i'j}) = \sigma_{ii'}$, $E(e_{ij}e_{i'j'}) = 0$ if $j \neq j'$, $b \geqslant t$, and e_{ij} is a normal variable. Then the quantity

$$u = \bar{\mathbf{Y}}'\mathbf{S}^{-1}\bar{\mathbf{Y}} \frac{(b - t + 1)b}{(b - 1)(t - 1)}$$

is distributed as $F(t - 1, b - t + 1)$ if $\tau_1 = \tau_2 = \cdots = \tau_t$.

To test the hypothesis $\tau_1 = \tau_2 = \cdots = \tau_t$, the following theorem can be used.

◆ **Theorem 15.2** Let the conditions of Theorem 15.1 be satisfied. Then the hypothesis $\tau_1 = \tau_2 = \cdots = \tau_t$ is rejected at the α level of significance if $u > F_\alpha(t-1, b-t+1)$.

Instead of testing the hypothesis $\tau_1 = \tau_2 = \cdots = \tau_t$, it may be desired to test the hypothesis that a contrast of the treatments is equal to zero; that is, H_0: $\Sigma c_i \tau_i = 0$, where $\Sigma c_i = 0$. If the distributional properties of (15.3) hold, then the conventional method of testing H_0 is not exact. An alternative method is the subject of the next three theorems.

◆ **Theorem 15.3** Let the distributional properties in (15.3) hold for the two-way classification model $y_{ij} = \mu + \tau_i + \beta_j + e_{ij}$ for

$$i = 1, 2, \ldots, t \qquad j = 1, 2, \ldots, b$$

Then the quantities

$$z_j = \sum_{i=1}^{t} c_i y_{ij} \qquad \Sigma c_i = 0$$

are independent and distributed $N(\Sigma c_i \tau_i, \sigma^2)$, where

$$\sigma^2 = \sum_i c_i^2 \sigma_{ii} + \sum_{i \neq i'} c_i c_{i'} \sigma_{ii'}$$

The proof will not be given.

◆ **Theorem 15.4** Using the assumptions of Theorem 15.3,

$$\frac{u^2}{\sigma^2} = \frac{b\bar{z}^2}{\sigma^2}$$

is distributed as $\chi'^2(1,\lambda)$, where

$$\lambda = \frac{b(\Sigma c_i \tau_i)^2}{2\sigma^2}$$

Also,

$$\frac{w^2}{\sigma^2} = \frac{\Sigma(z_j - \bar{z})^2}{\sigma^2}$$

is distributed as $\chi^2(b-1)$, and u and w are independent.

◆ **Theorem 15.5** Let u^2 and w^2 be as given in Theorem 15.4. Then

$$v = (b-1)u^2/w^2 = b(b-1)\bar{z}^2 / \sum_j (z_j - \bar{z})^2$$

is distributed as $F'(1, b-1, \lambda)$, where

$$\lambda = \frac{b(\Sigma c_i \tau_i)^2}{2\sigma^2}$$

Theorem 15.5 can be used to test H_0: $\Sigma c_i \tau_i = 0$.

15.3.1 Example. Suppose that the data of Table 15.1 are from the model $y_{ij} = \mu + \tau_i + \beta_j + e_{ij}$, where the assumptions of (15.3) hold. Suppose we wish to test the hypothesis $\tau_1 = \tau_2 = \tau_3 = \tau_4$.

TABLE 15.1 DATA FOR EXAMPLE OF ART. 15.3.1

β \ τ	1	2	3	4
1	19.41	43.60	24.05	19.47
2	23.84	40.40	21.76	16.61
3	16.08	18.08	14.19	16.69
4	18.29	19.57	18.61	17.78
5	30.08	45.20	29.33	20.19
6	27.04	25.87	25.60	23.31
7	39.95	55.20	38.77	21.15
8	25.12	55.32	34.19	18.56
9	22.45	19.79	21.65	23.31
10	29.28	46.24	31.52	22.48
11	22.56	14.88	15.68	19.79
12	22.08	7.52	4.69	20.53
13	43.95	41.17	32.59	29.25

Since we assume that (15.3) holds on the e_{ij}, we cannot use the conventional method of analysis. Instead we can use Theorem 15.2. By Theorem 15.2 we need

$$\mathbf{Y}_j' = (y_{2j} - y_{1j}, y_{3j} - y_{1j}, y_{4j} - y_{1j}) \qquad j = 1, 2, \ldots, 13$$

$$\mathbf{Y}_1' = (24.19, 4.64, .06)$$
$$\mathbf{Y}_2' = (16.56, -2.08, -7.23)$$
$$\mathbf{Y}_3' = (2.00, -1.89, .61)$$
$$\mathbf{Y}_4' = (1.28, .32, -.51)$$
$$\mathbf{Y}_5' = (15.12, -.75, -9.89)$$
$$\mathbf{Y}_6' = (-1.17, -1.44, -3.73)$$
$$\mathbf{Y}_7' = (15.25, -1.18, -18.80)$$
$$\mathbf{Y}_8' = (30.20, 9.07, -6.56)$$
$$\mathbf{Y}_9' = (-2.66, -.80, .86)$$
$$\mathbf{Y}_{10}' = (16.96, 2.24, -6.80)$$
$$\mathbf{Y}_{11}' = (-7.68, -6.88, -2.77)$$
$$\mathbf{Y}_{12}' = (-14.56, -17.39, -1.55)$$
$$\mathbf{Y}_{13}' = (-2.78, -11.36, -14.70)$$

From this we obtain

$$\bar{\mathbf{Y}}' = (7.13, -2.12, -5.46)$$

and $12\mathbf{S} = \sum_{j=1}^{13} (\mathbf{Y}_j - \bar{\mathbf{Y}})(\mathbf{Y}_j - \bar{\mathbf{Y}})' = \begin{pmatrix} 2{,}151.8706 & 894.5435 & -274.1924 \\ 894.5435 & 542.1045 & 36.3449 \\ -274.1924 & 36.3449 & 446.3738 \end{pmatrix}$

The inverse is

$$[\Sigma(\mathbf{Y}_j - \bar{\mathbf{Y}})(\mathbf{Y}_j - \bar{\mathbf{Y}})']^{-1} = \begin{pmatrix} .0023572760 & -.0040087773 & .0017744001 \\ -.0040087773 & .0086721040 & -.0031685618 \\ .0017744001 & -.0031685618 & .0035882215 \end{pmatrix}$$

This gives $u = 7.59$, which we compare with Snedecor's F value with 3 and 10 degrees of freedom. This is significant at the 1 per cent level; so we have evidence to reject the hypothesis $\tau_1 = \tau_2 = \tau_3 = \tau_4$. Notice that the Doolittle procedure and Theorem 7.1 can be used to obtain $\bar{\mathbf{Y}}'\mathbf{S}^{-1}\bar{\mathbf{Y}}$.

15.4 Test for Additivity

In a two-way classification model $y_{ij} = \mu + \tau_i + \beta_j + (\tau\beta)_{ij} + e_{ij}$, it may be desirable to have a method for testing H_0: $(\tau\beta)_{ij} = 0$ for all i and j. If these quantities are zero, then the conventional method given in Sec. 12.3 can be used to test $\tau_1 = \tau_2 = \cdots = \tau_t$. Also, the error mean square will be an unbiased estimate of σ^2 in this case.

In this section a method for testing the hypothesis $(\tau\beta)_{ij} = 0$ will be exhibited. This test is due to Tukey [14]. In brief, the method consists in partitioning the error sum of squares $\sum_{ij}(y_{ij} - y_{i.} - y_{.j} + y_{..})^2$ into two parts:

1. Sum of squares due to nonadditivity:

$$N_{ss} = \frac{\left[\sum_{ij} y_{ij}(y_{i.} - y_{..})(y_{.j} - y_{..})\right]^2}{\sum_i (y_{i.} - y_{..})^2 \sum_j (y_{.j} - y_{..})^2}$$

2. Sum of squares due to balance (remainder):

$$R_{ss} = \sum_{ij} (y_{ij} - y_{i.} - y_{.j} + y_{..})^2 - N_{ss}$$

If $(\tau\beta)_{ij} = 0$ for all i and j, then N_{ss}/σ^2 is distributed as $\chi^2(1)$, R_{ss}/σ^2 is distributed as $\chi^2[(b-1)(t-1)-1]$, and the two are independent.

Hence, if $(\tau\beta)_{ij} = 0$ for all i and j, the quantity $(N_{ss}/R_{ss})[(b-1)(t-1)-1]$ is distributed as $F[1, (b-1)(t-1)-1]$. The proof of these distributional properties will be the subject of the next few theorems.

◆ **Theorem 15.6** Let $y_{ij} = \mu + \tau_i + \beta_j + (\tau\beta)_{ij} + e_{ij}$, where $\Sigma\tau_i = \Sigma\beta_j = \sum_i(\tau\beta)_{ij} = \sum_j(\tau\beta)_{ij} = 0$. Let e_{ij} be independently distributed $N(0,\sigma^2)$. Let the quantities x, u_i, v_j, w_{pq} denote

$$x = y_{..} \qquad \ldots\ldots\ldots\ldots$$

$$u_i = y_{i.} - y_{..} \qquad i = 1, 2, \ldots, t$$

$$v_j = y_{.j} - y_{..} \qquad j = 1, 2, \ldots, b$$

$$w_{pq} = y_{pq} - y_{p.} - y_{.q} + y_{..} \qquad \begin{cases} p = 1, 2, \ldots, t \\ q = 1, 2, \ldots, b \end{cases}$$

Then

(1) x is independent of u_i, v_j, w_{pq} for all i, j, p, and q.
(2) u_i and v_j are independent for all i and j.
(3) u_i and w_{pq} are independent for all i, p, and q.
(4) v_j and w_{pq} are independent for all j, p, and q.

Proof: Since each of the quantities x, u_i, v_j, w_{pq} has a normal distribution, we need only show that they are uncorrelated, since this implies independence. We shall prove (3), and the rest will follow by similar reasoning. We must show that $\text{cov}(u_i,w_{pq}) = 0$ for all i, p, and q. But

$$\begin{aligned}
\text{cov}(u_i,w_{pq}) &= E\{[u_i - E(u_i)][w_{pq} - E(w_{pq})]\} \\
&= E\{[(y_{i.} - y_{..}) - E(y_{i.} - y_{..})][(y_{pq} - y_{p.} - y_{.q} + y_{..}) \\
&\qquad - E(y_{pq} - y_{p.} - y_{.q} + y_{..})]\} \\
&= E[(\bar{e}_{i.} - \bar{e}_{..})(e_{pq} - \bar{e}_{p.} - \bar{e}_{.q} + \bar{e}_{..})] \\
&= E[\bar{e}_{i.}(e_{pq} - \bar{e}_{p.} - \bar{e}_{.q} + \bar{e}_{..})] - E[\bar{e}_{..}(e_{pq} - \bar{e}_{p.} - \bar{e}_{.q} + \bar{e}_{..})]
\end{aligned}$$

Using the fact that the e_{ij} are independent, we get

$$\text{cov}(u_i,w_{pq}) = \sigma^2\left(\frac{\delta_{ip}}{b} - \frac{\delta_{ip}}{b} - \frac{1}{tb} + \frac{1}{tb}\right) - \sigma^2\left(\frac{1}{tb} - \frac{1}{tb} - \frac{1}{tb} + \frac{1}{tb}\right) = 0$$

where $\delta_{ip} = 1$ if $i = p$, $\delta_{ip} = 0$ if $i \neq p$.

A similar proof holds for (1), (2), and (3). Notice that the theorem does not say that u_i and u_j are independent. This is, of course, impossible, since $\Sigma u_i = 0$. Similarly for v_j and w_{pq}.

Next we shall state a theorem about the distribution of these quantities.

◆ **Theorem 15.7** Let the notation be the same as in the preceding theorem. Then
(1) x is distributed $N(\mu, \sigma^2/bt)$.
(2) u_i is distributed $N[\tau_i, \sigma^2(t-1)/bt]$.
(3) v_j is distributed $N[\beta_j, \sigma^2(b-1)/bt]$.
(4) w_{pq} is distributed $N[(\tau\beta)_{pq}, \sigma^2(b-1)(t-1)/bt]$.

The proof of this theorem will be left for the reader.

◆ **Theorem 15.8** Let the distributional properties and notation be the same as in Theorem 15.6. Then
(1) $\text{cov}(u_i,u_j) = -\sigma^2/bt$ when $i \neq j$.
(2) $\text{cov}(v_i,v_j) = -\sigma^2/bt$ when $i \neq j$.
(3) $\text{cov}(w_{pq},w_{ij}) = \sigma^2(\delta_{pi}\delta_{qj} + 1/bt - \delta_{pi}/b - \delta_{qj}/t)$, where $\delta_{rs} = 0$ if $r \neq s$; $\delta_{rs} = 1$ if $r = s$.

The proof will be left for the reader.

◆ **Theorem 15.9** Let the distributional properties and notation be the same as in Theorem 15.6, except that $(\tau\beta)_{ij} = 0$ for all i and j. Then the quantity

$$z = \frac{\sum_{ij} w_{ij}u_iv_j}{\sqrt{\Sigma u_i^2 \Sigma v_j^2}}$$

is distributed $N(0,\sigma^2)$.

Proof: First we shall show that z is distributed normally. We shall look at the conditional distribution of z given $u_i = a_i$ and $v_j = c_j$ for $i = 1, 2, \ldots, t; j = 1, 2, \ldots, b$. But, if u_i and v_j are fixed, z is just a linear combination of w_{ij}, that is, a linear combination of normal variables, and, hence, is normal. That is to say, the conditional distribution of z, given u_i and v_j, is normal. Now we want the mean and variance of this conditional distribution. For the mean we have

$$E(z \mid u_i = a_i, v_j = c_j) = E\left(\frac{\sum_{ij} w_{ij}a_ic_j}{\sqrt{\sum_{ij} a_i^2c_j^2}}\right) = \frac{\sum_{ij} E(w_{ij})a_ic_j}{\sqrt{\Sigma a_i^2c_j^2}}$$

But this is zero, since, by (4) of Theorem 15.7, $E(w_{ij}) = 0$ if $(\tau\beta)_{ij} = 0$. The conditional variance of z is

$$\begin{aligned}
\operatorname{var}(z \mid u_i = a_i, v_j = c_j) &= \operatorname{var}\left(\frac{\sum_{ij} w_{ij}a_ic_j}{\sqrt{\Sigma a_i^2c_j^2}}\right) \\
&= E\left(\frac{\sum_{ij} w_{ij}a_ic_j}{\sqrt{\Sigma a_i^2c_j^2}}\right)^2 \\
&= \frac{1}{\Sigma a_i^2c_j^2}\left[\sum_{ij} E(w_{ij}a_ic_j)^2\right. \\
&\quad + \sum_i \sum_{\substack{jj' \\ j\neq j'}} E(w_{ij}w_{ij'}a_i^2c_jc_{j'}) \\
&\quad + \sum_{\substack{ii' \\ i\neq i'}} \sum_j E(w_{ij}w_{i'j}a_ia_{i'}c_j^2) \\
&\quad \left. + \sum_{\substack{ii' \\ i\neq i'}} \sum_{\substack{jj' \\ j\neq j'}} E(w_{ij}w_{i'j'}a_ia_{i'}c_jc_{j'})\right]
\end{aligned}$$

If we use (3) of Theorem 15.8, this expression becomes

$$\begin{aligned}
\frac{\sigma^2}{\sum_{ij} a_i^2c_j^2}&\left[\sum_{ij} \frac{(b-1)(t-1)}{bt}\, a_i^2c_j^2 + \sum_i \sum_{\substack{jj' \\ j\neq j'}} \frac{(1-t)}{bt}\, a_i^2c_jc_{j'}\right. \\
&\left. + \sum_{\substack{ii' \\ i\neq i'}} \sum_j \frac{1-b}{bt}\, a_ia_{i'}c_j^2 + \sum_{\substack{ii' \\ i\neq i'}} \sum_{\substack{jj' \\ j\neq j'}} \frac{1}{bt}\, a_ia_{i'}c_jc_{j'}\right]
\end{aligned}$$

We can use the fact that $\Sigma a_i = \Sigma c_j = 0$ to obtain

$$\sum_{\substack{i \\ i\neq i'}} a_i = -a_{i'} \qquad \text{and} \qquad \sum_{\substack{j \\ j\neq j'}} c_j = -c_{j'}$$

The variance then becomes

$$\begin{aligned}
\frac{\sigma^2}{\sum_{ij} a_i^2c_j^2}&\left[\frac{(b-1)(t-1)}{bt} \sum_{ij} a_i^2c_j^2 + \frac{t-1}{bt} \sum_{ij} a_i^2c_j^2\right. \\
&\left. + \frac{b-1}{bt} \sum_{ij} a_i^2c_j^2 + \frac{1}{bt} \sum_{ij} a_i^2c_j^2\right] = \sigma^2
\end{aligned}$$

We have proved that the conditional distribution of z, given that $u_i = a_i$ and $v_j = c_j$, is normal with mean 0 and variance σ^2. But, since this conditional distribution is $N(0,\sigma^2)$, it is the same for every value of u_i and v_j; hence, the distribution of z is independent of u_i

and v_j, and the conditional distribution of z given u_i and v_j is equal to the marginal distribution of z, which is $N(0,\sigma^2)$. This completes the proof of the theorem.

Next we want to show that the two quantities $h = z^2$ and $g = \sum_{ij} w_{ij}^2 - z^2$ are independent. To do this, let us consider

$$\begin{aligned} g &= \sum_{ij} w_{ij}^2 - z^2 \\ &= \sum_{ij}\left[w_{ij} - \frac{u_i v_j}{\sqrt{\sum_{mn} u_m^2 v_n^2}}\left(\sum_{pq} w_{pq} \frac{u_p v_q}{\sqrt{\sum_{rs} u_r^2 v_s^2}}\right)\right]^2 \\ &= \sum_{ij}\left(w_{ij} - \frac{u_i v_j}{\sqrt{\sum_{mn} u_m^2 v_n^2}} z\right)^2 \\ &= \sum_{ij} g_{ij}^2 \text{ (say)} \end{aligned}$$

Therefore, if we can show that z is independent of g_{ij} for every i and j, it follows that z is independent of every function of g_{ij}, and, specifically, of Σg_{ij}^2. Now the conditional distribution of g_{ij} given $u_i = a_i$ and $v_j = c_j$ is normal, since, for fixed u_i and v_j, the quantity g_{ij} is a linear combination of w_{ij}, which is a normal variable. The mean (in the conditional distribution) of g_{ij} is zero; that is, $E(w_{ij}) = 0$ for all i and j if $(\tau\beta)_{ij} = 0$. The variance (in the conditional distribution) of g_{ij} is

$$\begin{aligned} E(g_{ij}^2 \mid u_i = a_i, v_j = c_j) &= E\left(w_{ij} - \frac{a_i c_j z}{\sqrt{\Sigma a_p^2 \Sigma c_q^2}}\right)^2 \\ &= E(w_{ij}^2) - 2E\left(\frac{w_{ij} a_i c_j z}{\sqrt{\Sigma a_p^2 \Sigma c_q^2}}\right) + E\left(\frac{a_i^2 c_j^2 z^2}{\Sigma a_p^2 \Sigma c_q^2}\right) \end{aligned}$$

If we use Theorem 15.8, this becomes

$$\sigma^2\left[\frac{(b-1)(t-1)}{bt} - \frac{a_i^2 c_j^2}{\Sigma a_i^2 \Sigma c_j^2}\right] = \operatorname{var}(g_{ij} \mid u_i = a_i, v_j = c_j)$$

The covariance of g_{ij}, g_{pq} in the conditional distribution given $u_i = a_i$, $v_j = c_j$ is

$$\operatorname{cov}(g_{ij}, g_{pq} \mid u_i = a_i, v_j = c_j) = \sigma^2\left(\delta_{pi}\delta_{qj} + \frac{1}{bt} - \frac{\delta_{pi}}{b} - \frac{\delta_{qj}}{t} - \frac{a_i a_p c_j c_q}{\Sigma a_i^2 \Sigma c_j^2}\right)$$

From this it follows that the conditional distribution of g/σ^2 given $u_i = a_i$ and $v_j = c_j$ is $\chi^2[(b-1)(t-1) - 1]$. Notice that the covariance matrix of the g_{ij} is idempotent when divided by σ^2. Hence we use Theorem 4.9. But, since this is chi-square for every value of a_i and c_j,

g is independent of u_i and v_j, and the *marginal* distribution of g/σ^2 is $\chi^2[(b-1)(t-1)-1]$.

Next we shall examine $E(zg_{ij} \mid u_i = a_i, v_j = c_j)$. By using Theorem 15.8 we see that this quantity is zero. This tells us that, in the conditional distribution of z and g_{ij} given $u_i = a_i$ and $v_j = c_j$, z and g_{ij} are independent (since they are uncorrelated normal variables). But, by Theorem 15.9, z is independent of u_i and v_j; hence, z is also independent of g_{ij} in the marginal distribution of z, g_{ij}, u_i, v_j. We have proved the following theorem.

◆ **Theorem 15.10** Let the distributional properties and notation be as given in Theorems 15.6 and 15.9. Then the quantity z is independent of the quantities

$$g_{ij} = w_{ij} - \frac{u_i v_j}{\sqrt{\Sigma u_p^2 \Sigma v_q^2}} z$$

Also $g/\sigma^2 = \Sigma g_{ij}^2/\sigma^2$ is distributed as $\chi^2[(b-1)(t-1)-1]$.

Notice that

$$\Sigma\, w_{ij}^2 = \Sigma\, g_{ij}^2 + z^2$$

and that z^2/σ^2 is distributed as $\chi^2(1)$. Therefore, we have the following.

◆ **Theorem 15.11** Let the distributional properties and notation be as given in Theorems 15.6 and 15.9. Then

(*a*)
$$\frac{\sum_{ij} (y_{i.} - y_{..})^2}{\sigma^2}$$

is distributed as $\chi'^2(t-1, \lambda_1)$;

(*b*)
$$\frac{\sum_{ij} (y_{.j} - y_{..})^2}{\sigma^2}$$

is distributed as $\chi'^2(b-1, \lambda_2)$;

(*c*)
$$\frac{z^2}{\sigma^2}$$

is distributed as $\chi^2(1)$ if $(\tau\beta)_{ij} = 0$ for all i and j; and

(*d*)
$$\frac{\sum_{ij} (y_{ij} - y_{i.} - y_{.j} + y_{..})^2}{\sigma^2} - \frac{z^2}{\sigma^2}$$

is distributed as $\chi^2[(b-1)(t-1)-1]$ if $(\tau\beta)_{ij} = 0$ for all i and j.

(*e*) (*a*), (*b*), (*c*), (*d*) are independent

Now we state the final theorem of this section.

◆ **Theorem 15.12** Let the notation and distributional properties be as given in Theorem 15.6. Then, to test $H_0: (\tau\beta)_{ij} = 0$ for all i and j, the quantity

$$u = \frac{z^2[(b-1)(t-1)-1]}{\sum_{ij} (y_{ij} - y_{i.} - y_{.j} + y_{..})^2 - z^2}$$

is distributed as $F[1, (b-1)(t-1)-1]$. If $u > F_\alpha$, the hypothesis $(\tau\beta)_{ij} = 0$ can be rejected at the α level of significance.

Proof: The proof is immediate from Theorem 15.11. Notice that this can be put into an AOV table, as shown in Table 15.2.

TABLE 15.2 ANALYSIS OF VARIANCE FOR NONADDITIVITY

SV	DF	SS	F
Total	bt	$\sum_{ij} y_{ij}^2$	
Mean	1	$M_{ss} = bty_{..}^2$	
Treatments	$t-1$	$T_{ss} = \sum_{ij}(y_{i.} - y_{..})^2$	
Blocks	$b-1$	$B_{ss} = \sum_{ji}(y_{.j} - y_{..})^2$	
Nonadditivity	1	$N_{ss} = \dfrac{\left[\sum_{ij} y_{ij}(y_{i.} - y_{..})(y_{.j} - y_{..})\right]^2}{\sum_i (y_{i.} - y_{..})^2 \sum_j (y_{.j} - y_{..})^2}$	$\dfrac{N_{ms}}{R_{ms}}$
Balance	$(b-1)(t-1)-1$	R_{ss} = subtraction	

Notice also that $N_{ss} = z^2$ can be simplified for computing as follows:

$$N_{ss} = z^2 = \frac{\left[\sum_{ij} (y_{ij} - y_{i.} - y_{.j} + y_{..})(y_{i.} - y_{..})(y_{.j} - y_{..})\right]^2}{\sum_i (y_{i.} - y_{..})^2 \sum_j (y_{.j} - y_{..})^2}$$

$$= \frac{\left[\sum_{ij} y_{ij}(y_{i.} - y_{..})(y_{.j} - y_{..})\right]^2}{\sum_i (y_{i.} - y_{..})^2 \sum_j (y_{.j} - y_{..})^2}$$

$$= \frac{\left[\sum_{ij} y_{ij} Y_{i.} Y_{.j} - Y_{..}(T_{ss} + B_{ss} + M_{ss})\right]^2}{btT_{ss}B_{ss}}$$

The quantities T_{ss}, B_{ss}, and M_{ss} are computed by the conventional method of computing treatment sum of squares, block sum of squares, and the mean sum of squares, respectively. To compute N_{ss} the only new quantity needed is $\sum_{ij} y_{ij} Y_{i.} Y_{.j}$. A convenient method of computing this quantity will be given in the next section.

15.4.1 Example of Test for Nonadditivity. Suppose that the data in Table 15.3 are assumed to satisfy a two-way classification model,

TABLE 15.3 DATA FOR EXAMPLE OF ART. 15.4.1

β	τ 1	2	3	4	Sum (A)	Sum of cross product row i with total row (B)
1	8	2	1	3	14	146
2	4	0	3	1	8	76
3	2	1	0	4	7	63
	14	3	4	8	29	

as given in Theorem 15.6. We desire to test the hypothesis $H_0: (\tau\beta)_{ij} = 0$ for all i and j.

The only new quantity is $\sum_{ij} y_{ij} Y_{i.} Y_{.j}$. To obtain this, column B must be computed. For example, the first element in column B is

$$\sum_j y_{1j} Y_{.j} = (8)(14) + (2)(3) + (1)(4) + (3)(8) = 146$$

the second element in column B is

$$\sum_j y_{2j} Y_{.j} = (4)(14) + (0)(3) + (3)(4) + (1)(8) = 76$$

the third element in column B is

$$\sum_j y_{3j} Y_{.j} = (2)(14) + (1)(3) + (0)(4) + (4)(8) = 63$$

Then $$\sum_i \left(\sum_j y_{ij} Y_{.j} \right) Y_{i.} = (146)(14) + (76)(8) + (63)(7) = 3{,}093$$

Also, $$M_{ss} = \frac{(29)^2}{12} = 70.08$$

$$B_{ss} = \frac{(14)^2 + (8)^2 + (7)^2}{4} - M_{ss} = 7.17$$

$$T_{ss} = \frac{(14)^2 + (3)^2 + (4)^2 + (8)^2}{3} - M_{ss} = 24.92$$

Then $$\left[\sum_{ij} y_{ij}Y_{i.}Y_{.j} - Y_{..}(T_{ss} + B_{ss} + M_{ss})\right] = [3{,}093$$
$$- (29)(102.17)] = 130.07$$

Then $$N_{ss} = \frac{(130.07)^2}{btT_{ss}B_{ss}} = \frac{(130.07)^2}{(12)(178.68)} = 7.89$$

The AOV is given in Table 15.4. The F value is 2.64; so there is no indication of interaction in these data.

TABLE 15.4 ANALYSIS OF VARIANCE FOR NONADDITIVITY

SV	DF	SS	MS	F
Total	12	125.00		
Mean	1	70.08		
Treatments	3	24.92		
Blocks	2	7.17		
Nonadditivity	1	7.89	7.89	2.64
Balance	5	14.94	2.99	

15.5 Transformation

If conditions A to E on model (15.1), listed in Sec. 15.2, are not met, then, as explained in previous sections, the conventional tests of hypotheses and methods of defining confidence intervals may not be strictly valid. In this case there are various things that might be done:

1. We might use nonparametric procedures that are valid for very general assumptions (along this line the reader should investigate "randomization" procedures).

2. We might ignore the fact that conditions A to E are not met and proceed as if they were.

3. We might transform the observed random variables y_{ij} in (15.1) so as to meet conditions A to E.

Suppose we are interested in testing a certain hypothesis H_0 in model 4. If the test function is insensitive to the conditions A to E, it will be said to be *robust*. If a test is *robust*, then procedure 2 above will be useful.

If enough information is available about the random variables y_{ij}, it may be possible to transform them so as to meet conditions A to E (procedure 3). For example, suppose $y_{ij} = \mu\tau_i e_{ij}$, where $\log e_{ij}$ is a normal variable. Then

$$\log y_{ij} = \log \mu + \log \tau_i + \log e_{ij}$$

and the conditions may be satisfied if $x_{ij} = \log y_{ij}$ is used instead of y_{ij}. There may be many reasons for using transformations, but we shall discuss only one.

Let y be a random variable, and suppose that we want it to satisfy conditions A to E. Suppose it is known that the mean of y is related to the variance of y by the function $f(t)$. That is to say, if $E(y) = \mu$ and $\text{var}(y) = \sigma^2$, $\sigma^2 = f(\mu)$. For example, if y is a Poisson variable with parameter m, $\mu = m$ and $\sigma^2 = m$; so $\mu = \sigma^2$. If y is a binomial variable with parameter p, $\mu = p$ and $\sigma^2 = p(1 - p)$; so $\sigma^2 = \mu(1 - \mu)$. Now, if conditions A to E are to be satisfied, y must be a normal variable, and in general we expect the mean and variance to be unrelated. Therefore, if it is known that $\sigma^2 = f(\mu)$, we shall try to find a transformation $h(y)$ such that $\text{var}[h(y)]$ is not related to $E[h(y)]$. Suppose that $x = h(y)$ is such a transformation. The problem is to find a function $h(y)$ such that the variance of x is a constant unrelated to $E(x)$. Suppose that $h(y)$ can be expanded about the point μ by a Taylor series. This gives

$$x = h(y) = h(\mu) + (y - \mu)h'(\mu) + \cdots$$

If we ignore all but the first two terms, we get

$$x = h(\mu) + (y - \mu)h'(\mu)$$

where $h'(\mu)$ is the derivative of $h(y)$ evaluated at $y = \mu$. Now

$$E(x) = E[h(\mu) + (y - \mu)h'(\mu)] = h(\mu)$$

since $E(y) = \mu$. Also,

$$\text{var}(x) = E[x - E(x)]^2 = E[x - h(\mu)]^2 = E[(y - \mu)h'(\mu)]^2 = \sigma^2[h'(\mu)]^2$$

since $\text{var}(y) = E(y - \mu)^2 = \sigma^2$. We have assumed that $\sigma^2 = f(\mu)$, where $f(\mu)$ is known; hence,

$$\text{var}(x) = f(\mu)[h'(\mu)]^2$$

Since $\text{var}(x)$ is to be independent of μ, we set $\text{var}(x)$ equal to a constant c^2. We have

$$c^2 = f(\mu)[h'(\mu)]^2$$

or

$$h'(\mu) = \frac{c}{\sqrt{f(\mu)}}$$

which gives

$$h(\mu) = c\int \frac{dt}{\sqrt{f(t)}} = cG(\mu) + k$$

where $G(\mu)$ is the indefinite integral of the function

$$\frac{1}{\sqrt{f(t)}}$$

and k is the constant of integration. Actually, the constants c^2 and k

are immaterial, since they do not depend on μ. They can be given any convenient value (so long as they are not related to μ).

For example, suppose $f(t) = t$; that is, the mean and variance of y are equal; then

$$h(\mu) = c \int \frac{dt}{\sqrt{t}} = 2c\sqrt{\mu} + k$$

So we could use the transformation $x = \sqrt{y}$.

This chapter is far from complete. For more complete information on the consequences that follow when the assumptions underlying model 4 are not satisfied, the reader is referred to the material in the bibliography.

Problems

15.1 Suppose the data of Table 15.5 satisfy the model given in Eq. (15.2) and the assumptions in (15.3). Test the hypothesis $\tau_1 = \tau_2 = \tau_3$ with a type I error of 5 per cent.

TABLE 15.5

Treatment	Block							
	1	2	3	4	5	6	7	8
1	52.3	54.1	54.3	55.1	56.4	59.2	60.2	53.1
2	53.2	53.0	56.3	36.9	54.4	58.5	60.1	54.1
3	30.6	54.7	55.2	32.4	55.7	58.3	59.7	54.7

15.2 Prove Theorem 15.3.
15.3 Prove Theorem 15.4.
15.4 Prove Theorem 15.5.
15.5 Show how the abbreviated Doolittle method can be used to find u in Theorem 15.1.
15.6 Use the data in Prob. 15.1 to test the hypothesis $\tau_1 - 2\tau_2 + \tau_3 = 0$. Use Theorem 15.5.
15.7 Test the data of Table 15.6 to see whether there is interaction (use a type I error probability of 5 per cent).

TABLE 15.6

Block	Treatment				
	1	2	3	4	5
1	4	3	10	5	1
2	6	2	19	12	7
3	1	1	7	2	1
4	8	7	14	9	4
5	1	4	9	6	2

15.8 Prove Theorem 15.7.
15.9 Prove Theorem 15.8.
15.10 Prove Theorem 15.10.
15.11 Prove Theorem 15.11.
15.12 The Poisson distribution is

$$P(x) = \frac{e^{-\lambda}\lambda^x}{x!} \qquad x = 0, 1, \ldots$$

Find the mean and variance of this distribution. What transformation makes the mean and variance independent?

15.13 The binomial distribution is

$$B(x) = \binom{n}{x} p^x q^{n-x} \qquad x = 0, 1, \ldots, n$$

Find the mean and variance of this distribution. What transformation makes the mean and variance independent?

15.14 Find the mean and variance of a chi-square distribution with n degrees of freedom. What transformation leaves the mean and variance independent?

15.15 Prove that the $(t - 1) \times (t - 1)$ matrix $\mathbf{V}$ with elements v_{pq} in Eq. (15.4) is positive definite if the $t \times t$ matrix with elements σ_{ii}' in Eq. (15.3) is positive definite.

15.16 In Theorem 15.1, prove that u is the same if y_{2j} instead of y_{1j} is subtracted from the other y_{ij} to form $\mathbf{Y}_j$.

Further Reading

1 F. N. David and N. L. Johnson: A Method of Investigating the Effect of Non Normality and Heterogeneity of Variance on Tests of the General Linear Hypothesis, *Ann. Math. Statist.*, vol. 22, pp. 382–392, 1951.

2 G. Horsnell: The Effect of Unequal Group Variances on the F-Test for the Homogeneity of Group Means, *Biometrika*, vol. 40, pp. 128–136, 1953.

3 G. E. P. Box: Some Theorems on Quadratic Forms Applied in the Study of Analysis of Variance Problems, I, II, *Ann. Math. Statist.*, vol. 25, pp. 290–302, pp. 448–498, 1954.

4 P. G. Moore: Transformations to Normality Using Fractional Powers of the Variable, *J. Am. Statist. Assoc.*, vol. 52, pp. 237–246, 1957.

5 M. S. Bartlett: The Use of Transformations, *Biometrics*, vol. 3, pp. 39–52, 1947.

6 F. J. Anscombe: The Transformation of Poisson, Binomial and Negative-Binomial Data, *Biometrika*, vol. 35, pp. 246–254, 1948.

7 J. H. Curtiss: Transformations Used in the A.O.V., *Ann. Math. Statist.*, vol. 14, pp. 107–122, 1943.

8 G. Beall: The Transformation of Data from Entomological Field Experiments so that the Analysis of Variance Becomes Applicable, *Biometrika*, vol. 32, pp. 243–262, 1941–1942.

9 H. Scheffé: Alternative Models for the Analysis of Variance, *Ann. Math. Statist.*, vol. 27, pp. 251–271, 1956.

10 C. Eisenhart: The Assumptions Underlying the A.O.V., *Biometrics*, vol. 3, pp. 1–21, 1947.

11 W. G. Cochran: Some Consequences when the Assumptions for the A.O.V. Are Not Satisfied, *Biometrics*, vol. 3, pp. 22–38, 1947.

12 G. W. Snedecor: "Statistical Methods," Iowa State College Press, Ames, Iowa, 1958.

13 O. Kempthorne: "Design and Analysis of Experiments," John Wiley & Sons, Inc., New York, 1952.

14 J. W. Tukey: One Degree of Freedom for Nonadditivity, *Biometrics*, vol. 5, pp. 232–242, 1949.

16

Model 5: Variance Components; Point Estimation

In this chapter we shall discuss the subject of variance components. Suppose there are two populations designated by π_1 and π_2. Let the mean and variance of π_1 be 0 and σ_a^2, respectively. Similarly, let 0 and σ_b^2 represent the mean and variance of π_2. Further, let $a_1, a_2, \ldots$ represent elements of population π_1 and $b_{11}, b_{12}, \ldots$ represent elements of π_2. The elements from population π_2 have two subscripts; this is solely for identification. Now let an observable random variable y be such that

$$y_{ij} = \mu + a_i + b_{ij}$$

where μ is an unknown constant. The thing to notice is the structure of the observed random variable y_{ij}. The object in this model is to observe the y_{ij} and, on the basis of this observation, to estimate or to test hypotheses about μ, σ_a^2, and σ_b^2.

Another way to look at this is as follows: The variance of an observable random variable y is

$$\operatorname{var}(y) = \sigma_b^2 + \sigma_a^2$$

Now suppose that we can stratify the random variables represented by y so that the variance of the random variables in a given stratum is σ_b^2. Then let y_{ij} be the jth observed random variable in the ith stratum, and y_{ij} can be written

$$y_{ij} = \mu + a_i + b_{ij}$$

For example, suppose a horticulturist wants to determine how a certain treatment has affected the nitrogen in the foliage of the trees in an orchard. He cannot examine every leaf on every tree; so he selects at random a group of trees for study. He cannot examine every leaf on the trees he has selected; so he chooses a set of leaves at random from each selected tree. Let y_{ij} be the observed nitrogen content of the jth leaf from the ith tree, and the structure is assumed to be of the form

$$y_{ij} = \mu + a_i + b_{ij}$$

where a_i is a random variable with variance σ_a^2, and b_{ij} is a random variable with variance σ_b^2.

Other models could be written

$$y_{ij} = \mu + a_i + b_j + c_{ij}$$
$$y_{ijk} = \mu + a_i + b_j + c_{ij} + d_{ijk}$$
etc.

Notice that the structure of the observation is similar to model 4, the only change being that the a_i, b_{ij}, etc., are random variables. Now we shall define model 5.

◆ **Definition 16.1** Let an observable random variable $y_{ij\ldots m}$ be such that

$$y_{ij\ldots m} = \mu + a_i + b_{ij} + \cdots + e_{ij\ldots m} \quad (16.1)$$

where μ is a constant; a_i is a random variable with mean 0 and variance σ_a^2; b_{ij} is a random variable with mean 0 and variance $\sigma_b^2, \ldots$; and $e_{ij\ldots m}$ is a random variable with mean 0 and variance σ_e^2. Let all the random variables be uncorrelated and let the covariance matrix of the $y_{ij\ldots m}$ be positive definite. When these conditions are satisfied, the relationship in Eq. (16.1) will be called model 5.

This is what Eisenhart has termed Model II in his classification [9]. In this chapter we shall discuss various models within model 5. We shall also devote a section to a general situation.

For point estimation two cases will be considered:

Case A. All random variables are independent and normally distributed.

Case B. All random variables are uncorrelated but not necessarily normal.

16.1 One-way Classification: Equal Subclass Numbers

Consider the model

$$y_{ij} = \mu + a_i + b_{ij} \qquad j = 1, 2, \ldots, s; \quad i = 1, 2, \ldots, r \quad (16.2)$$

The assumptions are:

Case A. The a_i are distributed $N(0,\sigma_a^2)$; the b_{ij} are distributed $N(0,\sigma_b^2)$; all the random variables a_i, b_{ij} are independent.

Case B. The a_i have mean 0 and variance σ_a^2; the b_{ij} have mean 0 and variance σ_b^2; all the random variables a_i, b_{ij} are uncorrelated.

16.1.1 Case A. In this case y_{ij} and $y_{ij'}$ are correlated even if $j \neq j'$. This is given by

$$\text{cov}(y_{ij},y_{ij'}) = \begin{cases} \sigma_a^2 & \text{if } j \neq j' \\ \sigma_b^2 + \sigma_a^2 & \text{if } j = j' \end{cases} \tag{16.3}$$

Also notice that

$$\text{cov}(y_{ij},y_{pq}) = 0 \qquad \text{if } i \neq p \tag{16.4}$$

The likelihood function for the random variables y_{ij} is

$$\begin{aligned} g(\mathbf{Y}) &= g(y_{11},y_{12},\ldots,y_{1s},y_{21},\ldots,y_{2s},\ldots,y_{r1},\ldots,y_{rs}) \\ &= f(y_{11},y_{12},\ldots,y_{1s})f(y_{21},y_{22},\ldots,y_{2s})\cdots f(y_{r1},y_{r2},\ldots,y_{rs}) \\ &= f(\mathbf{Y}_1)f(\mathbf{Y}_2)\cdots f(\mathbf{Y}_r) \end{aligned}$$

where $\mathbf{Y}_i' = (y_{i1}, y_{i2}, \ldots, y_{is})$ is an $s \times 1$ vector. $\mathbf{Y}_i$ is distributed $N(\mathbf{1}\mu,\mathbf{V})$, where $\mathbf{1}$ is a vector of 1's, and, if we let v_{pq} be the pqth element of $\mathbf{V}$, we have

$$v_{pq} = \begin{cases} \sigma_b^2 + \sigma_a^2 & \text{if } p = q \\ \sigma_a^2 & \text{if } p \neq q \end{cases}$$

or, combining,

$$v_{pq} = \sigma_a^2 + \delta_{pq}\sigma_b^2 \tag{16.5}$$

where $\delta_{pq} = 1$ if $p = q$ and $\delta_{pq} = 0$ otherwise. Therefore,

$$f(\mathbf{Y}_i) = \frac{1}{(2\pi)^{s/2}|\mathbf{V}|^{1/2}} \exp\left[-\tfrac{1}{2}(\mathbf{Y}_i - \boldsymbol{\mu}^*)'\mathbf{V}^{-1}(\mathbf{Y}_i - \boldsymbol{\mu}^*)\right]$$

where $\boldsymbol{\mu}^* = \mathbf{1}\mu$ is an $s \times 1$ vector with each element equal to μ. The likelihood function is

$$g(\mathbf{Y}) = \frac{1}{(2\pi)^{rs/2}|\mathbf{V}|^{r/2}} \exp\left[-\frac{1}{2}\sum_{i=1}^{r}(\mathbf{Y}_i - \boldsymbol{\mu}^*)'\mathbf{V}^{-1}(\mathbf{Y}_i - \boldsymbol{\mu}^*)\right] \tag{16.6}$$

Before taking derivatives of the likelihood function, we shall elaborate upon some of the quantities in (16.6). Notice that the matrix $\mathbf{V}$ has diagonal elements equal to $\sigma_b^2 + \sigma_a^2$ and off-diagonal elements equal to σ_a^2. Two lemmas that follow will be useful.

◆ **Lemma 16.1** The determinant of $\mathbf{V}$ is equal to

$$(\sigma_b^2)^{s-1}(\sigma_b^2 + s\sigma_a^2)$$

The determinant of $\mathbf{V}^{-1}$ is equal to

$$(\sigma_b^2)^{1-s}(\sigma_b^2 + s\sigma_a^2)^{-1}$$

The proof will be left for the reader.

Next we shall prove a lemma on the inverse of $\mathbf{V}$. First notice that

$$\mathbf{V} = \sigma_b^2\mathbf{I} + \sigma_a^2\mathbf{J}$$

where $\mathbf{J}$ is an $s \times s$ matrix each element of which is equal to 1. (Throughout this chapter $\mathbf{J}$ will be a matrix with each element equal to unity. The dimension of $\mathbf{J}$ will generally be clear from the context.)

◆ **Lemma 16.2**

$$\mathbf{V}^{-1} = \alpha_1\mathbf{I} + \frac{\alpha_2}{s}\mathbf{J}$$

where $\alpha_1 = \dfrac{1}{\sigma_b^2}$ and $\alpha_2 = -\dfrac{s\sigma_a^2}{\sigma_b^2}(\sigma_b^2 + s\sigma_a^2)^{-1}$

Proof: Notice that, since $\mathbf{JJ} = s\mathbf{J}$, we can write

$$\mathbf{V} = \sigma_b^2\mathbf{I} + s\sigma_a^2\left(\frac{1}{s}\mathbf{J}\right)$$

or, if we let $(1/s)\mathbf{J} = \mathbf{A}$, then $\mathbf{A}$ is idempotent and

$$\mathbf{V} = \sigma_b^2\mathbf{I} + s\sigma_a^2\mathbf{A}$$

Now let

$$\mathbf{V}^{-1} = \alpha_1\mathbf{I} + \alpha_2\mathbf{A}$$

where α_1 and α_2 are to be determined. So

$$\begin{aligned}\mathbf{I} = \mathbf{VV}^{-1} &= (\sigma_b^2\mathbf{I} + s\sigma_a^2\mathbf{A})(\alpha_1\mathbf{I} + \alpha_2\mathbf{A})\\ &= \sigma_b^2\alpha_1\mathbf{I} + s\alpha_1\sigma_a^2\mathbf{A} + \alpha_2\sigma_b^2\mathbf{A} + s\alpha_2\sigma_a^2\mathbf{A}\\ &= \sigma_b^2\alpha_1\mathbf{I} + (s\alpha_1\sigma_a^2 + \alpha_2\sigma_b^2 + s\alpha_2\sigma_a^2)\mathbf{A}\end{aligned}$$

Since this equation is to hold for all values of σ_b^2 and σ_a^2, we set the coefficient of $\mathbf{A}$ equal to zero and the coefficient of $\mathbf{I}$ equal to unity. This gives

$$\sigma_b^2\alpha_1 = 1$$

$$s\alpha_1\sigma_a^2 + \alpha_2\sigma_b^2 + s\alpha_2\sigma_a^2 = 0$$

This gives

$$\alpha_1 = \frac{1}{\sigma_b^2}$$

$$\alpha_2 = -\frac{s\sigma_a^2}{\sigma_b^2}(\sigma_b^2 + s\sigma_a^2)^{-1} = \frac{1}{\sigma_b^2 + s\sigma_a^2} - \frac{1}{\sigma_b^2} \tag{16.7}$$

This completes the proof.

If we examine the exponent (ignore the $-\frac{1}{2}$) of the likelihood, we get

$$Q = \sum_i (\mathbf{Y}_i - \boldsymbol{\mu}^*)'\mathbf{V}^{-1}(\mathbf{Y}_i - \boldsymbol{\mu}^*) = \sum_i \left[\sum_{pq} (y_{ip} - \mu)(y_{iq} - \mu)w_{pq} \right] \quad (16.8)$$

where y_{ip} is the pth element of $\mathbf{Y}_i$ and w_{pq} is the pqth element of $\mathbf{V}^{-1}$. But, by Lemma 16.2,

$$w_{pq} = + \frac{\alpha_2}{s} = - \frac{\sigma_a^2}{\sigma_b^2}(\sigma_b^2 + s\sigma_a^2)^{-1} \qquad \text{if } p \neq q$$

$$w_{pp} = \alpha_1 + \frac{\alpha_2}{s}$$

So (in what follows the ranges of the subscripts are: $i = 1, 2, \ldots, r$; $p, q = 1, 2, \ldots, s$) Eq. (16.8) becomes

$$Q = \sum_i \sum_p (y_{ip} - \mu)^2 w_{pp} + \sum_i \sum_p \sum_{\substack{q \\ q \neq p}} (y_{ip} - \mu)(y_{iq} - \mu)w_{pq}$$

$$= \left(\alpha_1 + \frac{\alpha_2}{s}\right) \sum_i \sum_p (y_{ip} - \mu)^2 + \left(\frac{\alpha_2}{s}\right) \sum_i \sum_{\substack{p \\ p \neq q}} \sum_q (y_{ip} - \mu)(y_{iq} - \mu) \quad (16.9)$$

$$= \alpha_1 \sum_i \sum_p (y_{ip} - \mu)^2 + \frac{\alpha_2}{s} \sum_i \left[\sum_p (y_{ip} - \mu) \right]^2$$

We shall examine this in more detail. This quantity can be written

$$Q = \alpha_1 \sum_{ip} (y_{ip} - \mu)^2 + \frac{\alpha_2}{s} \sum_i (sy_{i.} - s\mu)^2$$

$$= \alpha_1 \sum_{ip} (y_{ip} - \mu)^2 + \alpha_2 s \sum_i (y_{i.} - \mu)^2 = \alpha_1 \sum_{ip} (y_{ip} - \mu)^2 + \alpha_2 \sum_{ip} (y_{i.} - \mu)^2$$

If we use the definitions of α_1 and α_2 from (16.7), this becomes

$$Q = \frac{1}{\sigma_b^2} \sum_{ip} (y_{ip} - \mu)^2 + \frac{1}{\sigma_b^2 + s\sigma_a^2} \sum_{ip} (y_{i.} - \mu)^2 - \frac{1}{\sigma_b^2} \sum_{ip} (y_{i.} - \mu)^2$$

But
$$\sum_{ip} (y_{ip} - \mu)^2 - \sum_{ip} (y_{i.} - \mu)^2 = \sum_{ip} (y_{ip} - y_{i.})^2$$

so we get
$$Q = \frac{1}{\sigma_b^2} \sum_{ip} (y_{ip} - y_{i.})^2 + \frac{1}{\sigma_b^2 + s\sigma_a^2} \sum_{ip} (y_{i.} - \mu)^2$$

The quantity

$$\sum_{ip} (y_{i.} - \mu)^2 = \sum_{ip} [(y_{i.} - y_{..}) + (y_{..} - \mu)]^2 = \sum_{ip} (y_{i.} - y_{..})^2 + \sum_{ip} (y_{..} - \mu)^2$$

So finally

$$Q = \frac{1}{\sigma_b^2} \sum_{ip} (y_{ip} - y_{i.})^2 + \frac{1}{\sigma_b^2 + s\sigma_a^2} \sum_{ip} (y_{i.} - y_{..})^2 + \frac{rs(y_{..} - \mu)^2}{\sigma_b^2 + s\sigma_a^2}$$

Now let us return to the likelihood function $f(\mathbf{Y})$ and substitute (16.7) and (16.8). This gives us

$$f(\mathbf{Y}) = \frac{\exp\left\{-\frac{1}{2}\left[\frac{1}{\sigma_b^2}\sum_{ip}(y_{ip} - y_{i.})^2 + \frac{1}{\sigma_b^2 + s\sigma_a^2}\sum_{ip}(y_{i.} - y_{..})^2 + \frac{rs(y_{..} - \mu)^2}{\sigma_b^2 + s\sigma_a^2}\right]\right\}}{(2\pi)^{rs/2}(\sigma_b^2)^{\frac{1}{2}r(s-1)}(\sigma_b^2 + s\sigma_a^2)^{r/2}} \tag{16.10}$$

We see from (16.10) that the three quantities $\sum_{ip}(y_{ip} - y_{i.})^2$, $\sum_{ip}(y_{i.} - y_{..})^2$, and $y_{..}$ are a set of sufficient statistics. It can also be shown that they are complete. If the derivatives of $f(\mathbf{Y})$ are taken with respect to σ_b^2, σ_a^2, and μ, the maximum-likelihood estimators (corrected for bias) are

$$\hat{\mu} = y_{..}$$

$$\hat{\sigma}_b^2 = \frac{\sum_{ip}(y_{ip} - y_{i.})^2}{r(s-1)} \tag{16.11}$$

$$\hat{\sigma}_a^2 = \frac{\sum_{ip}(y_{i.} - y_{..})^2}{s(r-1)} - \frac{\sum_{ip}(y_{ip} - y_{i.})^2}{rs(s-1)}$$

This gives the following theorem.

◆ **Theorem 16.1** Under the assumptions of case A for the one-way classification given in Eq. (16.2), the estimators given in Eq. (16.11) are minimum-variance unbiased.

It is interesting to notice that the three sufficient statistics $\sum_{ip}(y_{ip} - y_{i.})^2$, $\sum_{ip}(y_{i.} - y_{..})^2$, and $\sqrt{rs}y_{..}$ are exactly the three quantities appearing in the AOV Table 12.1 for model 4. The expected mean squares are

TABLE 16.1 ANALYSIS OF VARIANCE FOR ONE-WAY CLASSIFICATION, MODEL 5

SV	DF	SS	MS	EMS
Total	rs	$\sum_{ip} y_{ip}^2$		
Mean	1	$rsy_{..}^2$		
Between	$r - 1$	$B_{ss} = \sum_{ip}(y_{i.} - y_{..})^2$	B_{ms}	$\sigma_b^2 + s\sigma_a^2$
Within	$r(s - 1)$	$W_{ss} = \sum_{ip}(y_{ip} - y_{i.})^2$	W_{ms}	σ_b^2

given in Table 16.1. The method of obtaining EMS from "between" will be demonstrated.

$$E\left[\frac{1}{r-1}\sum_{ip}(y_{i.}-y_{..})^2\right] = \frac{1}{r-1}E\left[\sum_{ip}(\mu + a_i + \bar{b}_{i.} - \mu - \bar{a}_. - \bar{b}_{..})^2\right]$$

$$= \frac{1}{r-1}E\left[\sum_{ip}(a_i - \bar{a}_.)^2 + \sum_{ip}(\bar{b}_{i.} - \bar{b}_{..})^2\right] \quad (16.12)$$

since the a_i and b_{ip} are independent. To find the expected values we shall use the following lemma.

◆ **Lemma 16.3** Let $z_1, z_2, \ldots, z_n$ be uncorrelated random variables with mean μ and variance σ^2. Then

$$E\left[\sum_{i=1}^{n}(z_i - \bar{z})^2\right] = (n-1)\operatorname{var}(z_i) = (n-1)\sigma^2$$

The proof will be left for the reader.

By using this lemma we get

$$\frac{1}{r-1}E\left[\sum_{ip}(a_i - \bar{a}_.)^2\right] = \frac{1}{r-1}\sum_{p}\left\{E\left[\sum_{i=1}^{r}(a_i - \bar{a}_.)^2\right]\right\} = \sum_{p}\sigma_a^2 = s\sigma_a^2$$

Similarly,

$$\frac{1}{r-1}E\left[\sum_{ip}(\bar{b}_{i.} - \bar{b}_{..})^2\right] = \frac{1}{r-1}\sum_{p}\left\{E\left[\sum_{i=1}^{r}(\bar{b}_{i.} - \bar{b}_{..})^2\right]\right\}$$

$$= \frac{1}{r-1}\sum_{p}(r-1)\operatorname{var}(\bar{b}_{i.})$$

But $\operatorname{var}(\bar{b}_{i.}) = \sigma_b^2/s$. So we get

$$\frac{1}{r-1}E\left[\sum_{ip}(\bar{b}_{i.} - \bar{b}_{..})^2\right] = \frac{1}{r-1}\sum_{p=1}^{s}(r-1)\frac{\sigma_b^2}{s} = \sigma_b^2$$

So the EMS for "between" is $\sigma_b^2 + s\sigma_a^2$. The "within" EMS can be found by a similar process.

Next we shall turn our attention to finding the distributions of the "between" and "within" sums of squares. Again we shall prove a lemma to help decide the distributional properties of these quantities.

◆ **Lemma 16.4** Let z_i be distributed $N(\mu_0,\sigma^2)$ for $i = 1, 2, \ldots, n$. Let $\operatorname{cov}(z_i,z_j) = \rho\sigma^2$ for all $i \neq j$ $(-1 < \rho < 1)$. Let

$$u = \frac{\sum_{i=1}^{n}(z_i - \bar{z})^2}{\sigma^2(1-\rho)}$$

Then u is distributed as $\chi^2(n-1)$.

Proof: Let the $n \times 1$ vector $\mathbf{Z}$ have z_i for its ith element. Then $\mathbf{Z}$ is distributed $N(\boldsymbol{\mu},\mathbf{V})$, where the diagonal elements of $\mathbf{V}$ are σ^2 and the off-diagonal elements are $\rho\sigma^2(-1 < \rho < 1)$. Thus $\mathbf{V}$ can be written

$$\mathbf{V} = (\sigma^2 - \rho\sigma^2)\mathbf{I} + \rho\sigma^2\mathbf{J}$$

Now let

$$u = \frac{\mathbf{Z}'\mathbf{A}\mathbf{Z}}{\sigma^2(1 - \rho)}$$

where the ith diagonal element of $\mathbf{A}$ is $(n - 1)/n$ and each off-diagonal element is $-1/n$. So

$$\mathbf{A} = \mathbf{I} - \frac{1}{n}\mathbf{J}$$

Let us examine the product $\mathbf{VA}$. We have

$$\begin{aligned}\mathbf{VA} &= [(\sigma^2 - \rho\sigma^2)\mathbf{I} + \rho\sigma^2\mathbf{J}]\left(\mathbf{I} - \frac{1}{n}\mathbf{J}\right) = (\sigma^2 - \rho\sigma^2)\mathbf{I} \\ &\quad - \frac{1}{n}(\sigma^2 - \rho\sigma^2)\mathbf{J} \\ &= (\sigma^2 - \rho\sigma^2)\left(\mathbf{I} - \frac{1}{n}\mathbf{J}\right) = (\sigma^2 - \rho\sigma^2)\mathbf{A}\end{aligned}$$

But $\mathbf{A}$ is idempotent of rank $n - 1$; hence, $\mathbf{VA}/\sigma^2(1 - \rho)$ is idempotent. Therefore, by Theorem 4.9, u is distributed as $\chi'^2(n - 1, \lambda)$, where

$$\lambda = \tfrac{1}{2}[E(\mathbf{Z})]'\mathbf{A}[E(\mathbf{Z})] = \tfrac{1}{2}\boldsymbol{\mu}'\mathbf{A}\boldsymbol{\mu}$$

But $\boldsymbol{\mu}$ has each element equal to μ_0; so $\boldsymbol{\mu} = \mathbf{1}\mu_0$. But $\mathbf{1}'\mathbf{A} = \mathbf{0}$; so $\lambda = 0$, and the lemma is proved.

We shall exhibit the use of this lemma to find the distribution of

$$v = \frac{\sum_{ip} (y_{i.} - y_{..})^2}{\sigma_b^2 + s\sigma_a^2}$$

Let us examine $z_i = y_{i.}$.

$$\begin{aligned}E(y_{i.}) &= \mu \\ \operatorname{var}(y_{i.}) &= E(y_{i.} - \mu)^2 \\ &= E(a_i + \bar{b}_{i.})^2 \\ &= E(a_i)^2 + E(\bar{b}_{i.})^2 \\ &= \sigma_a^2 + \frac{\sigma_b^2}{s} \\ &= \frac{\sigma_b^2 + s\sigma_a^2}{s}\end{aligned}$$

$$\begin{aligned}\operatorname{cov}(y_{i.}, y_{j.}) &= E[(y_{i.} - \mu)(y_{j.} - \mu)] \\ &= E[(a_i + \bar{b}_{i.})(a_j + \bar{b}_{j.})] \\ &= 0 \qquad \text{if } i \neq j\end{aligned}$$

So the quantities in the lemma are

$$\sigma^2 = \frac{\sigma_b^2 + s\sigma_a^2}{s} \qquad \rho = 0$$

So $y_{i.}$ is distributed

$$N\left(\mu, \frac{\sigma_b^2 + s\sigma_a^2}{s}\right)$$

and $$\text{cov}\,(y_{i.},y_{i'.}) = 0 \qquad \text{for } i \neq i'$$

Hence,

$$v_0 = \sum_i \frac{(y_{i.} - y_{..})^2}{(\sigma_b^2 + s\sigma_a^2)/s}$$

is distributed as $\chi^2(r - 1)$. But $v = v_0$; so v is distributed as $\chi^2(r - 1)$.

Now the "within" sum of squares

$$w = \sum_{ij} (y_{ij} - y_{i.})^2$$

gives us $$w = \sum_{ij} (b_{ij} - \bar{b}_{i.})^2$$

But, since the b_{ij} are distributed independently as $N(0,\sigma_b^2)$, we see that

$$w_i = \frac{\sum_j (b_{ij} - \bar{b}_{i.})^2}{\sigma_b^2}$$

is distributed as $\chi^2(s - 1)$. But w_i is independent of $w_{i'}$ if $i \neq i'$; hence, by the reproductive property of chi-square, we see that

$$\frac{w}{\sigma_b^2} = \sum_{i=1}^{r} w_i$$

is distributed as $\chi^2[r(s - 1)]$.

Next we want to show that v and w are independent. Let $v_i = y_{i.} - y_{..}$ and $w_{pq} = y_{pq} - y_{p.}$. These are each normal variates with zero means. Hence, if we can show that the covariance of v_i and w_{pq} is zero for all i, p, and q, then they are independent. Also, since any functions of independent variables are independent, it follows that, in particular, Σv_i^2 is independent of $\sum_{pq} w_{pq}^2$. To find $\text{cov}(v_i,w_{pq})$, we get

$$\begin{aligned}
\text{cov}(v_i, w_{pq}) &= E(v_i w_{pq}) \\
&= E[(y_{i.} - y_{..})(y_{pq} - y_{p.})] \\
&= E\{[(a_i - \bar{a}_.) + (\bar{b}_{i.} - \bar{b}_{..})][b_{pq} - \bar{b}_{p.}]\} \\
&= E[(\bar{b}_{i.} - \bar{b}_{..})(b_{pq} - \bar{b}_{p.})] \\
&= E(\bar{b}_{i.}b_{pq} - b_{pq}\bar{b}_{..} + \bar{b}_{..}\bar{b}_{p.} - \bar{b}_{i.}\bar{b}_{p.}) \\
&= \frac{\delta_{pi}\sigma_b^2}{s} - \frac{\sigma_b^2}{rs} + \frac{\sigma_b^2}{rs} - \frac{\delta_{pi}\sigma_b^2}{s} \\
&= 0
\end{aligned}$$

So the "between" mean square and the "within" mean square are independent. By a similar method it can be shown that

$$\frac{rsy_{..}^2}{\sigma_b^2 + s\sigma_a^2}$$

is distributed as $\chi'^2(1,\lambda)$, where

$$\lambda = \frac{rs\mu^2}{2(\sigma_b^2 + s\sigma_a^2)}$$

The distributional properties can be summed up in the following theorem.

◆ **Theorem 16.2** Under the assumptions of case A for the balanced one-way classification given in Eq. (16.2), the following are true:

(1) $$\frac{\sum_{ip} (y_{ip} - y_{i.})^2}{\sigma_b^2}$$

is distributed as $\chi^2[r(s - 1)]$.

(2) $$\frac{\sum_{ip} (y_{i.} - y_{..})^2}{\sigma_b^2 + s\sigma_a^2}$$

is distributed as $\chi^2(r - 1)$.

(3) $$\frac{rsy_{..}^2}{\sigma_b^2 + s\sigma_a^2}$$

is distributed as $\chi'^2(1,\lambda)$, where

$$\lambda = \frac{rs\mu^2}{2(\sigma_b^2 + s\sigma_a^2)}$$

(4) The quantities (1), (2), and (3) are mutually independent.

16.1.2 Case B. The assumptions for case B are that the a_i are uncorrelated with mean 0 and variance σ_a^2, the b_{ij} are uncorrelated with mean 0 and variance σ_b^2, and the a_i and b_{ij} are uncorrelated. Since the distribution of the random variables is not known, maximum-likelihood estimates cannot be obtained. Therefore, to obtain estimates of σ_a^2 and σ_b^2, the *method of analysis of variance* will be used. This method consists of assuming that the a_i are fixed constants, not random variables; that is, we assume model 4 instead of model 5 and form an AOV table, then find the expected mean squares assuming model 5. Then the mean squares will be set equal to the expected mean squares in the table and the resulting equations solved for the σ^2's. These solutions will be taken as the estimates.

The results are given in Table 16.1. The expected mean squares for this table were derived under the assumptions for case A. It is easy to verify that the same results hold for case B. It is clear that the estimates of σ_a^2 and σ_b^2 are unbiased. However, they do not possess the optimum property given in Theorem 16.1 for the model for case A.

Although the estimators (case B) are not minimum-variance unbiased estimators in the class of all functions of the observations, in the class restricted to quadratic functions only, they do have this property. This is the sense of the next theorem.

◆ **Theorem 16.3** Let the balanced one-way classification model satisfy the assumptions given for case B. In the class of quadratic estimators (quadratic functions of the observations) the estimators of the variance components σ_a^2 and σ_b^2 given by the *analysis of variance* are unbiased and have minimum variance.

The proof of this theorem will be omitted [17].

16.2 The General Balanced Case of Model 5

In the previous section the method of estimation by maximum likelihood was demonstrated for the balanced one-way classification model when the assumptions for case A were satisfied. The property of the estimators was given in Theorem 16.1. The method of estimation by the analysis of variance was demonstrated for the balanced one-way classification under the assumptions for case B. The optimum properties were given in Theorem 16.3.

Before proceeding to other special models, we shall state two general theorems for balanced cases of model 5. Notice that the theorems say nothing about maximum-likelihood estimators. Even in the one-way balanced classification, the maximum-likelihood estimators are tedious to obtain. We shall prefer the analysis-of-variance method, since for case A the estimators so derived are simple functions of the maximum-likelihood estimators.

◆ **Theorem 16.4** Let the model be model 5 and let it be balanced. Also let the assumptions for case A hold. Then the estimators of the variance components obtained by the analysis-of-variance method of estimation are minimum-variance unbiased.

Next we shall state a theorem concerning the distributional properties of each line in an AOV table.

◆ **Theorem 16.5** Let the model be model 5 and let it be balanced. Also let the assumptions for case A hold. Let the ith line (see Table 16.2) in the MS column of the AOV table for this model be

TABLE 16.2 TABLE FOR THEOREM 16.5

SV	DF	MS	EMS	Line
Total				
Mean	1	s_0^2	σ_0^2	0
1st class	f_1	s_1^2	σ_1^2	1
2nd class	f_2	s_2^2	σ_2^2	2
.	.	.	.	.
.	.	.	.	.
.	.	.	.	.
kth class	f_k	s_k^2	σ_k^2	k

denoted by s_i^2 ($i > 0$). Let s_i^2 have f_i degrees of freedom and let $E(s_i^2) = \sigma_i^2$ (generally σ_i^2 is a linear combination of the σ_a^2, σ_b^2, etc.). Then $u_i = f_i s_i^2/\sigma_i^2$ is distributed as $\chi^2(f_i)$, and $u_1, u_2, \ldots, u_k$ are mutually independent.

The proof of this theorem will not be given, but a particular case has been proved in Theorem 16.2.

◆ **Theorem 16.6** Let the model be model 5 and let it be balanced. Also let the assumptions for case B hold. Then the estimators of the variance components obtained by the analysis-of-variance method of estimation are best (minimum-variance) quadratic unbiased estimators.

The models that are balanced are all the models with equal numbers in all subclasses, such as one-way models, two-way models, latin-square, n-way, twofold nested, etc. The theorems do not hold for balanced incomplete block models or for partially balanced incomplete block models.

16.3 Two-way Classification

Consider the model for the two-way balanced cross classification:

$$y_{ij} = \mu + a_i + b_j + e_{ij} \qquad i = 1, 2, \ldots, r; \quad j = 1, 2, \ldots, s \tag{16.13}$$

If the a_i and b_j are fixed constants, the model fits the definition of model 4. The AOV is given in Table 12.2. This table is restated in Table 16.3. If the a_i are random variables with variance σ_a^2, if the b_j are random variables with variance σ_b^2, if the e_{ij} are random variables with variance σ_e^2, and if all random variables are independent, the expected mean squares can be shown to be as indicated.

TABLE 16.3 ANALYSIS OF VARIANCE FOR TWO-WAY BALANCED CLASSIFICATION, MODEL 5

SV	DF	SS	MS	EMS
Total	rs	$\sum_{ij} y_{ij}^2$		
Mean	1	$rsy_{..}^2$		
a classification	$r - 1$	$\sum_{ij} (y_{i.} - y_{..})^2$	A_{ms}	$\sigma_e^2 + s\sigma_a^2$
b classification	$s - 1$	$\sum_{ij} (y_{.j} - y_{..})^2$	B_{ms}	$\sigma_e^2 + r\sigma_b^2$
Error	$(r - 1)(s - 1)$	$\sum_{ij} (y_{ij} - y_{i.} - y_{.j} + y_{..})^2$	E_{ms}	σ_e^2

If the mean squares A_{ms}, B_{ms}, E_{ms} are equated to their respective expected mean squares, the following equations are obtained:

$$A_{\text{ms}} = \hat{\sigma}_e^2 + s\hat{\sigma}_a^2 \qquad B_{\text{ms}} = \hat{\sigma}_e^2 + r\hat{\sigma}_b^2 \qquad E_{\text{ms}} = \hat{\sigma}_e^2$$

The estimators are obtained by solving these equations. This gives

$$\hat{\sigma}_e^2 = E_{\text{ms}} \qquad \hat{\sigma}_a^2 = \frac{A_{\text{ms}} - E_{\text{ms}}}{s} \qquad \hat{\sigma}_b^2 = \frac{B_{\text{ms}} - E_{\text{ms}}}{r}$$

Estimates of variance components for the three-way, four-way, . . . , n-way classifications are obtained by similar methods.

16.4 The Balanced Twofold and Fourfold Nested Classification of Model 5

The balanced twofold nested classification components-of-variance model is defined by

$$y_{ijk} = \mu + a_i + b_{ij} + c_{ijk} \qquad \begin{cases} i = 1, 2, \ldots, r \\ j = 1, 2, \ldots, s \\ k = 1, 2, \ldots, t \end{cases} \tag{16.14}$$

where a_i has mean 0 and variance σ_a^2, b_{ij} has mean 0 and variance σ_b^2, c_{ijk} has mean 0 and variance σ_c^2, and all random variables are uncorrelated.

This is a very important type in model 5. For example, in a chicken-breeding experiment, suppose rs dams are mated to r sires (s to each sire) and t offsprings result from each sire-dam mating. If the 12-week body weight is the important characteristic, suppose y_{ijk} represents the weight of the kth offspring of the ijth sire-dam mating. Suppose the structure of y_{ijk} is

$$y_{ijk} = \mu + a_i + b_{ij} + c_{ijk}$$

where a_i is the contribution due to the ith sire, b_{ij} is the contribution due to the jth dam mated to the ith sire, and c_{ijk} is the effect due to the kth offspring of the ijth dam-sire mating. The object is to estimate σ_a^2, σ_b^2, and σ_c^2, since these components have special meaning in genetics.

For another example, suppose a company that manufactures transistors has various production lines and each production line has many machines that make the product. Suppose that the product is variable and that it is desired to study all sources of variation in the final product. Suppose t transistors are taken from each of s machines on each of r production lines. Let the observation of the kth transistor from the jth machine on the ith line have the structure

$$y_{ijk} = \mu + a_i + b_{ij} + c_{ijk}$$

where a_i is the contribution of the ith production line, b_{ij} is the contribution of the jth machine in the ith line, and c_{ijk} is the contribution of the kth transistor from the jth machine in the ith line. The object is to study the variances σ_a^2, σ_b^2, and σ_c^2.

The thing to notice in the nested classification model is that each successive contribution is imbedded or "nested" within the preceding.

This model is also important in many sample survey investigations. In many experiments the random variables a_i (or others) cannot conceivably be random variables from infinite populations. If this is true, the situation may not exactly fit the definition of model 5. However, only the case in which the random variables are from infinite populations will be considered.

Let us consider the model given in (16.14). We shall make AOV estimates of σ_a^2, σ_b^2, and σ_c^2. To do this, we consider the a_i and b_{ij} as fixed unknown parameters, construct an AOV table, and find the expected mean squares under the assumption of (16.14). This AOV is given in Table 16.4.

The expected mean squares are obtained by taking the expectations of the respective mean squares under the assumption that the a_i, b_{ij}, c_{ijk} are random variables. The estimates are obtained by equating mean squares to expected mean squares. The solution is

$$\hat{\sigma}_c^2 = C_{\text{ms}} = \frac{\sum_{ijk} (y_{ijk} - y_{ij.})^2}{rs(t-1)}$$

$$\hat{\sigma}_b^2 = \frac{B_{\text{ms}} - C_{\text{ms}}}{t} = \frac{1}{t}\left[\frac{\sum_{ijk} (y_{ij.} - y_{i..})^2}{r(s-1)} - \frac{\sum_{ijk} (y_{ijk} - y_{ij.})^2}{rs(t-1)}\right]$$

$$\hat{\sigma}_a^2 = \frac{A_{\text{ms}} - B_{\text{ms}}}{st} = \frac{1}{st}\left[\frac{\sum_{ijk} (y_{i..} - y_{...})^2}{r-1} - \frac{\sum_{ijk} (y_{ij.} - y_{i..})^2}{r(s-1)}\right]$$

TABLE 16.4 ANALYSIS OF VARIANCE FOR BALANCED TWOFOLD NESTED CLASSIFICATION

SV	DF	SS	MS	EMS
Total	rst	$\sum_{ijk} y_{ijk}^2$		
Mean	1	$\frac{Y_{...}^2}{rst}$		
Between a classes	$r - 1$	$\sum_{ijk} (y_{i..} - y_{...})^2$	A_{ms}	$\sigma_c^2 + t\sigma_b^2 + st\sigma_a^2$
Between b classes within a classes	$r(s - 1)$	$\sum_{ijk} (y_{ij.} - y_{i..})^2$	B_{ms}	$\sigma_c^2 + t\sigma_b^2$
Between c classes within b classes	$rs(t - 1)$	$\sum_{ijk} (y_{ijk} - y_{ij.})^2$	C_{ms}	σ_c^2

These estimates are minimum-variance unbiased (by Theorem 16.4) or minimum-variance quadratic unbiased (by Theorem 16.6), depending on whether the conditions of case A or those of case B are met.

Next we shall examine a fourfold classification model. Suppose the model is

$$y_{ijkmn} = \mu + a_i + b_{ij} + c_{ijk} + d_{ijkm} + e_{ijkmn} \qquad \begin{cases} i = 1, 2, \ldots, I \\ j = 1, 2, \ldots, J \\ k = 1, 2, \ldots, K \\ m = 1, 2, \ldots, M \\ n = 1, 2, \ldots, N \end{cases}$$

Let a_i, b_{ij}, c_{ijk}, d_{ijkm}, e_{ijkmn} have zero means and variances σ_a^2, σ_b^2, σ_c^2, σ_d^2, and σ_e^2, respectively. Let all random variables be uncorrelated. Table 16.5 contains the mean squares that are needed for estimates.

16.5 One-way Classification with Unequal Numbers in the Groups

Although it is usually desirable for each group to have an equal number of observations, this is not always possible. Suppose there are A groups and the ith group contains n_i elements. The model is

$$y_{ij} = \mu + a_i + b_{ij} \qquad \begin{cases} j = 1, 2, \ldots, n_i \\ i = 1, 2, \ldots, A \end{cases}$$

TABLE 16.5 ANALYSIS OF VARIANCE FOR FOURFOLD CLASSIFICATION IN MODEL 5

SV	DF	SS	MS	EMS
Total	$IJKMN$	Σy^2_{ijkmn}		
Mean	1	$\dfrac{Y^2_{.....}}{rst}$		
a classes	$I - 1$	$\Sigma(y_{i....} - y_{.....})^2$	A_{ms}	$\sigma_e^2 + N\sigma_d^2 + MN\sigma_c^2 + MNK\sigma_b^2 + MNKJ\sigma_a^2$
b classes	$I(J - 1)$	$\Sigma(y_{ij...} - y_{i....})^2$	B_{ms}	$\sigma_e^2 + N\sigma_d^2 + MN\sigma_c^2 + MNK\sigma_b^2$
c classes	$IJ(K - 1)$	$\Sigma(y_{ijk..} - y_{ij...})^2$	C_{ms}	$\sigma_e^2 + N\sigma_d^2 + MN\sigma_c^2$
d classes	$IJK(M - 1)$	$\Sigma(y_{ijkm.} - y_{ijk..})^2$	D_{ms}	$\sigma_e^2 + N\sigma_d^2$
e classes	$IJKM(N - 1)$	$\Sigma(y_{ijkmn} - y_{ijkm.})^2$	E_{ms}	σ_e^2

Note: Each summation is over all subscripts.

Let
$$\sum_{i=1}^{A} n_i = n$$

The AOV is given in Table 12.1 and is restated in Table 16.6 with the

TABLE 16.6 ANALYSIS OF VARIANCE FOR UNBALANCED ONEFOLD CLASSIFICATION OF MODEL 5

SV	DF	SS	MS	EMS
Total	n	$\sum_{ij} y_{ij}^2$		
Mean	1	$\frac{Y_{..}^2}{n}$		
a classes	$A-1$	$\sum_i \left(\frac{Y_{i.}^2}{n_i}\right) - \frac{Y_{..}^2}{n}$	A_{ms}	$\sigma_b^2 + k_0\sigma_a^2$
b classes	$n-A$	$\sum_{ij} y_{ij}^2 - \sum_i \left(\frac{Y_{i.}^2}{n_i}\right)$	B_{ms}	σ_b^2

expected mean squares. The expected mean squares are obtained as follows.

$$E(B_{\text{ms}}) = \frac{1}{n-A} E\left(\sum_{ij} y_{ij}^2 - \sum_i \frac{Y_{i.}^2}{n_i}\right)$$
$$= \frac{1}{n-A} E\left[\sum_{ij} (y_{ij} - y_{i.})^2\right]$$
$$= \frac{1}{n-A} E\left[\sum_{ij} (\mu + a_i + b_{ij} - \mu - a_i - \bar{b}_{i.})^2\right]$$
$$= \frac{1}{n-A} E\left[\sum_{ij} (b_{ij} - \bar{b}_{i.})^2\right]$$

Using Lemma 16.3, we get

$$\frac{1}{n-A} \sum_i \left\{E\left[\sum_j (b_{ij} - \bar{b}_{i.})^2\right]\right\} = \frac{1}{n-A} \sum_i (n_i - 1)\sigma_b^2 = \sigma_b^2$$

Next we find

$$E(A_{\text{ms}}) = \frac{1}{A-1} E\left(\sum_i \frac{Y_{i.}^2}{n_i} - \frac{Y_{..}^2}{n}\right) = \frac{1}{A-1} E\left[\sum_{ij} (y_{i.} - y_{..})^2\right]$$

where
$$y_{..} = \frac{1}{n} \sum_i n_i y_{i.} = \frac{1}{n} \sum_i n_i(\mu + a_i + \bar{b}_{i.})$$

This gives

$$E(A_{\text{ms}}) = \frac{1}{A-1} E\left[\sum_{ij}\left(\mu + a_i + \bar{b}_{i.} - \mu - \frac{1}{n}\sum_p n_p a_p - \frac{1}{n}\sum_{ij} b_{ij}\right)\right]^2$$

$$= \frac{1}{A-1}\left\{E\left[\sum_{ij}\left(a_i - \frac{1}{n}\sum_p n_p a_p\right)^2\right] + E\left[\sum_{ij}\left(\bar{b}_{i.} - \frac{1}{n}\sum_{pq} b_{pq}\right)^2\right]\right\}$$

$$= \frac{1}{A-1}\left(\sum_{ij} E\left[a_i^2 - \frac{2}{n} a_i \sum_p n_p a_p + \frac{1}{n^2}\left(\sum_p n_p a_p\right)^2\right]\right.$$

$$\left. + E\left\{\sum_{ij}\left[\bar{b}_{i.}^2 - \frac{2}{n}\bar{b}_{i.}\sum_{pq} b_{pq} + \frac{1}{n^2}\left(\sum_{pq} b_{pq}\right)^2\right]\right\}\right)$$

$$= \frac{1}{A-1}\left[\sum_{ij}\left(\sigma_a^2 - \frac{2}{n} n_i \sigma_a^2 + \frac{\Sigma n_p^2}{n^2}\sigma_a^2\right) + \sum_{ij}\left(\frac{\sigma_b^2}{n_i} - \frac{2n_i\sigma_b^2}{nn_i} + \frac{\sigma_b^2}{n}\right)\right]$$

$$= \frac{1}{A-1}\left[\sum_i\left(n_i\sigma_a^2 - \frac{2n_i^2}{n}\sigma_a^2 + n_i\frac{\Sigma n_p^2}{n^2}\sigma_a^2\right) + \sum_i\left(\sigma_b^2 - \frac{n_i}{n}\sigma_b^2\right)\right]$$

$$= \frac{1}{A-1}\left[\sigma_a^2\left(n - \frac{\Sigma n_i^2}{n}\right) + \sigma_b^2(A-1)\right]$$

$$= \sigma_b^2 + \frac{n^2 - \Sigma n_i^2}{n(A-1)}\sigma_a^2$$

So, in Table 16.6,

$$k_0 = \frac{n^2 - \Sigma n_i^2}{n(A-1)}$$

The estimates obtained from Table 16.6 do not satisfy the optimum properties of Theorems 16.4 and 16.6.

16.6 Twofold Nested Classification in Model 5 with Unequal Numbers in the Subclasses

The model is

$$y_{ijk} = \mu + a_i + b_{ij} + c_{ijk} \qquad \begin{cases} k = 1, 2, \ldots, n_{ij} \\ j = 1, 2, \ldots, B_i \\ i = 1, 2, \ldots, A \end{cases} \tag{16.15}$$

where

$$\sum_j n_{ij} = n_i$$

$$\sum_{ij} n_{ij} = \sum_i n_i = n$$

$$\Sigma B_i = B$$

This implies that there are A groups of the a_i. The a_i unit contains B_i

units of the b_{ij} (or the a_i unit contains n_i units of c_{ijk}), according to the pattern shown in Table 16.7.

TABLE 16.7 BLOCK PATTERN FOR UNBALANCED TWOFOLD NESTED CLASSIFICATION

a_1	a_2	$\cdots$	a_A
b_{11} b_{12} $\cdots$ b_{1B_1}	$b_{21}b_{22}\cdots b_{2B_2}$		$b_{A1}b_{A2}\cdots b_{AB_A}$
c_{111} c_{121} c_{112} c_{122} $\cdots$ c_{113} c_{123} $\cdots$ $\vdots$ $c_{11n_{11}}c_{12n_{12}}\cdots$			

TABLE 16.8 ANALYSIS OF VARIANCE FOR THE UNBALANCED TWOFOLD NESTED CLASSIFICATION IN MODEL 5

SV	DF	SS	MS	EMS
Total	n	$\sum_{ijk} y_{ijk}^2$		
Mean	1	$\frac{Y_{\ldots}^2}{n}$		
a classes	$A - 1$	$\sum_{ijk} (y_{i..} - y_{...})^2$	A_{ms}	$\sigma_c^2 + q_1\sigma_b^2 + q_2\sigma_a^2$
b classes	$B - A$	$\sum_{ijk} (y_{ij.} - y_{i..})^2$	B_{ms}	$\sigma_c^2 + q_0\sigma_b^2$
c classes	$n - B$	$\sum_{ijk} (y_{ijk} - y_{ij.})^2$	C_{ms}	σ_c^2

An AOV is given in Table 16.8, where

$$
\begin{aligned}
q_0 &= \frac{n - \sum_i \frac{\sum_j n_{ij}^2}{n_i}}{B - A} = \sum_{ij} n_{ij}^2 f_{ij} \\
q_1 &= \frac{\sum_{ij} \left(\frac{n_{ij}^2}{n_i} - \frac{n_{ij}^2}{n} \right)}{A - 1} = \sum_{ij} n_{ij}^2 f_i \\
q_2 &= \frac{n - \frac{\sum_i n_i^2}{n}}{A - 1} = \sum_i n_i^2 f_i
\end{aligned} \tag{16.16}
$$

where

$$f_i = \frac{\dfrac{1}{n_i} - \dfrac{1}{n}}{A - 1} \qquad f_{ij} = \frac{\dfrac{1}{n_{ij}} - \dfrac{1}{n_i}}{B - A}$$

The thing to notice is the EMS column and especially the coefficients q_0, q_1, and q_2. Next we shall show how to obtain them.

First consider

$$E(B_{\text{ms}}) = \frac{1}{B-A} E\left[\sum_{ijk} (y_{ij.} - y_{i..})^2\right]$$

$$= \frac{1}{B-A} E\left(\sum_{ij} \frac{Y_{ij.}^2}{n_{ij}} - \sum_{i} \frac{Y_{i..}^2}{n_i}\right)$$

$$= \frac{1}{B-A} E\left[\sum_{ij} \frac{\left(n_{ij}\mu + n_{ij}a_i + n_{ij}b_{ij} + \sum_k c_{ijk}\right)^2}{n_{ij}}\right.$$

$$\left. - \sum_i \frac{\left(n_i\mu + n_i a_i + \sum_t n_{it}b_{it} + \sum_{ts} c_{its}\right)^2}{n_i}\right]$$

Taking expectations [remembering that $E(a_i) = E(b_{ij}) = E(c_{ijk}) = 0$ and that all random variables are uncorrelated] gives

$$E(B_{\text{ms}}) = \frac{1}{B-A}\left(\sum_{ij} \frac{n_{ij}^2\mu^2 + n_{ij}^2\sigma_a^2 + n_{ij}^2\sigma_b^2 + n_{ij}\sigma_c^2}{n_{ij}}\right.$$

$$\left. - \sum_i \frac{n_i^2\mu^2 + n_i^2\sigma_a^2 + \sum_t n_{it}^2\sigma_b^2 + n_i\sigma_c^2}{n_i}\right)$$

$$= \frac{1}{B-A}\left(n\mu^2 + n\sigma_a^2 + n\sigma_b^2 + B\sigma_c^2 - n\mu^2 - n\sigma_a^2 - \sum_{it} \frac{n_{it}^2}{n_i}\sigma_b^2 - A\sigma_c^2\right)$$

$$= \sigma_c^2 + \frac{n - \sum_{it} \dfrac{n_{it}^2}{n_i}}{B-A}\sigma_b^2$$

So

$$q_0 = \frac{n - \sum_{it} \dfrac{n_{it}^2}{n_i}}{B-A} = \sum_{ij} \frac{1}{B-A}\left(\frac{1}{n_{ij}} - \frac{1}{n_i}\right) n_{ij}^2$$

as given in (16.16).

Next consider

$$E(A_{\text{ms}}) = \frac{1}{A-1} E\left[\sum_{ijk} (y_{i..} - y_{...})^2\right] = \frac{1}{A-1} E\left(\sum_i \frac{Y_{i..}^2}{n_i} - \frac{Y_{...}^2}{n}\right)$$

$$= \frac{1}{A-1}\left(\sum_i \frac{n_i^2\mu^2 + n_i^2\sigma_a^2 + \sum_t n_{it}^2\sigma_b^2 + n_i\sigma_c^2}{n_i}\right.$$

$$\left. - \frac{n^2\mu^2 + \sum_i n_i^2\sigma_a^2 + \sum_{ij} n_{ij}^2\sigma_b^2 + n\sigma_c^2}{n}\right)$$

$$= \frac{1}{A-1}\left[\sum_i n_i^2\left(\frac{1}{n_i} - \frac{1}{n}\right)\sigma_a^2 + \sum_{ij} n_{ij}^2\left(\frac{1}{n_i} - \frac{1}{n}\right)\sigma_b^2 + (A-1)\sigma_c^2\right]$$

The coefficients of σ_b^2 and σ_a^2 correspond to q_1 and q_2 in (16.16). If expected mean squares are equated to mean squares in Table 16.8 and if the resulting equations are solved for σ_a^2, σ_b^2, and σ_c^2, these σ^2's are the analysis-of-variance estimates and are unbiased. The general unbalanced n-fold classification is an extension of the twofold.

The analysis-of-variance estimators for the unbalanced classification do not lead to the same estimates as maximum-likelihood estimators. The maximum-likelihood equations are very difficult to solve in unbalanced classifications; so we shall rely completely on analysis-of-variance estimators.

16.6.1 Example: Twofold Nested Model. Suppose the data in Table 16.9 are assumed to satisfy a twofold nested model. The data are artificial, but we assume that it is a breeding experiment. The number of sires is $A = 4$, the number of dams is $B = 12$, and $n = 52$.

All the quantities needed for an AOV table can easily be computed on an automatic desk calculator. Quantities like

$$\sum_i \frac{Y_{i..}^2}{n_i} \quad \text{and} \quad \sum_{ij} \frac{n_{ij}^2}{n_i}$$

are easily computed. The results are

$$Y_{...} = 1{,}423$$

$$n = 52$$

$$\sum_{ij} \frac{Y_{ij.}^2}{n_{ij}} = 40{,}861.201$$

$$\sum_{ijk} y_{ijk}^2 = 41{,}811$$

TABLE 16.9 DATA FOR EXAMPLE OF ART. 16.6.1

Sires	i	1			2				3		4		
Dams	j	1	2	3	4	5	6	7	8	9	10	11	12
		32	30	34	26	22	23	21	16	14	31	42	26
		31	26	30	20	31	21	21	20	18	34	43	25
		23	29	26	18	20	24	30	32	16	41	40	29
		26	28	34		21	26			17	40	35	40
			18	32			18					29	37
				31									
				26									
Total of jth dam in ith sire	$Y_{ij.}$	112	131	213	64	94	112	72	68	65	146	189	157
Number in jth dam in ith sire	n_{ij}	4	5	7	3	4	5	3	3	4	4	5	5
Total of ith sire	$Y_{i..}$	456			342				133		492		
Number in ith sire	n_i	16			15				7		14		

$$\sum_i \frac{Y_{i..}^2}{n_i} = 40{,}610.885$$

$$\sum_{ij} \frac{n_{ij}^2}{n_i} = 17.844$$

$$\sum_{ij} \frac{n_{ij}^2}{n} = 4.615$$

$$\sum_i \frac{n_i^2}{n} = 13.961$$

$$\frac{Y_{...}^2}{n} = 38{,}940.942$$

So $$q_0 = \frac{52 - 17.844}{8} = 4.270 \qquad q_1 = \frac{17.844 - 4.615}{3} = 4.410$$

$$q_2 = \frac{52 - 13.961}{3} = 12.679$$

The AOV is given in Table 16.10. The estimators are

$$\hat{\sigma}_c^2 = 23.744 \qquad \hat{\sigma}_b^2 = 1.767 \qquad \hat{\sigma}_a^2 = 41.416$$

TABLE 16.10 ANALYSIS OF VARIANCE FOR ARTIFICIAL BREEDING DATA

SV	DF	SS	MS	EMS
Total	52	41,811.000		
Mean	1	38,940.942		
a classes	3	1,669.943	556.648	$\sigma_c^2 + 4.410\sigma_b^2 + 12.679\sigma_a^2$
b classes	8	250.316	31.289	$\sigma_c^2 + 4.270\sigma_b^2$
c classes	40	949.799	23.744	σ_c^2

16.7 The Unbalanced Two-way Classification in Model 5

The model for this case is

$$y_{ijk} = \mu + a_i + b_j + c_{ijk} \qquad \begin{cases} k = 0, 1, \ldots, n_{ij} \\ j = 1, 2, \ldots, B \\ i = 1, 2, \ldots, A \end{cases}$$

where $\quad \sum_j n_{ij} = n_{i.} \qquad \sum_i n_{ij} = n_{.j} \qquad \sum_{ij} n_{ij} = n$

The a_i, b_j, c_{ijk} are random variables with zero means and variances σ_a^2, σ_b^2, and σ_c^2, respectively. We find that the maximum-likelihood equations are very difficult to solve. Hence we shall rely on analysis-of-variance estimators. This, however, presents a difficulty. In the unbalanced cross classification discussed in Chap. 13 we saw that

$$R(\mu,\tau,\beta) = R(\tau \mid \mu,\beta) + R(\mu,\beta)$$

and

$$R(\mu,\tau,\beta) = R(\beta \mid \mu,\tau) + R(\mu,\tau)$$

So we can partition the total sum of squares in two ways:

$$\Sigma y_{ijk}^2 = \frac{Y_{...}^2}{n} + R(\tau \mid \mu,\beta) + R(\mu,\beta) + (\text{error})$$

and

$$\Sigma y_{ijk}^2 = \frac{Y_{...}^2}{n} + R(\beta \mid \mu,\tau) + R(\mu,\tau) + (\text{error})$$

If this is done for the model in this section and if mean squares are set equal to expected mean squares, we shall obtain two sets of estimators. Each set is unbiased.

Another method that is sometimes used is to partition the total sum of squares as follows:

$$\Sigma y_{ijk}^2 = \left(\frac{Y_{...}^2}{n}\right) + \left[\sum_i \left(\frac{Y_{i..}^2}{n_{i.}}\right) - \frac{Y_{...}^2}{n}\right] + \left[\sum_j \left(\frac{Y_{.j.}^2}{n_{.j}}\right) - \frac{Y_{...}^2}{n}\right] + (\text{remainder})$$

This can be put in an AOV table such as Table 16.11.

TABLE 16.11 ANALYSIS OF VARIANCE FOR UNBALANCED TWO-WAY CLASSIFICATION MODEL

SV	DF	SS	MS	EMS
Total	n	$\sum_{ijk} y_{ijk}^2$		
Mean	1	$\frac{Y_{...}^2}{n}$		
a classes	$A - 1$	$\sum_i \frac{Y_{i..}^2}{n_{i.}} - \frac{Y_{...}^2}{n}$	A_{ms}	$\sigma_c^2 + r_5\sigma_b^2 + r_6\sigma_a^2$
b classes	$B - 1$	$\sum_j \frac{Y_{.j.}^2}{n_{.j}} - \frac{Y_{...}^2}{n}$	B_{ms}	$\sigma_c^2 + r_3\sigma_b^2 + r_4\sigma_a^2$
Remainder	$n - A - B + 1$	Subtraction	R_{ms}	$\sigma_c^2 + r_1\sigma_b^2 + r_2\sigma_a^2$

To find the estimates, we equate observed mean squares to expected mean squares and solve the resulting set of three equations for the three unknowns $\hat{\sigma}_a^2$, $\hat{\sigma}_b^2$, and $\hat{\sigma}_c^2$.

It should be pointed out that the remainder mean square obtained by subtraction can be negative.

To find the expected mean squares and the r_i, we proceed as follows:

$$E(A_{\text{ms}}) = \frac{1}{A-1} E\left(\sum_i \frac{Y_{i..}^2}{n_{i.}} - \frac{Y_{...}^2}{n}\right)$$

$$= \frac{1}{A-1} E\left[\sum_i \frac{\left(n_{i.}\mu + n_{i.}a_i + \sum_j n_{ij}b_j + \sum_{jk} c_{ijk}\right)^2}{n_{i.}} - \frac{\left(n\mu + \sum_i n_{i.}a_i + \sum_j n_{.j}b_j + \sum_{ijk} c_{ijk}\right)^2}{n}\right]$$

$$= \frac{1}{A-1}\left[\left(n\mu^2 + n\sigma_a^2 + \sum_{ij} \frac{n_{ij}^2}{n_{i.}} \sigma_b^2 + A\sigma_c^2\right) - \left(n\mu^2 + \sum_i \frac{n_{i.}^2}{n} \sigma_a^2 + \sum_j \frac{n_{.j}^2}{n} \sigma_b^2 + \sigma_c^2\right)\right]$$

$$= \frac{1}{A-1}\left(n - \sum_i \frac{n_{i.}^2}{n}\right)\sigma_a^2 + \frac{1}{A-1}\left(\sum_{ij} \frac{n_{ij}^2}{n_{i.}} - \sum_j \frac{n_{.j}^2}{n}\right)\sigma_b^2 + \sigma_c^2$$

The values of r_5 and r_6 can be obtained from this expression. Also, $E(B_{\text{ms}})$ can be written down with symmetry, replacing i with j and interchanging σ_a^2 and σ_b^2 in $E(A_{\text{ms}})$. We get

$$E(B_{\text{ms}}) = \frac{1}{B-1}\left(\sum_{ij}\frac{n_{ij}^2}{n_{.j}} - \sum_i \frac{n_{i.}^2}{n}\right)\sigma_a^2 + \frac{1}{B-1}\left(n - \sum_j \frac{n_{.j}^2}{n}\right)\sigma_b^2 + \sigma_c^2$$

from which r_3 and r_4 can be obtained.

Since the sums of squares add to Σy_{ijk}^2, this gives

$$\Sigma y_{ijk}^2 = \frac{Y_{...}^2}{n} + (A-1)A_{\text{ms}} + (B-1)B_{\text{ms}} + (n-A-B+1)R_{\text{ms}}$$

Taking expected values of both sides and simplifying, we get

$$E(R_{\text{ms}}) = \frac{1}{n-A-B+1}\left[E(\Sigma y_{ijk}^2) - E\left(\frac{Y_{...}^2}{n}\right) - (A-1)E(A_{\text{ms}}) - (B-1)E(B_{\text{ms}})\right]$$

The quantities on the right have been computed, except for $E(\Sigma y_{ijk}^2)$. This is

$$E\left(\sum_{ijk} y_{ijk}^2\right) = E\left[\sum_{ijk}(\mu + a_i + b_j + c_{ijk})^2\right] = n(\mu^2 + \sigma_a^2 + \sigma_b^2 + \sigma_c^2)$$

Hence,

$$\begin{aligned} E(R_{\text{ms}}) &= \frac{1}{n-A-B+1} \\ &\quad\times \left[n(\mu^2 + \sigma_a^2 + \sigma_b^2 + \sigma_c^2) - n\mu^2 - \sum_i \frac{n_{i.}^2}{n}\sigma_a^2 - \sum_j \frac{n_{.j}^2}{n}\sigma_b^2\right. \\ &\quad - \sigma_c^2 - \left(n - \sum_i \frac{n_{i.}^2}{n}\right)\sigma_a^2 - \left(\sum_{ij}\frac{n_{ij}^2}{n_{i.}} - \sum_j \frac{n_{.j}^2}{n}\right)\sigma_b^2 - (A-1)\sigma_c^2 \\ &\quad \left. - \left(\sum_{ij}\frac{n_{ij}^2}{n_{.j}} - \sum_i \frac{n_{i.}^2}{n}\right)\sigma_a^2 - \left(n - \sum_j \frac{n_{.j}^2}{n}\right)\sigma_b^2 - (B-1)\sigma_c^2\right] \\ &= \sigma_c^2 - \frac{1}{n-A-B+1}\left(\sum_{ij}\frac{n_{ij}^2}{n_{i.}} - \sum_j \frac{n_{.j}^2}{n}\right)\sigma_b^2 \\ &\quad - \frac{1}{n-A-B+1}\left(\sum_{ij}\frac{n_{ij}^2}{n_{.j}} - \sum_i \frac{n_{i.}^2}{n}\right)\sigma_a^2 \end{aligned}$$

This gives r_1 and r_2. We have

$$r_1 = \frac{1}{n - A - B + 1}\left(\sum_j \frac{n_{.j}^2}{n} - \sum_{ij} \frac{n_{ij}^2}{n_{i.}}\right)$$

$$r_2 = \frac{1}{n - A - B + 1}\left(\sum_i \frac{n_{i.}^2}{n} - \sum_{ij} \frac{n_{ij}^2}{n_{.j}}\right)$$

$$r_3 = \frac{1}{B - 1}\left(n - \sum_j \frac{n_{.j}^2}{n}\right)$$

$$r_4 = \frac{1}{B - 1}\left(\sum_{ij} \frac{n_{ij}^2}{n_{.j}} - \sum_i \frac{n_{i.}^2}{n}\right)$$

$$r_5 = \frac{1}{A - 1}\left(\sum_{ij} \frac{n_{ij}^2}{n_{i.}} - \sum_j \frac{n_{.j}^2}{n}\right)$$

$$r_6 = \frac{1}{A - 1}\left(n - \sum_i \frac{n_{i.}^2}{n}\right)$$

16.8 The Balanced Incomplete Block in Model 5

The balanced incomplete block classification in model 4 has been defined and discussed in Chap. 14. The definition of this classification in model 5 is the same, except that the block β_j and treatment τ_i components are now random variables. Therefore, we shall write the model as

$$y_{ijm} = \mu + t_i + b_j + e_{ijm} \qquad i = 1, 2, \ldots, t; j = 1, 2, \ldots, b$$

where $m = n_{ij}$ and $n_{ij} = 0$ or 1. Also, there are k treatments per block, and each treatment is repeated r times. Each pair of treatments appears together in exactly λ blocks. Also, as in Chap. 13, when $m = 0$ the corresponding observation is not present; that is, y_{ij0} does not exist. Now t_i, b_j, e_{ijm} are uncorrelated random variables with zero means and variances σ_t^2, σ_b^2, σ_e^2, respectively.

Since the maximum-likelihood estimators are difficult to obtain, we shall use analysis-of-variance estimators. However, this presents the same problem as did the general two-way classification model discussed in Sec. 16.7.

There are at least three ways to find analysis-of-variance estimators, just as there were for the model in the preceding section.

16.8.1 Method 1. If we use the method of Table 16.11, the coefficients r_i simplify somewhat. This AOV is given in Table 16.12. The correspondence in notation is as follows: A becomes t; $B = b$;

TABLE 16.12 ANALYSIS OF VARIANCE (METHOD 1) FOR BALANCED INCOMPLETE BLOCK IN MODEL 5

SV	DF	SS	MS	EMS
Total	$bk = rt$	$\Sigma\, y_{ijm}^2$		
Mean	1	$\dfrac{Y_{...}^2}{bk}$		
Treatments	$t - 1$	$\dfrac{\sum_i Y_{i..}^2}{r} - \dfrac{Y_{...}^2}{bk}$	T_{ms}	$\sigma_e^2 + \dfrac{t-k}{t-1}\sigma_b^2 + r\sigma_t^2$
Blocks	$b - 1$	$\dfrac{\sum_j Y_{.j.}^2}{k} - \dfrac{Y_{...}^2}{bk}$	B_{ms}	$\sigma_e^2 + k\sigma_b^2 + \dfrac{b-r}{b-1}\sigma_t^2$
Error	$bk - t - b + 1$	Subtraction	R_{ms}	$\sigma_e^2 - \dfrac{(t-k)\sigma_b^2 + (b-r)\sigma_t^2}{bk - b - t + 1}$

$n_{i.} = r$; $n_{.j} = k$; $n = bk$. Also, $\sigma_c^2 = \sigma_e^2$; $\sigma_a^2 = \sigma_t^2$; and σ_b^2 is the same in both tables. For example,

$$r_1 = \frac{1}{n - A - B + 1}\left(-\sum_{ij}\frac{n_{ij}^2}{n_{i.}} + \sum_j \frac{n_{.j}^2}{n}\right)$$

$$= \frac{1}{bk - t - b + 1}\left(-\sum_{ij}\frac{n_{ij}}{r} + \sum_j \frac{k^2}{bk}\right)$$

$$= \frac{1}{bk - t - b + 1}\left(-\frac{rt}{r} + k\right)$$

$$= -\frac{t - k}{bk - t - b + 1}$$

$$r_2 = \frac{1}{bk - t - b + 1}\left(-\sum_{ij}\frac{n_{ij}^2}{n_{.j}} + \sum_i \frac{n_{i.}^2}{n}\right)$$

$$= \frac{1}{bk - t - b + 1}\left(-\sum_{ij}\frac{n_{ij}}{k} + \sum_i \frac{r^2}{bk}\right)$$

$$= \frac{1}{bk - t - b + 1}(-b + r)$$

$$= -\frac{b - r}{bk - t - b + 1}$$

$$r_3 = \frac{1}{b-1}\left(n - \sum_j \frac{n_{.j}^2}{n}\right)$$

$$= \frac{1}{b-1}\left(bk - \sum_j \frac{k^2}{bk}\right)$$

$$= \frac{bk-k}{b-1} = k$$

$$r_4 = \frac{1}{b-1}\left(\sum_{ij} \frac{n_{ij}^2}{n_{.j}} - \sum_i \frac{n_{i.}^2}{n}\right)$$

$$= \frac{b-r}{b-1}$$

$$r_5 = \frac{1}{t-1}\left(\sum_{ij} \frac{n_{ij}^2}{n_{i.}} - \sum_j \frac{n_{.j}^2}{n}\right)$$

$$= \frac{t-k}{t-1}$$

$$r_6 = \frac{1}{t-1}\left(n - \sum_i \frac{n_{i.}^2}{n}\right)$$

$$= \frac{1}{t-1}\left(rt - \sum_i \frac{r^2}{rt}\right)$$

$$= \frac{rt-r}{t-1} = r$$

16.8.2 Method 2. Another method for constructing an AOV for the balanced incomplete block classification is that used in Chap. 14.

TABLE 16.13 ANALYSIS OF VARIANCE (METHOD 2) FOR BALANCED INCOMPLETE BLOCK IN MODEL 5

SV	DF	SS	MS	EMS
Total	$bk = rt$	$\sum_{ijm} y_{ijm}^2$		
Mean	1	$\frac{Y_{...}^2}{bk}$		
Treatments adj for blocks	$t-1$	$\frac{k}{\lambda t}\sum_i q_i^2$	T_{ms}	$\sigma_e^2 + \frac{\lambda t}{k}\sigma_t^2$
Blocks unadj	$b-1$	$\frac{\sum_j Y_{.j.}^2}{k} - \frac{Y_{...}^2}{bk}$	B_{ms}	$\sigma_e^2 + k\sigma_b^2 + \frac{b-r}{b-1}\sigma_t^2$
Error	$bk - t - b + 1$	Subtraction	E_{ms}	σ_e^2

This method is demonstrated in Table 16.13. As in Chap. 14,

$$q_i = Y_{i..} - \frac{1}{k}\sum_j n_{ij} Y_{.j.}$$

and

$$E(T_{\text{ms}}) = E\left(\frac{k\Sigma q_i^2}{\lambda t}\right) = \sigma_e^2 + \frac{\lambda t}{k}\sigma_t^2$$

The other EMS values can easily be shown to be as given.

Problems

16.1 Evaluate the following determinant by any method, and verify Lemma 16.1.

$$\begin{vmatrix} 8 & 2 & 2 & 2 \\ 2 & 8 & 2 & 2 \\ 2 & 2 & 8 & 2 \\ 2 & 2 & 2 & 8 \end{vmatrix}$$

16.2 Evaluate the inverse of the following matrix by any method, and verify Lemma 16.2.

$$\begin{pmatrix} 3 & 1 & 1 \\ 1 & 3 & 1 \\ 1 & 1 & 3 \end{pmatrix}$$

16.3 Prove Lemma 16.1.

16.4 Prove Lemma 16.3.

16.5 Assume that the data shown in Table 16.14 satisfy the assumptions for case A given in Eq. (16.2). Find the estimators of σ_a^2 and σ_b^2 given in Eq. (16.11).

TABLE 16.14

a effects					
1	2	3	4	5	6
16.1	19.3	15.3	20.1	13.7	18.3
16.5	17.2	15.7	19.1	13.1	18.9
15.3	18.3	14.1	19.4	12.6	19.1
17.0	18.5	16.2	18.9	12.8	19.3
16.8	17.3	17.1	18.6	13.1	17.9
16.7	17.9	17.3	20.5	13.4	17.8

16.6 Prove Theorem 16.3.

16.7 Go through the details of finding the expected mean squares in Table 16.1.

16.8 Assume that the data of Table 16.15 satisfy the conditions of a one-way classification model with unequal numbers in the subclasses. Find k_0.

16.9 In Prob. 16.8, compute the quantities needed in an AOV table.

16.10 In Prob. 16.8, find estimates of σ_a^2 and σ_b^2.

TABLE 16.15

a effect			
1	2	3	4
30.2	29.1	35.1	27.2
31.1	28.7	35.3	28.3
31.8	27.5	36.4	29.4
32.5	27.9		28.5
30.6	28.3		
31.3			
32.0			

16.11 Go through the details of finding q_0, q_1, and q_2 in Table 16.8.

16.12 Assume that the data of Table 16.16 satisfy the conditions of a two-way cross classification in model 5 with unequal numbers in the cells. Find the quantities r_1, r_2, r_3, r_4, r_5, and r_6 in Table 16.11.

TABLE 16.16

b \ *a*	1	2	3	4	5
1	1.3 1.6		7.2 7.5 7.6	9.1	11.2
2		4.2 4.3	10.1 10.3	12.2 12.4 12.6	14.1 14.3
3	6.1 6.3	8.4 8.2	12.1		16.1 16.3 17.2

16.13 From the data in Prob. 16.12, estimate σ_a^2, σ_b^2, and σ_c^2 by constructing Table 16.11.

16.14 Assume that the data of Table 16.17 satisfy a balanced incomplete

TABLE 16.17

Block	Treatments			
1	(1)	1.2	(2)	2.3
2	(1)	2.9	(3)	8.1
3	(1)	5.1	(4)	11.9
4	(2)	10.0	(3)	14.1
5	(2)	11.7	(4)	17.8
6	(3)	17.0	(4)	20.1

block in model 5. Estimate σ_a^2, σ_b^2, and σ_c^2 by using method 1 and Table 16.12. The numbers in parentheses are treatment numbers.

16.15 In Prob. 16.14, estimate σ_a^2, σ_b^2, and σ_c^2 by using method 2 and Table 16.13.

Further Reading

1 R. A. Fisher: "Statistical Methods for Research Workers," Oliver and Boyd, Ltd., London, 1946.

2 G. W. Snedecor: "Statistical Methods," Iowa State College Press, Ames, Iowa, 1958.

3 O. Kempthorne: "Design and Analysis of Experiments," John Wiley & Sons, Inc., New York, 1952.

4 C. H. Goulden: "Methods of Statistical Analysis," John Wiley & Sons, Inc., New York, 1939.

5 O. L. Davies: "Design and Analysis of Industrial Experiments," Oliver and Boyd, Ltd., London, 1954.

6 R. L. Anderson and T. A. Bancroft: "Statistical Theory in Research," McGraw-Hill Book Company, Inc., New York, 1952.

7 P. L. Hsu: On the Best Unbiased Quadratic Estimate of the Variance, *Statist. Research Mem.*, vol. 2, pp. 91–104, 1938.

8 F. A. Graybill and A. W. Wortham: A Note on Uniformly Best Unbiased Estimators for Variance Components, *J. Am. Statist. Assoc.*, vol. 51, pp. 266–268, 1956.

9 C. Eisenhart: The Assumptions Underlying the A.O.V., *Biometrics*, vol. 3, pp. 1–21, 1947.

10 S. L. Crump: The Present Status of Variance Component Analysis, *Biometrics*, vol. 7, pp. 1–16, 1951.

11 E. F. Schultz, Jr.: Rules of Thumb for Determining Expectations of Mean Squares in A.O.V., *Biometrics*, vol. 11, pp. 123–135, 1955.

12 H. Naglor: On the Best Unbiased Quadratic Estimate of the Variance, *Biometrika*, vol. 37, pp. 444–445, 1950.

13 J. A. Nelder: The Interpretation of Negative Components of Variance, *Biometrika*, vol. 41, pp. 544–548, 1954.

14 J. W. Tukey: Variances of Variance Components: I, Balanced Designs, *Ann. Math. Statist.*, vol. 27, pp. 722–736, 1956.

15 S. R. Searle: Matrix Methods in Components of Variance and Covariance Analysis, *Ann. Math. Statist.*, vol. 27, pp. 737–748, 1956.

16 C. R. Henderson: Estimation of Variance and Covariance Components, *Biometrics*, vol. 9, pp. 226–252, 1953.

17 F. A. Graybill: On Quadratic Estimates of Variance Components, *Ann. Math. Statist.*, vol. 25, pp. 367–372, 1954.

17

Model 5: Variance Components; Interval Estimation and Tests of Hypotheses

Throughout this chapter the following assumptions will be made:

1. The model used is model 5, the variance-components model.
2. There are an equal number of observations in all subclasses.
3. The errors satisfy the distributional properties of case A; i.e., all random variables are independently and normally distributed.

First some general theorems will be given and then these will be applied to certain special cases.

17.1 Distribution of a Linear Combination of Chi-square Variates

Let n_1x_1/σ_1^2 be distributed as $\chi^2(n_1)$, let n_2x_2/σ_2^2 be distributed as $\chi^2(n_2), \ldots$, let n_kx_k/σ_k^2 be distributed as $\chi^2(n_k)$, and let $x_1, x_2, \ldots, x_k$ be independent. The problem we shall consider is that of setting confidence limits on

$$\gamma = g_1\sigma_1^2 + g_2\sigma_2^2 + \cdots + g_k\sigma_k^2 \tag{17.1}$$

where $\gamma > 0$ and $g_1, g_2, \ldots, g_k$ are known constants. To do this we shall consider the distribution of $g = g_1x_1 + g_2x_2 + \cdots + g_kx_k$. Since the x_i are independent, the moment-generating function of g can easily be found. However, the distribution of g is very complicated, and attempts to use it to set exact confidence limits on (17.1) have not been successful.

Various approximate methods for setting confidence limits on functions of variances such as (17.1) have been given [9]. We shall demonstrate a method that has been described by Welch [7]. Before examining different models we shall state some theorems that will be useful.

◆ **Theorem 17.1** Let $n_i x_i/\sigma_i^2$ $(i = 1, 2, \ldots, k)$ be independently distributed as $\chi^2(n_i)$. Let

$$\gamma = \sum_{i=1}^{k} g_i\sigma_i^2 \qquad \gamma > 0$$

Let
$$g = \sum_{i=1}^{k} g_i x_i$$

Then $u = ng/\gamma$ is approximately distributed as $\chi^2(n)$, where

$$n = \frac{(\Sigma g_i\sigma_i^2)^2}{\Sigma(g_i^2\sigma_i^4/n_i)}$$

Proof: We shall not calculate the accuracy of this approximation, but shall merely evaluate n. If u is distributed as $\chi^2(n)$, then $\text{var}(u) = 2n$. But

$$\text{var}(u) = \frac{n^2}{\gamma^2}\,\text{var}(g) = \frac{n^2}{\gamma^2}\,[g_1^2\,\text{var}(x_1) + \cdots + g_k^2\,\text{var}(x_k)]$$

$$= \frac{n^2}{\gamma^2}\left(\frac{2g_1^2\sigma_1^4}{n_1} + \frac{2g_2^2\sigma_2^4}{n_2} + \cdots + \frac{2g_k^2\sigma_k^4}{n_k}\right)$$

Equating this to $2n$ gives

$$n = \frac{\gamma^2}{\Sigma(g_i^2\sigma_i^4/n_i)} = \frac{(\Sigma g_i\sigma_i^2)^2}{\Sigma(g_i^2\sigma_i^4/n_i)}$$

Ordinarily the σ_i^2 are not known, so they could be replaced by their estimators x_i in the formula for n. Another point should be emphasized: this approximation should not be used unless g is positive. However, since $E(g) = \gamma$ and since we have assumed that γ is positive, a negative value of g would not be very useful in estimating γ.

To set confidence limits on γ, Theorem 17.1 can be used, and the result is as follows.

◆ **Theorem 17.2** Since ng/γ is approximately distributed $\chi^2(n)$, an approximate confidence interval on γ with coefficient $1 - \alpha$ is given by

$$P\left(\frac{ng}{\chi^2_{\alpha/2}(n)} \leqslant \gamma \leqslant \frac{ng}{\chi^2_{1-\alpha/2}(n)}\right) = 1 - \alpha$$

where
$$n = \frac{(\Sigma g_i x_i)^2}{\Sigma(g_i^2 x_i^2/n_i)}$$

and $\chi^2_\beta(n)$ is the upper β percentage point of a chi-square with n degrees of freedom.

Proof: The theorem can be proved by straightforward application of the method of using a chi-square variate to determine a confidence interval.

Although Theorem 17.2 can be used to set approximate confidence intervals on variances, a better method in most experimental situations has been given by Welch; this is the text of the next theorem.

◆ **Theorem 17.3** Let $n_i x_i/\sigma_i^2$ $(i = 1, 2, \ldots, k)$ be independently distributed as $\chi^2(n_i)$, and let

$$g = \sum_{i=1}^{k} g_i x_i$$

and

$$\gamma = \sum_{i=1}^{k} g_i \sigma_i^2 \qquad \gamma > 0$$

Then the probability statement

$$P\left(\frac{ng}{\gamma} \leqslant \chi_\beta^2(n) - \frac{2}{3}(2z_\beta^2 + 1)(gs^{-2}t - 1)\right) = 1 - \beta$$

is approximately true. This can be used to give the $1 - \alpha$ confidence interval

$$P\left(\frac{ng}{A_{\alpha/2}} \leqslant \gamma \leqslant \frac{ng}{A_{1-\alpha/2}}\right) = 1 - \alpha$$

$A_{\alpha/2}$ and $A_{1-\alpha/2}$ are computed by using

$$A_\beta = \chi_\beta^2(n) - C_\beta = \chi_\beta^2(n) - \tfrac{2}{3}(2z_\beta^2 + 1)(gs^{-2}t - 1)$$

with $\alpha/2$ and $1 - \alpha/2$ respectively substituted for β. C_β will be called the *correction factor*. z_β is such that

$$\int_{-\infty}^{z_\beta} h(y)\, dy = 1 - \beta$$

where y is distributed $N(0,1)$. Also,

$$t = \sum_{i=1}^{k} \frac{g_i^3 x_i^3}{n_i^2}$$

$$s = \sum \frac{g_i^2 x_i^2}{n_i}$$

For a proof of this theorem, see Welch [7]. If $n > 30$, the confidence interval given in Theorem 17.2 is probably adequate. If $n < 30$, the improved approximation given in Theorem 17.3 can be used. If the correction term

$$C_\beta = \tfrac{2}{3}(2z_\beta^2 + 1)(gs^{-2}t - 1)$$

is small, the method described in Theorem 17.2 is adequate.

17.1.1 Example. To illustrate the theory in this section, we shall work an example of a twofold classification. The model is

$$y_{ijk} = \mu + a_i + b_{ij} + c_{ijk} \qquad \begin{cases} k = 1, 2, 3 \\ j = 1, 2, 3, 4, 5 \\ i = 1, 2, \ldots, 12 \end{cases}$$

The distributional properties are those given for case A in Chap. 16, where the variances of a_i, b_{ij}, and c_{ijk} are σ_a^2, σ_b^2, and σ_c^2, respectively. The original data will not be given, but the AOV is shown in Table 17.1.

TABLE 17.1 ANALYSIS OF VARIANCE OF TWOFOLD CLASSIFICATION MODEL

SV	DF (n_i)	MS (x_i)	EMS
Total	180		
Mean	1		
a classes	$11 = n_1$	$3.5629 = x_1$	$\sigma_1^2 = \sigma_c^2 + 3\sigma_b^2 + 15\sigma_a^2$
b classes	$48 = n_2$	$1.2055 = x_2$	$\sigma_2^2 = \sigma_c^2 + 3\sigma_b^2$
c classes	$120 = n_3$	$.6113 = x_3$	$\sigma_3^2 = \sigma_c^2$

The estimates are

$$\hat{\sigma}_c^2 = .6113 \qquad \hat{\sigma}_b^2 = .1981 \qquad \hat{\sigma}_a^2 = .1572$$

Suppose we want the following:

1. A 95 per cent confidence interval on σ_b^2
2. A 90 per cent confidence interval on $\sigma_a^2 + \sigma_b^2 + \sigma_c^2$

To find interval 1, Theorem 17.2 will be used. We have

$$\gamma = \sigma_b^2 = \tfrac{1}{3}\sigma_2^2 - \tfrac{1}{3}\sigma_3^2$$

so

$$g_1 = 0 \qquad g_2 = \tfrac{1}{3} \qquad g_3 = -\tfrac{1}{3}$$

Table 17.2 will be useful for computing the confidence limits.

Now, $(\Sigma g_i x_i)^2 = [(0)(3.5629) + (.3333)(1.2055) - (.3333)(.6113)]^2$

$$= (.1981)^2 = .03924$$

and

$$\Sigma \frac{g_i^2 x_i^2}{n_i} = \frac{(0)^2}{11} + \frac{(.40180)^2}{48} + \frac{(-.20375)^2}{120} = .003709$$

TABLE 17.2 TABLE OF QUANTITIES FOR INTERVAL 1, EXAMPLE OF ART. 17.1.1

i	n_i	g_i	x_i	$g_i x_i$	$g_i^2 x_i^2/n_i$	$g_i^3 x_i^3/n_i^2$
1	11	.0000	3.5629	.00000	.00000000	.000000000
2	48	.3333	1.2055	.40180	.0033630	.000028150
3	120	−.3333	.6113	−.20375	.0003459	−.000000587
				.1981 = g	.003709 = s	.0000276 = t

So

$$n = \frac{.03924}{.003709} = 10.58$$

and we shall use $n = 11$. Also,

$$g = g_1x_1 + g_2x_2 + g_3x_3$$
$$= 0 + .4018 - .20375 = .1981$$

and

$$ng = 2.1791$$

Since $1 - \alpha = .95$, we have $1 - \alpha/2 = .975$ and $\alpha/2 = .025$. So, for 11 degrees of freedom, we get

$$\chi^2_{.025} = 21.92 \qquad \text{and} \qquad \chi^2_{.975} = 3.82$$

Using the formula in Theorem 17.2, we get

$$P\left(\frac{ng}{\chi^2_{.025}} \leqslant \gamma \leqslant \frac{ng}{\chi^2_{.975}}\right) = .95$$

which gives

$$P\left(\frac{2.1791}{21.92} \leqslant \sigma_b^2 \leqslant \frac{2.1791}{3.82}\right) = .95$$

or

$$P(.0994 \leqslant \sigma_b^2 \leqslant .570) = .95$$

Next we shall compute the more exact confidence limits, by Theorem 17.3. We can compute C_β as follows:

$$C_{.975} = (+.6666)[(2)(-1.96)^2 + 1][(.1981)(.003709)^{-2}(.00002763) - 1]$$
$$= -3.4877$$

$$C_{.025} = (+.6666)[(2)(1.96)^2 + 1][(.1981)(.003709)^{-2}(.00002763) - 1]$$
$$= -3.4877$$

and $A_{.025} = 21.92 + 3.49 = 25.41 \qquad A_{.975} = 3.82 + 3.49 = 7.31$

The limits are

$$P\left(\frac{2.179}{25.41} \leqslant \sigma_b^2 \leqslant \frac{2.179}{7.31}\right) = .95$$

or

$$P(.086 \leqslant \sigma_b^2 \leqslant .298) = .95$$

There seems to be an improvement resulting from use of Theorem 17.3 instead of Theorem 17.2.

To compute interval 2, the confidence interval for

$$\gamma = \sigma_a^2 + \sigma_b^2 + \sigma_c^2 = \tfrac{1}{15}\sigma_1^2 + \tfrac{4}{15}\sigma_2^2 + \tfrac{10}{15}\sigma_3^2$$

we get

$$g = \tfrac{1}{15}x_1 + \tfrac{4}{15}x_2 + \tfrac{10}{15}x_3$$

The quantities in Table 17.3 can easily be computed.

TABLE 17.3 TABLE OF QUANTITIES FOR INTERVAL 2, EXAMPLE OF ART. 17.1.1

i	n_i	g_i	x_i	g_ix_i	$g_i^2x_i^2/n_i$
1	11	.06666	3.5629	.23753	.0051291
2	48	.26666	1.2055	.32147	.0021529
3	120	.66666	.6113	.40753	.0013840
				.96653	.0086660

The estimate is

$$\hat{\gamma} = g = \Sigma g_i x_i = \hat{\sigma}_a^2 + \hat{\sigma}_b^2 + \hat{\sigma}_c^2 = .9665$$

To demonstrate confidence limits on γ, we shall use only Theorem 17.2. This gives

$$n = \frac{(\Sigma g_i x_i)^2}{\Sigma(g_i^2 x_i^2/n_i)} = 107.8 \cong 108$$

$$ng = 104.39$$

Since $1 - \alpha = .90$, we get $\chi^2_{.05}(108) = 125$ and $\chi^2_{.95}(108) = 78$. Thus

$$P\left(\frac{104.39}{125} < \gamma < \frac{104.39}{78}\right) = .90$$

and

$$P(.835 < \sigma_a^2 + \sigma_b^2 + \sigma_c^2 < 1.338) = .90$$

The correction term C_β in Theorem 17.3 can be computed to give improved limits, if desired.

17.2 Tests of Hypotheses

In this section we shall discuss two general tests. One is a test that a variance component is zero, for example, that $\sigma_a^2 = 0$. The other is that a linear combination of $\sigma_a^2, \sigma_b^2, \sigma_c^2, \ldots$, is equal to zero. The notation in Chap. 16 will be used, and references to various special models and AOV tables will be made.

First we shall consider a general analysis-of-variance model and the quantities in Table 16.2.

In many cases, the difference $\sigma_i^2 - \sigma_j^2$ in Table 16.2 is equal to a multiple of one of the variance components $\sigma_a^2, \sigma_b^2, \ldots$. For example, in Table 16.1, we get $\sigma_1^2 - \sigma_2^2 = s\sigma_a^2$; in Table 16.3, we have $\sigma_1^2 - \sigma_3^2 = s\sigma_a^2$; in Table 16.4, we have $\sigma_2^2 - \sigma_3^2 = t\sigma_b^2$; etc.

A test of the hypothesis $H_0: \sigma_i^2 = \sigma_j^2$ will be given in the next theorem.

◆ **Theorem 17.4** Let the model satisfy the conditions of Theorem 16.5, and let the AOV be represented by Table 16.2. To test the hypothesis $H_0: \sigma_i^2 = \sigma_j^2$ $(i \neq j)$ against the alternative hypothesis $H_A: \sigma_j^2 > \sigma_i^2$, the quantity

$$u = \frac{s_j^2}{s_i^2}$$

can be used, and H_0 is rejected at the α level of significance if $u > F_\alpha$, where F_α is the upper percentage point of Snedecor's F distribution with n_j and n_i degrees of freedom. That is to say, F_α is such that

$$\alpha = \int_{F_\alpha}^{\infty} h(u)\, du$$

where $h(u)$ is the $F(n_j,n_i)$ distribution. The power of the test depends on the quantity

$$\lambda = \frac{\sigma_i^2}{\sigma_j^2} < 1$$

and is
$$\int_{\lambda F_\alpha}^{\infty} h(u)\, du = \beta(\lambda)$$

Proof: Let $g(u;\lambda)$ be the frequency function of u for various values of λ. If $\lambda = 1$, u is distributed as $F(n_j,n_i)$; so $g(u;1)$ is Snedecor's F distribution with n_j and n_i degrees of freedom. Of course, if $\lambda \neq 1$, then $g(u;\lambda)$ is not the F distribution. The test procedure is to reject H_0 if $u > F_\alpha$, where F_α is the upper α point of the F distribution; that is, where F_α is given by

$$\alpha = \int_{F_\alpha}^{\infty} g(u;1)\, du$$

The power of the test, denoted by $\beta(\lambda)$, is

$$\beta(\lambda) = \int_{F_\alpha}^{\infty} g(u;\lambda)\, du$$

To evaluate $\beta(\lambda)$, we make the transformation $w = \lambda u$ and substitute this into $g(u;\lambda)$. Now $g(u;\lambda)$ becomes the density function of w, say $h(w)$. But w is distributed as $F(n_j,n_i)$ for every value of λ $(0 \leqslant \lambda < \infty)$; so $h(w)$ is the central F distribution. To evaluate the limits on the integral, we note that, when $u = F_\alpha$, $w = \lambda F_\alpha$, and when $u = \infty$, $w = \infty$; so

$$\beta(\lambda) = \int_{\lambda F_\alpha}^{\infty} h(w)\, dw$$

and the power can be evaluated by using only the central F tables.

17.2.1 Example: One-way Classification. For this example we refer to Table 16.1. The correspondence in notation is

$$\sigma_1^2 = \sigma_b^2 + s\sigma_a^2$$
$$s_1^2 = B_{\text{ms}}$$
$$\sigma_2^2 = \sigma_b^2$$
$$s_2^2 = W_{\text{ms}}$$

We shall test $\text{H}_0\colon \sigma_1^2 = \sigma_2^2$; that is, $\sigma_a^2 = 0$, and H_A is $\sigma_a^2 > 0$. The test criterion is to reject H_0 at the α level of significance if $u > F_\alpha$, where

$$u = \frac{B_{\text{ms}}}{W_{\text{ms}}}$$

Also
$$\lambda = \frac{\sigma_b^2}{\sigma_b^2 + s\sigma_a^2}$$

As another example, suppose we wish to test $\text{H}_0\colon \sigma_a^2 = 0$ using the data in Table 17.1. We get

$$u = \frac{3.5629}{1.2055} = 2.956$$

The tabulated F value for 11 and 48 degrees of freedom for $\alpha = .05$ is $F_{.05} = 2.00$. Hence we reject H_0.

17.2.2 Test of a Linear Combination of σ_i^2. In the previous article a test criterion was devised for testing the hypothesis that $\sigma_i^2 - \sigma_j^2 = 0$. This is equivalent to testing the hypothesis that $\sigma_i^2/\sigma_j^2 = 1$. This test will satisfy many experimenter's requirements. However, a more general test may be desired. That is to say, we may need to test H_0: $\Sigma g_i \sigma_i^2 = 0$, where the g_i are known constants that are either plus or minus unity. Of course, a more general test would be one in which g_i were any known constants, but the case in which they are plus or minus unity covers many important situations. In most cases an exact test for this hypothesis has not been found, but an approximate test will be the subject of the next theorem.

◆ **Theorem 17.5** Let the model satisfy the conditions of Theorem 16.5, and let the AOV be represented by Table 16.2. To test the hypothesis $\sigma_p^2 + \cdots + \sigma_q^2 = \sigma_r^2 + \cdots + \sigma_t^2$ against the alternative hypothesis $\sigma_p^2 + \cdots + \sigma_q^2 > \sigma_r^2 + \cdots + \sigma_t^2$, the quantity

$$u = \frac{s_p^2 + \cdots + s_q^2}{s_r^2 + \cdots + s_t^2}$$

can be considered to be approximately distributed as $F(n,m)$, where

$$n = \frac{(s_p^2 + \cdots + s_q^2)^2}{s_p^4/f_p + \cdots + s_q^4/f_q}$$

$$m = \frac{(s_r^2 + \cdots + s_t^2)^2}{s_r^4/f_r + \cdots + s_t^4/f_t}$$

and the hypothesis will be rejected at the approximate α significance level if $u > F_\alpha$.

This theorem is just an extension of Theorem 17.1; i.e., we are assuming that

$$v = \frac{(s_p^2 + \cdots + s_q^2)n}{\sigma_p^2 + \cdots + \sigma_q^2}$$

is approximately distributed as $\chi^2(n)$. The first two moments of v are equated to the first two moments of $\chi^2(n)$ to evaluate n. Similarly, the quantity

$$w = \frac{(s_r^2 + \cdots + s_t^2)m}{\sigma_r^2 + \cdots + \sigma_t^2}$$

is approximately distributed as $\chi^2(m)$, where m is similarly found by equating the first two moments of w to those of $\chi^2(m)$. Then since v and w are independent, the quantity

$$\frac{v/n}{w/m} = u$$

is approximately distributed as $F(n,m)$. The quantities n and m will depend on the σ_i^2, but these are replaced by s_i^2, which introduces a further approximation. Also, m and n will not be integers, but their values can be approximated in any F table.

It should be pointed out that some care should be exercised in using any of these approximate methods where no bounds on the error of approximation are known.

17.2.3 Example: Testing a Linear Combination of Variance Components. The details will not be given, but suppose a three-way classification model with equal numbers and all interaction is given by

$$y_{ijk} = \mu + a_i + b_j + (ab)_{ij} + c_k + (ac)_{ik} + (bc)_{jk} + e_{ijk} \qquad \begin{aligned} i &= 1, 2, \ldots, 10 \\ j &= 1, 2, \ldots, 8 \\ k &= 1, 2, \ldots, 4 \end{aligned}$$

Let the distributional properties satisfy the requirements for case A in Chap. 16. Let the variances be σ_a^2, σ_b^2, σ_{ab}^2, σ_c^2, σ_{ac}^2, σ_{bc}^2, and σ^2. The AOV will be as given in Table 17.4. This should be compared with the notation introduced in Table 16.2.

TABLE 17.4 ANALYSIS OF VARIANCE FOR THREE-WAY CLASSIFICATION

SV	DF	MS	EMS
Total	320		
Mean	1		
a class	$f_1 = 9$	$s_1^2 = 47.54$	$\sigma_1^2 = \sigma^2 + 4\sigma_{ab}^2 + 8\sigma_{ac}^2 + 32\sigma_a^2$
b class	$f_2 = 7$	$s_2^2 = 74.54$	$\sigma_2^2 = \sigma^2 + 4\sigma_{ab}^2 + 10\sigma_{bc}^2 + 40\sigma_b^2$
ab interaction	$f_3 = 63$	$s_3^2 = 14.31$	$\sigma_3^2 = \sigma^2 + 4\sigma_{ab}^2$
c class	$f_4 = 3$	$s_4^2 = 99.26$	$\sigma_4^2 = \sigma^2 + 8\sigma_{ac}^2 + 10\sigma_{bc}^2 + 80\sigma_c^2$
ac interaction	$f_5 = 27$	$s_5^2 = 7.29$	$\sigma_5^2 = \sigma^2 + 8\sigma_{ac}^2$
bc interaction	$f_6 = 21$	$s_6^2 = 18.01$	$\sigma_6^2 = \sigma^2 + 10\sigma_{bc}^2$
Remainder	$f_7 = 189$	$s_7^2 = 6.01$	$\sigma_7^2 = \sigma^2$

For any of the hypotheses

$$\sigma_{bc}^2 = 0 \qquad \sigma_{ab}^2 = 0 \qquad \sigma_{ac}^2 = 0$$

Theorem 17.4 provides an exact test. However, for any of the following hypotheses:

$$\sigma_a^2 = 0 \qquad \sigma_b^2 = 0 \qquad \sigma_c^2 = 0$$

Theorem 17.4 provides no test, since $\sigma_a^2 = 0$ is not equivalent to $\sigma_i^2 - \sigma_j^2 = 0$ for any $i, j = 1, 2, \ldots, 7$. Theorem 17.5, on the other hand, provides an approximate test. We shall demonstrate by testing the hypothesis H_0: $\sigma_a^2 = 0$ against the alternative hypothesis H_A: $\sigma_a^2 > 0$ at the 5 per cent level of significance.

First notice that $\sigma_a^2 = 0$ if and only if $\sigma_1^2 + \sigma_7^2 = \sigma_3^2 + \sigma_5^2$. So we can use Theorem 17.5 for an approximate test of the hypothesis $H_0: \sigma_1^2 + \sigma_7^2 = \sigma_3^2 + \sigma_5^2$ against the alternative hypothesis $H_A: \sigma_1^2 + \sigma_7^2 > \sigma_3^2 + \sigma_5^2$, that is, $H_A: \sigma_a^2 > 0$.

The test function is

$$u = \frac{s_1^2 + s_7^2}{s_3^2 + s_5^2} = \frac{53.55}{21.60} = 2.48$$

Also,

$$n = \frac{(s_1^2 + s_7^2)^2}{s_1^4/f_1 + s_7^4/f_7} = \frac{2{,}867.6025}{251.3074} = 11.41$$

and

$$m = \frac{(s_3^2 + s_5^2)^2}{s_3^4/f_3 + s_5^4/f_5} = \frac{466.5600}{5.2187} = 89.40$$

The 5 per cent significance level of Snedecor's F for 11.41 and 89.40 degrees of freedom is approximately 1.88. Since $u > 1.88$, the hypothesis is rejected.

17.3 Ratio of Variances

A problem of some importance in applied work such as genetics, breeding, and certain industrial work is estimation of the *ratio* of variance components. For example, in the one-way model given in Table 16.1, it may be desirable to estimate σ_a^2/σ_b^2 and to set confidence limits on this quantity or on $\sigma_b^2/(\sigma_a^2 + \sigma_b^2)$. This will be the subject of the next theorem.

◆ **Theorem 17.6** Let the model be a one-way classification with equal numbers in the subclasses. Let the distributional properties satisfy the requirements for case A in Chap. 16. The AOV is given in Table 16.1. The best (minimum-variance) unbiased estimate of σ_a^2/σ_b^2 is u, where

$$u = \frac{B_{\text{ms}}}{W_{\text{ms}}} \frac{r(s-1) - 2}{rs(s-1)} - \frac{1}{s}$$

Proof: Since $y_{..}^2$, B_{ms}, and W_{ms} is a set of complete sufficient statistics (see Theorem 16.1), it is necessary only to show that u is an unbiased estimate of σ_a^2/σ_b^2. The minimum-variance unbiased property then follows. Let

$$w = \frac{B_{\text{ms}}/(\sigma_b^2 + s\sigma_a^2)}{W_{\text{ms}}/\sigma_b^2} = \frac{\sigma_b^2}{\sigma_b^2 + s\sigma_a^2} \frac{B_{\text{ms}}}{W_{\text{ms}}}$$

Then w is distributed as $F[r-1, r(s-1)]$, and

$$E(w) = \frac{r(s-1)}{r(s-1)-2}$$

So
$$E\left(\frac{B_{ms}}{W_{ms}}\right) = \frac{\sigma_b^2 + s\sigma_a^2}{\sigma_b^2} E(w) = \left(1 + s\frac{\sigma_a^2}{\sigma_b^2}\right)\frac{r(s-1)}{r(s-1)-2}$$

Solving for σ_a^2/σ_b^2 gives

$$E\left[\frac{B_{ms}}{W_{ms}}\frac{r(s-1)-2}{r(s-1)s} - \frac{1}{s}\right] = \frac{\sigma_a^2}{\sigma_b^2}$$

so
$$\frac{B_{ms}}{W_{ms}}\frac{r(s-1)-2}{r(s-1)s} - \frac{1}{s}$$

is an unbiased estimate of σ_a^2/σ_b^2.

Next we shall give a theorem concerning confidence limits on $\sigma_b^2/(\sigma_a^2 + \sigma_b^2)$.

◆ **Theorem 17.7** Let the conditions and the model be the same as in Theorem 17.6. Then a confidence interval on $\sigma_b^2/(\sigma_a^2 + \sigma_b^2)$ with coefficient $1-\alpha$ is given by

$$\frac{W_{ms}sF_{1-\alpha/2}}{B_{ms} + W_{ms}(s-1)F_{1-\alpha/2}} \leqslant \frac{\sigma_b^2}{\sigma_a^2 + \sigma_b^2} \leqslant \frac{W_{ms}sF_{\alpha/2}}{B_{ms} + W_{ms}(s-1)F_{\alpha/2}}$$

Proof: By Theorem 17.6, the quantity w is distributed as $F[r-1, r(s-1)]$; so

$$P(F_{1-\alpha/2} \leqslant w \leqslant F_{\alpha/2}) = 1 - \alpha$$

or
$$P\left(F_{1-\alpha/2} \leqslant \frac{\sigma_b^2}{\sigma_b^2 + s\sigma_a^2}\frac{B_{ms}}{W_{ms}} \leqslant F_{\alpha/2}\right) = 1 - \alpha$$

After manipulating the quantities within the parentheses, we arrive at

$$P\left[\frac{W_{ms}sF_{1-\alpha/2}}{B_{ms} + W_{ms}(s-1)F_{1-\alpha/2}} \leqslant \frac{\sigma_b^2}{\sigma_a^2 + \sigma_b^2} \leqslant \frac{W_{ms}sF_{\alpha/2}}{B_{ms} + W_{ms}(s-1)F_{\alpha/2}}\right] = 1 - \alpha$$

which is the desired confidence statement.

This theorem can be extended to many other models as well.

Problems

17.1 If x_1 and x_2 are independent, if n_1x_1/σ_1^2 is distributed as $\chi^2(n_1)$, and if n_2x_2/σ_2^2 is distributed as $\chi^2(n_2)$, find the moment-generating function of $x_1 + x_2$.

17.2 In Theorem 17.3, show that $C_\beta = C_{1-\beta}$.

17.3 If x and y are independent, if x is distributed as $\chi^2(n)$, and if y is distributed as $\chi^2(m)$, prove that $x + y$ and x/y are independent.

17.4 In Prob. 17.1, find the moment-generating function of $a_1x_1 + a_2x_2$, where a_1 and a_2 are constants.

17.5 Use the results of Prob. 17.1 to find the mean and variance of $x_1 + x_2$.

17.6 If the $p \times 1$ vector $\mathbf{Y}$ is distributed $N(\mathbf{0},\mathbf{V})$, show that the expected value of $\mathbf{Y}'\mathbf{AY}$ is $\text{tr}(\mathbf{VA})$.

17.7 If the $p \times 1$ vector $\mathbf{Y}$ is distributed $N(\mathbf{0},\mathbf{V})$, show that the variance of $\mathbf{Y}'\mathbf{AY}$ is $2\ \text{tr}(\mathbf{AV})^2$.

17.8 An analysis of variance is run on data that satisfy the model

$$y_{ij} = \mu + a_i + b_{ij} \qquad j = 1, 2, \ldots, 10; \qquad i = 1, 2, 3, 4$$

The data are assumed to meet the conditions for case A in Chap. 16. The AOV is given in Table 17.5.

TABLE 17.5

SV	DF	MS (x_i)	EMS
Total	40		
Mean	1		
a class	3	4.3	$\sigma_b^2 + 10\sigma_a^2$
b class	36	1.7	σ_b^2

Set a 90 per cent confidence interval on σ_b^2.

17.9 In Prob. 17.8, set a 95 per cent confidence interval on $\sigma_b^2 + 10\sigma_a^2$.

17.10 In Prob. 17.8, set a 90 per cent confidence interval on σ_a^2.

17.11 In Prob. 17.8, set a 90 per cent confidence interval on $\sigma_a^2 + \sigma_b^2$.

17.12 In Prob. 17.8, set a 90 per cent confidence interval on σ_a^2/σ_b^2.

17.13 In Prob. 17.8, set a 90 per cent confidence interval on $\sigma_a^2/(\sigma_a^2 + \sigma_b^2)$.

TABLE 17.6

SV	DF	MS	EMS
Total	960		
Mean	1		
a effect	7	236.1	$\sigma^2 + 12\sigma_{ab}^2 + 10\sigma_{ac}^2 + 120\sigma_a^2$
b effect	9	347.8	$\sigma^2 + 8\sigma_{bc}^2 + 12\sigma_{ab}^2 + 96\sigma_b^2$
ab interaction	63	30.2	$\sigma^2 + 12\sigma_{ab}^2$
c effect	11	108.8	$\sigma^2 + 10\sigma_{ac}^2 + 8\sigma_{bc}^2 + 80\sigma_c^2$
ac interaction	77	34.1	$\sigma^2 + 10\sigma_{ac}^2$
bc interaction	99	18.5	$\sigma^2 + 8\sigma_{bc}^2$
Remainder	693	8.1	σ^2

17.14 In Prob. 17.8, find the best (minimum-variance) unbiased estimate of σ_a^2/σ_b^2.

17.15 In Prob. 17.8, test the hypothesis $\sigma_a^2 = 0$ with a type I error probability of 5 per cent. Plot the power as a function of $\lambda = \sigma_b^2/(\sigma_b^2 + 10\sigma_a^2)$.

17.16 The AOV of Table 17.6 is from data that satisfy the following model:

$$y_{ijm} = \mu + a_i + b_j + (ab)_{ij} + c_m + (ac)_{im} + (bc)_{jm} + e_{ijm} \quad \begin{cases} i = 1, 2, \ldots, 8 \\ j = 1, 2, \ldots, 10 \\ m = 1, 2, \ldots, 12 \end{cases}$$

Use Theorem 17.4 to test the hypothesis $\sigma_{bc}^2 = 0$ with a type I error probability of 5 per cent.

17.17 Use the data in Prob. 17.16 and Theorem 17.5 to test the hypothesis $\sigma_c^2 = 0$ with a type I error probability of 5 per cent.

Further Reading

1 G. W. Snedecor: "Statistical Methods," Iowa State College Press, Ames, Iowa, 1958.

2 O. Kempthorne: "Design and Analysis of Experiments," John Wiley & Sons, Inc., New York, 1952.

3 C. H. Goulden: "Methods of Statistical Analysis," John Wiley & Sons, Inc., New York, 1939.

4 O. L. Davies: "Design and Analysis of Industrial Experiments," Oliver and Boyd, Ltd., London, 1954.

5 R. L. Anderson and T. A. Bancroft: "Statistical Theory in Research," McGraw-Hill Book Company, New York, 1952.

6 C. R. Rao: "Advanced Statistical Methods in Biometric Research," John Wiley & Sons, Inc., New York, 1952.

7 B. L. Welch: On Linear Combinations of Several Variances, *J. Am. Statist. Assoc.*, vol. 51, pp. 132–148, 1956.

8 W. G. Cochran: Some Consequences when the Assumptions for the A.O.V. Are Not Satisfied, *Biometrics*, vol. 3, pp. 22–38, 1947.

9 I. Bross: Fiducial Intervals for Variance Components, *Biometrics*, vol. 6, pp. 137–144, 1951.

10 S. L. Crump: The Present Status of Variance Component Analysis, *Biometrics*, vol. 7, pp. 1–16, 1951.

11 W. A. Hendricks: Variance Components as a Tool for the Analysis of Sample Data, *Biometrics*, vol. 7, pp. 97–101, 1951.

12 W. G. Cochran: Testing a Linear Relation among Variances, *Biometrics*, vol. 7, pp. 17–32, 1951.

13 J. R. Green: A Confidence Interval for Variance Components, *Ann. Math. Statist.*, vol. 25, pp. 671–685, 1954.

14 N. L. Johnson: Comparison of Analysis of Variance Power Functions in the Parametric and Random Models, *Biometrika*, vol. 39, pp. 427–429, 1952.

15 A. Huitson: A Method of Assigning Confidence Limits to Linear Combinations of Variances, *Biometrika*, vol. 42, pp. 471–479, 1955.

16 C. R. Henderson: Estimation of Variance and Covariance Components, *Biometrics*, vol. 9, pp. 226–252, 1953.

17 F. Yates and I. Zacopanay: The Estimation of the Efficiency of Sampling with Special Reference to Sampling for Yield in Cereal Experiments, *J. Agr. Sci.*, vol. 25, pp. 545–577, 1935.

18 H. E. Daniels: The Estimation of Components of Variance, *J. Roy. Statist. Soc.*, suppl., vol. 6, pp. 186–197, 1939.

19 F. E. Satterthwaite: An Approximate Distribution of Estimates of Variance Components, *Biometrics Bull.*, vol. 2, pp. 110–114, 1946.

20 M. Ganguli: A Note on Nested Sampling, *Sankhyā*, vol. 5, pp. 449–452, 1941.

21 M. G. Bulmer: Approximate Confidence Limits for Components of Variance, *Biometrika*, vol. 44, pp. 159–167, 1957.

18

Mixed Models

In the previous chapters five models were defined and discussed in some detail. In many scientific investigations these will form an adequate basis for the models needed to describe experimental situations. However, in many instances a combination of these models will be necessary. There are many possible combinations, but only three will be discussed in this chapter. These should form a nucleus from which the reader can obtain a basic understanding of "mixed" models.

Those which will be discussed are:

1. A mixture of model 1 and model 4
   (*a*) A covariance model
2. A mixture of model 4 and model 5
   (*a*) A two-way classification model with fixed and random effects
   (*b*) The balanced incomplete block model with blocks random and treatments fixed, i.e., with recovery of interblock information

18.1 Covariance

Covariance can be used with any case of model 4. However, only the one-way classification will be discussed here.

The model is

$$y_{ij} = \mu + \tau_i + \beta x_{ij} + e_{ij} \qquad \begin{cases} j = 1, 2, \ldots, r \\ i = 1, 2, \ldots, t \end{cases} \tag{18.1}$$

where y_{ij} is an observed random variable; x_{ij} is an observed fixed quantity, usually called a *concomitant variable*; μ is a constant; τ_i is the ith treatment constant; β is the regression coefficient (unknown); and the e_{ij} are unobserved normal variables that are independent and have means 0 and constant variances σ^2.

The likelihood equation is

$$L = f(\mathbf{e}) = \frac{1}{(2\pi\sigma^2)^{rt/2}} \exp\left[-\frac{1}{2\sigma^2}\sum_{ij}(y_{ij} - \mu - \tau_i - \beta x_{ij})^2\right]$$

and $$\log L = -\frac{rt}{2}\log 2\pi - \frac{rt}{2}\log\sigma^2 - \frac{1}{2\sigma^2}\sum_{ij}(y_{ij} - \mu - \tau_i - \beta x_{ij})^2$$

To find maximum-likelihood estimators, we can set the partial derivatives equal to zero. This gives

$$\begin{aligned}
\frac{\partial \log L}{\partial \mu} &= \frac{1}{\tilde{\sigma}^2}\sum_{ij}(y_{ij} - \hat{\mu} - \hat{\tau}_i - \hat{\beta}x_{ij}) = 0 \\
\frac{\partial \log L}{\partial \tau_p} &= \frac{1}{\tilde{\sigma}^2}\sum_{j}(y_{pj} - \hat{\mu} - \hat{\tau}_p - \hat{\beta}x_{pj}) = 0 \qquad p = 1, 2, \ldots, t \\
\frac{\partial \log L}{\partial \beta} &= \frac{1}{\tilde{\sigma}^2}\sum_{ij}(y_{ij} - \hat{\mu} - \hat{\tau}_i - \hat{\beta}x_{ij})x_{ij} = 0 \\
\frac{\partial \log L}{\partial \sigma^2} &= -\frac{rt}{2\tilde{\sigma}^2} + \frac{1}{2\tilde{\sigma}^4}\sum_{ij}(y_{ij} - \hat{\mu} - \hat{\tau}_i - \hat{\beta}x_{ij})^2 = 0
\end{aligned} \tag{18.2}$$

Notice that we have used $\tilde{\sigma}^2$. This is because we reserve $\hat{\sigma}^2$ for the unbiased estimator.

The first three equations can be used to give the following normal equations:

$$\begin{aligned}
\mu: &\quad rt\hat{\mu} + r\Sigma\hat{\tau}_i + \hat{\beta}X_{..} = Y_{..} \\
\tau_p: &\quad r\hat{\mu} + r\hat{\tau}_p + \hat{\beta}X_{p.} = Y_{p.} \qquad p = 1, 2, \ldots, t \\
\beta: &\quad X_{..}\hat{\mu} + \sum_i X_{i.}\hat{\tau}_i + \hat{\beta}\sum_{ij} x_{ij}^2 = \sum_{ij} y_{ij}x_{ij}
\end{aligned} \tag{18.3}$$

There are $t + 2$ equations, and the rank is $t + 1$; so there is one linear dependency. The nonestimable function $\Sigma\hat{\tau}_i = 0$ can be used to solve the normal equations.

The object in this model is to determine what functions of the τ_i are estimable, estimate these functions, estimate β, find the standard errors of the estimates, and test hypotheses about the τ_i and β.

18.1.1 Point Estimation. The τ_i equation in (18.3) can be used to obtain

$$\hat{\mu} + \hat{\tau}_i = \frac{1}{r}(Y_{i.} - \hat{\beta}X_{i.})$$

The β equation can be written

$$\sum_i (\hat{\mu} + \hat{\tau}_i)X_{i.} + \hat{\beta} \sum_{ij} x_{ij}^2 = \sum_{ij} x_{ij}y_{ij}$$

Substituting the value of $\hat{\mu} + \hat{\tau}_i$ into this equation gives

$$\sum_i \frac{1}{r}(Y_{i.} - \hat{\beta}X_{i.})X_{i.} + \hat{\beta} \sum_{ij} x_{ij}^2 = \sum_{ij} x_{ij}y_{ij}$$

This can be solved for $\hat{\beta}$. This gives

$$\hat{\beta} = \frac{\sum_{ij} x_{ij}y_{ij} - \frac{1}{r}\sum_i Y_{i.}X_{i.}}{\sum_{ij} x_{ij}^2 - \frac{1}{r}\sum_i X_{i.}^2} = \frac{\sum_{ij}(x_{ij} - x_{i.})(y_{ij} - y_{i.})}{\sum_{ij}(x_{ij} - x_{i.})^2}$$

To assist in simplifying the results the following notation will be used:

$$\begin{aligned}
E_{xx} &= \sum_{ij} (x_{ij} - x_{i.})^2 \\
E_{xy} &= \sum_{ij} (x_{ij} - x_{i.})(y_{ij} - y_{i.}) \\
E_{yy} &= \sum_{ij} (y_{ij} - y_{i.})^2 \\
T_{xx} &= \sum_{ij} (x_{i.} - x_{..})^2 \\
T_{xy} &= \sum_{ij} (x_{i.} - x_{..})(y_{i.} - y_{..}) \\
T_{yy} &= \sum_{ij} (y_{i.} - y_{..})^2
\end{aligned} \qquad (18.4)$$

These quantities will be quite useful, and they can be tabulated in the form shown in Table 18.1.

TABLE 18.1 COMPUTATIONS FOR COVARIANCE MODEL

SV	DF	SS:x^2	SS:xy	SS:y^2
Total	rt	Σx_{ij}^2	$\Sigma x_{ij}y_{ij}$	Σy_{ij}^2
Mean	1	$\frac{X_{..}^2}{rt}$	$\frac{X_{..}Y_{..}}{rt}$	$\frac{Y_{..}^2}{rt}$
Treatment	$t - 1$	T_{xx}	T_{xy}	T_{yy}
Error	$t(r - 1)$	E_{xx}	E_{xy}	E_{yy}
Treatment plus error	$rt - 1$	$S_{xx} = T_{xx} + E_{xx}$	$S_{xy} = T_{xy} + E_{xy}$	$S_{yy} = T_{yy} + E_{yy}$

Notice the identity in the SS:x^2 column,

$$\sum_{ij} x_{ij}^2 - \frac{X_{..}^2}{rt} = T_{xx} + E_{xx}$$

Similar identities hold for the other two columns. Hence E_{xx}, E_{xy}, and E_{yy} can be obtained by subtraction. The treatment-plus-error row will play an important part in covariance models. This will be explained later.

If this notation is used, we write

$$\hat{\beta} = \frac{E_{xy}}{E_{xx}} \tag{18.5}$$

If this is substituted into the τ_p equation, we get

$$\hat{\mu} + \hat{\tau}_p = y_{p.} - \hat{\beta} x_{p.} = y_{p.} - \frac{E_{xy}}{E_{xx}} x_{p.} \tag{18.6}$$

From this we notice that $\Sigma\lambda_i\tau_i$ is estimable if and only if $\Sigma\lambda_i = 0$; that is to say, only treatment *contrasts* are estimable. This gives us

$$\Sigma\lambda_i\hat{\tau}_i = \Sigma\lambda_i(y_{i.} - \hat{\beta}x_{i.}) \qquad \text{if } \Sigma\lambda_i = 0 \tag{18.7}$$

It can be shown that

$$E(\Sigma\lambda_i\hat{\tau}_i) = \Sigma\lambda_i\tau_i \qquad \text{and} \qquad E(\hat{\beta}) = \beta$$

The variance of $\hat{\beta}$ is σ^2/E_{xx}, and

$$\text{var}(\Sigma\lambda_i\hat{\tau}_i) = \sigma^2\left[\frac{\Sigma\lambda_i^2}{r} + \frac{(\Sigma\lambda_i x_{i.})^2}{E_{xx}}\right]$$

In particular, the variance of $\hat{\tau}_i - \hat{\tau}_j$ is

$$\sigma^2\left[\frac{2}{r} + \frac{(x_{i.} - x_{j.})^2}{E_{xx}}\right]$$

Since $x_{i.}$ and $x_{j.}$ enter into this variance, a different standard error is required for the estimate of each treatment difference. The average value of the variance of $\hat{\tau}_i - \hat{\tau}_j$, averaged over all the $t(t-1)$ treatment differences, is

$$\frac{1}{t(t-1)} \sum_{\substack{ij \\ i \neq j}} \left[\frac{2}{r} + \frac{(x_{i.} - x_{j.})^2}{E_{xx}}\right] \sigma^2$$

$$= \frac{2\sigma^2}{r} + \frac{\sigma^2}{t(t-1)E_{xx}} \sum_{\substack{ij \\ i \neq j}} [(x_{i.} - x_{..}) - (x_{j.} - x_{..})]^2$$

$$= \frac{2\sigma^2}{r} + \frac{\sigma^2}{t(t-1)E_{xx}}$$

$$\times \left[\sum_{\substack{ij \\ i \neq j}} (x_{i.} - x_{..})^2 + \sum_{\substack{ij \\ i \neq j}} (x_{j.} - x_{..})^2 - 2 \sum_{\substack{ij \\ i \neq j}} (x_{i.} - x_{..})(x_{j.} - x_{..})\right]$$

$$= \frac{2\sigma^2}{r} + \frac{\sigma^2}{t(t-1)E_{xx}}$$

$$\times \left[(t-1) \sum_{i} (x_{i.} - x_{..})^2 + (t-1) \sum_{j} (x_{j.} - x_{..})^2 + 2 \sum_{i} (x_{i.} - x_{..})^2\right]$$

$$= \sigma^2 \left[\frac{2}{r} + \frac{2T_{xx}}{r(t-1)E_{xx}}\right]$$

Also, from the last derivative of Eq. (18.2), we get

$$\tilde{\sigma}^2 = \frac{1}{rt} \sum_{ij} (y_{ij} - \hat{\mu} - \hat{\tau}_i - \hat{\beta}x_{ij})^2$$

To this equation let us add and subtract $y_{i.}$; we get

$$\tilde{\sigma}^2 = \frac{1}{rt} \sum_{ij} [(y_{ij} - y_{i.}) + (y_{i.} - \hat{\mu} - \hat{\tau}_i - \hat{\beta}x_{ij})]^2$$

From Eq. (18.6) we get

$$y_{i.} - \hat{\mu} - \hat{\tau}_i = \hat{\beta}x_{i.}$$

and this gives us

$$\tilde{\sigma}^2 = \frac{1}{rt} \sum_{ij} [(y_{ij} - y_{i.}) - \hat{\beta}(x_{ij} - x_{i.})]^2$$

$$= \frac{1}{rt}\left[\sum_{ij} (y_{ij} - y_{i.})^2 - 2\hat{\beta} \sum_{ij} (y_{ij} - y_{i.})(x_{ij} - x_{i.}) + \hat{\beta}^2 \sum_{ij} (x_{ij} - x_{i.})^2\right]$$

$$= \frac{1}{rt} [E_{yy} - 2\hat{\beta}E_{xy} + \hat{\beta}^2 E_{xx}]$$

By the definition of $\hat{\beta}$, we get

$$\tilde{\sigma}^2 = \frac{1}{rt}\left(E_{yy} - \frac{E_{xy}^2}{E_{xx}}\right)$$

We shall omit the proof, but it can easily be shown that

$$E(\tilde{\sigma}^2) = \frac{t(r-1)-1}{rt}\sigma^2$$

Therefore, we shall define

$$\hat{\sigma}^2 = \frac{1}{t(r-1)-1}\left(E_{yy} - \frac{E_{xy}^2}{E_{xx}}\right)$$

so that

$$E(\hat{\sigma}^2) = \sigma^2$$

Next let us examine the exponent of the likelihood equation

$$-\frac{1}{2\sigma^2}\sum_{ij}(y_{ij} - \mu - \tau_i - \beta x_{ij})^2$$

If we add and subtract the quantity $\hat{\mu} + \hat{\tau}_i + \hat{\beta}x_{ij}$, this becomes

$$-\frac{1}{2\sigma^2}\sum_{ij}[(y_{ij} - \hat{\mu} - \hat{\tau}_i - \hat{\beta}x_{ij}) + (\hat{\mu} + \hat{\tau}_i + \hat{\beta}x_{ij} - \mu - \tau_i - \beta x_{ij})]^2$$

$$= -\frac{1}{2\sigma^2}\left[\sum_{ij}(y_{ij} - \hat{\mu} - \hat{\tau}_i - \hat{\beta}x_{ij})^2 + \sum_{ij}(\hat{\mu} + \hat{\tau}_i + \hat{\beta}x_{ij} - \mu - \tau_i - \beta x_{ij})^2\right]$$

$$= -\frac{1}{2\sigma^2}\left[\{t(r-1)-1\}\hat{\sigma}^2 + \sum_{ij}(\hat{\mu} + \hat{\tau}_i + \hat{\beta}x_{ij} - \mu - \tau_i - \beta x_{ij})^2\right]$$

The cross-product term is zero, by virtue of the normal equations. The quantity $\beta x_{i.}$ will be added and subtracted to the second term, and $\hat{\mu} + \hat{\tau}_i$ will be replaced by $y_{i.} - \hat{\beta}x_{i.}$. This gives us

$$-\frac{1}{2\sigma^2}\left\{[t(r-1)-1]\hat{\sigma}^2 + \sum_{ij}[(\hat{\beta}-\beta)(x_{ij} - x_{i.}) + (y_{i.} - \mu - \tau_i - \beta x_{i.})]^2\right\}$$

$$= -\frac{\hat{\sigma}^2}{2\sigma^2}[t(r-1)-1]$$

$$-\frac{(\hat{\beta}-\beta)^2}{2\sigma^2}\sum_{ij}(x_{ij} - x_{i.})^2 - \frac{1}{2\sigma^2}\sum_{ij}(y_{i.} - \mu - \tau_i - \beta x_{i.})^2$$

Therefore, the $t + 2$ quantities $\hat{\sigma}^2, \hat{\beta}, y_{1.}, y_{2.}, \ldots, y_{t.}$ are a set of sufficient statistics. They are also complete. The foregoing can be summed up in the following theorem.

◆ **Theorem 18.1** Let the covariance model be given by Eq. (18.1) with the distributional properties noted there. Then the statistics $\hat{\sigma}^2, \hat{\beta}, y_{1.}, y_{2.}, \ldots, y_{t.}$, are a sufficient complete set. Also, the quantities $\hat{\sigma}^2, \hat{\beta}, \Sigma\lambda_i(y_{i.} - \hat{\beta}x_{i.})$ for $\Sigma\lambda_i = 0$ are unbiased estimators of $\sigma^2, \beta, \Sigma\lambda_i\tau_i$, respectively; hence, they are the best (minimum-variance) unbiased estimators.

Next we state an important theorem on the distributional properties of the above estimators.

◆ **Theorem 18.2** Let the covariance model be given by Eq. (18.1) with the distributional properties noted there. Then

(1) $\hat{\beta}$ is distributed $N(\beta, \sigma^2/E_{xx})$.

(2) $\Sigma\lambda_i\hat{\tau}_i$ is distributed $N(\Sigma\lambda_i\tau_i, \sigma_1^2)$, where

$$\sigma_1^2 = \sigma^2\left[\frac{\Sigma\lambda_i^2}{r} + \frac{(\Sigma\lambda_i x_{i.})^2}{E_{xx}}\right]$$

if $\Sigma\lambda_i = 0$.

(3) $\dfrac{[t(r-1)-1]\hat{\sigma}^2}{\sigma^2}$ is distributed as $\chi^2[t(r-1)-1]$.

(4) $\hat{\sigma}^2$ is independent of $\hat{\beta}$ and $\Sigma\lambda_i\hat{\tau}_i$.

Proof: The proofs of (1) and (2) are clear, since $\hat{\beta}$ and $\Sigma\lambda_i\hat{\tau}_i$ are each equal to a linear function of normal variables y_{ij}. The mean and variance of both have been given elsewhere in this article. The proofs for (3) and (4) follow from general theorems in Chap. 11.

We are now in a position to state a theorem about confidence intervals on σ^2, β, and $\Sigma\lambda_i\tau_i$.

18.1.2 Confidence Intervals. From Theorem 18.2 it is clear that confidence intervals can be put on the parameters $\sigma^2, \beta, \Sigma\lambda_i\tau_i$.

◆ **Theorem 18.3** A confidence interval on σ^2 with confidence coefficient $1-\alpha$ is

$$\frac{[t(r-1)-1]\hat{\sigma}^2}{\chi^2_{\alpha/2}} \leqslant \sigma^2 \leqslant \frac{[t(r-1)-1]\hat{\sigma}^2}{\chi^2_{1-\alpha/2}}$$

A confidence interval on β with confidence coefficient $1-\alpha$ is

$$\hat{\beta} - t_{\alpha/2}\frac{\hat{\sigma}}{\sqrt{E_{xx}}} \leqslant \beta \leqslant \hat{\beta} + t_{\alpha/2}\frac{\hat{\sigma}}{\sqrt{E_{xx}}}$$

A confidence interval on $\Sigma\lambda_i\tau_i$, where $\Sigma\lambda_i = 0$, with confidence coefficient $1-\alpha$ is

$$\Sigma\lambda_i\hat{\tau}_i - t_{\alpha/2}\hat{\sigma}\left[\frac{\Sigma\lambda_i^2}{r} + \frac{(\Sigma\lambda_i x_{i.})^2}{E_{xx}}\right]^{\frac{1}{2}} \leqslant \Sigma\lambda_i\tau_i \leqslant \Sigma\lambda_i\hat{\tau}_i + t_{\alpha/2}\hat{\sigma}\left[\frac{\Sigma\lambda_i^2}{r} + \frac{(\Sigma\lambda_i x_{i.})^2}{E_{xx}}\right]^{\frac{1}{2}}$$

The proof is straightforward and will be left for the reader.

18.1.3 Tests of Hypotheses. From Theorem 18.2 it follows that $\hat{\beta}\sqrt{E_{xx}}/\sigma$ is distributed $N(\beta\sqrt{E_{xx}}/\sigma, 1)$. Then $u = \hat{\beta}^2 E_{xx}/\sigma^2$ is distributed as $\chi'^2(1,\lambda)$, where

$$\lambda = \frac{\beta^2 E_{xx}}{2\sigma^2}$$

If we let
$$v = \frac{[t(r-1)-1]\hat{\sigma}^2}{\sigma^2}$$

then, by (3) and (4) of Theorem 18.2, v is distributed as $\chi^2[t(r-1)-1]$ and is independent of u. So

$$w = \frac{u}{v}[t(r-1)-1] = \frac{\hat{\beta}^2 E_{xx}}{\hat{\sigma}^2}$$

is distributed as $F'(1, t(r-1)-1, \lambda)$. This is summed up in the next theorem.

◆ **Theorem 18.4** Let the covariance model be given by Eq. (18.1) with the distributional properties noted there. To test $H_0: \beta = 0$ at the α level of significance, a test criterion is

$$w = \frac{\hat{\beta}^2 E_{xx}}{\hat{\sigma}^2}$$

and H_0 is rejected if $w > F_\alpha$, where F_α is the upper α percentage level of the F distribution with 1 and $t(r-1)-1$ degrees of freedom. The power of the test is

$$\int_{F_\alpha}^{\infty} g(w)\,dw$$

where w is distributed as $F'[1, t(r-1)-1, \lambda]$, where

$$\lambda = \frac{\beta^2 E_{xx}}{2\sigma^2}$$

Next we shall prove a theorem that can be used to test the very important hypothesis $H_0: \tau_1 = \tau_2 = \cdots = \tau_t$.

This hypothesis is estimable, since $\Sigma\lambda_i\tau_i$ is estimable if $\Sigma\lambda_i = 0$. We can use Theorem 11.15. Using this theorem and the normal equations (18.3), we get

$$\begin{aligned}\hat{\boldsymbol{\beta}}'\mathbf{X}'\mathbf{Y} &= \hat{\mu}Y_{..} + \sum_{i=1}^{t}\hat{\tau}_i Y_{i.} + \hat{\beta}\sum_{ij} y_{ij}x_{ij} \\ &= \sum_i (\hat{\mu} + \hat{\tau}_i)Y_{i.} + \hat{\beta}\sum_{ij} y_{ij}x_{ij}\end{aligned}$$

If the value $y_{i.} - \hat{\beta}x_{i.}$ is substituted for $\hat{\mu} + \hat{\tau}_i$, this becomes

$$\hat{\boldsymbol{\beta}}'\mathbf{X}'\mathbf{Y} = \sum_i (y_{i.} - \hat{\beta}x_{i.})Y_{i.} + \hat{\beta}\sum_{ij} y_{ij}x_{ij}$$

$$= \frac{\sum_i Y_{i.}^2}{r} + \hat{\beta}\left(\sum_{ij} y_{ij}x_{ij} - \frac{\sum_i Y_{i.}X_{i.}}{r}\right)$$

$$= \frac{\sum_i Y_{i.}^2}{r} + \hat{\beta}E_{xy}$$

$$= \frac{\sum_i Y_{i.}^2}{r} + \frac{E_{xy}^2}{E_{xx}}$$

Also, $\mathbf{Y}'\mathbf{Y} - \hat{\boldsymbol{\beta}}'\mathbf{X}'\mathbf{Y}$ of Theorem 11.15 is

$$\sum_{ij} y_{ij}^2 - \frac{\sum_i Y_{i.}^2}{r} - \hat{\beta}E_{xy} = \sum_{ij}(y_{ij} - y_{i.})^2 - \frac{E_{xy}^2}{E_{xx}} = E_{yy} - \frac{E_{xy}^2}{E_{xx}} = [t(r-1) - 1]\hat{\sigma}^2$$

The reduced model (substitute $\tau_1 = \tau_2 = \cdots = \tau_t = \tau$) can be written

$$y_{ij} = \mu^* + \beta x_{ij} + e_{ij}$$

where we have written μ^* in place of $\mu + \tau$. The normal equations can be obtained directly from this model and from the normal equations (18.3). They are

$$\mu^*: \qquad rt\tilde{\mu}^* + \tilde{\beta}X_{..} = Y_{..}$$

$$\beta: \qquad X_{..}\tilde{\mu}^* + \tilde{\beta}\sum_{ij} x_{ij}^2 = \sum_{ij} y_{ij}x_{ij}$$

The solution is

$$\tilde{\beta} = \frac{rt\Sigma y_{ij}x_{ij} - X_{..}Y_{..}}{rt\sum_{ij} x_{ij}^2 - X_{..}^2} = \frac{\sum_{ij} y_{ij}x_{ij} - X_{..}Y_{..}/rt}{\sum_{ij} x_{ij}^2 - X_{..}^2/rt}$$

If the quantities in Table 18.1 are used, we obtain

$$\tilde{\beta} = \frac{T_{xy} + E_{xy}}{T_{xx} + E_{xx}}$$

Further, $$\tilde{\mu}^* = y_{..} - \tilde{\beta}x_{..}$$

So, referring to Theorem 11.15,

$$\tilde{\boldsymbol{\gamma}}'\mathbf{Z}_2'\mathbf{Y} = \tilde{\beta}\sum_{ij} y_{ij}x_{ij} + \tilde{\mu}^* Y_{..}$$

$$= \tilde{\beta}\Sigma y_{ij}x_{ij} + \frac{Y_{..}^2}{rt} - \frac{\tilde{\beta}Y_{..}X_{..}}{rt}$$

$$= \frac{Y_{..}^2}{rt} + \tilde{\beta}\left(\Sigma y_{ij}x_{ij} - \frac{X_{..}Y_{..}}{rt}\right)$$

$$= \frac{Y_{..}^2}{rt} + \tilde{\beta}(T_{xy} + E_{xy})$$

$$= \frac{Y_{..}^2}{rt} + \frac{(T_{xy} + E_{xy})^2}{T_{xx} + E_{xx}}$$

So, noting that s of Theorem 11.15 is $t - 1$, that $k = t + 1$, and that $n - k$ is the degrees of freedom for error $t(r - 1) - 1$, we find that the quantity v in Theorem 11.15 here is

$$v = \frac{1}{(t-1)\hat{\sigma}^2}\left[\frac{\sum_i Y_{i.}^2}{r} + \frac{E_{xy}^2}{E_{xx}} - \frac{Y_{..}^2}{rt} - \frac{(T_{xy} + E_{xy})^2}{T_{xx} + E_{xx}}\right]$$

Since

$$\frac{\sum_i Y_{i.}^2}{r} - \frac{Y_{..}^2}{rt} = T_{yy}$$

we can write v as

$$v = \frac{1}{(t-1)\hat{\sigma}^2}\left[T_{yy} + E_{yy} - \frac{(T_{xy} + E_{xy})^2}{T_{xx} + E_{xx}} - \left(E_{yy} - \frac{E_{xy}^2}{E_{xx}}\right)\right] \tag{18.8}$$

This is the text of the following theorem.

◆ **Theorem 18.5** Let the covariance model be given by Eq. (18.1) with the distributional properties noted there. To test the hypothesis $H_0: \tau_1 = \tau_2 = \cdots = \tau_t$, the function v in (18.8) can be used. v is distributed as $F'[t - 1, t(r - 1) - 1, \lambda]$, where

$$\lambda = \frac{r\sum_i (\tau_i - \bar{\tau}_.)^2}{2\sigma^2} - \frac{\left[\sum_i (\tau_i - \bar{\tau}_.)X_{i.}\right]^2}{2\sigma^2(T_{xx} + E_{xx})}$$

So H_0 is rejected if $v > F_\alpha$. A computational procedure is given in Tables 18.1 and 18.2.

TABLE 18.2 AUXILIARY TABLE FOR TESTING $\tau_1 = \tau_2 = \cdots = \tau_t$

SV	DF	$SS{:}y^2 - \dfrac{(SS{:}xy)^2}{SS{:}x^2}$	MS	F
Treatment plus error	$rt - 2$	$S_{yy} - \dfrac{S_{xy}^2}{S_{xx}} = A$		
Error	$rt - t - 1$	$E_{yy} - \dfrac{E_{xy}^2}{E_{xx}} = B$	$\dfrac{B}{t(r-1)-1}$	$v = \dfrac{A-B}{B}\,\dfrac{t(r-1)-1}{t-1}$
Difference	$t - 1$	$A - B$	$\dfrac{A-B}{t-1}$	

Proof: The proof has been given above with the aid of Theorem 11.15.

The noncentrality parameter λ can be found by using Theorem 11.16. We need the expected value of the treatment mean square, i.e., the treatment mean square $T_{\text{ms}} = (A - B)/(t - 1)$. This gives

$$E(T_{\text{ms}}) = E\,\frac{1}{t-1}\left[T_{yy} + E_{yy} - \frac{(T_{xy} + E_{xy})^2}{T_{xx} + E_{xx}} - E_{yy} + \frac{E_{xy}^2}{E_{xx}}\right] \tag{18.9}$$

But, by the definition of $\hat{\sigma}^2$, we get

$$[t(r-1) - 1]E(\hat{\sigma}^2) = E\left(E_{yy} - \frac{E_{xy}^2}{E_{xx}}\right) = [t(r-1) - 1]\sigma^2 \tag{18.10}$$

So we need only evaluate

$$\frac{1}{t-1}\,E\left[T_{yy} + E_{yy} - \frac{(T_{xy} + E_{xy})^2}{T_{xx} + E_{xx}}\right] \tag{18.11}$$

Using the notation of (18.4) and Table 18.1, this can be written

$$\frac{1}{t-1}\,E\left\{\sum_{ij}(y_{ij} - y_{..})^2 - \frac{\left[\sum_{ij}(y_{ij} - y_{..})(x_{ij} - x_{..})\right]^2}{T_{xx} + E_{xx}}\right\}$$

Substituting by model (18.1), we get

$$\frac{1}{t-1} E\left\{\sum_{ij} (\mu + \tau_i + \beta x_{ij} + e_{ij} - \mu - \bar{\tau}_{.} - \beta x_{..} - \bar{e}_{..})^2 - \frac{\left[\sum_{ij} (\mu + \tau_i + \beta x_{ij} + e_{ij} - \mu - \bar{\tau}_{.} - \beta x_{..} - \bar{e}_{..})(x_{ij} - x_{..})\right]^2}{T_{xx} + E_{xx}}\right\}$$

$$= \frac{1}{t-1}\left\{\sum_{ij} (\tau_i - \bar{\tau}_{.} + \beta x_{ij} - \beta x_{..})^2 + (rt-1)\sigma^2 - \frac{E\left[\sum_{ij} (\tau_i - \bar{\tau}_{.})x_{ij} + \beta \sum_{ij} (x_{ij} - x_{..})^2 + \sum_{ij} e_{ij}(x_{ij} - x_{..})\right]^2}{T_{xx} + E_{xx}}\right\}$$

$$= \frac{1}{t-1}\left\{r \sum_{i} (\tau_i - \bar{\tau}_{.})^2 + \beta^2 \sum_{ij} (x_{ij} - x_{..})^2 + 2\beta \sum_{ij} (\tau_i - \bar{\tau}_{.})(x_{ij} - x_{..}) + (rt-1)\sigma^2 - \frac{\left[\sum_{i}(\tau_i - \bar{\tau}_{.})X_{i.}\right]^2}{T_{xx} + E_{xx}} - \frac{\beta^2\left[\sum_{ij} (x_{ij} - x_{..})^2\right]^2}{T_{xx} + E_{xx}} - \frac{2\beta \sum_{ij}[(\tau_i - \bar{\tau}_{.})x_{ij}] \sum_{pq}(x_{pq} - x_{..})^2}{T_{xx} + E_{xx}} - \frac{\sigma^2 \sum_{ij} (x_{ij} - x_{..})^2}{T_{xx} + E_{xx}}\right\}$$

If we use the fact that $\sum_{ij} (x_{ij} - x_{..})^2 = T_{xx} + E_{xx}$, this can be written

$$\frac{1}{t-1}\left\{r \sum_{i} (\tau_i - \bar{\tau}_{.})^2 - \frac{\left[\sum_{i} (\tau_i - \bar{\tau}_{.})X_{i.}\right]^2}{T_{xx} + E_{xx}} + (rt-2)\sigma^2\right\}$$

Putting (18.10) and (18.11) together, we get for (18.9)

$$E(T_{\text{ms}}) = \frac{1}{t-1}\left\{r \sum_{i} (\tau_i - \bar{\tau}_{.})^2 - \frac{\left[\sum_{i} (\tau_i - \bar{\tau}_{.})X_{i.}\right]^2}{T_{xx} + E_{xx}} + (t-1)\sigma^2\right\}$$

By using Theorem 11.16, the value of λ can be evaluated.

A covariance model for a one-way classification has been presented in some detail. Other covariance models can be analyzed in similar fashion.

18.1.4 Example. The model for the following artificial data is Eq. (18.1). The problem is to estimate β, estimate contrasts of the treatment effects τ_i, and test the hypothesis $\tau_1 = \tau_2 = \tau_3 = \tau_4$.

TABLE 18.3 DATA FOR EXAMPLE OF COVARIANCE MODEL

1		2		3		4	
y	x	y	x	y	x	y	x
5.0	2.0	6.2	2.1	9.4	2.7	13.5	2.3
5.3	3.0	6.7	3.3	11.5	6.3	13.4	3.1
5.2	3.5	7.0	3.5	10.9	6.4	14.0	4.0
7.2	6.1	7.1	4.1	13.0	8.3	14.5	5.1
8.5	8.3	6.9	4.2	13.1	10.2	15.0	6.0
31.2	22.9	33.9	17.2	57.9	33.9	70.4	20.5

$\Sigma x_{ij} = 94.5 \qquad \Sigma y_{ij} = 193.4$

From the row of totals in Table 18.3 and from $\hat{\beta}$ as calculated in Table 18.4, the quantities $\hat{\mu} + \hat{\tau}_i = y_{i.} - \hat{\beta}x_{i.}$ can be computed. The results are as follows:

$$\hat{\mu} + \hat{\tau}_1 = 3.77 \qquad \hat{\mu} + \hat{\tau}_2 = 4.92 \qquad \hat{\mu} + \hat{\tau}_3 = 7.93$$
$$\hat{\mu} + \hat{\tau}_4 = 11.87$$

The average variance is $.461\sigma^2$. From these values any contrast of the τ_i can be estimated. The auxiliary table for testing

TABLE 18.4 ANALYSIS OF VARIANCE FOR THE DATA IN TABLE 18.3

SV	DF	SS:x^2	SS:xy	SS:y^2
Total	20	547.13	977.94	2,107.66
Mean	1	446.51	913.82	1,870.18
Treatments	3	31.43 = T_{xx}	26.89 = T_{xy}	216.06 = T_{yy}
Error	16	69.19 = E_{xx}	37.23 = E_{xy}	21.42 = E_{yy}
Treatment plus error	19	100.62 = S_{xx}	64.12 = S_{xy}	237.48 = S_{yy}

$$\hat{\beta} = \frac{E_{xy}}{E_{xx}} = \frac{37.23}{69.19} = .538$$

$$H_0: \tau_1 = \tau_2 = \tau_3 = \tau_4$$

corresponding to Table 18.2 is Table 18.5.

TABLE 18.5 AUXILIARY TABLE FOR TESTING $\tau_1 = \tau_2 = \tau_3 = \tau_4$

SV	DF	$\text{SS}:y^2 - \dfrac{(\text{SS}:xy)^2}{\text{SS}:x^2}$	MS	F
Treatment plus error	19	196.62		
Error	16	1.39	.087	748 +
Difference	3	195.23	65.077	

The result indicates that the treatment effects are certainly different; i.e., we reject H_0.

18.2 Two-way Classification Model with Interaction and with Fixed and Random Effects

In Chap. 12 the two-way classification model with interaction was discussed. In that chapter model 4 was assumed; that is to say, μ, τ_i, β_j, and $(\tau\beta)_{ij}$ were assumed to be fixed unknown constants. In Chap. 17 various types of model 5 were discussed. In these models all effects were random variables except the over-all mean μ.

In many experiments the model that fits the situation is neither 4 nor 5 but a combination of the two. In a mixed model such as this, there could be many specific types; that is to say, in an n-way classification model, any of the main effects could be fixed and any could be random. However, in a realistic model, the interaction terms are fixed or random depending on the main effects that occur in them.

Since so many different types are possible, all cannot be described. Only one will be discussed: the two-way classification model. From this model the general procedure can be inferred.

The distributional properties for model 4 and model 5 were fairly straightforward, but this is not true of the mixed model. If any element in an interaction term is random, it may be realistic to assume that the interaction term is random also. However, in that case we must postulate the distributional properties of the interaction terms. The proper error term to use to test certain hypotheses and set certain confidence intervals depends on what distributional properties are assumed to hold.[1] The model and distributional properties given in Eq. (18.12) will be assumed, and an AOV for this situation will be

[1] There are many sets of distributional properties that have been advanced as realistic for various situations [3, 4, 5, 6]. A discussion of these various assumptions is beyond the scope of this book. The concern here is to explain methods of analysis when various assumptions hold.

discussed. If an experimenter desires distributional properties other than those given in Eq. (18.12), tests and confidence intervals must be altered accordingly.

The two-way classification mixed model with interaction and with equal numbers in the subclasses will be defined by

$$y_{ijk} = \mu + \tau_i + \beta_j + (\tau\beta)_{ij} + e_{ijk} \qquad \begin{cases} i = 1, 2, \ldots, t \\ j = 1, 2, \ldots, b \\ k = 1, 2, \ldots, m \end{cases} \tag{18.12}$$

where μ and τ_i are assumed to be fixed unknown parameters such that

$$\sum_{i=1}^{t} \tau_i = 0$$

and where β_j, $(\tau\beta)_{ij}$, and e_{ijk} are random variables such that

$$E(\beta_j) = E(\tau\beta)_{ij} = E(e_{ijk}) = 0$$

the e_{ijk} are independent and distributed $N(0,\sigma_e^2)$; the e_{ijk} are independent of the β_j and the $(\tau\beta)_{ij}$; the β_j are independent and distributed $N(0,\sigma_\beta^2)$; the β_j are independent of the $(\tau\beta)_{ij}$; the $(\tau\beta)_{ij}$ are distributed $N(0, \frac{t-1}{t}\sigma_{\tau\beta}^2)$;

$$E[(\tau\beta)_{ij}(\tau\beta)_{i'j}] = -\frac{1}{t}\sigma_{\tau\beta}^2 \qquad \text{if } i \neq i'$$

$$E[(\tau\beta)_{ij}(\tau\beta)_{ij'}] = 0 \qquad \text{if } j \neq j'$$

and $\sum_i (\tau\beta)_{ij} = 0$ for all j.

If all the terms were fixed except e_{ijk}, the AOV would be as given in Chap. 12. This AOV is also given in Table 18.6, except for the EMS column, which is different for the mixed model (18.12). The object in this model is to test H_0: $\sigma_\beta^2 = 0$, to test H_0: $\sigma_{\tau\beta}^2 = 0$, to test H_0: $\tau_1 = \tau_2 = \cdots = \tau_t$, and to set confidence intervals on contrasts of the τ_i.

To accomplish this, we shall find the distribution of the sums of squares in Table 18.6.

The following theorem will be proved concerning the distributional properties of the sums of squares in Table 18.6.

◆ **Theorem 18.6** Let the model and distributional properties be as given by Eq. (18.12), i.e., a mixed two-way classification model. Then

(1) $$\frac{T_{ss}}{\sigma_e^2 + m\sigma_{\tau\beta}^2} = \frac{\sum_{ijk} (y_{i..} - y_{...})^2}{\sigma_e^2 + m\sigma_{\tau\beta}^2}$$

is distributed as $\chi'^2[(t-1), \lambda]$, where

$$\lambda = \frac{mb \sum_i (\tau_i - \bar{\tau}_.)^2}{2(\sigma_e^2 + m\sigma_{\tau\beta}^2)}$$

(2)
$$\frac{B_{\text{ss}}}{\sigma_e^2 + mt\sigma_\beta^2} = \frac{\sum_{ijk} (y_{.j.} - y_{...})^2}{\sigma_e^2 + mt\sigma_\beta^2}$$

is distributed as $\chi^2(b-1)$.

(3)
$$\frac{I_{\text{ss}}}{\sigma_e^2 + m\sigma_{\tau\beta}^2} = \frac{\sum_{ijk} (y_{ij.} - y_{i..} - y_{.j.} + y_{...})^2}{\sigma_e^2 + m\sigma_{\tau\beta}^2}$$

is distributed as $\chi^2(bt - b - t + 1)$.

(4)
$$\frac{E_{\text{ss}}}{\sigma_e^2} = \frac{\sum_{ijk} (y_{ijk} - y_{ij.})^2}{\sigma_e^2}$$

is distributed as $\chi^2[bt(m-1)]$.

(5) The quantities (1), (2), (3), and (4) are pairwise independent.

Table 18.6 Analysis of Variance for Two-way Classification

SV	DF	SS	EMS
Total	btm	$\sum_{ijk} y_{ijk}^2$	
Mean	1	$\frac{Y_{...}^2}{btm}$	
Due to τ	$t-1$	$T_{\text{ss}} = \sum_{ijk} (y_{i..} - y_{...})^2$	$\sigma_e^2 + m\sigma_{\tau\beta}^2 + \frac{mb}{t-1} \Sigma (\tau_i - \bar{\tau}_.)^2$
Due to β	$b-1$	$B_{\text{ss}} = \sum_{ijk} (y_{.j.} - y_{...})^2$	$\sigma_e^2 + mt\sigma_\beta^2$
Interaction	$(t-1)(b-1)$	$I_{\text{ss}} = \sum_{ijk} (y_{ij.} - y_{i..} - y_{.j.} + y_{...})^2$	$\sigma_e^2 + m\sigma_{\tau\beta}^2$
Error	$bt(m-1)$	$E_{\text{ss}} = \sum_{ijk} (y_{ijk} - y_{ij.})^2$	σ_e^2

Proof: First we shall prove (1). If we let $x_i = y_{i..}$, then x_i is distributed $N(\mu + \tau_i, \sigma_{ii})$, where

$$\sigma_{ii} = \text{var}(y_{i..}) = E[\bar{\beta}_. + (\overline{\tau\beta})_{i.} + \bar{e}_{i..}]^2$$
$$= \frac{\sigma_\beta^2}{b} + \frac{t-1}{bt}\sigma_{\tau\beta}^2 + \frac{\sigma_e^2}{mb}$$
$$= \frac{1}{mb}\left[\sigma_e^2 + m\sigma_\beta^2 + \frac{m(t-1)}{t}\sigma_{\tau\beta}^2\right]$$

Further, for $i \neq i'$,

$$\begin{aligned}\operatorname{cov}(x_i, x_{i'}) = \sigma_{ii'} &= E[\bar{\beta}_{.} + (\overline{\tau\beta})_{i.} + \bar{e}_{i..}][\bar{\beta}_{.} + (\overline{\tau\beta})_{i'.} + \bar{e}_{i'..}] \\ &= \frac{\sigma_\beta^2}{b} - \frac{1}{tb}\,\sigma_{\tau\beta}^2 \\ &= \frac{1}{tb}\,(t\sigma_\beta^2 - \sigma_{\tau\beta}^2)\end{aligned}$$

Now let $\mathbf{X}$ be a $t \times 1$ vector whose ith element is x_i; then the vector $\mathbf{X}$ is distributed $N(\boldsymbol{\mu}^*, \mathbf{V})$, where $\boldsymbol{\mu}^*$ is a $t \times 1$ vector whose ith element is $\mu + \tau_i$ and where $\mathbf{V}$ is a $t \times t$ covariance matrix whose ii'th element is $\sigma_{ii'}$. Also, we can write

$$\sum_i (y_{i..} - y_{...})^2 = \sum_{i=1}^{t} (x_i - x_.)^2 = \mathbf{X}'\mathbf{A}\mathbf{X}$$

where $\mathbf{A} = \mathbf{I} - (1/t)\mathbf{J}$.

Now, by Theorem 4.9, $\mathbf{X}'\mathbf{B}\mathbf{X}$ is distributed as $\chi'^2(n,\lambda)$ if and only if $\mathbf{BVB} = \mathbf{B}$, where n is the rank of $\mathbf{B}$ and $\lambda = \frac{1}{2}(\boldsymbol{\mu}^*)'\mathbf{B}(\boldsymbol{\mu}^*)$. Let us set

$$\frac{bm\mathbf{A}}{\sigma_e^2 + m\sigma_{\tau\beta}^2} = \mathbf{B}$$

Then
$$\mathbf{BVB} = \frac{(mb)^2}{(\sigma_e^2 + m\sigma_{\tau\beta}^2)^2}\,\mathbf{AVA}$$

But
$$\mathbf{AVA} = \left(\mathbf{I} - \frac{1}{t}\mathbf{J}\right)\mathbf{V}\left(\mathbf{I} - \frac{1}{t}\mathbf{J}\right) = \mathbf{V} - \frac{1}{t}\left(\mathbf{JV} + \mathbf{VJ} - \frac{1}{t}\mathbf{JVJ}\right)$$

Let $\mathbf{C} = \mathbf{JV}$; then

$$\begin{aligned}c_{pq} &= \sum_{s=1}^{t} \sigma_{sq} = \sum_{\substack{s \\ s \neq q}} \sigma_{sq} + \sigma_{qq} \\ &= \sum_{\substack{s=1 \\ s \neq q}}^{t} \left(\frac{\sigma_\beta^2}{b} - \frac{1}{tb}\,\sigma_{\tau\beta}^2\right) + \frac{1}{mb}\left[\sigma_e^2 + m\sigma_\beta^2 + \frac{m(t-1)}{t}\,\sigma_{\tau\beta}^2\right] \\ &= \frac{1}{mb}\,(\sigma_e^2 + mt\sigma_\beta^2)\end{aligned}$$

So every element of $\mathbf{C}$ equals

$$\frac{1}{mb}\,(\sigma_e^2 + mt\sigma_\beta^2)$$

or
$$\mathbf{C} = \mathbf{J}\,\frac{\sigma_e^2 + mt\sigma_\beta^2}{mb}$$

Also, $$\frac{1}{t}\,\mathbf{JVJ} = \frac{1}{t}\,\mathbf{JJ}\,\frac{\sigma_e^2 + mt\sigma_\beta^2}{mb} = \mathbf{J}\,\frac{\sigma_e^2 + mt\sigma_\beta^2}{mb}$$

Therefore, $$\mathbf{AVA} = \mathbf{V} - \mathbf{J}\,\frac{\sigma_e^2 + mt\sigma_\beta^2}{mbt} = \mathbf{D}\ \text{(say)}$$

Then $$d_{ii} = \sigma_{ii} - \frac{\sigma_e^2 + mt\sigma_\beta^2}{mtb} = \frac{t-1}{tmb}\,\sigma_e^2 + \frac{m(t-1)}{mtb}\,\sigma_{\tau\beta}^2$$

$$= \frac{t-1}{mtb}\,(\sigma_e^2 + m\sigma_{\tau\beta}^2)$$

and, if $i \neq i'$,

$$d_{ii'} = \sigma_{ii'} - \frac{\sigma_e^2 + mt\sigma_\beta^2}{mtb} = -\,\frac{1}{mtb}\,(\sigma_e^2 + m\sigma_{\tau\beta}^2)$$

So we see that

$$\mathbf{D} = \left(\mathbf{I} - \frac{1}{t}\,\mathbf{J}\right)\frac{\sigma_e^2 + m\sigma_{\tau\beta}^2}{mb} = \mathbf{A}\,\frac{\sigma_e^2 + m\sigma_{\tau\beta}^2}{mb}$$

Therefore, $$\mathbf{AVA} = \mathbf{A}\,\frac{\sigma_e^2 + m\sigma_{\tau\beta}^2}{mb}$$

and $$\mathbf{BVB} = \mathbf{B}$$

We can now invoke Theorem 4.9 and conclude that $\mathbf{X'BX}$ is a noncentral chi-square variate. But

$$\mathbf{X'BX} = \frac{mb}{\sigma_e^2 + m\sigma_{\tau\beta}^2}\,\mathbf{X'AX} = \frac{mb}{\sigma_e^2 + m\sigma_{\tau\beta}^2}\sum_i (y_{i..} - y_{...})^2 = \frac{\sum_{ijk} (y_{i..} - y_{...})^2}{\sigma_e^2 + m\sigma_{\tau\beta}^2}$$

The rank of $\mathbf{B}$ equals the rank of $\mathbf{A}$, which is $t - 1$. Also,

$$\lambda = \tfrac{1}{2}(\boldsymbol{\mu}^*)'\mathbf{B}(\boldsymbol{\mu}^*) = \frac{mb}{2(\sigma_e^2 + m\sigma_{\tau\beta}^2)}\,(\boldsymbol{\mu}^*)'\mathbf{A}(\boldsymbol{\mu}^*)$$

But

$$(\boldsymbol{\mu}^*)'\mathbf{A}(\boldsymbol{\mu}^*) = (\boldsymbol{\mu} + \boldsymbol{\tau})'\left(\mathbf{I} - \frac{1}{t}\,\mathbf{J}\right)(\boldsymbol{\mu} + \boldsymbol{\tau})$$

where $\boldsymbol{\tau}$ is a $t \times 1$ vector with elements τ_i and where $\boldsymbol{\mu}$ is a $t \times 1$ vector with elements μ. So

$$(\boldsymbol{\mu}^*)'\mathbf{A}(\boldsymbol{\mu}^*) = \boldsymbol{\tau}'\left(\mathbf{I} - \frac{1}{t}\,\mathbf{J}\right)\boldsymbol{\tau} = \sum_i (\tau_i - \bar{\tau}_.)^2$$

and

$$\lambda = \frac{mb \sum (\tau_i - \bar{\tau}_.)^2}{2(\sigma_e^2 + m\sigma_{\tau\beta}^2)}$$

This completes the proof of (1).

To prove (2), substitute the model into the equation for B_{ss} and obtain

$$B_{ss} = \sum_{ijk} (y_{.j.} - y_{...})^2 = \sum_{ijk} [\beta_j - \bar{\beta}_. + \bar{e}_{.j.} - \bar{e}_{...}]^2$$
$$= \sum_{ijk} [(\beta_j + \bar{e}_{.j.}) - (\bar{\beta}_. + \bar{e}_{...})]^2$$

Now let $\beta_j + \bar{e}_{.j.} = x_j$, and x_j is distributed independently and $N(0, \sigma_\beta^2 + \sigma_e^2/mt)$. If we set

$$u = \frac{\sum_{j=1}^{b} (x_j - x_.)^2}{\sigma_\beta^2 + \sigma_e^2/mt}$$

then, clearly, u is distributed as $\chi^2(b - 1)$; but

$$u = mt \frac{\sum_{j=1}^{b} (x_j - x_.)^2}{\sigma_e^2 + mt\sigma_\beta^2} = \sum_{ijk} \frac{(y_{.j.} - y_{...})^2}{\sigma_e^2 + mt\sigma_\beta^2}$$

To prove (3) we shall use the fact that, if random independent quantities x_{ij} are distributed $N(0,\sigma^2)$, then

$$\frac{\sum_{i=1}^{t} \sum_{j=1}^{b} (x_{ij} - x_{i.} - x_{.j} + x_{..})^2}{\sigma^2}$$

is distributed as $\chi^2[(b - 1)(t - 1)]$. This was proved in Chap. 12. But let $x_{ij} - x_{.j} = z_{ij}$; then

$$\sum_{ij} (x_{ij} - x_{i.} - x_{.j} + x_{..})^2 = \sum_{ij} (z_{ij} - z_{i.})^2$$

and z_{ij} is distributed $N\left(0, \frac{t-1}{t} \sigma^2\right)$, $E(z_{ij}z_{ij'}) = 0$ if $j \neq j$ and $E(z_{ij}z_{i'j}) = -\sigma^2/t$ if $i \neq i'$. This says that, if z_{ij} has these distributional properties, then

$$v = \frac{\sum_{ij} (z_{ij} - z_{i.})^2}{\sigma^2}$$

and v is distributed as $\chi^2[(b-1)(t-1)]$. This result will be used shortly. Now let us consider

$$\frac{I_{\text{ss}}}{\sigma_e^2 + m\sigma_{\tau\beta}^2} = \frac{\sum_{ijk} (y_{ij.} - y_{i..} - y_{.j.} + y_{...})^2}{\sigma_e^2 + m\sigma_{\tau\beta}^2}$$

If we substitute model (18.12) into this, we get

$$I_{\text{ss}} = \sum_{ijk} [(\tau\beta)_{ij} - (\overline{\tau\beta})_{i.} + \bar{e}_{ij.} - \bar{e}_{i..} - \bar{e}_{.j.} + \bar{e}_{...}]^2$$

which can be written

$$\sum_{ijk} \{[(\tau\beta)_{ij} + \bar{e}_{ij.} - \bar{e}_{.j.}] - [(\overline{\tau\beta})_{i.} + \bar{e}_{i..} - \bar{e}_{...}]\}^2 = \sum_{ijk} (w_{ij} - w_{i.})^2$$

where

$$(\tau\beta)_{ij} + \bar{e}_{ij.} - \bar{e}_{.j.} = w_{ij}$$

So w_{ij} is distributed normally with zero mean, and

$$E(w_{ij})^2 = \frac{t-1}{t}\sigma_{\tau\beta}^2 + \frac{t-1}{mt}\sigma_e^2$$

$$= \frac{t-1}{t}\frac{\sigma_e^2 + m\sigma_{\tau\beta}^2}{m}$$

$$= \frac{t-1}{t}\sigma^2 \text{ (say)}$$

Also, clearly, $E(w_{ij}w_{ij'}) = 0$ if $j \neq j'$, and

$$E(w_{ij}w_{i'j}) = -\frac{1}{t}\sigma_{\tau\beta}^2 - \frac{1}{mt}\sigma_e^2 = -\frac{1}{t}\frac{\sigma_e^2 + m\sigma_{\tau\beta}^2}{m} = -\frac{1}{t}\sigma^2 \qquad \text{if } i' \neq i$$

The w_{ij} satisfy the distributional properties of z_{ij} above; hence,

$$u = \frac{\sum_{ij} (w_{ij} - w_{i.})^2}{\sigma^2} = \frac{\sum_{ijk} (y_{ij.} - y_{i..} - y_{.j.} + y_{...})^2}{\sigma_e^2 + m\sigma_{\tau\beta}^2}$$

is distributed as $\chi^2[(b-1)(t-1)]$.

The proof for (4) is straightforward and will be left for the reader. Also, the proof for (5) will not be given.

From Theorem 18.6 we have the following theorem.

◆ **Theorem 18.7** (1) To test $H_0: \sigma_\beta^2 = 0$, use $B_{\text{ms}}/E_{\text{ms}}$ as an F statistic with $b-1$ and $bt(m-1)$ degrees of freedom.
(2) To test $H_0: \sigma_{\tau\beta}^2 = 0$, use $I_{\text{ms}}/E_{\text{ms}}$ as an F statistic with $(b-1) \times (t-1)$ and $bt(m-1)$ degrees of freedom.

◆ **Theorem 18.8** To test H_0: $\tau_1 = \tau_2 = \cdots = \tau_t$, use $T_{\text{ms}}/I_{\text{ms}}$ as an F statistic with $t - 1$ and $(b - 1)(t - 1)$ degrees of freedom.

The power of these tests can be found by using previous theorems. It can be shown that the best (minimum-variance) unbiased estimate of the treatment contrast $\Sigma\lambda_i\tau_i$ is $\Sigma\lambda_i y_{i..}$ and that the variance of this estimate is

$$\frac{1}{mb}(\sigma_e^2 + m\sigma_{\tau\beta}^2)\Sigma\lambda_i^2$$

So a confidence interval on $\Sigma\lambda_i\tau_i$ with coefficient $1 - \alpha$ is

$$\Sigma\lambda_i y_{i..} - t_{\alpha/2}\sqrt{\frac{I_{\text{ms}}\Sigma\lambda_i^2}{mb}} \leqslant \Sigma\lambda_i\tau_i \leqslant \Sigma\lambda_i y_{i..} + t_{\alpha/2}\sqrt{\frac{I_{\text{ms}}\Sigma\lambda_i^2}{mb}}$$

It is beyond the scope of this book to argue for or against the distributional properties of (18.12). The object here is merely to give methods for finding tests of hypotheses, etc., for this particular set of properties. If other sets are desired, then similar methods can be used to ascertain proper tests, etc.

18.3 Balanced Incomplete Block with Blocks Random: Recovery of Interblock Information

In Chap. 14 the balanced incomplete block model was discussed as a special case of model 4. It is written

$$y_{ijm} = \mu + \tau_i + \beta_j + e_{ijm}$$

and the elements μ, τ_i, and β_j were assumed to be unknown constants. However, in many experimental situations it is quite realistic to assume that the block effects, β_j, are random variables. If the β_j are random, it may be possible to obtain more information about the treatment constants τ_i than when the β_j are fixed.

18.3.1 Definition and Notation. The balanced incomplete block model with blocks random will be defined as follows: Each treatment appears r times, and there are k plots in each block; every pair of treatments appears together in λ blocks. The model can be written

$$y_{ijm} = \mu + \tau_i + b_j + e_{ijm} \qquad \begin{cases} i = 1, 2, \ldots, t \\ j = 1, 2, \ldots, b \end{cases} \tag{18.13}$$

where $m = n_{ij}$ and where $n_{ij} = 1$ if treatment i appears in block j and $n_{ij} = 0$ otherwise; y_{ij0} does not exist. The μ and τ_i are fixed unknown parameters; b_j is distributed $N(0,\sigma_b^2)$; e_{ijm} is distributed $N(0,\sigma^2)$; and all random variables are independent.

18.3.2 Point Estimates of $\tau_p - \bar{\tau}_.$. If the model in Chap. 14 with b_j fixed is used, the normal equations are as given in Eq. (14.3). From these an unbiased estimate of $\tau_p - \bar{\tau}_.$ was formed. This is given in Eq. (14.9) and is

$$\frac{k}{\lambda t} q_p \tag{18.14}$$

where q_p is defined as

$$q_p = Y_{p..} - \frac{1}{k} \sum_{j=1}^{b} n_{pj} Y_{.j.} \tag{18.15}$$

The variance of q_p is given in Eq. (14.10), and is

$$\frac{r(k-1)}{k} \sigma^2 \tag{18.16}$$

From these results it follows that $(\hat{\tau}_p - \hat{\bar{\tau}}_.) = (k/\lambda t)q_p$ is distributed

$$N\left(\tau_p - \bar{\tau}_., \frac{k}{\lambda t} \frac{t-1}{t} \sigma^2\right)$$

and

$$\operatorname{cov}(\hat{\tau}_p - \hat{\bar{\tau}}_., \hat{\tau}_{p'} - \hat{\bar{\tau}}_.) = -\frac{k}{\lambda t} \frac{1}{t} \sigma^2 \tag{18.17}$$

The estimator of a treatment contrast $\Sigma c_i \tau_i$ based on the q_p is given by Theorem 14.1 and is

$$\frac{k}{\lambda t} \Sigma c_i q_i$$

So

$$E(\Sigma c_i \hat{\tau}_i) = E\left(\frac{k}{\lambda t} \Sigma c_i q_i\right) = \Sigma c_i \tau_i$$

and

$$\operatorname{var}(\Sigma c_i \hat{\tau}_i) = \frac{k}{\lambda t} (\Sigma c_i^2) \sigma^2 \tag{18.18}$$

Equation (14.8) shows that q_p does not depend on β_j; hence, Eq. (18.18) holds even when the blocks are random variables. Next consider Table 14.2. The quantity E_{ms} is an unbiased estimate of σ^2, and, if the β_j are fixed,

$$\frac{E_{\text{ss}}}{\sigma^2} \text{ is distributed as } \chi^2(bk - b - t + 1) \tag{18.19}$$

However, by Table 14.2,

$$E_{ss} = \sum_{ijm} y_{ijm}^2 - \frac{\sum_j Y_{.j.}^2}{k} - \frac{k}{\lambda t}\Sigma q_p^2 = \sum_{ijm}(y_{ijm} - y_{.j.})^2 - \frac{k}{\lambda t}\Sigma q_p^2$$

If we substitute the model into $\Sigma(y_{ijm} - y_{.j.})^2$, we get

$$E_{ss} = \sum_{ijm}[\tau_i - \bar{\tau}_. + e_{ijm} - \bar{e}_{.j.}]^2 - \frac{k}{\lambda t}\Sigma q_p^2$$

But, by Eq. (14.8), it was shown that q_p does not involve β_j; hence E_{ss} does not involve β_j. Thus E_{ss}/σ^2 is distributed as $\chi^2(bk - b - t + 1)$ under the assumptions of model (18.13), that is, where the b_j are random variables. Therefore, the following theorem has been proved.

◆ **Theorem 18.9** Let the model be a balanced incomplete block model with blocks random, i.e., model (18.13). Then

(1) $$\Sigma c_p\hat{\tau}_p = \frac{k}{\lambda t}\Sigma c_p\left(Y_{p..} - \frac{1}{k}\sum_j n_{pj}Y_{.j.}\right) = \frac{k}{\lambda t}\Sigma c_p q_p$$

is an unbiased estimator of $\Sigma c_i \tau_i$ if $\Sigma c_i = 0$.

(2) $\Sigma c_p \hat{\tau}_p$ is distributed

$$N\left(\Sigma c_p \tau_p, \frac{k\sigma^2}{\lambda t}\Sigma c_p^2\right)$$

(3) E_{ss}/σ^2 is distributed as $\chi^2(bk - b - t + 1)$, where (see Table 14.2)

$$E_{ss} = \sum_{ijm} y_{ijm}^2 - \frac{\sum_j Y_{.j.}^2}{k} - \frac{k}{\lambda t}\Sigma q_p^2$$

(4) E_{ss} and $\Sigma c_p \hat{\tau}_p$ are independent.

The quantities $\frac{k}{\lambda t}\Sigma c_p q_p$ and E_{ss} will be called *intrablock estimators* of $\Sigma c_p \tau_p$ and σ^2, respectively.

Next we shall show that there exists another unbiased estimator of $\Sigma c_p \tau_p$ independent of the intrablock estimators $\Sigma c_p \hat{\tau}_p$ and E_{ss}. This estimator will be denoted by $\Sigma c_p \tilde{\tau}_p$ and will be a function of the block totals $Y_{.j.}$. The symbol B_p will be used to denote the total of all blocks that contain treatment p; this notation was used also in Sec. 14.3. This gives

$$B_p = \sum_j n_{pj} Y_{.j.}$$

Also we shall use $\bar{B}_. = \frac{1}{t}\sum_{i=1}^{t} B_i$

Now

$$Y_{.j.} = \sum_{im} y_{ijm}$$

$$= \sum_{im} (\mu + \tau_i + b_j + e_{ijm})$$

$$= k\mu + \sum_i n_{ij}\tau_i + kb_j + e_{.j.}$$

and

$$B_p = \sum_j n_{pj}\left(k\mu + \sum_i n_{ij}\tau_i + kb_j + e_{.j.}\right)$$

Also, $E(B_p) = \sum_j n_{pj}\left(k\mu + \sum_i n_{ij}\tau_i\right) = rk(\mu + \bar{\tau}_.) + (r - \lambda)(\tau_p - \bar{\tau}_.)$

So

$$\frac{E(\Sigma c_p B_p)}{r - \lambda} = \Sigma c_p \tau_p$$

and

$$\frac{\Sigma c_p B_p}{r - \lambda} \tag{18.20}$$

is an unbiased estimator of the treatment contrast $\Sigma c_p \tau_p$ where $\Sigma c_p = 0$. The variance of this estimator will be found, but first $\text{var}(B_p)$ and $\text{cov}(B_p, B_{p'})$ will be evaluated.

$$\text{var}(B_p) = E[B_p - E(B_p)]^2 = E\left(k \sum_j n_{pj}b_j + \sum_j \sum_{im} n_{pj}e_{ijm}\right)^2$$

$$= k^2 E\left(\sum_j n_{pj}b_j\right)^2 + E\left(\sum_j \sum_{im} n_{pj}e_{ijm}\right)^2$$

But, since b_j is independent of $b_{j'}$ if $j \neq j'$ and since the e_{ijm} are independent, this becomes ($n_{pj}^2 = n_{pj}$)

$$k^2 E\left(\sum_j n_{pj}^2 b_j^2\right) + E\left(\sum_j \sum_{im} n_{pj}^2 e_{ijm}^2\right) = k^2\sigma_b^2 \sum_j n_{pj} + \sigma^2 \sum_{ijm} n_{pj}$$

$$= k^2 r\sigma_b^2 + \sigma^2 \sum_{ij} n_{pj}n_{ij}$$

$$= k^2 r\sigma_b^2 + rk\sigma^2$$

$$= kr(\sigma^2 + k\sigma_b^2)$$

The covariance of B_p and $B_{p'}$ is needed.

$$\text{cov}(B_p, B_{p'}) = E\left[\left(k \sum_j n_{pj}b_j + \sum_j \sum_{im} n_{pj}e_{ijm}\right)\left(k \sum_{j'} n_{p'j'}b_{j'} + \sum_{j'} \sum_{i'm'} n_{p'j'}e_{i'j'm'}\right)\right] \quad p' \neq p$$

$$= k^2\sigma_b^2 \sum_j n_{pj}n_{p'j} + \sigma^2 \sum_{ji} n_{pj}n_{p'j}n_{ij}$$

$$= k^2\lambda\sigma_b^2 + k\lambda\sigma^2$$

$$= k\lambda(\sigma^2 + k\sigma_b^2)$$

Now the variance of $\Sigma c_p B_p/(r - \lambda)$ can be found. It is

$$\operatorname{var}\left(\frac{\Sigma c_p B_p}{r - \lambda}\right) = \frac{1}{(r - \lambda)^2}\left[\sum_{p'} c_p^2 \operatorname{var}(B_p) + \sum_{p \neq p'} c_p c_{p'} \operatorname{cov}(B_p, B_{p'})\right]$$

$$= \frac{1}{(r - \lambda)^2}\left[kr \sum_p c_p^2(\sigma^2 + k\sigma_b^2) - \sum_p c_p^2 k\lambda(\sigma^2 + k\sigma_b^2)\right]$$

$$= \frac{kr - k\lambda}{(r - \lambda)^2} \Sigma c_p^2(\sigma^2 + k\sigma_b^2)$$

$$= \frac{k}{r - \lambda}(\sigma^2 + k\sigma_b^2)\Sigma c_p^2$$

To prove that the estimator based on block totals is independent of the intrablock estimator, we shall show that B_j is independent of q_i for all i and j. Since B_j and q_i are each normally distributed, their independence will follow if they can be shown to have zero covariance. Thus we have

$$\operatorname{cov}(B_j, q_i) = E\{[B_j - E(B_j)][q_i - E(q_i)]\}$$

$$= E\left\{\left[\sum_s n_{js} Y_{.s.} - E\left(\sum_s n_{js} Y_{.s.}\right)\right]\right.$$

$$\left. \times \left[Y_{i..} - \frac{1}{k}\sum_v n_{iv} Y_{.v.} - E(Y_{i..}) + \frac{1}{k} E\left(\sum_v n_{iv} Y_{.v.}\right)\right]\right\}$$

$$= E\left[\left(k \sum_s n_{js} b_s + \sum_u \sum_{sm} n_{js} e_{usm}\right)\left(\sum_{vu} e_{ivu} - \frac{1}{k}\sum_{pv}\sum_{\omega} n_{iv} e_{pvw}\right)\right]$$

$$= E\left(\sum_{usm} n_{js} e_{usm} \sum_{vu} e_{ivu}\right) - \frac{1}{k} E\left(\sum_u \sum_{sm} n_{js} e_{usm} \sum_p \sum_{v\omega} n_{iv} e_{pvw}\right)$$

$$= \sigma^2 \sum_s n_{js} n_{is} - \frac{\sigma^2}{k} \sum_{pv} n_{jv} n_{iv} n_{pv} = 0$$

for every i and j.

To prove that B_j is independent of E_{ss}, we can argue as follows: In the balanced incomplete model with block effects fixed, the B_j are clearly independent of the intrablock error E_{ss}, and the block sum of squares B_{ss} is a function of B_j. Now, even if the block effects b_j are random, they do not appear in E_{ss}. So in this case the B_j and E_{ss} are still independent.

This proves (1) and (2) of the following theorem; the proof for (3) and (4) will be omitted.

◆ **Theorem 18.10** Let the model be a balanced incomplete block model with blocks random, i.e., model (18.13). Then

(1) $$\Sigma c_p \tilde{\tau}_p = \frac{\Sigma c_p B_p}{r - \lambda} = \sum_p c_p \frac{\sum_j n_{pj} Y_{.j.}}{r - \lambda}$$

is an unbiased estimator of $\Sigma c_p \tau_p$ if $\Sigma c_p = 0$.

(2) $\Sigma c_p \tilde{\tau}_p$ is distributed

$$N\left(\Sigma c_p \tau_p, \Sigma c_p^2 k \frac{\sigma^2 + k\sigma_b^2}{r - \lambda}\right)$$

(3) $$\frac{R_{\text{ss}}}{\sigma^2 + k\sigma_b^2} = \frac{\sum_j \frac{Y_{.j.}^2}{k} - \frac{Y_{...}^2}{bk} - \left[\sum_j \frac{B_j^2}{k(r - \lambda)} - \frac{B_.^2}{tk(r - \lambda)}\right]}{\sigma^2 + k\sigma_b^2}$$

is distributed as $\chi^2(b - t)$ if $b > t$.

(4) $\Sigma c_p \hat{\tau}_p$, $\Sigma c_p \tilde{\tau}_p$, R_{ss}, and E_{ss} are mutually independent.

The estimator $\Sigma c_p B_p/(r - \lambda)$ will be termed the *interblock estimator* of the treatment contrast $\Sigma c_p \tau_p$.

Sufficient machinery has now been developed so that methods of utilizing interblock estimators can be explored. These will be discussed in the next two articles.

18.3.3 Combining Interblock and Intrablock Estimators. Instead of working with estimators of general contrasts $\Sigma c_p \tau_p$, we shall consider estimators of $\tau_i - \bar{\tau}_.$. From these any contrast can easily be estimated. From Theorems 18.9 and 18.10, the intrablock and interblock estimators of $\tau_i - \bar{\tau}_.$ can easily be found by setting $c_i = (t - 1)/t$ and the remaining c_j equal to $-1/t$. Then

$$\Sigma c_p^2 = \frac{t - 1}{t}$$

This gives us

Intrablock:

$$\hat{\tau}_i - \hat{\bar{\tau}}_. = \frac{k}{\lambda t} q_i \qquad \text{var}(\hat{\tau}_i - \hat{\bar{\tau}}_.) = \frac{k(t - 1)}{\lambda t^2} \sigma^2$$

Interblock:

$$\tilde{\tau}_i - \tilde{\bar{\tau}}_. = \frac{B_i - \bar{B}_.}{r - \lambda} \qquad \text{var}(\tilde{\tau}_i - \tilde{\bar{\tau}}_.) = \frac{(t - 1)k}{t(r - \lambda)} (\sigma^2 + k\sigma_b^2) \qquad (18.21)$$

Before discussing methods for combining the interblock and intrablock estimators in the balanced incomplete block, we shall prove a general theorem about combining two independent unbiased estimators of a parameter.

◆ **Theorem 18.11** Let $\hat{\theta}_1$ be an unbiased estimator of θ, and let the variance of $\hat{\theta}_1$ be denoted by σ_1^2. Let $\hat{\theta}_2$ be another unbiased estimator of θ, and let the variance of $\hat{\theta}_2$ be denoted by σ_2^2. Let $\hat{\theta}_1$ and $\hat{\theta}_2$ be uncorrelated. Then the best (minimum-variance) linear unbiased estimator of θ is

$$\hat{\theta} = (\sigma_2^2\hat{\theta}_1 + \sigma_1^2\hat{\theta}_2)(\sigma_1^2 + \sigma_2^2)^{-1}$$

Proof: By a linear estimator of θ we mean a linear combination of $\hat{\theta}_1$ and $\hat{\theta}_2$; that is, $y = \alpha_1\hat{\theta}_1 + \alpha_2\hat{\theta}_2$. The problem is to determine α_1 and α_2 such that y is unbiased $[E(y) = \theta]$ and such that $\text{var}(y)$ is a minimum.

If $\theta = E(y)$, this gives

$$\theta = E(y) = E(\alpha_1\hat{\theta}_1 + \alpha_2\hat{\theta}_2) = \alpha_1\theta + \alpha_2\theta = (\alpha_1 + \alpha_2)\theta$$

So
$$\alpha_1 + \alpha_2 = 1 \qquad \text{or} \qquad \alpha_1 = 1 - \alpha_2$$

Now
$$\begin{aligned}\text{var}(y) &= \text{var}(\alpha_1\hat{\theta}_1 + \alpha_2\hat{\theta}_2) = \alpha_1^2\,\text{var}(\hat{\theta}_1) + \alpha_2^2\,\text{var}(\hat{\theta}_2)\\ &= \alpha_1^2\sigma_1^2 + (1 - \alpha_1)^2\sigma_2^2\end{aligned}$$

The only unknown in $\text{var}(y)$ is α_1. The value of α_1 that minimizes $\text{var}(y)$ is found by

$$\frac{d\,\text{var}(y)}{d\alpha_1} = 2\alpha_1\sigma_1^2 - 2(1 - \alpha_1)\sigma_2^2 = 0$$

This gives

$$\alpha_1 = \sigma_2^2(\sigma_1^2 + \sigma_2^2)^{-1} \qquad \text{and} \qquad \alpha_2 = \sigma_1^2(\sigma_1^2 + \sigma_2^2)^{-1}$$

So
$$y = (\sigma_2^2\hat{\theta}_1 + \sigma_1^2\hat{\theta}_2)(\sigma_1^2 + \sigma_2^2)^{-1}$$

and $y = \hat{\theta}$.

It is important to notice that if the variances σ_1^2 and σ_2^2 are *not known*, $\hat{\theta}$ cannot be constructed. In the balanced incomplete block model, where we shall make use of this theorem, the variances are *not known*. Therefore, the theorem cannot be used directly, but a slight variation will be employed. The variation consists in replacing the variances σ_1^2 and σ_2^2 with their unbiased estimators. The full impact of this variation has not been studied in great detail at the present time. In order to get some insight into this, the following theorem will be useful.

◆ **Theorem 18.12** Let $\hat{\theta}_1$ and $\hat{\theta}_2$ be unbiased estimators of θ with variance σ_1^2 and σ_2^2, respectively. Let $\hat{\sigma}_1^2$ and $\hat{\sigma}_2^2$ be unbiased estimators of σ_1^2 and σ_2^2, respectively. Let $\hat{\theta}_1$, $\hat{\theta}_2$, $\hat{\sigma}_1^2$, $\hat{\sigma}_2^2$ be mutually independent. Then

$$\tilde{\theta} = (\hat{\sigma}_2^2\hat{\theta}_1 + \hat{\sigma}_1^2\hat{\theta}_2)(\hat{\sigma}_1^2 + \hat{\sigma}_2^2)^{-1}$$

is an unbiased estimator of θ.

Proof: The theorem can be proved by using the fact that, if the random variables x and y are independent, then $E(xy) = E(x)E(y)$. Now consider

$$E(\tilde{\theta}) = E\left(\frac{\hat{\sigma}_2^2\hat{\theta}_1}{\hat{\sigma}_1^2 + \hat{\sigma}_2^2} + \frac{\hat{\sigma}_1^2\hat{\theta}_2}{\hat{\sigma}_1^2 + \hat{\sigma}_2^2}\right)$$

$$= E\left(\hat{\theta}_1 \frac{\hat{\sigma}_2^2}{\hat{\sigma}_1^2 + \hat{\sigma}_2^2}\right) + E\left(\hat{\theta}_2 \frac{\hat{\sigma}_1^2}{\hat{\sigma}_1^2 + \hat{\sigma}_2^2}\right)$$

But, because $\hat{\theta}_1$ and $\hat{\sigma}_2^2/(\hat{\sigma}_1^2 + \hat{\sigma}_2^2)$ are independent, the first term can be written

$$E(\hat{\theta}_1)E\left(\frac{\hat{\sigma}_2^2}{\hat{\sigma}_1^2 + \hat{\sigma}_2^2}\right) = \theta E\left(\frac{\hat{\sigma}_2^2}{\hat{\sigma}_1^2 + \hat{\sigma}_2^2}\right)$$

Similarly, the second term can be written

$$E(\hat{\theta}_2)E\left(\frac{\hat{\sigma}_1^2}{\hat{\sigma}_1^2 + \hat{\sigma}_2^2}\right) = \theta E\left(\frac{\hat{\sigma}_1^2}{\hat{\sigma}_1^2 + \hat{\sigma}_2^2}\right)$$

So we get

$$E(\tilde{\theta}) = \theta E\left(\frac{\hat{\sigma}_2^2}{\hat{\sigma}_1^2 + \hat{\sigma}_2^2}\right) + \theta E\left(\frac{\hat{\sigma}_1^2}{\hat{\sigma}_1^2 + \hat{\sigma}_2^2}\right) = \theta E\left(\frac{\hat{\sigma}_1^2 + \hat{\sigma}_2^2}{\hat{\sigma}_1^2 + \hat{\sigma}_2^2}\right) = \theta E(1) = \theta$$

To find a combined estimator of $\tau_i - \bar{\tau}_.$, Theorem 18.11 and the estimators in Eq. (18.21) can be used. If we let τ_i^* denote the combined estimator of $\tau_i - \bar{\tau}_.$, this gives

$$\tau_i^* = \frac{\dfrac{k}{\lambda t} q_i \dfrac{(t-1)k}{t(r-\lambda)}(\sigma^2 + k\sigma_b^2) + \dfrac{\bar{B}_i - \bar{B}_.}{r - \lambda}\dfrac{k(t-1)}{\lambda t^2}\sigma^2}{\dfrac{k(t-1)}{\lambda t^2}\sigma^2 + \dfrac{k(t-1)}{t(r-\lambda)}(\sigma^2 + k\sigma_b^2)}$$

If each term is multiplied by $\lambda t^2(r - \lambda)/k(t - 1)$, this simplifies to

$$\tau_i^* = \frac{kq_i(\sigma^2 + k\sigma_b^2) + (B_i - \bar{B}_.)\sigma^2}{(r-\lambda)\sigma^2 + \lambda t(\sigma^2 + k\sigma_b^2)} \tag{18.22}$$

Of course, this quantity cannot be calculated, since the variances are not known. The estimators of σ^2 and σ_b^2 can be used; but, before this is discussed, various forms of τ_i^* will be examined.

Let T_i denote the total of the observations that receive the ith treatment; that is, $T_i = Y_{i..}$; and let $\bar{T}_. = \frac{1}{t}\sum_{i=1}^{t} T_i$. Then

$$q_i = T_i - \frac{1}{k} B_i = (T_i - \bar{T}_.) - \frac{1}{k}(B_i - \bar{B}_.)$$

If we substitute this into Eq. (18.22) and simplify, we have

$$\tau_i^* = \frac{T_i - \bar{T}_.}{r} + \frac{\nu}{r}[(t - k)(T_i - \bar{T}_.) - (t - 1)(B_i - \bar{B}_.)] \quad (18.23)$$

where

$$\nu = \frac{w - w'}{t(k - 1)w + (t - k)w'} \quad (18.24)$$

and

$$w = \frac{1}{\sigma^2} \qquad w' = \frac{1}{\sigma^2 + k\sigma_b^2} \quad (18.25)$$

τ_i^* in Eq. (18.23) is the best (minimum-variance) linear unbiased combined estimator of $\tau_i - \bar{\tau}_.$. However, since ν is unknown, this quantity cannot be used. Therefore, w and w' (and hence ν) will be estimated from an AOV table, and these estimates will be used to find a combined estimator of $\tau_i - \bar{\tau}_.$. This estimator will be denoted by

$$\hat{\tau}_i^* = \frac{T_i - \bar{T}_.}{r} + \frac{\hat{\nu}}{r}[(t - k)(T_i - \bar{T}_.) - (t - 1)(B_i - \bar{B}_.)] \quad (18.26)$$

where

$$\hat{\nu} = \frac{\hat{w} - \hat{w}'}{t(k - 1)\hat{w} + (t - k)\hat{w}'}$$

and where $\hat{w}$ and $\hat{w}'$ are estimators of $1/\sigma^2$ and $1/(\sigma^2 + k\sigma_b^2)$, respectively, obtained from an AOV table.

There are many ways of estimating $1/\sigma^2$ and $1/(\sigma^2 + k\sigma_b^2)$, but we shall present the method given by Yates [11], who was the first to consider the incomplete block model with blocks random.

To obtain an analysis of variance, the model will be considered with blocks fixed in order to use the theory in Chap. 14. Then, when an AOV table is obtained, the model will be considered with blocks random, and the appropriate mean squares and expected mean squares will be equated, so as to give rise to estimators of σ^2 and σ_b^2.

Table 18.7 Analysis of Variance for Incomplete Block Model with Blocks Random

Method A			Method B		
SV	DF	SS	SS	DF	SV
Total	bk	$A = \Sigma y_{ijm}^2$	$B = \Sigma y_{ijm}^2$	bk	Total
Mean	1	$A_1 = \frac{Y_{...}^2}{bk}$	$B_1 = \frac{Y_{...}^2}{bk}$	1	Mean
Treatments (adj)	$t - 1$	$A_2 = \frac{k}{\lambda t} \Sigma q_i^2$	$B_2 = \Sigma \frac{Y_{i..}^2}{r} - \frac{Y_{...}^2}{bk}$	$t - 1$	Treatments (unadj)
Blocks (unadj)	$b - 1$	$A_3 = \sum_j \frac{Y_{.j.}^2}{k} - \frac{Y_{...}^2}{bk}$	$B_3 = A_2 + A_3 - B_2$	$b - 1$	Blocks (adj)
[Treatment component	[$t - 1$	[$A_{31} = \sum_i \frac{(B_i - \bar{B}.)^2}{k(r - \lambda)}$			
Remainder]	$b - t$]	$A_3 - A_{31}$ = subtraction]			
Intrablock error	$bk - b - t + 1$	$A_4 = A - A_1 - A_2 - A_3$	$B_4 = B - B_1 - B_2 - B_3$	$bk - b - t + 1$	Intrablock error

There are two methods of analysis that could be used:

Method A. This is often called the *method of treatments eliminating* (adjusted for) *blocks.* The sum of squares due to τ and β is broken into

$$\begin{aligned} R(\tau,\beta) &= R(\tau \mid \beta) + R(\beta) \\ &= [\text{SS due to treatments (adj)}] + [\text{SS due to blocks (unadj)}] \end{aligned}$$

Method B. This is often called the *method of blocks eliminating* (adjusted for) *treatments.* The sum of squares due to τ and β is broken into

$$\begin{aligned} R(\tau,\beta) &= R(\beta \mid \tau) + R(\tau) \\ &= [\text{SS due to blocks (adj)}] + [\text{SS due to treatments (unadj)}] \end{aligned}$$

Method A is used in Table 14.1. From the above it follows that

$$R(\beta \mid \tau) + R(\tau) = R(\tau \mid \beta) + R(\beta)$$

or, in words, that the sum of squares due to blocks (adj) plus the sum of squares due to treatments (unadj) equals the sum of squares due to treatments (adj) plus the sum of squares due to blocks (unadj).

These two methods of analysis are given in Table 18.7. Method A in Table 18.7 is exactly the analysis given in Table 14.2 except that the sum of squares due to blocks (unadj) has been further partitioned into two parts (A_{31} and $A_3 - A_{31}$). However, these quantities are not used directly for estimating $\tau_i - \bar{\tau}_{..}$

To compute $\hat{\tau}_i^*$ in Eq. (18.26), Yates used the analysis given under method B. The procedure will be explained by rewriting method B of Table 18.7 in Table 18.8 and putting in the EMS column. The expected mean square for intrablock error is as shown because $A_4 = B_4$ and A_4 has the same expectation whether the model has blocks fixed or blocks random. The expectation of $B_3/(b-1)$ will be examined in some detail.

$$E(B_3) = E(A_2 + A_3 - B_2) = E\left(\frac{k}{\lambda t}\Sigma q_i^2 + \sum_j \frac{Y_{.j.}^2}{k} - \sum_i \frac{Y_{i..}^2}{r}\right) \tag{18.27}$$

By Eq. (18.16), $$\operatorname{var}(q_i) = \frac{\lambda t}{k}\frac{t-1}{t}\sigma^2$$

Also, $$E(q_i) = \frac{\lambda t}{k}(\tau_i - \bar{\tau}_.)$$

so $$E(q_i^2) = \operatorname{var}(q_i) + [E(q_i)]^2 = \frac{\lambda t}{k}\frac{t-1}{t}\sigma^2 + \frac{\lambda^2 t^2}{k^2}(\tau_i - \bar{\tau}_.)^2$$

and $$\frac{k}{\lambda t}E\left(\sum_{i=1} q_i^2\right) = (t-1)\sigma^2 + \frac{\lambda t}{k}\Sigma(\tau_i - \bar{\tau}_.)^2 \tag{18.28}$$

Now,

$$E(A_3 - B_2) = E\left(\sum_j \frac{Y_{.j.}^2}{k} - \sum_i \frac{Y_{i..}^2}{r}\right)$$

$$= E\left[\sum_j \frac{\left(k\mu + \sum_p n_{pj}\tau_p + kb_j + e_{.j.}\right)^2}{k} - \sum_i \frac{\left(r\mu + r\tau_i + \sum_q n_{iq}b_q + e_{i..}\right)^2}{r}\right]$$

$$= \sum_j \frac{\left(k\mu + \sum_p n_{pj}\tau_p\right)^2}{k} + kb\sigma_b^2 + b\sigma^2 - \sum_i \frac{(r\mu + r\tau_i)^2}{r} - t\sigma_b^2 - t\sigma^2$$

$$= \sum_j \frac{\left(\sum_p n_{pj}\tau_p\right)^2}{k} - r\sum_p \tau_p^2 + (b - t)\sigma^2 + t(r - 1)\sigma_b^2$$

$$= \sum_j \sum_p n_{pj}^2 \frac{\tau_p^2}{k} + \sum_j \sum_p \sum_{\substack{p' \\ p \neq p'}} n_{pj}n_{p'j}\tau_p \frac{\tau_{p'}}{k} - r\sum_p \tau_p^2 + (b - t)\sigma^2 + t(r - 1)\sigma_b^2$$

$$= \frac{r}{k}\Sigma\tau_p^2 + \lambda \sum_p \sum_{\substack{p' \\ p \neq p'}} \tau_p \frac{\tau_{p'}}{k} - r\Sigma\tau_p^2 + (b - t)\sigma^2 + t(r - 1)\sigma_b^2$$

$$= \frac{r}{k}\Sigma\tau_p^2 + \frac{\lambda}{k}\sum_p (t\bar{\tau}_. - \tau_p)\tau_p - r\Sigma\tau_p^2 + (b - t)\sigma^2 + t(r - 1)\sigma_b^2$$

$$= -\frac{\lambda t}{k}\sum_p (\tau_p - \bar{\tau}_.)^2 + (b - t)\sigma^2 + t(r - 1)\sigma_b^2$$

If this result is combined with Eq. (18.28), we get

$$E(B_3) = \frac{\lambda t}{k}\Sigma(\tau_i - \bar{\tau}_.)^2 + (t - 1)\sigma^2 - \frac{\lambda t}{k}\Sigma(\tau_i - \bar{\tau}_.)^2 + (b - t)\sigma^2 + t(r - 1)\sigma_b^2$$

$$= (b - 1)\sigma^2 + t(r - 1)\sigma_b^2$$

Hence the expected mean square is as given in Table 18.8.

Yates equated the observed mean square with the expected mean square of blocks (adj) and intrablock error in Table 18.8 to obtain estimates of σ^2 and σ_b^2. These estimates are used to obtain $\hat{\nu}$ in Eq. (18.26).

TABLE 18.8 ANALYSIS OF VARIANCE FOR METHOD B OF TABLE 18.7

SV	DF	SS	MS	EMS
Total	bk	$B = \Sigma y_{ijm}^2$		
Mean	1	$B_1 = \dfrac{Y_{...}^2}{bk}$		
Treatments (unadj)	$t - 1$	$B_2 = \sum_i \dfrac{Y_{i..}^2}{r} - \dfrac{Y_{...}^2}{bk}$		
Blocks (adj)	$b - 1$	$B_3 = A_2 + A_3 - B_2$	E_b	$\sigma^2 + \dfrac{t(r-1)}{b-1}\sigma_b^2$
Intrablock error	$bk - b - t + 1$	$B_4 = B - B_1 - B_2 - B_3$	E_e	σ^2

We get from the EMS column in Table 18.8

$$E_b = \hat{\sigma}^2 + \frac{t(r-1)}{b-1}\hat{\sigma}_b^2 \qquad E_e = \hat{\sigma}^2 \tag{18.29}$$

and

$$\hat{\sigma}^2 = E_e \qquad \hat{\sigma}_b^2 = (E_b - E_e)\frac{b-1}{t(r-1)} \tag{18.30}$$

and

$$\hat{w} = \frac{1}{E_e} \qquad \hat{w}' = \frac{t(r-1)}{k(b-1)E_b - (t-k)E_e}$$

If $E_b \leqslant E_e$, then $\hat{\sigma}_b^2$ is taken to be zero, and $\hat{w} = \hat{w}'$. From (18.30) we get

$$\hat{\nu} = \begin{cases} \dfrac{(b-1)(E_b - E_e)}{t(k-1)(b-1)E_b + (t-k)(b-t)E_e} & \text{if } E_b > E_e \\ 0 & \text{if } E_b \leqslant E_e \end{cases} \tag{18.31}$$

The preceding can be summed up in the following theorem.

◆ **Theorem 18.13** Let the model be given by Eq. (18.13), i.e., a balanced incomplete block model with blocks random. An estimator of $\tau_i - \bar{\tau}_.$ using both interblock and intrablock estimators is given by

$$\hat{\tau}_i^* = \begin{cases} \dfrac{T_i - \bar{T}_.}{r} + \dfrac{\hat{\nu}}{r}[(t-k)(T_i - \bar{T}_.) - (t-1)(B_i - \bar{B}_.)] & \text{if } E_b > E_e \\ \dfrac{T_i - \bar{T}_.}{r} & \text{if } E_b \leqslant E_e \end{cases}$$

where $$\hat{\nu} = \frac{(b-1)(E_b - E_e)}{t(k-1)(b-1)E_b + (t-k)(b-t)E_e}$$

TABLE 18.9 TABLE OF TREATMENT ESTIMATES

Treatment number i	Treatment total T_i	Total of blocks containing ith treatment B_i	$T_i - \frac{1}{k} B_i$ q_i	$(t - k)T_i - (t - 1)B_i$ W_i	$W_i - \overline{W}.$	$r\tilde{\tau}_i^* = T_i - \bar{T}. + \hat{v}(W_i - \overline{W}.)$
1	7.49	25.68	−1.0700	−124.12	−7.52	−.87
2	5.63	22.74	−1.9500	−113.92	2.68	−2.69
3	9.70	26.47	.8767	−120.02	−3.42	1.36
4	9.18	25.24	.7667	−114.72	1.88	.86
5	8.57	23.57	.7133	−107.14	9.46	.28
6	9.13	24.69	.9000	−111.62	4.98	.82
7	8.60	26.51	−.2367	−124.66	−8.06	.24

$$\hat{v} = \frac{(b - 1)(E_b - E_e)}{t(k - 1)(b - 1)E_b + (t - k)(b - t)E_e} = \frac{(6)(.2563 - .2424)}{(7)(2)(6)(.2563)} = \frac{.0834}{21.5292} = .00387$$

Before we proceed to computing instructions, we shall state two theorems on combining interblock and intrablock information for testing the hypothesis $H_0: \tau_1 = \tau_2 = \cdots = \tau_t$.

◆ **Theorem 18.14** Let the model be given by Eq. (18.13), i.e., a balanced incomplete block model with blocks random. In addition, let $b > t$. The notation will be that given in Table 18.7 for method A. If $\tau_1 = \tau_2 = \cdots = \tau_t$, the following distributional properties hold:

(1) $\dfrac{A_2}{\sigma^2}$ is distributed as $\chi^2(t - 1)$.

(2) $\dfrac{A_4}{\sigma^2}$ is distributed as $\chi^2(bk - b - t + 1)$.

(3) $\dfrac{A_{31}}{\sigma^2 + k\sigma_b^2}$ is distributed as $\chi^2(t - 1)$.

(4) $\dfrac{A_3 - A_{31}}{\sigma^2 + k\sigma_b^2}$ is distributed as $\chi^2(b - t)$.

(5) The quantities in (1), (2), (3), (4) are mutually independent.

From this we obtain the following theorem.

◆ **Theorem 18.15** Let the conditions of Theorem 18.14 hold. Then

(1) $u = \dfrac{(bk - b - t + 1)}{(t - 1)} \dfrac{A_2}{A_4}$ is distributed as $F(t - 1, bk - b - t + 1)$.

(2) $v = \dfrac{(b - t)}{(t - 1)} \dfrac{A_{31}}{A_3 - A_{31}}$ is distributed as $F(t - 1, b - t)$.

(3) u and v are independent.

(4) $z = -2 \log P_u P_v$ is distributed as $\chi^2(4)$, where P_u is the significance level of u and P_v is similarly defined for v. So the hypothesis $\tau_1 = \tau_2 = \cdots = \tau_t$ is rejected at the α level of significance if $z > \chi_\alpha^2(4)$.

For a complete discussion of this combined test the reader may consult Zelen [7] and Weeks [1].

18.3.4 Computing Instructions for the Balanced Incomplete Block Model with Recovery of Interblock Information. Computing instructions for $\hat{\tau}_i^*$ in Theorem 18.13 will be given. A format such as that of Table 18.9 will be useful. From this table an AOV can be computed as shown in Table 18.8. The following quantities will be needed:

$$\begin{aligned}
&\text{SS for treatments (unadj):} \quad && B_2 = \frac{\Sigma T_i^2}{r} - \frac{T_.^2}{bk} \\
&\text{SS for treatments (adj):} && A_2 = \frac{k}{\lambda t}\, \Sigma q_i^2 \\
&\text{SS for blocks (unadj):} && A_3 = \sum_j \frac{Y_{.j.}^2}{k} - \frac{Y_{...}^2}{bk} \\
&\text{SS for blocks (adj):} && B_3 = A_2 + A_3 - B_2
\end{aligned} \tag{18.32}$$

The example given in Art. 14.4.1 will be used to illustrate the above procedure. A_2 and A_3 can be obtained from Table 14.6; the other quantities are easily computed from (18.32).

$$B_2 = 3.7880 \qquad A_2 = 3.2908 \qquad A_3 = 2.0347 \qquad B_3 = 1.5375$$

The AOV is given in Table 18.10.

TABLE 18.10 ANALYSIS OF VARIANCE FOR EXAMPLE OF ART. 14.4.1

SV	DF	SS	MS
Total	21	169.1162	
Mean	1	161.8519	
Treatments (unadj)	6	3.7880	
Blocks (adj)	6	1.5375	$.2563 = E_b$
Intrablock error	8	1.9388	$.2424 = E_e$

Problems

18.1 Prove that $E(\Sigma\lambda_i\hat{\tau}_i) = \Sigma\lambda_i\tau_i$ in Eq. (18.7).

18.2 Show that the rank of the normal equations (18.3) is $t + 1$.

18.3 Show that $E(\hat{\beta}) = \beta$ in Eq. (18.5).

18.4 Show that $\text{var}(\hat{\beta}) = \sigma^2/E_{xx}$.

18.5 Show that

$$\text{var}(\Sigma\lambda_i\hat{\tau}_i) = \sigma^2\left[\frac{\Sigma\lambda_i^2}{r} + \frac{(\Sigma\lambda_i x_{i.})^2}{E_{xx}}\right]$$

18.6 Show that

$$E(\tilde{\sigma}^2) = \frac{t(r-1)-1}{rt}\,\sigma^2$$

18.7 The covariance model given in Eq. (18.1) is assumed to fit the data in Table 18.11. Find $\hat{\beta}$.

TABLE 18.11

Treatment

1		2		3	
y	x	y	x	y	x
4.0	1.0	7.1	1.3	7.0	.3
8.0	2.1	9.0	2.0	8.3	.8
9.4	3.0	13.2	3.3	9.2	1.1
12.8	4.2	12.8	3.4	10.4	1.5
15.7	5.0	15.5	4.1	11.5	1.8

18.8 In Prob. 18.7, find $\hat{\sigma}^2$ and set a 90 per cent confidence interval on σ^2.

18.9 In Prob. 18.7, find $\hat{\tau}_1 - \hat{\tau}_2$.

18.10 In Prob. 18.7, set a 95 per cent confidence interval on $\tau_1 - \tau_2$.

18.11 In Prob. 18.7, find the average variance of the estimates of the difference-of-treatment means.

18.12 In Prob. 18.7, test the hypothesis $\tau_1 = \tau_2 = \tau_3$ with a type I error probability of 5 per cent.

18.13 Suppose the data of Table 18.12 satisfy the assumptions for a balanced incomplete block with blocks random. Find the intrablock estimates of $\tau_i - \bar{\tau}_.$. The numbers in parentheses are treatment numbers.

TABLE 18.12

Block	Treatments	
1	(1) 1.2	(2) 2.2
2	(1) .8	(3) 1.8
3	(1) 1.1	(4) 7.1
4	(2) 2.2	(3) 4.2
5	(2) 1.6	(4) 6.7
6	(3) 4.5	(4) 6.3

18.14 In Prob. 18.13, find the interblock estimates of $\tau_i - \bar{\tau}_.$.

18.15 Find the variance of the estimates in Probs. 18.13 and 18.14.

18.16 Run an AOV on the data in Prob. 18.13 such as that given in Table 18.7.

18.17 Find combined estimators of $\tau_i - \bar{\tau}_.$ for the data in Prob. 18.13.

18.18 Use Theorem 18.15 and the data in Problem 18.13 to test the hypothesis $\tau_1 = \tau_2 = \tau_3 = \tau_4$ with a 5 per cent type I error.

18.19 Prove parts (3) and (4) of Theorem 18.10.

18.20 If $b > t$, show that R_{ss} and E_{ss} in Theorem 18.10 can be used to form a combined estimator of $\tau_i - \bar{\tau}_.$.

Further Reading

1 D. L. Weeks: An Exact Test of Significance in the Balanced Incomplete Block Design with Recovery of Inter-block Information, unpublished M.S. thesis, Oklahoma State University, Stillwater, Okla., 1957.

2 C. R. Rao: General Methods of Analysis for Incomplete Block Designs, *J. Am. Statist. Assoc.*, vol. 42, pp. 541–561, 1947.
3 M. B. Wilk: Linear Models and Randomized Experiments, Iowa Ph.D thesis, Iowa State College, Ames, Iowa, 1955.
4 M. B. Wilk and O. Kempthorne: Some Aspects of the Analysis of Factorial Experiments in a Completely Randomized Design, *Ann. Math. Statist.*, vol. 27, pp. 950–984, 1956.
5 H. Scheffé: A "Mixed Model" for the Analysis of Variance, *Ann. Math. Statist.*, vol. 27, pp. 23–36, 1956.
6 H. Scheffé: Alternative Models for Analysis of Variance, *Ann. Math. Statist.*, vol. 27, pp. 251–271, 1956.
7 M. Zelen: The Analysis of Incomplete Block Designs, *J. Am. Statist. Assoc.*, vol. 52, pp. 204–217, 1957.
8 D. A. Sprott: A Note on Combined Interblock and Intrablock Estimation in Incomplete Block Designs, *Ann. Math. Statist.*, vol. 27, pp. 633–641, 1956.
9 E. S. Pearson: A Note on Further Properties of Statistical Tests, *Biometrika*, vol. 32, pp. 59–61, 1941–42.
10 F. Yates: Incomplete Randomized Blocks, *Ann. Eugenics*, vol. 7, pp. 121–140, 1936.
11 F. Yates: The Recovery of Inter-block Information in Balanced Incomplete Block Designs, *Ann. Eugenics*, vol. 10, pp. 317–325, 1940.

Appendix

Four tables are presented: central chi-square, Student's t, central F, noncentral beta. (The cumulative normal distribution can be obtained from Table A.2 with $\nu = \infty$.)

Table A.1 Central Chi-square

The entries in this table are $\chi_p^2(\nu)$, where

$$p = \int_{\chi_p^2(\nu)}^{\infty} g(u)\,du = P[u \geqslant \chi_p^2(\nu)]$$

for probability values p = .0001, .001, .005, .01, .025, .05, .1, .25, .5, .75, .9, .95, .975, .99, .995, .999, .9999, and for degrees of freedom values ν = 1, 2, 3, 4, 5, 6, 7, 8, 9, 10, 12, 15, 20, 24, 30, 40, 60, 120.

Table A. 2 Student's t

The entries in this table are $t_p(\nu)$, where

$$p/2 = \int_{t_p(\nu)}^{\infty} g(t)\,dt = P[t \geqslant t_p(\nu)]$$

for probability values p = .0001, .001, .005, .01, .025, .05, .1, .25, .5, .75, .9, .95, .975, .99, .995, .999, .9999, and for degrees of freedom values $\nu = 1, 1.2, 1.5, 2, 3, 4, 5, 6, 7, 8, 9, 10, 12, 15, 20, 24, 30, 40, 60, 120, \infty$.

Table A.3 Central F

The entries in this table are $F_p(\nu_1, \nu_2)$, where

$$p = \int_{F_p(\nu_1, \nu_2)}^{\infty} g(F)\, dF = P[F \geqslant F_p(\nu_1, \nu_2)]$$

where ν_1 is the numerator degrees of freedom and ν_2 is the denominator degrees of freedom.

The values of p are .0001, .001, .005, .01, .025, .05, .1, .25, .5, .75, .9, .95, .975, .99, .995, .999, .9999; numerator degrees of freedom $\nu_1 = 1, 2, 3, 4, 5, 6, 7, 8, 9, 10, 12, 15, 20, 24, 30, 40, 60, 120, \infty$; and denominator degrees of freedom $\nu_2 = 1, 1.2, 1.5, 2, 3, 4, 5, 6, 7, 8, 9, 10, 12, 15, 20, 24, 30, 40, 60, 120, \infty$.

Table A.4 Noncentral Beta

The entries in this table, except for the E_p^2 column, are $\beta_p(f_1, f_2, \phi)$, where

$$p = \int_0^{\beta_p(f_1, f_2, \phi)} g(\beta)\, d\beta = P\,[\beta \leqslant \beta_p(f_1, f_2, \phi)]$$

where the values of p are .01 and .05; the values of f_1, the numerator degrees of freedom, are 1, 2, 3, 4, 5, 6, 7, 8; the values of f_2, the denominator degrees of freedom, are 2, 4, 6, 7, 8, 9, 10, 11, 12, 13, 14, 15, 16, 17, 18, 19, 20, 21, 22, 23, 24, 25, 26, 27, 28, 29, 30, 60, ∞; and the values for ϕ = 1, 1.5, 2, 2.5, 3, 4, 5, 6, 7, 8.

The entries in the E_p^2 column are $\beta_p(f_1, f_2, \phi = 0)$, where

$$p = \int_{\beta_p(f_1, f_2, \phi = 0)}^{1} g(\beta)\, d\beta = P\,[\beta \geqslant \beta_p(f_1, f_2, \phi = 0)]$$

Table A.1 Chi-square Table

Entries are $\chi_p^2(\nu)$, where $p = \int_{\chi_p^2(\nu)}^{\infty} g(\chi^2)\, d\chi^2$

ν	p = 0.0001	p = 0.001	p = 0.005	p = 0.01	p = 0.025	p = 0.05	p = 0.1	p = 0.25	p = 0.5	p = 0.75	p = 0.9	p = 0.95	p = 0.975	p = 0.99	p = 0.995	p = 0.999	p = 0.9999	ν
1	(+1) 1.5137	(+1) 1.0828	7.8794	6.6349	5.0239	3.8415	2.7055	1.3233	(-1) 4.5494	(-1) 1.0153	(-2) 1.5791	(-3) 3.9321	(-4) 9.8207	(-4) 1.5709	(-5) 3.9270	(-6) 1.5708	(-8) 1.5708	1
2	(+1) 1.8421	(+1) 1.3816	(+1) 1.0597	9.2103	7.3778	5.9915	4.6052	2.7726	1.3863	(-1) 5.7536	(-1) 2.1072	(-1) 1.0259	(-2) 5.0636	(-2) 2.0101	(-2) 1.0025	(-3) 2.0010	(-4) 2.0002	2
3	(+1) 2.1108	(+1) 1.6266	(+1) 1.2838	(+1) 1.1345	9.3484	7.8147	6.2514	4.1084	2.3660	1.2125	(-1) 5.8438	(-1) 3.5185	(-1) 2.1580	(-1) 1.1483	(-2) 7.1721	(-2) 2.4298	(-3) 5.2149	3
4	(+1) 2.3513	(+1) 1.8467	(+1) 1.4860	(+1) 1.3277	(+1) 1.1143	9.4877	7.7794	5.3853	3.3567	1.9226	1.0636	(-1) 7.1072	(-1) 4.8442	(-1) 2.9711	(-1) 2.0699	(-2) 9.0804	(-2) 2.8418	4
5	(+1) 2.5745	(+1) 2.0515	(+1) 1.6750	(+1) 1.5086	(+1) 1.2832	(+1) 1.1070	9.2364	6.6257	4.3515	2.6746	1.6103	1.1455	(-1) 8.3121	(-1) 5.5430	(-1) 4.1174	(-1) 2.1022	(-2) 8.2175	5
6	(+1) 2.7856	(+1) 2.2458	(+1) 1.8548	(+1) 1.6812	(+1) 1.4449	(+1) 1.2592	(+1) 1.0645	7.8408	5.3481	3.4546	2.2041	1.6354	1.2373	(-1) 8.7208	(-1) 6.7573	(-1) 3.8107	(-1) 1.7235	6
7	(+1) 2.9877	(+1) 2.4322	(+1) 2.0278	(+1) 1.8475	(+1) 1.6013	(+1) 1.4067	(+1) 1.2017	9.0372	6.3458	4.2548	2.8331	2.1674	1.6899	1.2390	(-1) 9.8926	(-1) 5.9849	(-1) 2.9996	7
8	(+1) 3.1828	(+1) 2.6125	(+1) 2.1955	(+1) 2.0090	(+1) 1.7535	(+1) 1.5507	(+1) 1.3362	(+1) 1.0219	7.3441	5.0706	3.4895	2.7326	2.1797	1.6465	1.3444	(-1) 8.5712	(-1) 4.6359	8
9	(+1) 3.3720	(+1) 2.7877	(+1) 2.3589	(+1) 2.1666	(+1) 1.9023	(+1) 1.6919	(+1) 1.4684	(+1) 1.1389	8.3428	5.8988	4.1682	3.3251	2.7004	2.0879	1.7349	1.1519	(-1) 6.6081	9
10	(+1) 3.5564	(+1) 2.9588	(+1) 2.5188	(+1) 2.3209	(+1) 2.0483	(+1) 1.8307	(+1) 1.5987	(+1) 1.2549	9.3418	6.7372	4.8652	3.9403	3.2470	2.5582	2.1558	1.4787	(-1) 8.8890	10
12	(+1) 3.9134	(+1) 3.2909	(+1) 2.8300	(+1) 2.6217	(+1) 2.3337	(+1) 2.1026	(+1) 1.8549	(+1) 1.4845	(+1) 1.1340	8.4384	6.3038	5.2260	4.4038	3.5706	3.0738	2.2142	1.4275	12
15	(+1) 4.4264	(+1) 3.7697	(+1) 3.2801	(+1) 3.0578	(+1) 2.7488	(+1) 2.4996	(+1) 2.2307	(+1) 1.8245	(+1) 1.4339	(+1) 1.1036	8.5468	7.2609	6.2621	5.2294	4.6009	3.4827	2.4082	15
20	(+1) 5.2386	(+1) 4.5315	(+1) 3.9997	(+1) 3.7566	(+1) 3.4170	(+1) 3.1410	(+1) 2.8412	(+1) 2.3828	(+1) 1.9337	(+1) 1.5452	(+1) 1.2443	(+1) 1.0851	9.5908	8.2604	7.4339	5.9210	4.3952	20
24	(+1) 5.8613	(+1) 5.1179	(+1) 4.5558	(+1) 4.2980	(+1) 3.9364	(+1) 3.6415	(+1) 3.3196	(-1) 2.8241	(+1) 2.3337	(+1) 1.9037	(+1) 1.5659	(+1) 1.3848	(+1) 1.2401	(+1) 1.0856	9.8862	8.0849	6.2230	24
30	(+1) 6.7632	(+1) 5.9703	(+1) 5.3672	(+1) 5.0892	(+1) 4.6979	(+1) 4.3773	(+1) 4.0256	(+1) 3.4800	(+1) 2.9336	(+1) 2.4478	(+1) 2.0599	(+1) 1.8493	(+1) 1.6791	(+1) 1.4954	(+1) 1.3787	(+1) 1.1588	9.2580	30
40	(+1) 8.2064	(+1) 7.3402	(+1) 6.6766	(+1) 6.3691	(+1) 5.9342	(+1) 5.5758	(+1) 5.1805	(+1) 4.5616	(+1) 3.9335	(+1) 3.3660	(+1) 2.9050	(+1) 2.6509	(+1) 2.4433	(+1) 2.2164	(+1) 2.0706	(+1) 1.7916	(+1) 1.4883	40
60	(+2) 1.0950	(+1) 9.9607	(+1) 9.1952	(+1) 8.8379	(+1) 8.3298	(+1) 7.9082	(+1) 7.4397	(+1) 6.6981	(+1) 5.9335	(+1) 5.2294	(+1) 4.6459	(+1) 4.3188	(+1) 4.0482	(+1) 3.7485	(+1) 3.5535	(+1) 3.1738	(+1) 2.7497	60
120	(+2) 1.8633	(+2) 1.7362	(+2) 1.6364	(+2) 1.5895	(+2) 1.5221	(+2) 1.4657	(+2) 1.4023	(+2) 1.3006	(+2) 1.1933	(+2) 1.0922	(+2) 1.0062	(+1) 9.5701	(+1) 9.1576	(+1) 8.6926	(+1) 8.3851	(+1) 7.7750	(+1) 7.0728	120

ν = degrees of freedom.

The numbers in parentheses indicate the power of ten by which the number following is to be multiplied; e.g., (-1) 1.2345 = .12345.

Table A.1 is reprinted from Lewis E. Vogler and Kenneth A. Norton, Graphs and Tables of the Significance Levels $F(\nu_1, \nu_2, p)$ for the Fisher-Snedecor Variance Ratio, from NBS report 5069, National Bureau of Standards, Boulder Laboratories, by the kind permission of the authors and the director of the Boulder Laboratory, F. W. Brown.

Table A.2 Student's t Table

Entries are $t_p(\nu)$, where $p/2 = \int_{t_p(\nu)}^{\infty} g(t)\,dt$

$t(\nu, p) = +\sqrt{F(1, \nu, p)}$

ν	p = 0.0001	p = 0.001	p = 0.005	p = 0.01	p = 0.025	p = 0.05	p = 0.1	p = 0.25	p = 0.5	p = 0.75	p = 0.9	p = 0.95	p = 0.975	p = 0.99	p = 0.995	p = 0.999	p = 0.9999	ν
1	(+3) 6.3662	(+2) 6.3662	(+2) 1.2732	(+1) 6.3657	(+1) 2.5452	(+1) 1.2706	6.3138	2.4142	1.0000	(-1) 4.1421	(-1) 1.5838	(-2) 7.8701	(-2) 3.9290	(-2) 1.5709	(-3) 7.8541	(-3) 1.5708	(-4) 1.5708	1
1.2	(+3) 1.5434	(+2) 2.2654	(+1) 5.9240	(+1) 3.3239	(+1) 1.5468	8.6488	4.7958	2.0897	(-1) 9.3358	(-1) 3.9768	(-1) 1.5305	(-2) 7.6117	(-2) 3.8008	(-2) 1.5198	(-3) 7.5984	(-3) 1.5197	(-4) 1.5197	1.2
1.5	(+2) 3.8458	(+1) 8.2847	(+1) 2.8317	(+1) 1.7820	9.6353	6.0166	3.7051	1.8230	(-1) 8.7259	(-1) 3.8131	(-1) 1.4763	(-2) 7.3481	(-2) 3.6699	(-2) 1.4675	(-3) 7.3372	(-3) 1.4674	(-4) 1.4674	1.5
2	(+1) 9.9992	(+1) 3.1599	(+1) 1.4089	9.9249	6.2053	4.3027	2.9200	1.6036	(-1) 8.1650	(-1) 3.6514	(-1) 1.4213	(-2) 7.0799	(-2) 3.5367	(-2) 1.4143	(-3) 7.0711	(-3) 1.4142	(-4) 1.4142	2
3	(+1) 2.8000	(+1) 1.2924	7.4533	5.8409	4.1765	3.1825	2.3534	1.4226	(-1) 7.6489	(-1) 3.4921	(-1) 1.3660	(-2) 6.8087	(-2) 3.4018	(-2) 1.3604	(-3) 6.8018	(-3) 1.3604	(-4) 1.3603	3
4	(+1) 1.5544	8.6103	5.5976	4.6041	3.4954	2.7764	2.1319	1.3444	(-1) 7.4070	(-1) 3.4138	(-1) 1.3383	(-2) 6.6729	(-2) 3.3341	(-2) 1.3334	(-3) 6.6666	(-3) 1.3333	(-4) 1.3333	4
5	(+1) 1.1178	6.8688	4.7734	4.0321	3.1634	2.5706	2.0150	1.3010	(-1) 7.2668	(-1) 3.3672	(-1) 1.3217	(-2) 6.5915	(-2) 3.2936	(-2) 1.3172	(-3) 6.5858	(-3) 1.3172	(-4) 1.3172	5
6	9.0823	5.9588	4.3168	3.7074	2.9687	2.4469	1.9432	1.2733	(-1) 7.1756	(-1) 3.3365	(-1) 1.3108	(-2) 6.5374	(-2) 3.2666	(-2) 1.3064	(-3) 6.5321	(-3) 1.3064	(-4) 1.3064	6
7	7.8846	5.4079	4.0294	3.4994	2.8412	2.3646	1.8946	1.2543	(-1) 7.1114	(-1) 3.3145	(-1) 1.3029	(-2) 6.4988	(-2) 3.2475	(-2) 1.2988	(-3) 6.4936	(-3) 1.2987	(-4) 1.2987	7
8	7.1200	5.0413	3.8325	3.3554	2.7515	2.3060	1.8595	1.2403	(-1) 7.0639	(-1) 3.2983	(-1) 1.2971	(-2) 6.4701	(-2) 3.2331	(-2) 1.2930	(-3) 6.4651	(-3) 1.2930	(-4) 1.2930	8
9	6.5937	4.7809	3.6897	3.2498	2.6850	2.2622	1.8331	1.2297	(-1) 7.0272	(-1) 3.2856	(-1) 1.2925	(-2) 6.4477	(-2) 3.2220	(-2) 1.2886	(-3) 6.4427	(-3) 1.2885	(-4) 1.2885	9
10	6.2110	4.5869	3.5813	3.1692	2.6338	2.2281	1.8125	1.2213	(-1) 6.9981	(-1) 3.2755	(-1) 1.2889	(-2) 6.4299	(-2) 3.2131	(-2) 1.2850	(-3) 6.4250	(-3) 1.2850	(-4) 1.2850	10
12	5.6945	4.3178	3.4284	3.0545	2.5600	2.1788	1.7823	1.2088	(-1) 6.9548	(-1) 3.2605	(-1) 1.2835	(-2) 6.4030	(-2) 3.1997	(-2) 1.2797	(-3) 6.3984	(-3) 1.2797	(-4) 1.2797	12
15	5.2391	4.0727	3.2860	2.9467	2.4899	2.1315	1.7531	1.1967	(-1) 6.9119	(-1) 3.2456	(-1) 1.2781	(-2) 6.3764	(-2) 3.1865	(-2) 1.2744	(-3) 6.3719	(-3) 1.2744	(-4) 1.2744	15
20	4.8373	3.8495	3.1534	2.8453	2.4231	2.0860	1.7247	1.1848	(-1) 6.8696	(-1) 3.2306	(-1) 1.2727	(-2) 6.3499	(-2) 3.1732	(-2) 1.2691	(-3) 6.3454	(-3) 1.2691	(-4) 1.2691	20
24	4.6544	3.7454	3.0905	2.7969	2.3910	2.0639	1.7109	1.1789	(-1) 6.8485	(-1) 3.2232	(-1) 1.2700	(-2) 6.3366	(-2) 3.1667	(-2) 1.2665	(-3) 6.3321	(-3) 1.2665	(-4) 1.2665	24
30	4.4824	3.6460	3.0298	2.7500	2.3596	2.0423	1.6973	1.1731	(-1) 6.8276	(-1) 3.2157	(-1) 1.2673	(-2) 6.3234	(-2) 3.1601	(-2) 1.2638	(-3) 6.3190	(-3) 1.2638	(-4) 1.2638	30
40	4.3206	3.5509	2.9712	2.7045	2.3289	2.0211	1.6839	1.1673	(-1) 6.8066	(-1) 3.2084	(-1) 1.2646	(-2) 6.3102	(-2) 3.1535	(-2) 1.2612	(-3) 6.3059	(-3) 1.2612	(-4) 1.2612	40
60	4.1686	3.4602	2.9145	2.6603	2.2991	2.0003	1.6707	1.1616	(-1) 6.7862	(-1) 3.2011	(-1) 1.2619	(-2) 6.2969	(-2) 3.1469	(-2) 1.2586	(-3) 6.2928	(-3) 1.2585	(-4) 1.2585	60
120	4.0254	3.3734	2.8599	2.6174	2.2699	1.9799	1.6576	1.1559	(-1) 6.7656	(-1) 3.1937	(-1) 1.2593	(-2) 6.2839	(-2) 3.1404	(-2) 1.2559	(-3) 6.2796	(-3) 1.2559	(-4) 1.2559	120
∞	3.8906	3.2905	2.8070	2.5758	2.2414	1.9600	1.6449	1.1503	(-1) 6.7449	(-1) 3.1864	(-1) 1.2566	(-2) 6.2707	(-2) 3.1338	(-2) 1.2533	(-3) 6.2666	(-3) 1.2533	(-4) 1.2533	∞

ν = degrees of freedom.

The numbers in parentheses indicate the power of ten by which the number following is to be multiplied; e.g., (-1) 1.2345 = .12345.

Table A.2 is reprinted from Lewis E. Vogler and Kenneth A. Norton, Graphs and Tables of the Significance Levels $F(\nu_1, \nu_2, p)$ for the Fisher-Snedecor Variance Ratio, from NBS report 5069, National Bureau of Standards, Boulder Laboratories, by the kind permission of the authors and the director of the Boulder Laboratory, F. W. Brown.

Entries are $F_p(\nu_1, \nu_2)$ where $p = \int_{F_p(\nu_1, \nu_2)}^{\infty} g(F)\, dF$

$\nu_1 = 1$

ν_2	p=0.0001	p=0.001	p=0.005	p=0.01	p=0.025	p=0.05	p=0.1	p=0.25	p=0.5	p=0.75	p=0.9	p=0.95	p=0.975	p=0.99	p=0.995	p=0.999	p=0.9999	ν_2
1	(+7) 4.0528	(+5) 4.0528	(+4) 1.6211	(+3) 4.0522	(+2) 6.4779	(+2) 1.6145	(+1) 3.9864	5.8285	1.0000	(-1) 1.7157	(-2) 2.5085	(-3) 6.1939	(-3) 1.5437	(-4) 2.4678	(-5) 6.1687	(-6) 2.4674	(-8) 2.4674	1
1.2	(6) 2.3821	(4) 5.1319	(3) 3.5094	(3) 1.1048	(2) 2.3927	(1) 7.4802	(1) 2.3000	4.3669	(-1) 8.7158	(-1) 1.5815	(-2) 2.3424	(-3) 5.7938	(-3) 1.4446	(-4) 2.3097	(-5) 5.7736	(-6) 2.3094	(-8) 2.3094	1.2
1.5	(5) 1.4790	(3) 6.8637	(2) 8.0184	(2) 3.1756	(1) 9.2839	(1) 3.6200	(1) 1.3728	3.3235	(-1) 7.6142	(-1) 1.4540	(-2) 2.1794	(-3) 5.3994	(-3) 1.3468	(-4) 2.1536	(-5) 5.3834	(-6) 2.1533	(-8) 2.1533	1.5
2	(+3) 9.9985	(+2) 9.9850	(+2) 1.9850	(+1) 9.8503	(+1) 3.8506	(+1) 1.8513	8.5263	2.5714	(-1) 6.6667	(-1) 1.3333	(-2) 2.0202	(-3) 5.0125	(-3) 1.2508	(-4) 2.0002	(-5) 5.0000	(-6) 2.0000	(-8) 2.0000	2
3	(+2) 7.8401	(+2) 1.6703	(+1) 5.5552	(+1) 3.4116	(+1) 1.7443	(+1) 1.0128	5.5383	2.0239	(-1) 5.8506	(-1) 1.2195	(-2) 1.8659	(-3) 4.6359	(-3) 1.1572	(-4) 1.8507	(-5) 4.6264	(-6) 1.8506	(-8) 1.8505	3
4	(+2) 2.4162	(+1) 7.4137	(+1) 3.1333	(+1) 2.1198	(+1) 1.2218	7.7086	4.5448	1.8074	(-1) 5.4863	(-1) 1.1654	(-2) 1.7911	(-3) 4.4528	(-3) 1.1116	(-4) 1.7779	(-5) 4.4444	(-6) 1.7778	(-8) 1.7778	4
5	(+2) 1.2494	(+1) 4.7181	(+1) 2.2785	(+1) 1.6258	(+1) 1.0007	6.6079	4.0604	1.6925	(-1) 5.2807	(-1) 1.1338	(-2) 1.7470	(-3) 4.3448	(-3) 1.0848	(-4) 1.7350	(-5) 4.3373	(-6) 1.7349	(-8) 1.7349	5
6	(+1) 8.2489	(+1) 3.5507	(+1) 1.8635	(+1) 1.3745	8.8131	5.9874	3.7760	1.6214	(-1) 5.1489	(-1) 1.1132	(-2) 1.7181	(-3) 4.2737	(-3) 1.0671	(-4) 1.7068	(-5) 4.2668	(-6) 1.7067	(-8) 1.7067	6
7	(+1) 6.2167	(+1) 2.9245	(+1) 1.6236	(+1) 1.2246	8.0727	5.5914	3.5894	1.5732	(-1) 5.0572	(-1) 1.0986	(-2) 1.6976	(-3) 4.2235	(-3) 1.0546	(-4) 1.6868	(-5) 4.2167	(-6) 1.6867	(-8) 1.6867	7
8	(+1) 5.0694	(+1) 2.5415	(+1) 1.4688	(+1) 1.1259	7.5709	5.3177	3.4579	1.5384	(-1) 4.9898	(-1) 1.0879	(-2) 1.6824	(-3) 4.1862	(-3) 1.0453	(-4) 1.6718	(-5) 4.1797	(-6) 1.6718	(-8) 1.6718	8
9	(+1) 4.3477	(+1) 2.2857	(+1) 1.3614	(+1) 1.0561	7.2093	5.1174	3.3603	1.5121	(-1) 4.9382	(-1) 1.0796	(-2) 1.6706	(-3) 4.1573	(-3) 1.0381	(-4) 1.6604	(-5) 4.1509	(-6) 1.6603	(-8) 1.6603	9
10	(+1) 3.8577	(+1) 2.1040	(+1) 1.2826	(+1) 1.0044	6.9367	4.9646	3.2850	1.4915	(-1) 4.8973	(-1) 1.0729	(-2) 1.6613	(-3) 4.1343	(-3) 1.0324	(-4) 1.6513	(-5) 4.1281	(-6) 1.6512	(-8) 1.6512	10
12	(+1) 3.2427	(+1) 1.8643	(+1) 1.1754	9.3302	6.5538	4.7472	3.1765	1.4613	(-1) 4.8369	(-1) 1.0631	(-2) 1.6473	(-3) 4.0999	(-3) 1.0238	(-4) 1.6377	(-5) 4.0940	(-6) 1.6376	(-8) 1.6376	12
15	(+1) 2.7448	(+1) 1.6587	(+1) 1.0798	8.6831	6.1995	4.5431	3.0732	1.4321	(-1) 4.7775	(-1) 1.0534	(-2) 1.6335	(-3) 4.0659	(-3) 1.0154	(-4) 1.6241	(-5) 4.0601	(-6) 1.6240	(-8) 1.6240	15
20	(+1) 2.3399	(+1) 1.4819	9.9439	8.0960	5.8715	4.3513	2.9747	1.4037	(-1) 4.7192	(-1) 1.0437	(-2) 1.6197	(-3) 4.0321	(-3) 1.0069	(-4) 1.6106	(-5) 4.0264	(-6) 1.6105	(-8) 1.6105	20
24	(+1) 2.1663	(+1) 1.4028	9.5513	7.8229	5.7167	4.2597	2.9271	1.3898	(-1) 4.6902	(-1) 1.0389	(-2) 1.6129	(-3) 4.0153	(-3) 1.0028	(-4) 1.6040	(-5) 4.0096	(-6) 1.6039	(-8) 1.6039	24
30	(+1) 2.0092	(+1) 1.3293	9.1797	7.5625	5.5675	4.1709	2.8807	1.3761	(-1) 4.6616	(-1) 1.0341	(-2) 1.6060	(-3) 3.9986	(-4) 9.9860	(-4) 1.5973	(-5) 3.9930	(-6) 1.5972	(-8) 1.5972	30
40	(+1) 1.8668	(+1) 1.2609	8.8278	7.3141	5.4239	4.0848	2.8354	1.3626	(-1) 4.6330	(-1) 1.0294	(-2) 1.5993	(-3) 3.9818	(-4) 9.9443	(-4) 1.5906	(-5) 3.9765	(-6) 1.5906	(-8) 1.5906	40
60	(+1) 1.7377	(+1) 1.1973	8.4946	7.0771	5.2857	4.0012	2.7914	1.3493	(-1) 4.6053	(-1) 1.0247	(-2) 1.5925	(-3) 3.9651	(-4) 9.9030	(-4) 1.5840	(-5) 3.9599	(-6) 1.5839	(-8) 1.5839	60
120	(+1) 1.6204	(+1) 1.1380	8.1790	6.8510	5.1524	3.9201	2.7478	1.3362	(-1) 4.5774	(-1) 1.0200	(-2) 1.5858	(-3) 3.9487	(-4) 9.8619	(-4) 1.5774	(-5) 3.9434	(-6) 1.5774	(-8) 1.5774	120
∞	(+1) 1.5137	(+1) 1.0828	7.8794	6.6349	5.0239	3.8415	2.7055	1.3233	(-1) 4.5494	(-1) 1.0153	(-2) 1.5791	(-3) 3.9321	(-4) 9.8203	(-4) 1.5708	(-5) 3.9270	(-6) 1.5708	(-8) 1.5708	∞

ν_1 = numerator degrees of freedom; ν_2 = denominator degrees of freedom.

The numbers in parentheses indicate the power of ten by which the number following is to be multiplied, e.g., (-1) 1.2345 = .12345.

Table A.3 is reprinted from Lewis E. Vogler and Kenneth A. Norton, Graphs and Tables of the Significance Levels $F(\nu_1, \nu_2, p)$ for the Fisher-Snedecor Variance Ratio, from NBS report 5069, National Bureau of Standards, Boulder Laboratories, by the kind permission of the authors and the director of the Boulder Laboratory, F. W. Brown.

$\nu_1 = 2$

ν_2	p=0.0001	p=0.001	p=0.005	p=0.01	p=0.025	p=0.05	p=0.1	p=0.25	p=0.5	p=0.75	p=0.9	p=0.95	p=0.975	p=0.99	p=0.995	p=0.999	p=0.9999	ν_2
1	(+7) 5.0000	(+5) 5.0000	(+4) 2.0000	(+3) 4.9995	(+2) 7.9950	(+2) 1.9950	(+1) 4.9500	7.5000	1.5000	(-1) 3.8889	(-1) 1.1728	(-2) 5.4016	(-2) 2.5970	(-2) 1.0152	(-3) 5.0378	(-3) 1.0015	(-4) 1.0002	1
1.2	(6) 2.7850	(4) 5.9999	(3) 4.1033	(3) 1.2921	(2) 2.8011	(1) 8.7817	(1) 2.7250	5.4476	1.3049	(-1) 3.6913	(-1) 1.1518	(-2) 5.3550	(-2) 2.5860	(-2) 1.0135	(-3) 5.0335	(-3) 1.0013	(-4) 1.0001	1.2
1.5	(5) 1.6158	(3) 7.4992	(2) 8.7646	(2) 3.4737	(2) 1.0185	(1) 3.9966	(1) 1.5408	4.0122	1.1399	(-1) 3.5064	(-1) 1.1312	(-2) 5.3088	(-2) 2.5750	(-2) 1.0118	(-3) 5.0293	(-3) 1.0012	(-4) 1.0001	1.5
2	(+3) 9.9990	(+2) 9.9900	(+2) 1.9900	(+1) 9.9000	(+1) 3.9000	(+1) 1.9000	9.0000	3.0000	1.0000	(-1) 3.3333	(-1) 1.1111	(-2) 5.2632	(-2) 2.5641	(-2) 1.0101	(-3) 5.0251	(-3) 1.0010	(-4) 1.0001	2
3	(+2) 6.9474	(+2) 1.4850	(+1) 4.9799	(+1) 3.0817	(+1) 1.6044	9.5521	5.4624	2.2798	(-1) 8.8110	(-1) 3.1712	(-1) 1.0915	(-2) 5.2181	(-2) 2.5533	(-2) 1.0084	(-3) 5.0208	(-3) 1.0008	(-4) 1.0001	3
4	(+2) 1.9800	(+1) 6.1246	(+1) 2.6284	(+1) 1.8000	(+1) 1.0649	6.9443	4.3246	2.0000	(-1) 8.2843	(-1) 3.0941	(-1) 1.0819	(-2) 5.1956	(-2) 2.5479	(-2) 1.0076	(-3) 5.0188	(-3) 1.0008	(-4) 1.0001	4
5	(+1) 9.7027	(+1) 3.7122	(+1) 1.8314	(+1) 1.3274	8.4336	5.7861	3.7797	1.8528	(-1) 7.9877	(-1) 3.0489	(-1) 1.0761	(-2) 5.1824	(-2) 2.5447	(-2) 1.0071	(-3) 5.0176	(-3) 1.0007	(-4) 1.0001	5
6	(+1) 6.1633	(+1) 2.7000	(+1) 1.4544	(+1) 1.0925	7.2598	5.1433	3.4633	1.7622	(-1) 7.7976	(-1) 3.0192	(-1) 1.0723	(-2) 5.1733	(-2) 2.5425	(-2) 1.0067	(-3) 5.0168	(-3) 1.0007	(-4) 1.0001	6
7	(+1) 4.5132	(+1) 2.1689	(+1) 1.2404	9.5466	6.5415	4.7374	3.2574	1.7010	(-1) 7.6655	(-1) 2.9983	(-1) 1.0696	(-2) 5.1672	(-2) 2.5410	(-2) 1.0065	(-3) 5.0161	(-3) 1.0006	(-4) 1.0001	7
8	(+1) 3.6000	(+1) 1.8494	(+1) 1.1042	8.6491	6.0595	4.4590	3.1131	1.6569	(-1) 7.5683	(-1) 2.9828	(-1) 1.0676	(-2) 5.1624	(-2) 2.5398	(-2) 1.0063	(-3) 5.0158	(-3) 1.0006	(-4) 1.0001	8
9	(+1) 3.0342	(+1) 1.6387	(+1) 1.0107	8.0215	5.7147	4.2565	3.0065	1.6236	(-1) 7.4938	(-1) 2.9708	(-1) 1.0660	(-2) 5.1586	(-2) 2.5389	(-2) 1.0062	(-3) 5.0153	(-3) 1.0006	(-4) 1.0001	9
10	(+1) 2.6548	(+1) 1.4905	9.4270	7.5594	5.4564	4.1028	2.9245	1.5975	(-1) 7.4349	(-1) 2.9612	(-1) 1.0648	(-2) 5.1557	(-2) 2.5382	(-2) 1.0060	(-3) 5.0150	(-3) 1.0006	(-4) 1.0001	10
12	(+1) 2.1850	(+1) 1.2974	8.5096	6.9266	5.0959	3.8853	2.8068	1.5595	(-1) 7.3477	(-1) 2.9469	(-1) 1.0629	(-2) 5.1512	(-2) 2.5371	(-2) 1.0059	(-3) 5.0145	(-3) 1.0006	(-4) 1.0001	12
15	(+1) 1.8109	(+1) 1.1339	7.7008	6.3589	4.7650	3.6823	2.6952	1.5227	(-1) 7.2619	(-1) 2.9327	(-1) 1.0610	(-2) 5.1469	(-2) 2.5361	(-2) 1.0057	(-3) 5.0143	(-3) 1.0006	(-4) 1.0001	15
20	(+1) 1.5119	9.9526	6.9865	5.8489	4.4613	3.4928	2.5893	1.4870	(-1) 7.1773	(-1) 2.9186	(-1) 1.0592	(-2) 5.1424	(-2) 2.5350	(-2) 1.0055	(-3) 5.0138	(-3) 1.0006	(-4) 1.0001	20
24	(+1) 1.3853	9.3394	6.6610	5.6136	4.3187	3.4028	2.5383	1.4695	(-1) 7.1356	(-1) 2.9116	(-1) 1.0582	(-2) 5.1403	(-2) 2.5345	(-2) 1.0054	(-3) 5.0135	(-3) 1.0005	(-4) 1.0001	24
30	(+1) 1.2718	8.7734	6.3547	5.3904	4.1821	3.3158	2.4887	1.4524	(-1) 7.0941	(-1) 2.9046	(-1) 1.0573	(-2) 5.1382	(-2) 2.5339	(-2) 1.0054	(-3) 5.0133	(-3) 1.0005	(-4) 1.0001	30
40	(+1) 1.1698	8.2508	6.0664	5.1785	4.0510	3.2317	2.4404	1.4355	(-1) 7.0531	(-1) 2.8976	(-1) 1.0564	(-2) 5.1358	(-2) 2.5334	(-2) 1.0053	(-3) 5.0133	(-3) 1.0005	(-4) 1.0001	40
60	(+1) 1.0781	7.7678	5.7950	4.9774	3.9253	3.1504	2.3932	1.4188	(-1) 7.0122	(-1) 2.8907	(-1) 1.0555	(-2) 5.1337	(-2) 2.5329	(-2) 1.0052	(-3) 5.0130	(-3) 1.0005	(-4) 1.0001	60
120	9.9549	7.3211	5.5393	4.7865	3.8046	3.0718	2.3473	1.4024	(-1) 6.9717	(-1) 2.8838	(-1) 1.0545	(-2) 5.1316	(-2) 2.5323	(-2) 1.0051	(-3) 5.0128	(-3) 1.0005	(-4) 1.0001	120
∞	9.2103	6.9078	5.2983	4.6052	3.6889	2.9957	2.3026	1.3863	(-1) 6.9315	(-1) 2.8768	(-1) 1.0536	(-2) 5.1293	(-2) 2.5318	(-2) 1.0050	(-3) 5.0123	(-3) 1.0005	(-4) 1.0001	∞

$\nu_1 = 3$

ν_2	p = 0.0001	p = 0.001	p = 0.005	p = 0.01	p = 0.025	p = 0.05	p = 0.1	p = 0.25	p = 0.5	p = 0.75	p = 0.9	p = 0.95	p = 0.975	p = 0.99	p = 0.995	p = 0.999	p = 0.9999	ν_2
1	(+7) 5.4038	(+5) 5.4038	(+4) 2.1615	(+3) 5.4033	(+2) 8.6416	(+2) 2.1571	(+1) 5.3593	8.1999	1.7092	(-1) 4.9410	(-1) 1.8056	(-2) 9.8736	(-2) 5.7330	(-2) 2.9312	(-2) 1.8001	(-3) 5.9868	(-3) 1.2755	1
1.2	(6) 2.9549	(4) 6.3660	(3) 4.3538	(3) 1.3710	(2) 2.9731	(1) 9.3286	(1) 2.9023	5.8883	1.4842	(-1) 4.7352	(-1) 1.8079	(-1) 1.0030	(-2) 5.8709	(-2) 3.0192	(-2) 1.8586	(-3) 6.1978	(-3) 1.3218	1.2
1.5	(5) 1.6727	(3) 7.7635	(2) 9.0745	(2) 3.5973	(2) 1.0557	(1) 4.1506	(1) 1.6083	4.2806	1.2947	(-1) 4.5502	(-1) 1.8158	(-1) 1.0225	(-2) 6.0349	(-2) 3.1222	(-2) 1.9269	(-3) 6.4432	(-3) 1.3756	1.5
2	(+3) 9.9992	(+2) 9.9917	(+2) 1.9917	(+1) 9.9166	(+1) 3.9165	(+1) 1.9164	9.1618	3.1534	1.1349	(-1) 4.3863	(-1) 1.8307	(-1) 1.0469	(-2) 6.2329	(-2) 3.2450	(-2) 2.0081	(-3) 6.7340	(-3) 1.4394	2
3	(+2) 6.5934	(+2) 1.4111	(+1) 4.7467	(+1) 2.9457	(+1) 1.5439	9.2766	5.3908	2.3555	1.0000	(-1) 4.2454	(-1) 1.8550	(-1) 1.0780	(-2) 6.4771	(-2) 3.3948	(-2) 2.1067	(-3) 7.0868	(-3) 1.5167	3
4	(+2) 1.8102	(+1) 5.6177	(+1) 2.4259	(+1) 1.6694	9.9792	6.5914	4.1908	2.0467	(-1) 9.4054	(-1) 4.1839	(-1) 1.8717	(-1) 1.0968	(-2) 6.6221	(-2) 3.4831	(-2) 2.1647	(-3) 7.2939	(-3) 1.5620	4
5	(+1) 8.6292	(+1) 3.3202	(+1) 1.6530	(+1) 1.2060	7.7636	5.4095	3.6195	1.8843	(-1) 9.0715	(-1) 4.1502	(-1) 1.8835	(-1) 1.1094	(-2) 6.7182	(-2) 3.5415	(-2) 2.2030	(-3) 7.4305	(-3) 1.5919	5
6	(+1) 5.3680	(+1) 2.3703	(+1) 1.2917	9.7795	6.5988	4.7571	3.2888	1.7844	(-1) 8.8578	(-1) 4.1292	(-1) 1.8923	(-1) 1.1185	(-2) 6.7866	(-2) 3.5828	(-2) 2.2303	(-3) 7.5275	(-3) 1.6132	6
7	(+1) 3.8676	(+1) 1.8772	(+1) 1.0882	8.4513	5.8898	4.3468	3.0741	1.7169	(-1) 8.7095	(-1) 4.1149	(-1) 1.8989	(-1) 1.1253	(-2) 6.8381	(-2) 3.6138	(-2) 2.2505	(-3) 7.5998	(-3) 1.6290	7
8	(+1) 3.0456	(+1) 1.5829	9.5965	7.5910	5.4160	4.0662	2.9238	1.6683	(-1) 8.6004	(-1) 4.1044	(-1) 1.9041	(-1) 1.1306	(-2) 6.8776	(-2) 3.6378	(-2) 2.2662	(-3) 7.6559	(-3) 1.6413	8
9	(+1) 2.5404	(+1) 1.3902	8.7171	6.9919	5.0781	3.8626	2.8129	1.6315	(-1) 8.5168	(-1) 4.0967	(-1) 1.9084	(-1) 1.1348	(-2) 6.9094	(-2) 3.6570	(-2) 2.2788	(-3) 7.7006	(-3) 1.6511	9
10	(+1) 2.2038	(+1) 1.2553	8.0807	6.5523	4.8256	3.7083	2.7277	1.6028	(-1) 8.4508	(-1) 4.0905	(-1) 1.9119	(-1) 1.1382	(-2) 6.9353	(-2) 3.6726	(-2) 2.2891	(-3) 7.7371	(-3) 1.6590	10
12	(+1) 1.7899	(+1) 1.0804	7.2258	5.9526	4.4742	3.4903	2.6055	1.5609	(-1) 8.3530	(-1) 4.0816	(-1) 1.9173	(-1) 1.1436	(-2) 6.9750	(-2) 3.6966	(-2) 2.3048	(-3) 7.7933	(-3) 1.6713	12
15	(+1) 1.4635	9.3353	6.4760	5.4170	4.1528	3.2874	2.4898	1.5202	(-1) 8.2569	(-1) 4.0730	(-1) 1.9230	(-1) 1.1490	(-2) 7.0161	(-2) 3.7213	(-2) 2.3210	(-3) 7.8509	(-3) 1.6839	15
20	(+1) 1.2050	8.0984	5.8177	4.9382	3.8587	3.0984	2.3801	1.4808	(-1) 8.1621	(-1) 4.0647	(-1) 1.9288	(-1) 1.1547	(-2) 7.0587	(-2) 3.7467	(-2) 2.3377	(-3) 7.9103	(-3) 1.6969	20
24	(+1) 1.0964	7.5545	5.5190	4.7181	3.7211	3.0088	2.3274	1.4615	(-1) 8.1153	(-1) 4.0607	(-1) 1.9318	(-1) 1.1576	(-2) 7.0801	(-2) 3.7597	(-2) 2.3462	(-3) 7.9406	(-3) 1.7036	24
30	9.9942	7.0545	5.2388	4.5097	3.5894	2.9223	2.2761	1.4426	(-1) 8.0689	(-1) 4.0568	(-1) 1.9349	(-1) 1.1606	(-2) 7.1018	(-2) 3.7729	(-2) 2.3548	(-3) 7.9714	(-3) 1.7103	30
40	9.1278	6.5945	4.9759	4.3126	3.4633	2.8387	2.2261	1.4239	(-1) 8.0228	(-1) 4.0528	(-1) 1.9381	(-1) 1.1635	(-2) 7.1240	(-2) 3.7863	(-2) 2.3636	(-3) 8.0026	(-3) 1.7171	40
60	8.3526	6.1712	4.7290	4.1259	3.3425	2.7581	2.1774	1.4055	(-1) 7.9770	(-1) 4.0491	(-1) 1.9413	(-1) 1.1666	(-2) 7.1469	(-2) 3.8000	(-2) 2.3725	(-3) 8.0343	(-3) 1.7241	60
120	7.6584	5.7814	4.4973	3.9493	3.2270	2.6802	2.1300	1.3873	(-1) 7.9314	(-1) 4.0453	(-1) 1.9446	(-1) 1.1697	(-2) 7.1700	(-2) 3.8137	(-2) 2.3816	(-3) 8.0665	(-3) 1.7311	120
∞	7.0358	5.4221	4.2794	3.7816	3.1161	2.6049	2.0838	1.3694	(-1) 7.8866	(-1) 4.0417	(-1) 1.9479	(-1) 1.1728	(-2) 7.1932	(-2) 3.8278	(-2) 2.3907	(-3) 8.0992	(-3) 1.7383	∞

$\nu_1 = 4$

ν_2	p = 0.0001	p = 0.001	p = 0.005	p = 0.01	p = 0.025	p = 0.05	p = 0.1	p = 0.25	p = 0.5	p = 0.75	p = 0.9	p = 0.95	p = 0.975	p = 0.99	p = 0.995	p = 0.999	p = 0.9999	ν_2
1	(+7) 5.6250	(+5) 5.6250	(+4) 2.2500	(+3) 5.6246	(+2) 8.9958	(+2) 2.2458	(+1) 5.5833	8.5810	1.8227	(-1) 5.5328	(-1) 2.2003	(-1) 1.2973	(-2) 8.1846	(-2) 4.7174	(-2) 3.1915	(-2) 1.3488	(-3) 4.1387	1
1.2	(6) 3.0478	(4) 6.5663	(3) 4.4908	(3) 1.4142	(2) 3.0671	(1) 9.6274	(1) 2.9990	6.1265	1.5810	(-1) 5.3271	(-1) 2.2249	(-1) 1.3347	(-2) 8.5083	(-2) 4.9438	(-2) 3.3572	(-2) 1.4256	(-3) 4.3849	1.2
1.5	(5) 1.7038	(3) 7.9077	(2) 9.2437	(2) 3.6648	(2) 1.0759	(1) 4.2343	(1) 1.6446	4.4235	1.3780	(-1) 5.1487	(-1) 2.2608	(-1) 1.3811	(-2) 8.9012	(-2) 5.2170	(-2) 3.5571	(-2) 1.5181	(-3) 4.6821	1.5
2	(+3) 9.9992	(+2) 9.9925	(+2) 1.9925	(+1) 9.9249	(+1) 3.9248	(+1) 1.9247	9.2434	3.2320	1.2071	(-1) 5.0000	(-1) 2.3124	(-1) 1.4400	(-2) 9.3906	(-2) 5.5555	(-2) 3.8046	(-2) 1.6328	(-3) 5.0505	2
3	(+2) 6.4019	(+2) 1.3710	(+1) 4.6195	(+1) 2.8710	(+1) 1.5101	9.1172	5.3427	2.3901	1.0632	(-1) 4.8859	(-1) 2.3862	(-1) 1.5171	(-1) 1.0021	(-2) 5.9902	(-2) 4.1222	(-2) 1.7801	(-3) 5.5243	3
4	(+2) 1.7187	(+1) 5.3436	(+1) 2.3155	(+1) 1.5977	9.6045	6.3883	4.1073	2.0642	1.0000	(-1) 4.8445	(-1) 2.4347	(-1) 1.5654	(-1) 1.0412	(-2) 6.2590	(-2) 4.3187	(-2) 1.8714	(-3) 5.8183	4
5	(+1) 8.0527	(+1) 3.1085	(+1) 1.5556	(+1) 1.1392	7.3879	5.1922	3.5202	1.8927	(-1) 9.6456	(-1) 4.8256	(-1) 2.4688	(-1) 1.5985	(-1) 1.0679	(-2) 6.4425	(-2) 4.4532	(-2) 1.9338	(-3) 6.0193	5
6	(+1) 4.9419	(+1) 2.1924	(+1) 1.2028	9.1483	6.2272	4.5337	3.1808	1.7872	(-1) 9.4191	(-1) 4.8156	(-1) 2.4939	(-1) 1.6226	(-1) 1.0873	(-2) 6.5759	(-2) 4.5506	(-2) 1.9792	(-3) 6.1657	6
7	(+1) 3.5222	(+1) 1.7198	(+1) 1.0050	7.8467	5.5226	4.1203	2.9605	1.7157	(-1) 9.2619	(-1) 4.8100	(-1) 2.5132	(-1) 1.6409	(-1) 1.1020	(-2) 6.6774	(-2) 4.6249	(-2) 2.0138	(-3) 6.2771	7
8	(+1) 2.7493	(+1) 1.4392	8.8051	7.0060	5.0526	3.8378	2.8064	1.6642	(-1) 9.1464	(-1) 4.8065	(-1) 2.5285	(-1) 1.6554	(-1) 1.1136	(-2) 6.7572	(-2) 4.6834	(-2) 2.0410	(-3) 6.3648	8
9	(+1) 2.2766	(+1) 1.2560	7.9559	6.4221	4.7181	3.6331	2.6927	1.6253	(-1) 9.0580	(-1) 4.8045	(-1) 2.5408	(-1) 1.6670	(-1) 1.1230	(-2) 6.8217	(-2) 4.7306	(-2) 2.0629	(-3) 6.4357	9
10	(+1) 1.9630	(+1) 1.1283	7.3428	5.9943	4.4683	3.4780	2.6053	1.5949	(-1) 8.9882	(-1) 4.8031	(-1) 2.5511	(-1) 1.6766	(-1) 1.1307	(-2) 6.8747	(-2) 4.7694	(-2) 2.0811	(-3) 6.4941	10
12	(+1) 1.5793	9.6327	6.5211	5.4119	4.1212	3.2592	2.4801	1.5503	(-1) 8.8848	(-1) 4.8017	(-1) 2.5671	(-1) 1.6916	(-1) 1.1427	(-2) 6.9570	(-2) 4.8298	(-2) 2.1092	(-3) 6.5849	12
15	(+1) 1.2783	8.2527	5.8029	4.8932	3.8043	3.0556	2.3614	1.5071	(-1) 8.7830	(-1) 4.8010	(-1) 2.5847	(-1) 1.7071	(-1) 1.1552	(-2) 7.0432	(-2) 4.8928	(-2) 2.1385	(-3) 6.6796	15
20	(+1) 1.0415	7.0960	5.1743	4.4307	3.5147	2.8661	2.2489	1.4652	(-1) 8.6830	(-1) 4.8012	(-1) 2.6013	(-1) 1.7234	(-1) 1.1682	(-2) 7.1327	(-2) 4.9586	(-2) 2.1692	(-3) 6.7786	20
24	9.4246	6.5892	4.8898	4.2184	3.3794	2.7763	2.1949	1.4447	(-1) 8.6335	(-1) 4.8015	(-1) 2.6103	(-1) 1.7318	(-1) 1.1750	(-2) 7.1793	(-2) 4.9925	(-2) 2.1850	(-3) 6.8298	24
30	8.5437	6.1245	4.6233	4.0179	3.2499	2.6896	2.1422	1.4244	(-1) 8.5844	(-1) 4.8019	(-1) 2.6196	(-1) 1.7404	(-1) 1.1819	(-2) 7.2265	(-2) 5.0271	(-2) 2.2013	(-3) 6.8822	30
40	7.7592	5.6981	4.3738	3.8283	3.1261	2.6060	2.0909	1.4045	(-1) 8.5357	(-1) 4.8028	(-1) 2.6291	(-1) 1.7492	(-1) 1.1889	(-2) 7.2754	(-2) 5.0628	(-2) 2.2179	(-3) 6.9358	40
60	7.0599	5.3067	4.1399	3.6491	3.0077	2.5252	2.0410	1.3848	(-1) 8.4873	(-1) 4.8038	(-1) 2.6388	(-1) 1.7581	(-1) 1.1961	(-2) 7.3249	(-2) 5.0992	(-2) 2.2349	(-3) 6.9907	60
120	6.4357	4.9472	3.9207	3.4796	2.8943	2.4472	1.9923	1.3654	(-1) 8.4392	(-1) 4.8049	(-1) 2.6488	(-1) 1.7674	(-1) 1.2035	(-2) 7.3757	(-2) 5.1366	(-2) 2.2523	(-3) 7.0470	120
∞	5.8782	4.6167	3.7151	3.3192	2.7858	2.3719	1.9449	1.3463	(-1) 8.3918	(-1) 4.8063	(-1) 2.6591	(-1) 1.7768	(-1) 1.2110	(-2) 7.4278	(-2) 5.1746	(-2) 2.2701	(-3) 7.1046	∞

$\nu_1 = 5$

ν_2	p = 0.0001	p = 0.001	p = 0.005	p = 0.01	p = 0.025	p = 0.05	p = 0.1	p = 0.25	p = 0.5	p = 0.75	p = 0.9	p = 0.95	p = 0.975	p = 0.99	p = 0.995	p = 0.999	p = 0.9999	ν_2
1	(+7) 5.7640	(+5) 5.7640	(+4) 2.3056	(+3) 5.7637	(+2) 9.2185	(+2) 2.3016	(+1) 5.7241	8.8198	1.8937	(-1) 5.9084	(-1) 2.4628	(-1) 1.5133	(-2) 9.9930	(-2) 6.1508	(-2) 4.3889	(-2) 2.1195	(-3) 8.0038	1
1.2	(6) 3.1062	(4) 6.6921	(3) 4.5769	(3) 1.4413	(2) 3.1262	(1) 9.8152	(1) 3.0596	6.2753	1.6415	(-1) 5.7046	(-1) 2.5048	(-1) 1.5689	(-1) 1.0483	(-2) 6.5149	(-2) 4.6708	(-2) 2.2701	(-3) 8.6059	1.2
1.5	(5) 1.7233	(3) 7.9984	(2) 9.3500	(2) 3.7072	(2) 1.0887	(1) 4.2867	(1) 1.6673	4.5121	1.4298	(-1) 5.5327	(-1) 2.5636	(-1) 1.6385	(-1) 1.1087	(-2) 6.9633	(-2) 5.0181	(-2) 2.4561	(-3) 9.3515	1.5
2	(+3) 9.9993	(+2) 9.9930	(+2) 1.9930	(+1) 9.9299	(+1) 3.9298	(+1) 1.9296	9.2926	3.2799	1.2519	(-1) 5.3972	(-1) 2.6457	(-1) 1.7283	(-1) 1.1857	(-2) 7.5335	(-2) 5.4603	(-2) 2.6938	(-2) 1.0306	2
3	(+2) 6.2817	(+2) 1.3458	(+1) 4.5392	(+1) 2.8237	(+1) 1.4885	9.0135	5.3092	2.4095	1.1024	(-1) 5.3070	(-1) 2.7628	(-1) 1.8486	(-1) 1.2881	(-2) 8.2919	(-2) 6.0496	(-2) 3.0118	(-2) 1.1589	3
4	(+2) 1.6613	(+1) 5.1712	(+1) 2.2456	(+1) 1.5522	9.3645	6.2560	4.0506	2.0723	1.0367	(-1) 5.2835	(-1) 2.8407	(-1) 1.9260	(-1) 1.3536	(-2) 8.7781	(-2) 6.4284	(-2) 3.2170	(-2) 1.2418	4
5	(+1) 7.6911	(+1) 2.9752	(+1) 1.4940	(+1) 1.0967	7.1464	5.0503	3.4530	1.8947	1.0000	(-1) 5.2779	(-1) 2.8960	(-1) 1.9801	(-1) 1.3993	(-2) 9.1183	(-2) 6.6934	(-2) 3.3611	(-2) 1.3002	5
6	(+1) 4.6747	(+1) 2.0803	(+1) 1.1464	8.7459	5.9876	4.3874	3.1075	1.7852	(-1) 9.7654	(-1) 5.2784	(-1) 2.9373	(-1) 2.0201	(-1) 1.4331	(-2) 9.3703	(-2) 6.8904	(-2) 3.4681	(-2) 1.3436	6
7	(+1) 3.3056	(+1) 1.6206	9.5221	7.4604	5.2852	3.9715	2.8833	1.7111	(-1) 9.6026	(-1) 5.2812	(-1) 2.9692	(-1) 2.0509	(-1) 1.4592	(-2) 9.5639	(-2) 7.0423	(-2) 3.5508	(-2) 1.3772	7
8	(+1) 2.5635	(+1) 1.3485	8.3018	6.6318	4.8173	3.6875	2.7265	1.6575	(-1) 9.4831	(-1) 5.2846	(-1) 2.9946	(-1) 2.0754	(-1) 1.4799	(-2) 9.7191	(-2) 7.1628	(-2) 3.6167	(-2) 1.4040	8
9	(+1) 2.1112	(+1) 1.1714	7.4711	6.0569	4.4844	3.4817	2.6106	1.6170	(-1) 9.3916	(-1) 5.2879	(-1) 3.0154	(-1) 2.0953	(-1) 1.4968	(-2) 9.8445	(-2) 7.2611	(-2) 3.6705	(-2) 1.4259	9
10	(+1) 1.8120	(+1) 1.0481	6.8723	5.6363	4.2361	3.3258	2.5216	1.5853	(-1) 9.3193	(-1) 5.2913	(-1) 3.0327	(-1) 2.1119	(-1) 1.5108	(-2) 9.9493	(-2) 7.3432	(-2) 3.7152	(-2) 1.4440	10
12	(+1) 1.4471	8.8921	6.0711	5.0643	3.8911	3.1059	2.3940	1.5389	(-1) 9.2124	(-1) 5.2975	(-1) 3.0598	(-1) 2.1378	(-1) 1.5327	(-1) 1.0113	(-2) 7.4716	(-2) 3.7853	(-2) 1.4726	12
15	(+1) 1.1621	7.5674	5.3721	4.5556	3.5764	2.9013	2.2730	1.4938	(-1) 9.1073	(-1) 5.3048	(-1) 3.0883	(-1) 2.1651	(-1) 1.5558	(-1) 1.0286	(-2) 7.6069	(-2) 3.8594	(-2) 1.5028	15
20	9.3880	6.4606	4.7616	4.1027	3.2891	2.7109	2.1582	1.4500	(-1) 9.0038	(-1) 5.3135	(-1) 3.1185	(-1) 2.1939	(-1) 1.5802	(-1) 1.0468	(-2) 7.7501	(-2) 3.9378	(-2) 1.5348	20
24	8.4578	5.9768	4.4857	3.8951	3.1548	2.6207	2.1030	1.4285	(-1) 8.9527	(-1) 5.3186	(-1) 3.1343	(-1) 2.2089	(-1) 1.5929	(-1) 1.0564	(-2) 7.8247	(-2) 3.9788	(-2) 1.5515	24
30	7.6322	5.5339	4.2276	3.6990	3.0265	2.5336	2.0492	1.4073	(-1) 8.9019	(-1) 5.3237	(-1) 3.1505	(-1) 2.2243	(-1) 1.6059	(-1) 1.0662	(-2) 7.9014	(-2) 4.0211	(-2) 1.5687	30
40	6.8987	5.1283	3.9860	3.5138	2.9037	2.4495	1.9968	1.3863	(-1) 8.8516	(-1) 5.3296	(-1) 3.1673	(-1) 2.2402	(-1) 1.6194	(-1) 1.0763	(-2) 7.9808	(-2) 4.0647	(-2) 1.5865	40
60	6.2465	4.7565	3.7600	3.3389	2.7863	2.3683	1.9457	1.3657	(-1) 8.8017	(-1) 5.3356	(-1) 3.1845	(-1) 2.2566	(-1) 1.6333	(-1) 1.0867	(-2) 8.0632	(-2) 4.1097	(-2) 1.6049	60
120	5.6661	4.4157	3.5482	3.1735	2.6740	2.2900	1.8959	1.3453	(-1) 8.7521	(-1) 5.3422	(-1) 3.2023	(-1) 2.2736	(-1) 1.6476	(-1) 1.0975	(-2) 8.1473	(-2) 4.1562	(-2) 1.6239	120
∞	5.1490	4.1030	3.3499	3.0173	2.5665	2.2141	1.8473	1.3251	(-1) 8.7029	(-1) 5.3493	(-1) 3.2206	(-1) 2.2910	(-1) 1.6624	(-1) 1.1086	(02) 8.2345	(-2) 4.2043	(-2) 1.6435	∞

$\nu_1 = 6$

ν_2	p=0.0001	p=0.001	p=0.005	p=0.01	p=0.025	p=0.05	p=0.1	p=0.25	p=0.5	p=0.75	p=0.9	p=0.95	p=0.975	p=0.99	p=0.995	p=0.999	p=0.9999	ν_2
1	(+7) 5.8594	(+5) 5.8594	(+4) 2.3437	(+3) 5.8590	(+2) 9.3711	(+2) 2.3399	(+1) 5.8204	8.9833	1.9422	(-1) 6.1675	(-1) 2.6483	(-1) 1.6702	(-1) 1.1347	(-2) 7.2754	(-2) 5.3662	(-2) 2.8163	(-2) 1.2123	1
1.2	(6) 3.1463	(4) 6.7785	(3) 4.6360	(3) 1.4599	(2) 3.1668	(1) 9.9439	(1) 3.1012	6.3770	1.6828	(-1) 5.9653	(-1) 2.7040	(-1) 1.7403	(-1) 1.1975	(-2) 7.7613	(-2) 5.7558	(-2) 3.0441	(-2) 1.3171	1.2
1.5	(5) 1.7367	(3) 8.0606	(2) 9.4230	(2) 3.7363	(2) 1.0974	(1) 4.3226	(1) 1.6829	4.5724	1.4652	(-1) 5.7991	(-1) 2.7806	(-1) 1.8288	(-1) 1.2760	(-2) 8.3679	)-2) 6.2430	(-2) 3.3302	(-2) 1.4492	1.5
2	(+3) 9.9993	(+2) 9.9933	(+2) 1.9933	(+1) 9.9332	(+1) 3.9331	(+1) 1.9330	9.3255	3.3121	1.2824	(-1) 5.6747	(-1) 2.8874	(-1) 1.9443	(-1) 1.3774	(-2) 9.1533	(-2) 6.8757	(-2) 3.7037	(-2) 1.6225	2
3	(+2) 6.1991	(+2) 1.3285	(+1) 4.4838	(+1) 2.7911	(+1) 1.4735	8.9406	5.2847	2.4218	1.1289	(-1) 5.6041	(-1) 3.0406	(-1) 2.1021	(-1) 1.5154	(-1) 1.0225	(-2) 7.7417	(-2) 4.2188	(-2) 1.8629	3
4	(+2) 1.6219	(+1) 5.0525	(+1) 2.1975	(+1) 1.5207	9.1973	6.1631	4.0098	2.0766	1.0617	(-1) 5.5953	(-1) 3.1439	(-1) 2.2057	(-1) 1.6059	(-1) 1.0931	(-2) 8.3139	(-2) 4.5613	(-2) 2.0235	4
5	(+1) 7.4426	(+1) 2.8834	(+1) 1.4513	(+1) 1.0672	6.9777	4.9503	3.4045	1.8945	1.0240	(-1) 5.6016	(-1) 3.2180	(-1) 2.2793	(-1) 1.6701	(-1) 1.1434	(-2) 8.7230	(-2) 4.8071	(-2) 2.1392	5
6	(+1) 4.4909	(+1) 2.0030	(+1) 1.1073	8.4661	5.8197	4.2839	3.0546	1.7821	1.0000	(-1) 5.6114	(-1) 3.2738	(-1) 2.3343	(-1) 1.7183	(-1) 1.1812	(-2) 9.0310	(-2) 4.9926	(-2) 2.2267	6
7	(+1) 3.1567	(+1) 1.5521	9.1554	7.1914	5.1186	3.8660	2.8274	1.7059	(-1) 9.8334	(-1) 5.6215	(-1) 3.3173	(-1) 2.3772	(-1) 1.7558	(-1) 1.2107	(-2) 9.2713	(-2) 5.1378	(-2) 2.2954	7
8	(+1) 2.4357	(+1) 1.2858	7.9520	6.3707	4.6517	3.5806	2.6683	1.6508	(-1) 9.7111	(-1) 5.6306	(-1) 3.3523	(-1) 2.4115	(-1) 1.7858	(-1) 1.2343	(-2) 9.4643	(-2) 5.2548	(-2) 2.3507	8
9	(+1) 1.9974	(+1) 1.1128	7.1338	5.8018	4.3197	3.3738	2.5509	1.6091	(-1) 9.6175	(-1) 5.6392	(-1) 3.3810	(-1) 2.4396	(-1) 1.8105	(-1) 1.2537	(-2) 9.6237	(-2) 5.3510	(-2) 2.3963	9
10	(+1) 1.7081	9.9256	6.5446	5.3858	4.0721	3.2172	2.4606	1.5765	(-1) 9.5436	(-1) 5.6472	(-1) 3.4050	(-1) 2.4631	(-1) 1.8311	(-1) 1.2700	(-2) 9.7561	(-2) 5.4316	(-2) 2.4345	10
12	(+1) 1.3560	8.3788	5.7570	4.8206	3.7283	2.9961	2.3310	1.5286	(-1) 9.4342	(-1) 5.6600	(-1) 3.4427	(-1) 2.5000	(-1) 1.8635	(-1) 1.2956	(-2) 9.9661	(-2) 5.5590	(-2) 2.4950	12
15	(+1) 1.0819	7.0917	5.0708	4.3183	3.4147	2.7905	2.2081	1.4820	(-1) 9.3267	(-1) 5.6750	(-1) 3.4829	(-1) 2.5393	(-1) 1.8980	(-1) 1.3229	(-1) 1.0190	(-2) 5.6952	(-2) 2.5597	15
20	8.6789	6.0186	4.4721	3.8714	3.1283	2.5990	2.0913	1.4366	(-1) 9.2210	(-1) 5.6918	(-1) 3.5257	(-1) 2.5812	(-1) 1.9348	(-1) 1.3521	(-1) 1.0429	(-2) 5.8411	(-2) 2.6291	20
24	7.7896	5.5504	4.2019	3.6667	2.9946	2.5082	2.0351	1.4143	(-1) 9.1687	(-1) 5.7013	(-1) 3.5482	(-1) 2.6031	(-1) 1.9542	(-1) 1.3675	(-1) 1.0555	(-2) 5.9181	(-2) 2.6658	24
30	7.0017	5.1223	3.9492	3.4735	2.8667	2.4205	1.9803	1.3923	(-1) 9.1169	(-1) 5.7110	(-1) 3.5714	(-1) 2.6259	(-1) 1.9743	(-1) 1.3834	(-1) 1.0686	(-2) 5.9980	(-2) 2.7039	30
40	6.3031	4.7306	3.7129	3.2910	2.7444	2.3359	1.9269	1.3706	(-1) 9.0654	(-1) 5.7218	(-1) 3.5956	(-1) 2.6495	(-1) 1.9950	(-1) 1.3999	(-1) 1.0822	(-2) 6.0811	(-2) 2.7435	40
60	5.6830	4.3721	3.4918	3.1187	2.6274	2.2540	1.8747	1.3491	(-1) 9.0144	(-1) 5.7330	(-1) 3.6206	(-1) 2.6739	(-1) 2.0166	(-1) 1.4171	(-1) 1.0963	(-2) 6.1674	(-2) 2.7847	60
120	5.1323	4.0437	3.2849	2.9559	2.5154	2.1750	1.8238	1.3278	(-1) 8.9637	(-1) 5.7448	(-1) 3.6466	(-1) 2.6993	(-1) 2.0389	(-1) 1.4349	(-1) 1.1109	(-2) 6.2574	(-2) 2.8277	120
∞	4.6427	3.7430	3.0913	2.8020	2.4082	2.0986	1.7741	1.3068	(-1) 8.9135	(-1) 5.7577	(-1) 3.6735	(-1) 2.7257	(-1) 2.0622	(-1) 1.4535	(-1) 1.1262	(-2) 6.3511	(-2) 2.8725	∞

$\nu_1 = 7$

ν_2	p = 0.0001	p = 0.001	p = 0.005	p = 0.01	p = 0.025	p = 0.05	p = 0.1	p = 0.25	p = 0.5	p = 0.75	p = 0.9	p = 0.95	p = 0.975	p = 0.99	p = 0.995	p = 0.999	p = 0.9999	ν_2
1	(+7) 5.9287	(+5) 5.9287	(+4) 2.3715	(+3) 5.9283	(+2) 9.4822	(+2) 2.3677	(+1) 5.8906	9.1021	1.9774	(-1) 6.3565	(-1) 2.7860	(-1) 1.7885	(-1) 1.2387	(-2) 8.1659	(-2) 6.1592	(-2) 3.4194	(-2) 1.6086	1
1.2	(6) 3.1755	(4) 6.8413	(3) 4.6790	(3) 1.4735	(2) 3.1963	(2) 1.0038	(1) 3.1314	6.4509	1.7128	(-1) 6.1560	(-1) 2.8523	(-1) 1.8702	(-1) 1.3130	(-2) 8.7554	(-2) 6.6435	(-2) 3.7203	(-2) 1.7609	1.2
1.5	(5) 1.7465	(3) 8.1059	(2) 9.4761	(2) 3.7574	(2) 1.1037	(1) 4.3487	(1) 1.6941	4.6161	1.4908	(-1) 5.9944	(-1) 2.9432	(-1) 1.9741	(-1) 1.4064	(-2) 9.4990	(-2) 7.2556	(-2) 4.1029	(-2) 1.9557	1.5
2	(+3) 9.9994	(+2) 9.9936	(+2) 1.9936	(+1) 9.9356	(+1) 3.9355	(+1) 1.9353	9.3491	3.3352	1.3045	(-1) 5.8789	(-1) 3.0699	(-1) 2.1109	(-1) 1.5287	(-1) 1.0475	(-2) 8.0619	(-2) 4.6106	(-2) 2.2157	2
3	(+2) 6.1388	(+2) 1.3158	(+1) 4.4434	(+1) 2.7672	(+1) 1.4624	8.8868	5.2662	2.4302	1.1482	(-1) 5.8245	(-1) 3.2530	(-1) 2.3005	(-1) 1.6979	(-1) 1.1832	(-2) 9.1895	(-2) 5.3270	(-2) 2.5856	3
4	(+2) 1.5931	(+1) 4.9658	(+1) 2.1622	(+1) 1.4976	9.0741	6.0942	3.9790	2.0790	1.0797	(-1) 5.8285	(-1) 3.3778	(-1) 2.4270	(-1) 1.8107	(-1) 1.2744	(-2) 9.9502	(-2) 5.8146	(-2) 2.8391	4
5	(+1) 7.2611	(+1) 2.8163	(+1) 1.4200	(+1) 1.0456	6.8531	4.8759	3.3679	1.8935	1.0414	(-1) 5.8442	(-1) 3.4682	(-1) 2.5179	(-1) 1.8921	(-1) 1.3404	(-1) 1.0502	(-2) 6.1706	(-2) 3.0251	5
6	(+1) 4.3566	(+1) 1.9463	(+1) 1.0786	8.2600	5.6955	4.2066	3.0145	1.7789	1.0169	(-1) 5.8620	(-1) 3.5368	(-1) 2.5867	(-1) 1.9537	(-1) 1.3905	(-1) 1.0923	(-2) 6.4430	(-2) 3.1679	6
7	(+1) 3.0477	(+1) 1.5019	8.8854	6.9928	4.9949	3.7870	2.7849	1.7011	1.0000	(-1) 5.8785	(-1) 3.5908	(-1) 2.6406	(-1) 2.0020	(-1) 1.4300	(-1) 1.1254	(-2) 6.6584	(-2) 3.2812	7
8	(+1) 2.3421	(+1) 1.2398	7.6942	6.1776	4.5286	3.5005	2.6241	1.6448	(-1) 9.8757	(-1) 5.8931	(-1) 3.6342	(-1) 2.6841	(-1) 2.0411	(-1) 1.4620	(-1) 1.1523	(-2) 6.8334	(-2) 3.3733	8
9	(+1) 1.9140	(+1) 1.0698	6.8849	5.6129	4.1971	3.2927	2.5053	1.6022	(-1) 9.7805	(-1) 5.9063	(-1) 3.6701	(-1) 2.7198	(-1) 2.0733	(-1) 1.4884	(-1) 1.1746	(-2) 6.9784	(-2) 3.4498	9
10	(+1) 1.6319	9.5175	6.3025	5.2001	3.9498	3.1355	2.4140	1.5688	(-1) 9.7054	(-1) 5.9179	(-1) 3.7003	(-1) 2.7499	(-1) 2.1004	(-1) 1.5106	(-1) 1.1933	(-2) 7.1006	(-2) 3.5143	10
12	(+1) 1.2892	8.0009	5.5245	4.6395	3.6065	2.9134	2.2828	1.5197	(-1) 9.5943	(-1) 5.9372	(-1) 3.7480	(-1) 2.7974	(-1) 2.1433	(-1) 1.5458	(-1) 1.2230	(-2) 7.2954	(-2) 3.6174	12
15	(+1) 1.0231	6.7408	4.8473	4.1415	3.2934	2.7066	2.1582	1.4718	(-1) 9.4850	(-1) 5.9591	(-1) 3.7991	(-1) 2.8484	(-1) 2.1892	(-1) 1.5837	(-1) 1.2551	(-2) 7.5054	(-2) 3.7287	15
20	8.1577	5.6920	4.2569	3.6987	3.0074	2.5140	2.0397	1.4252	(-1) 9.3776	(-1) 5.9837	(-1) 3.8540	(-1) 2.9032	(-1) 2.2388	(-1) 1.6246	(-1) 1.2897	(-2) 7.7330	(-2) 3.8495	20
24	7.2980	5.2349	3.9905	3.4959	2.8738	2.4226	1.9826	1.4022	(-1) 9.3245	(-1) 5.9970	(-1) 3.8830	(-1) 2.9321	(-1) 2.2650	(-1) 1.6463	(-1) 1.3080	(-2) 7.8541	(-2) 3.9139	24
30	6.5375	4.8173	3.7416	3.3045	2.7460	2.3343	1.9269	1.3795	(-1) 9.2719	(-1) 6.0114	(-1) 3.9131	(-1) 2.9623	(-1) 2.2923	(-1) 1.6689	(-1) 1.3272	(-2) 7.9806	(-2) 3.9813	30
40	5.8640	4.4355	3.5088	3.1238	2.6238	2.2490	1.8725	1.3571	(-1) 9.2197	(-1) 6.0266	(-1) 3.9446	(-1) 2.9937	(-1) 2.3208	(-1) 1.6925	(-1) 1.3473	(-2) 8.1129	(-2) 4.0518	40
60	5.2672	4.0864	3.2911	2.9530	2.5068	2.1665	1.8194	1.3349	(-1) 9.1679	(-1) 6.0430	(-1) 3.9774	(-1) 3.0264	(-1) 2.3505	(-1) 1.7172	(-1) 1.3682	(-2) 8.2516	(-2) 4.1258	60
120	4.7380	3.7670	3.0874	2.7918	2.3948	2.0867	1.7675	1.3128	(-1) 9.1164	(-1) 6.0599	(-1) 4.0116	(-1) 3.0605	(-1) 2.3816	(-1) 1.7430	(-1) 1.3902	(-2) 8.3970	(-2) 4.2035	120
∞	4.2682	3.4746	2.8968	2.6393	2.2875	2.0096	1.7167	1.2910	(-1) 9.0654	(-1) 6.0783	(-1) 4.0473	(-1) 3.0962	(-1) 2.4141	(-1) 1.7701	(-1) 1.4132	(-2) 8.5499	(-2) 4.2852	∞

$\nu_1 = 8$

ν_2	p = 0.0001	p = 0.001	p = 0.005	p = 0.01	p = 0.025	p = 0.05	p = 0.1	p = 0.25	p = 0.5	p = 0.75	p = 0.9	p = 0.95	p = 0.975	p = 0.99	p = 0.995	p = 0.999	p = 0.9999	ν_2
1	(+7) 5.9814	(+5) 5.9814	(+4) 2.3925	(+3) 5.9816	(+2) 9.5666	(+2) 2.3888	(+1) 5.9439	9.1922	2.0041	(-1) 6.5003	(-1) 2.8919	(-1) 1.8805	(-1) 1.3208	(-2) 8.8818	(-2) 6.8083	(-2) 3.9347	(-2) 1.9726	1
1.2	(6) 3.1977	(4) 6.8891	(3) 4.7117	(3) 1.4838	(2) 3.2187	(2) 1.0109	(1) 3.1544	6.5069	1.7355	(-1) 6.3014	(-1) 2.9668	(-1) 1.9717	(-1) 1.4045	(-2) 9.5596	(-2) 7.3741	(-2) 4.3021	(-2) 2.1719	1.2
1.5	(5) 1.7539	(3) 8.1403	(2) 9.5165	(2) 3.7736	(2) 1.1086	(1) 4.3685	(1) 1.7027	4.6492	1.5103	(-1) 6.1436	(-1) 3.0693	(-1) 2.0882	(-1) 1.5104	(-1) 1.0420	(-2) 8.0954	(-2) 4.7737	(-2) 2.4293	1.5
2	(+3) 9.9994	(+2) 9.9937	(+2) 1.9937	(+1) 9.9374	(+1) 3.9373	(+1) 1.9371	9.3668	3.3526	1.3213	(-1) 6.0354	(-1) 3.2122	(-1) 2.2427	(-1) 1.6503	(-1) 1.1562	(-2) 9.0563	(-2) 5.4072	(-2) 2.7778	2
3	(+2) 6.0929	(+2) 1.3062	(+1) 4.4126	(+1) 2.7489	(+1) 1.4540	8.8452	5.2517	2.4364	1.1627	(-1) 5.9941	(-1) 3.4202	(-1) 2.4593	(-1) 1.8464	(-1) 1.3173	(-1) 1.0420	(-2) 6.3173	(-2) 3.2834	3
4	(+2) 1.5711	(+1) 4.8996	(+1) 2.1352	(+1) 1.4799	8.9796	6.0410	3.9549	2.0805	1.0933	(-1) 6.0089	(-1) 3.5633	(-1) 2.6057	(-1) 1.9792	(-1) 1.4273	(-1) 1.1357	(-2) 6.9485	(-2) 3.6373	4
5	(+1) 7.1226	(+1) 2.7649	(+1) 1.3961	(+1) 1.0289	6.7572	4.8183	3.3393	1.8923	1.0545	(-1) 6.0332	(-1) 3.6677	(-1) 2.7119	(-1) 2.0759	(-1) 1.5079	(-1) 1.2046	(-2) 7.4158	(-2) 3.9009	5
6	(+1) 4.2541	(+1) 1.9030	(+1) 1.0566	8.1016	5.5996	4.1468	2.9830	1.7760	1.0298	(-1) 6.0577	(-1) 3.7477	(-1) 2.7928	(-1) 2.1498	(-1) 1.5697	(-1) 1.2575	(-2) 7.7773	(-2) 4.1057	6
7	(+1) 2.9644	(+1) 1.4634	8.6781	6.8401	4.8994	3.7257	2.7516	1.6969	1.0126	(-1) 6.0798	(-1) 3.8108	(-1) 2.8567	(-1) 2.2082	(-1) 1.6188	(-1) 1.2997	(-2) 8.0658	(-2) 4.2697	7
8	(+1) 2.2706	(+1) 1.2046	7.4960	6.0289	4.4332	3.4381	2.5893	1.6396	1.0000	(-1) 6.0990	(-1) 3.8620	(-1) 2.9086	(-1) 2.2557	(-1) 1.6587	(-1) 1.3340	(-2) 8.3019	(-2) 4.4042	8
9	(+1) 1.8503	(+1) 1.0368	6.6933	5.4671	4.1020	3.2296	2.4694	1.5961	(-1) 9.9037	(-1) 6.1162	(-1) 3.9044	(-1) 2.9515	(-1) 2.2951	(-1) 1.6919	(-1) 1.3627	(-2) 8.4987	(-2) 4.5166	9
10	(+1) 1.5736	9.2042	6.1159	5.0567	3.8549	3.0717	2.3772	1.5621	(-1) 9.8276	(-1) 6.1312	(-1) 3.9401	(-1) 2.9876	(-1) 2.3282	(-1) 1.7199	(-1) 1.3868	(-2) 8.6654	(-2) 4.6120	10
12	(+1) 1.2381	7.7104	5.3451	4.4994	3.5118	2.8486	2.2446	1.5120	(-1) 9.7152	(-1) 6.1561	(-1) 3.9968	(-1) 3.0451	(-1) 2.3811	(-1) 1.7647	(-1) 1.4255	(-2) 8.9330	(-2) 4.7653	12
15	9.7796	6.4707	4.6743	4.0045	3.1987	2.6408	2.1185	1.4631	(-1) 9.6046	(-1) 6.1843	(-1) 4.0581	(-1) 3.1071	(-1) 2.4383	(-1) 1.8132	(-1) 1.4675	(-2) 9.2240	(-2) 4.9325	15
20	7.7573	5.4400	4.0900	3.5644	2.9128	2.4471	1.9985	1.4153	(-1) 9.4959	(-1) 6.2158	(-1) 4.1244	(-1) 3.1743	(-1) 2.5003	(-1) 1.8660	(-1) 1.5133	(-2) 9.5423	(-2) 5.1159	20
24	6.9201	4.9912	3.8264	3.3629	2.7791	2.3551	1.9407	1.3918	(-1) 9.4422	(-1) 6.2332	(-1) 4.1596	(-1) 3.2101	(-1) 2.5334	(-1) 1.8942	(-1) 1.5378	(-2) 9.7131	(-2) 5.2145	24
30	6.1802	4.5814	3.5801	3.1726	2.6513	2.2662	1.8841	1.3685	(-1) 9.3889	(-1) 6.2516	(-1) 4.1964	(-1) 3.2474	(-1) 2.5681	(-1) 1.9238	(-1) 1.5635	(-2) 9.8925	(-2) 5.3182	30
40	5.5257	4.2070	3.3498	2.9930	2.5289	2.1802	1.8289	1.3455	(-1) 9.3361	(-1) 6.2716	(-1) 4.2348	(-1) 3.2864	(-1) 2.6043	(-1) 1.9548	(-1) 1.5905	(-1) 1.0081	(-2) 5.4275	40
60	4.9465	3.8648	3.1344	2.8233	2.4117	2.0970	1.7748	1.3226	(-1) 9.2838	(-1) 6.2925	(-1) 4.2751	(-1) 3.3275	(-1) 2.6424	(-1) 1.9874	(-1) 1.6189	(-1) 1.0280	(-2) 5.5430	60
120	4.4333	3.5519	2.9330	2.6629	2.2994	2.0164	1.7220	1.2999	(-1) 9.2318	(-1) 6.3147	(-1) 4.3174	(-1) 3.3705	(-1) 2.6825	(-1) 2.0218	(-1) 1.6488	(-1) 1.0491	(-2) 5.6652	120
∞	3.9785	3.2656	2.7444	2.5113	2.1918	1.9384	1.6702	1.2774	(-1) 9.1802	(-1) 6.3383	(-1) 4.3619	(-1) 3.4158	(-1) 2.7246	(-1) 2.0581	(-1) 1.6805	(-1) 1.0714	(-2) 5.7949	∞

$\nu_1 = 9$

ν_2	p = 0.0001	p = 0.001	p = 0.005	p = 0.01	p = 0.025	p = 0.05	p = 0.1	p = 0.25	p = 0.5	p = 0.75	p = 0.9	p = 0.95	p = 0.975	p = 0.99	p = 0.995	p = 0.999	p = 0.9999	ν_2
1	(+7) 6.0228	(+5) 6.0228	(+4) 2.4091	(+3) 6.0225	(+2) 9.6328	(+2) 2.4054	(+1) 5.9858	9.2631	2.0250	(-1) 6.6133	(-1) 2.9759	(-1) 1.9541	(-1) 1.3871	(-2) 9.4688	(-2) 7.3454	(-2) 4.3750	(-2) 2.3001	1
1.2	(6) 3.2151	(4) 6.9266	(3) 4.7374	(3) 1.4919	(2) 3.2363	(2) 1.0165	(1) 3.1725	6.5509	1.7533	(-1) 6.4159	(-1) 3.0579	(-1) 2.0531	(-1) 1.4785	(-1) 1.0221	(-2) 7.9818	(-2) 4.8018	(-2) 2.5438	1.2
1.5	(5) 1.7597	(3) 8.1674	(2) 9.5482	(2) 3.7862	(2) 1.1123	(1) 4.3841	(1) 1.7094	4.6752	1.5255	(-1) 6.2613	(-1) 3.1697	(-1) 2.1801	(-1) 1.5950	(-1) 1.1182	(-2) 8.7980	(-2) 5.3537	(-2) 2.8612	1.5
2	(+3) 9.9994	(+2) 9.9939	(+2) 1.9939	(+1) 9.9388	(+1) 3.9387	(+1) 1.9385	9.3805	3.3661	1.3344	(-1) 6.1592	(-1) 3.3261	(-1) 2.3493	(-1) 1.7499	(-1) 1.2466	(-2) 9.8941	(-2) 6.1024	(-2) 3.2958	2
3	(+2) 6.0567	(+2) 1.2986	(+1) 4.3882	(+1) 2.7345	(+1) 1.4473	8.8123	5.2400	2.4410	1.1741	(-1) 6.1293	(-1) 3.5550	(-1) 2.5889	(-1) 1.9692	(-1) 1.4302	(-1) 1.1472	(-2) 7.1933	(-2) 3.9364	3
4	(+2) 1.5538	(+1) 4.8475	(+1) 2.1139	(+1) 1.4659	8.9047	5.9988	3.9357	2.0814	1.1040	(-1) 6.1527	(-1) 3.7137	(-1) 2.7525	(-1) 2.1195	(-1) 1.5571	(-1) 1.2569	(-2) 7.9616	(-2) 4.3925	4
5	(+1) 7.0133	(+1) 2.7244	(+1) 1.3772	(+1) 1.0158	6.6810	4.7725	3.3163	1.8911	1.0648	(-1) 6.1843	(-1) 3.8305	(-1) 2.8722	(-1) 2.2300	(-1) 1.6510	(-1) 1.3385	(-2) 8.5370	(-2) 4.7366	5
6	(+1) 4.1732	(+1) 1.8688	(+1) 1.0391	7.9761	5.5234	4.0990	2.9577	1.7733	1.0398	(-1) 6.2147	(-1) 3.9202	(-1) 2.9640	(-1) 2.3150	(-1) 1.7236	(-1) 1.4018	(-2) 8.9863	(-2) 5.0065	6
7	(+1) 2.8987	(+1) 1.4330	8.5138	6.7188	4.8232	3.6767	2.7247	1.6931	1.0224	(-1) 6.2414	(-1) 3.9915	(-1) 3.0370	(-1) 2.3826	(-1) 1.7816	(-1) 1.4525	(-2) 9.3476	(-2) 5.2246	7
8	(+1) 2.2141	(+1) 1.1767	7.3386	5.9106	4.3572	3.3881	2.5612	1.6350	1.0097	(-1) 6.2653	(-1) 4.0496	(-1) 3.0964	(-1) 2.4378	(-1) 1.8291	(-1) 1.4940	(-2) 9.6451	(-2) 5.4046	8
9	(+1) 1.7999	(+1) 1.0107	6.5411	5.3511	4.0260	3.1789	2.4403	1.5909	1.0000	(-1) 6.2858	(-1) 4.0979	(-1) 3.1457	(-1) 2.4839	(-1) 1.8688	(-1) 1.5288	(-2) 9.8945	(-2) 5.5560	9
10	(+1) 1.5275	8.9558	5.9676	4.9424	3.7790	3.0204	2.3473	1.5563	(-1) 9.9232	(-1) 6.3040	(-1) 4.1386	(-1) 3.1875	(-1) 2.5228	(-1) 1.9024	(-1) 1.5583	(-1) 1.0107	(-2) 5.6851	10
12	(+1) 1.1976	7.4797	5.2021	4.3875	3.4358	2.7964	2.2135	1.5054	(-1) 9.8097	(-1) 6.3339	(-1) 4.2036	(-1) 3.2543	(-1) 2.5852	(-1) 1.9564	(-1) 1.6058	(-1) 1.0449	(-2) 5.8939	12
15	9.4218	6.2559	4.5364	3.8948	3.1227	2.5876	2.0862	1.4556	(-1) 9.6981	(-1) 6.3674	(-1) 4.2742	(-1) 3.3266	(-1) 2.6529	(-1) 2.0153	(-1) 1.6577	(-1) 1.0825	(-2) 6.1235	15
20	7.4394	5.2392	3.9564	3.4567	2.8365	2.3928	1.9649	1.4069	(-1) 9.5884	(-1) 6.4057	(-1) 4.3510	(-1) 3.4054	(-1) 2.7271	(-1) 2.0799	(-1) 1.7147	(-1) 1.1239	(-2) 6.3776	20
24	6.6197	4.7968	3.6949	3.2560	2.7027	2.3002	1.9063	1.3828	(-1) 9.5342	(-1) 6.4257	(-1) 4.3921	(-1) 3.4477	(-1) 2.7669	(-1) 2.1146	(-1) 1.7454	(-1) 1.1463	(-2) 6.5153	24
30	5.8960	4.3930	3.4505	3.0665	2.5746	2.2107	1.8490	1.3590	(-1) 9.4805	(-1) 6.4491	(-1) 4.4352	(-1) 3.4920	(-1) 2.8087	(-1) 2.1512	(-1) 1.7778	(-1) 1.1699	(-2) 6.6609	30
40	5.2564	4.0243	3.2220	2.8876	2.4519	2.1240	1.7929	1.3354	(-1) 9.4272	(-1) 6.4725	(-1) 4.4803	(-1) 3.5387	(-1) 2.8527	(-1) 2.1898	(-1) 1.8121	(-1) 1.1950	(-2) 6.8155	40
60	4.6907	3.6873	3.0083	2.7185	2.3344	2.0401	1.7380	1.3119	(-1) 9.3743	(-1) 6.4981	(-1) 4.5280	(-1) 3.5878	(-1) 2.8991	(-1) 2.2306	(-1) 1.8483	(-1) 1.2215	(-2) 6.9798	60
120	4.1901	3.3793	2.8083	2.5586	2.2217	1.9588	1.6843	1.2886	(-1) 9.3218	(-1) 6.5253	(-1) 4.5781	(-1) 3.6397	(-1) 2.9483	(-1) 2.2739	(-1) 1.8868	(-1) 1.2498	(-2) 7.1550	120
∞	3.7467	3.0975	2.6210	2.4073	2.1136	1.8799	1.6315	1.2654	(-1) 9.2698	(-1) 6.5344	(-1) 4.6313	(-1) 3.6945	(-1) 3.0004	(-1) 2.3199	(-1) 1.9277	(-1) 1.2799	(-2) 7.3423	∞

$\nu_1 = 10$

ν_2	p = 0.0001	p = 0.001	p = 0.005	p = 0.01	p = 0.025	p = 0.05	p = 0.1	p = 0.25	p = 0.5	p = 0.75	p = 0.9	p = 0.95	p = 0.975	p = 0.99	p = 0.995	p = 0.999	p = 0.9999	ν_2
1	(+7) 6.0562	(+5) 6.0562	(+4) 2.4224	(+3) 6.0558	(+2) 9.6863	(+2) 2.4188	(+1) 6.0195	9.3202	2.0419	(-1) 6.7047	(-1) 3.0441	(-1) 2.0143	(-1) 1.4416	(-2) 9.9562	(-2) 7.7967	(-2) 4.7529	(-2) 2.5922	1
1.2	(6) 3.2291	(5) 6.9569	(3) 4.7581	(3) 1.4984	(2) 3.2506	(2) 1.0210	(1) 3.1870	6.5863	1.7677	(-1) 6.5084	(-1) 3.1319	(-1) 2.1198	(-1) 1.5397	(-1) 1.0772	(-2) 8.4934	(-2) 5.2325	(-2) 2.8772	1.2
1.5	(5) 1.7644	(3) 8.1893	(2) 9.5739	(2) 3.7964	(2) 1.1154	(1) 4.3967	(1) 1.7148	4.6961	1.5378	(-1) 6.3565	(-1) 3.2516	(-1) 2.2555	(-1) 1.6651	(-1) 1.1819	(-2) 9.3921	(-2) 5.8562	(-2) 3.2508	1.5
2	(+3) 9.9994	(+2) 9.9940	(+2) 1.9940	(+1) 9.9399	(+1) 3.9398	(+1) 1.9396	9.3916	3.3770	1.3450	(-1) 6.2598	(-1) 3.4194	(-1) 2.4374	(-1) 1.8327	(-1) 1.3229	(-1) 1.0608	(-2) 6.7090	(-2) 3.7668	2
3	(+2) 6.0275	(+2) 1.2925	(+1) 4.3686	(+1) 2.7229	(+1) 1.4419	8.7855	5.2304	2.4447	1.1833	(-1) 6.2391	(-1) 3.6661	(-1) 2.6967	(-1) 2.0723	(-1) 1.5262	(-1) 1.2375	(-2) 7.9664	(-2) 4.5377	3
4	(+2) 1.5398	(+1) 4.8053	(+1) 2.0967	(+1) 1.4546	8.8439	5.9644	3.9199	2.0820	1.1126	(-1) 6.2700	(-1) 3.8383	(-1) 2.8752	(-1) 2.2380	(-1) 1.6683	(-1) 1.3619	(-2) 8.8630	(-2) 5.0942	4
5	(+1) 6.9250	(+1) 2.6917	(+1) 1.3618	(+1) 1.0051	6.6192	4.7351	3.2974	1.8899	1.0730	(-1) 6.3080	(-1) 3.9657	(-1) 3.0068	(-1) 2.3607	(-1) 1.7742	(-1) 1.4551	(-1) 9.5413	(-2) 5.5187	5
6	(+1) 4.1077	(+1) 1.8411	(+1) 1.0250	7.8741	5.4613	4.0600	2.9369	1.7708	1.0478	(-1) 6.3432	(-1) 4.0640	(-1) 3.1083	(-1) 2.4557	(-1) 1.8567	(-1) 1.5280	(-1) 1.0075	(-2) 5.8546	6
7	(+1)		8.3803	6.6201	4.7611	3.6365	2.7025	1.6898	1.0304	(-1) 6.3743	(-1) 4.1425	(-1) 3.1893	(-1) 2.5318	(-1) 1.9230	(-1) 1.5867	(-1) 1.0507	(-2) 6.1279	7
8	(+1) 2.1683	(+1) 1.1540	7.2107	5.8143	4.2951	3.3472	2.5380	1.6310	1.0175	(-1) 6.4016	(-1) 4.2066	(-1) 3.2555	(-1) 2.5941	(-1) 1.9776	(-1) 1.6351	(-1) 1.0865	(-2) 6.3549	8
9	(+1) 1.7590	9.8943	6.4171	5.2565	3.9639	3.1373	2.4163	1.5863	1.0077	(-1) 6.4255	(-1) 4.2602	(-1) 3.3108	(-1) 2.6462	(-1) 2.0233	(-1) 1.6757	(-1) 1.1166	(-2) 6.5467	9
10	(+1) 1.4901	8.7539	5.8467	4.8492	3.7168	2.9782	2.3226	1.5513	1.0000	(-1) 6.4462	(-1) 4.3055	(-1) 3.3577	(-1) 2.6905	(-1) 2.0622	(-1) 1.7104	(-1) 1.1424	(-2) 6.7111	10
12	(+1) 1.1647	7.2920	5.0855	4.2961	3.3736	2.7534	2.1878	1.4996	(-1) 9.8856	(-1) 6.4809	(-1) 4.3781	(-1) 3.4329	(-1) 2.7617	(-1) 2.1250	(-1) 1.7664	(-1) 1.1841	(-2) 6.9783	12
15	9.1309	6.0808	4.4236	3.8049	3.0602	2.5437	2.0593	1.4491	(-1) 9.7732	(-1) 6.5198	(-1) 4.4573	(-1) 3.5149	(-1) 2.8395	(-1) 2.1938	(-1) 1.8279	(-1) 1.2302	(-2) 7.2744	15
20	7.1805	5.0752	3.8470	3.3682	2.7737	2.3479	1.9367	1.3995	(-1) 9.6626	(-1) 6.5638	(-1) 4.5440	(-1) 3.6049	(-1) 2.9252	(-1) 2.2699	(-1) 1.8961	(-1) 1.2814	(-2) 7.6049	20
24	6.3750	4.6379	3.5870	3.1681	2.6396	2.2547	1.8775	1.3750	(-1) 9.6081	(-1) 6.5880	(-1) 4.5905	(-1) 3.6534	(-1) 2.9714	(-1) 2.3111	(-1) 1.9330	(-1) 1.3093	(-2) 7.7852	24
30	5.6641	4.2388	3.3440	2.9791	2.5112	2.1646	1.8195	1.3507	(-1) 9.5540	(-1) 6.6142	(-1) 4.6395	(-1) 3.7043	(-1) 3.0202	(-1) 2.3547	(-1) 1.9722	(-1) 1.3389	(-2) 7.9770	30
40	5.0363	3.8744	3.1167	2.8005	2.3882	2.0772	1.7627	1.3266	(-1) 9.5003	(-1) 6.6419	(-1) 4.6911	(-1) 3.7581	(-1) 3.0718	(-1) 2.4008	(-1) 2.0137	(-1) 1.3704	(-2) 8.1816	40
60	4.4815	3.5415	2.9042	2.6318	2.2702	1.9926	1.7070	1.3026	(-1) 9.4471	(-1) 6.6711	(-1) 4.7456	(-1) 3.8152	(-1) 3.1266	(-1) 2.4498	(-1) 2.0580	(-1) 1.4040	(-2) 8.4007	60
120	3.9907	3.2372	2.7052	2.4721	2.1570	1.9105	1.6524	1.2787	(-1) 9.3943	(-1) 6.7029	(-1) 4.8035	(-1) 3.8758	(-1) 3.1848	(-1) 2.5022	(-1) 2.1052	(-1) 1.4400	(-2) 8.6358	120
∞	3.5564	2.9588	2.5188	2.3209	2.0483	1.8307	1.5987	1.2549	(-1) 9.3418	(-1) 6.7372	(-1) 4.8652	(-1) 3.9403	(-1) 3.2470	(-1) 2.5582	(-1) 2.1559	(-1) 1.4787	(-2) 8.8891	∞

$\nu_1 = 12$

ν_2	p = 0.0001	p = 0.001	p = 0.005	p = 0.01	p = 0.025	p = 0.05	p = 0.1	p = 0.25	p = 0.5	p = 0.75	p = 0.9	p = 0.95	p = 0.975	p = 0.99	p = 0.995	p = 0.999	p = 0.9999	ν_2
1	(+7) 6.1067	(+5) 6.1067	(+4) 2.4426	(+3) 6.1063	(+2) 9.7671	(+2) 2.4391	(+1) 6.0705	9.4064	2.0674	(-1) 6.8432	(-1) 3.1481	(-1) 2.1065	(-1) 1.5258	(-1) 1.0718	(-2) 8.5077	(-2) 5.3638	(-2) 3.0838	1
1.2	(6) 3.2504	(4) 7.0027	(3) 4.7894	(3) 1.5083	(2) 3.2721	(2) 1.0278	(1) 3.2090	6.6399	1.7894	(-1) 6.6485	(-1) 3.2449	(-1) 2.2223	(-1) 1.6344	(-1) 1.1636	(-2) 9.3036	(-2) 5.9318	(-2) 3.4412	1.2
1.5	(5) 1.7715	(3) 8.2223	(2) 9.6126	(2) 3.8119	(2) 1.1200	(1) 4.4157	(1) 1.7230	4.7276	1.5563	(-1) 6.5009	(-1) 3.3771	(-1) 2.3719	(-1) 1.7741	(-1) 1.2824	(-1) 1.0338	(-2) 6.6769	(-2) 3.9143	1.5
2	(+3) 9.9994	(+2) 9.9942	(+2) 1.9942	(+1) 9.9416	(+1) 3.9415	(+1) 1.9413	9.4081	3.3934	1.3610	(-1) 6.4123	(-1) 3.5628	(-1) 2.5738	(-1) 1.9624	(-1) 1.4437	(-1) 1.1751	(-2) 7.7080	(-2) 4.5768	2
3	(+2) 5.9833	(+2) 1.2832	(+1) 4.3387	(+1) 2.7052	(+1) 1.4337	8.7446	5.2156	2.4500	1.1972	(-1) 6.4066	(-1) 3.8380	(-1) 2.8651	(-1) 2.2350	(-1) 1.6799	(-1) 1.3839	(-2) 9.2557	(-2) 5.5867	3
4	(+2) 1.5186	(+1) 4.7412	(+1) 2.0705	(+1) 1.4374	8.7512	5.9117	3.8955	2.0826	1.1255	(-1) 6.4504	(-1) 4.0321	(-1) 3.0682	(-1) 2.4265	(-1) 1.8478	(-1) 1.5335	(-1) 1.0381	(-2) 6.3321	4
5	(+1) 6.7908	(+1) 2.6418	(+1) 1.3384	9.8883	6.5246	4.6777	3.2682	1.8877	1.0855	(-1) 6.4981	(-1) 4.1771	(-1) 3.2197	(-1) 2.5700	(-1) 1.9746	(-1) 1.6471	(-2) 1.1246	(-2) 6.9104	5
6	(+1) 4.0081	(+1) 1.7989	(+1) 1.0034	7.7183	5.3662	3.9999	2.9047	1.7668	1.0600	(-1) 6.5419	(-1) 4.2900	(-1) 3.3377	(-1) 2.6822	(-1) 2.0744	(-1) 1.7370	(-1) 1.1935	(-2) 7.3745	6
7	(+1) 2.7644	(+1) 1.3707	8.1764	6.4691	4.6658	3.5747	2.6681	1.6843	1.0423	(-1) 6.5302	(-1) 4.3806	(-1) 3.4324	(-1) 2.7728	(-1) 2.1554	(-1) 1.8101	(-1) 1.2499	(-2) 7.7566	7
8	(+1) 2.0985	(+1) 1.1194	7.0149	5.6668	4.1997	3.2840	2.5020	1.6244	1.0293	(-1) 6.6138	(-1) 4.4551	(-1) 3.5105	(-1) 2.8475	(-1) 2.2225	(-1) 1.8709	(-1) 1.2970	(-2) 8.0771	8
9	(+1) 1.6967	9.5700	6.2274	5.1114	3.8682	3.0729	2.3789	1.5788	1.0194	(-1) 6.6428	(-1) 4.5177	(-1) 3.5760	(-1) 2.9105	(-1) 2.2792	(-1) 1.9223	(-1) 1.3369	(-2) 8.3503	9
10	(+1) 1.4330	8.4452	5.6613	4.7059	3.6209	2.9130	2.2841	1.5430	1.0116	(-1) 6.6684	(-1) 4.5708	(-1) 3.6319	(-1) 2.9642	(-1) 2.3277	(-1) 1.9664	(-1) 1.3714	(-2) 8.5861	10
12	(+1) 1.1144	7.0046	4.9063	4.1553	3.2773	2.6866	2.1474	1.4902	1.0000	(-1) 6.7105	(-1) 4.6568	(-1) 3.7222	(-1) 3.0513	(-1) 2.4066	(-1) 2.0382	(-1) 1.4276	(-2) 8.9732	12
15	8.6859	5.8121	4.2498	3.6662	2.9633	2.4753	2.0171	1.4383	(-1) 9.8863	(-1) 6.7586	(-1) 4.7508	(-1) 3.8213	(-1) 3.1474	(-1) 2.4940	(-1) 2.1180	(-1) 1.4905	(-2) 9.4075	15
20	6.7837	4.8229	3.6779	3.2311	2.6758	2.2776	1.8924	1.3873	(-1) 9.7746	(-1) 6.8129	(-1) 4.8551	(-1) 3.9314	(-1) 3.2544	(-1) 2.5917	(-1) 2.2076	(-1) 1.5613	(-2) 9.8996	20
24	5.9992	4.3929	3.4199	3.0316	2.5412	2.1834	1.8319	1.3621	(-1) 9.7194	(-1) 6.8432	(-1) 4.9116	(-1) 3.9912	(-1) 3.3127	(-1) 2.6452	(-1) 2.2566	(-1) 1.6003	(-1) 1.0171	24
30	5.3075	4.0006	3.1787	2.8431	2.4120	2.0921	1.7727	1.3369	(-1) 9.6647	(-1) 6.8757	(-1) 4.9714	(-1) 4.0547	(-1) 3.3746	(-1) 2.7021	(-1) 2.3090	(-1) 1.6421	(-1) 1.0464	30
40	4.6973	3.6425	2.9531	2.6648	2.2882	2.0035	1.7145	1.3119	(-1) 9.6104	(-1) 6.9104	(-1) 5.0350	(-1) 4.1222	(-1) 3.4408	(-1) 2.7630	(-1) 2.3651	(-1) 1.6870	(-1) 1.0778	40
60	4.1585	3.3153	2.7419	2.4961	2.1692	1.9174	1.6574	1.2870	(-1) 9.5566	(-1) 6.9478	(-1) 5.1028	(-1) 4.1943	(-1) 3.5115	(-1) 2.8285	(-1) 2.4255	(-1) 1.7354	(-1) 1.1119	60
120	3.6823	3.0161	2.5439	2.3363	2.0548	1.8337	1.6012	1.2621	(-1) 9.5032	(-1) 6.9881	(-1) 5.1752	(-1) 4.2717	(-1) 3.5876	(-1) 2.8991	(-1) 2.4907	(-1) 1.7879	(-1) 1.1490	120
∞	3.2612	2.7425	2.3583	2.1848	1.9447	1.7522	1.5458	1.2371	(-1) 9.4503	(-1) 7.0319	(-1) 5.2532	(-1) 4.3550	(-1) 3.6699	(-1) 2.9755	(-1) 2.5615	(-1) 1.8452	(-1) 1.1896	∞

$\nu_1 = 15$

ν_2	p = 0.0001	p = 0.001	p = 0.005	p = 0.01	p = 0.025	p = 0.05	p = 0.1	p = 0.25	p = 0.5	p = 0.75	p = 0.9	p = 0.95	p = 0.975	p = 0.99	p = 0.995	p = 0.999	p = 0.9999	ν_2
1	(+7) 6.1576	(+5) 6.1576	(+4) 2.4630	(+3) 6.1573	(+2) 9.8487	(+2) 2.4595	(+1) 6.1220	9.4934	2.0931	(-1) 6.9828	(-1) 3.2539	(-1) 2.2011	(-1) 1.6130	(-1) 1.1517	(-2) 9.2610	(-2) 6.0287	(-2) 3.6432	1
1.2	(6) 3.2718	(4) 7.0489	(3) 4.8210	(3) 1.5183	(2) 3.2938	(2) 1.0347	(1) 3.2312	6.6939	1.8112	(-1) 6.7903	(-1) 3.3603	(-1) 2.3278	(-1) 1.7328	(-1) 1.2546	(-1) 1.0166	(-2) 6.6968	(-2) 4.0868	1.2
1.5	(5) 1.7787	(3) 8.2557	(2) 9.6518	(2) 3.8275	(2) 1.1247	(1) 4.4349	(1) 1.7313	4.7594	1.5750	(-1) 6.6473	(-1) 3.5055	(-1) 2.4922	(-1) 1.8878	(-1) 1.3887	(-1) 1.1350	(-2) 7.5809	(-2) 4.6800	1.5
2	(+3) 9.9994	(+2) 9.9943	(+2) 1.9943	(+1) 9.9432	(+1) 3.9431	(+1) 1.9429	9.4247	3.4098	1.3771	(-1) 6.5673	(-1) 3.7103	(-1) 2.7157	(-1) 2.0986	(-1) 1.5726	(-1) 1.2986	(-2) 8.8190	(-2) 5.5221	2
3	(+2) 5.9384	(+2) 1.2737	(+1) 4.3085	(+1) 2.6872	(+1) 1.4253	8.7029	5.2003	2.4552	1.2111	(-1) 6.5781	(-1) 4.0164	(-1) 3.0419	(-1) 2.4080	(-1) 1.8460	(-1) 1.5442	(-1) 1.0712	(-2) 6.8329	3
4	(+2) 1.4971	(+1) 4.6761	(+1) 2.0438	(+1) 1.4198	8.6565	5.8578	3.8689	2.0829	1.1386	(-1) 6.6353	(-1) 4.2348	(-1) 3.2727	(-1) 2.6286	(-1) 2.0437	(-1) 1.7233	(-1) 1.2117	(-2) 7.8228	4
5	(+1) 6.6544	(+1) 2.5911	(+1) 1.3146	9.7222	6.4277	4.6188	3.2380	1.8851	1.0980	(-1) 6.6943	(-1) 4.3995	(-1) 3.4467	(-1) 2.7961	(-1) 2.1951	(-1) 1.8615	(-1) 1.3215	(-2) 8.6050	5
6	(+1) 3.9068	(+1) 1.7559	9.8140	7.5590	5.2687	3.9381	2.8712	1.7621	1.0722	(-1) 6.7476	(-1) 4.5288	(-1) 3.5836	(-1) 2.9285	(-1) 2.3157	(-1) 1.9721	(-1) 1.4101	(-2) 9.2427	6
7	(+1) 2.6819	(+1) 1.3324	7.9678	6.3143	4.5678	3.5108	2.6322	1.6781	1.0543	(-1) 6.7944	(-1) 4.6335	(-1) 3.6947	(-1) 3.0364	(-1) 2.4146	(-1) 2.0630	(-1) 1.4835	(-2) 9.7743	7
8	(+1) 2.0274	(+1) 1.0841	6.8143	5.5151	4.1012	3.2184	2.4642	1.6170	1.0412	(-1) 6.8348	(-1) 4.7203	(-1) 3.7867	(-1) 3.1263	(-1) 2.4972	(-2) 2.1394	(-1) 1.5454	(-1) 1.0225	8
9	(+1) 1.6331	9.2381	6.0325	4.9621	3.7694	3.0061	2.3396	1.5705	1.0311	(-1) 6.8700	(-1) 4.7934	(-1) 3.8646	(-1) 3.2024	(-1) 2.5675	(-1) 2.2044	(-1) 1.5985	(-1) 1.0614	9
10	(+1) 1.3747	8.1288	5.4707	4.5582	3.5217	2.8450	2.2435	1.5338	1.0232	(-1) 6.9008	(-1) 4.8560	(-1) 3.9313	(-1) 3.2678	(-1) 2.6282	(-1) 2.2606	(-1) 1.6445	(-1) 1.0952	10
12	(+1) 1.0630	6.7092	4.7214	4.0096	3.1772	2.6169	2.1049	1.4796	1.0115	(-1) 6.9527	(-1) 4.9576	(-1) 4.0399	(-1) 3.3746	(-1) 2.7276	(-1) 2.3531	(-1) 1.7206	(-1) 1.1513	12
15	8.2290	5.5351	4.0698	3.5222	2.8621	2.4035	1.9722	1.4263	1.0000	(-1) 7.0111	(-1) 5.0705	(-1) 4.1606	(-1) 3.4939	(-1) 2.8391	(-1) 2.4571	(-1) 1.8067	(-1) 1.2152	15
20	6.3748	4.5618	3.5020	3.0880	2.5731	2.2033	1.8449	1.3736	(-1) 9.8870	(-1) 7.0786	(-1) 5.1967	(-1) 4.2965	(-1) 3.6286	(-1) 2.9657	(-1) 2.5756	(-1) 1.9053	(-1) 1.2890	20
24	5.6112	4.1387	3.2456	2.8887	2.4374	2.1077	1.7831	1.3474	(-1) 9.8312	(-1) 7.1164	(-1) 5.2659	(-1) 4.3710	(-1) 3.7029	(-1) 3.0358	(-1) 2.6414	(-1) 1.9604	(-1) 1.3304	24
30	4.9385	3.7527	3.0057	2.7002	2.3072	2.0148	1.7223	1.3213	(-1) 9.7759	(-1) 7.1567	(-1) 5.3396	(-1) 4.4508	(-1) 3.7826	(-1) 3.1113	(-1) 2.7125	(-1) 2.0201	(-1) 1.3754	30
40	4.3455	3.4003	2.7811	2.5216	2.1819	1.9245	1.6624	1.2952	(-1) 9.7211	(-1) 7.2005	(-1) 5.4189	(-1) 4.5366	(-1) 3.8685	(-1) 3.1929	(-1) 2.7894	(-1) 2.0851	(-1) 1.4246	40
60	3.8221	3.0781	2.5705	2.3523	2.0613	1.8364	1.6034	1.2691	(-1) 9.6667	(-1) 7.2485	(-1) 5.5042	(-1) 4.6294	(-1) 3.9617	(-1) 3.2818	(-1) 2.8733	(-1) 2.1562	(-1) 1.4787	60
120	3.3600	2.7833	2.3727	2.1915	1.9450	1.7505	1.5450	1.2428	(-1) 9.6128	(-1) 7.3003	(-1) 5.5969	(-1) 4.7301	(-1) 4.0632	(-1) 3.3789	(-1) 2.9654	(-1) 2.2347	(-1) 1.5386	120
∞	2.9509	2.5132	2.1868	2.0385	1.8326	1.6664	1.4871	1.2163	(-1) 9.5593	(-1) 7.3578	(-1) 5.6977	(-1) 4.8407	(-1) 4.1748	(-1) 3.4863	(-1) 3.0673	(-1) 2.3218	(-1) 1.6055	∞

$\nu_1 = 20$

ν_2	p=0.0001	p=0.001	p=0.005	p=0.01	p=0.025	p=0.05	p=0.1	p=0.25	p=0.5	p=0.75	p=0.9	p=0.95	p=0.975	p=0.99	p=0.995	p=0.999	p=0.9999	ν_2
1	(+7) 6.2091	(+5) 6.2091	(+4) 2.4836	(+3) 6.2087	(+2) 9.9310	(+2) 2.4801	(+1) 6.1740	9.5813	2.1190	(-1) 7.1240	(-1) 3.3617	(-1) 2.2982	(-1) 1.7031	(-1) 1.2352	(-1) 1.0056	(-2) 6.7482	(-2) 4.2736	1
1.2	(6) 3.2935	(4) 7.0956	(3) 4.8530	(3) 1.5283	(2) 3.3157	(2) 1.0417	(1) 3.2536	6.7484	1.8333	(-1) 6.9338	(-1) 3.4780	(-1) 2.4362	(-1) 1.8348	(-1) 1.3502	(-1) 1.1080	(-2) 7.5289	(-2) 4.8187	1.2
1.5	(5) 1.7860	(3) 8.2895	(2) 9.6914	(2) 3.8433	(2) 1.1294	(1) 4.4543	(1) 1.7396	4.7914	1.5939	(-1) 6.7958	(-1) 3.6370	(-1) 2.6164	(-1) 2.0063	(-1) 1.5009	(-1) 1.2429	(-2) 8.5708	(-2) 5.5549	1.5
2	(+3) 9.9995	(+2) 9.9945	(+2) 1.9945	(+1) 9.9449	(+1) 3.9448	(+1) 1.9446	9.4413	3.4263	1.3933	(-1) 6.7249	(-1) 3.8620	(-1) 2.8630	(-1) 2.2415	(-1) 1.7097	(-1) 1.4313	(-1) 1.0048	(-2) 6.6142	2
3	(+2) 5.8930	(+2) 1.2642	(+1) 4.2778	(+1) 2.6690	(+1) 1.4167	8.6602	5.1845	2.4602	1.2252	(-1) 6.7531	(-1) 4.2015	(-1) 3.2275	(-1) 2.5915	(-1) 2.0250	(-1) 1.7189	(-1) 1.2348	(-2) 8.2988	3
4	(+2) 1.4752	(+1) 4.6100	(+1) 2.0167	(+1) 1.4020	8.5599	5.8025	3.8443	2.0828	1.1517	(-1) 6.8250	(-1) 4.4466	(-1) 3.4891	(-1) 2.8452	(-1) 2.2570	(-1) 1.9326	(-1) 1.4092	(-2) 9.6018	4
5	(+1) 6.5157	(+1) 2.5395	(+1) 1.2903	9.5527	6.3285	4.5581	3.2067	1.8820	1.1106	(-1) 6.8966	(-1) 4.6335	(-1) 3.6888	(-1) 3.0403	(-1) 2.4374	(-1) 2.1001	(-1) 1.5479	(-1) 1.0652	5
6	(+1) 3.8036	(+1) 1.7120	9.5888	7.3958	5.1684	3.8742	2.8363	1.7569	1.0845	(-1) 6.9609	(-1) 4.7817	(-1) 3.8476	(-1) 3.1966	(-1) 2.5830	(-1) 2.2361	(-1) 1.6615	(-1) 1.1522	6
7	(+1) 2.5977	(+1) 1.2932	7.7540	6.1554	4.4667	3.4445	2.5947	1.6712	1.0664	(-1) 7.0166	(-1) 4.9027	(-1) 3.9777	(-1) 3.3251	(-1) 2.7037	(-1) 2.3491	(-1) 1.7569	(-1) 1.2258	7
8	(+1) 1.9547	(+1) 1.0480	6.6082	5.3591	3.9995	3.1503	2.4246	1.6088	1.0531	(-1) 7.0656	(-1) 5.0038	(-1) 4.0865	(-1) 3.4331	(-1) 2.8055	(-1) 2.4450	(-1) 1.8382	(-1) 1.2891	8
9	(+1) 1.5680	8.8976	5.8318	4.8080	3.6669	2.9365	2.2983	1.5611	1.0429	(-1) 7.1078	(-1) 5.0893	(-1) 4.1792	(-1) 3.5255	(-1) 2.8929	(-1) 2.5276	(-1) 1.9087	(-1) 1.3442	9
10	(+1) 1.3150	7.8037	5.2740	4.4054	3.4186	2.7740	2.2007	1.5235	1.0349	(-1) 7.1454	(-1) 5.1634	(-1) 4.2591	(-1) 3.6053	(-1) 2.9689	(-1) 2.5994	(-1) 1.9703	(-1) 1.3927	10
12	(+1) 1.0101	6.4048	4.5299	3.8584	3.0728	2.5436	2.0597	1.4678	1.0231	(-1) 7.2082	(-1) 5.2843	(-1) 4.3906	(-1) 3.7372	(-1) 3.0949	(-1) 2.7189	(-1) 2.0734	(-1) 1.4741	12
15	7.7582	5.2484	3.8826	3.3719	2.7559	2.3275	1.9243	1.4127	1.0114	(-1) 7.2801	(-1) 5.4203	(-1) 4.5386	(-1) 3.8864	(-1) 3.2383	(-1) 2.8555	(-1) 2.1921	(-1) 1.5687	15
20	5.9516	4.2900	3.3178	2.9377	2.4645	2.1242	1.7938	1.3580	1.0000	(-1) 7.3638	(-1) 5.5748	(-1) 4.7077	(-1) 4.0576	(-1) 3.4040	(-1) 3.0140	(-1) 2.3310	(-1) 1.6802	20
24	5.2084	3.8732	3.0624	2.7380	2.3273	2.0267	1.7302	1.3307	(-1) 9.9436	(-1) 7.4107	(-1) 5.6603	(-1) 4.8019	(-1) 4.1535	(-1) 3.4972	(-1) 3.1037	(-1) 2.4100	(-1) 1.7441	24
30	4.5540	3.4928	2.8230	2.5487	2.1952	1.9317	1.6673	1.3033	(-1) 9.8877	(-1) 7.4621	(-1) 5.7531	(-1) 4.9041	(-1) 4.2579	(-1) 3.5991	(-1) 3.2016	(-1) 2.4969	(-1) 1.8147	30
40	3.9772	3.1450	2.5984	2.3689	2.0677	1.8389	1.6052	1.2758	(-1) 9.8323	(-1) 7.5182	(-1) 5.8538	(-1) 5.0155	(-1) 4.3720	(-1) 3.7110	(-1) 3.3096	(-1) 2.5931	(-1) 1.8933	40
60	3.4681	2.8266	2.3872	2.1978	1.9445	1.7480	1.5435	1.2481	(-1) 9.7773	(-1) 7.5798	(-1) 5.9637	(-1) 5.1377	(-1) 4.4976	(-1) 3.8348	(-1) 3.4295	(-1) 2.7005	(-1) 1.9817	60
120	3.0180	2.5344	2.1881	2.0346	1.8249	1.6587	1.4821	1.2200	(-1) 9.7228	(-1) 7.6488	(-1) 6.0853	(-1) 5.2734	(-1) 4.6378	(-1) 3.9733	(-1) 3.5640	(-1) 2.8218	(-1) 2.0820	120
∞	2.6193	2.2657	1.9998	1.8783	1.7085	1.5705	1.4206	1.1914	(-1) 9.6687	(-1) 7.7262	(-1) 6.2212	(-1) 5.4253	(-1) 4.7955	(-1) 4.1302	(-1) 3.7169	(-1) 2.9605	(-1) 2.1976	∞

$\nu_1 = 24$

ν_2	p=0.0001	p=0.001	p=0.005	p=0.01	p=0.025	p=0.05	p=0.1	p=0.25	p=0.5	p=0.75	p=0.9	p=0.95	p=0.975	p=0.99	p=0.995	p=0.999	p=0.9999	ν_2
1	(+7) 6.2350	(+5) 6.2350	(+4) 2.4940	(+3) 6.2346	(+2) 9.9725	(+2) 2.4905	(+1) 6.2002	9.6255	2.1321	(-1) 7.1953	(-1) 3.4164	(-1) 2.3476	(-1) 1.7493	(-1) 1.2783	(-1) 1.0470	(-2) 7.1286	(-2) 4.6161	1
1.2	(6) 3.3044	(4) 7.1192	(3) 4.8691	(3) 1.5334	(2) 3.3267	(2) 1.0452	(1) 3.2649	6.7759	1.8444	(-1) 7.0062	(-1) 3.5378	(-1) 2.4916	(-1) 1.8872	(-1) 1.3996	(-1) 1.1556	(-2) 7.9704	(-2) 5.2180	1.2
1.5	(5) 1.7897	(3) 8.3065	(2) 9.7113	(2) 3.8512	(2) 1.1318	(1) 4.4641	(1) 1.7438	4.8075	1.6034	(-1) 6.8707	(-1) 3.7039	(-1) 2.6799	(-1) 2.0673	(-1) 1.5592	(-1) 1.2994	(-2) 9.0987	(-2) 6.0349	1.5
2	(+3) 9.9995	(+2) 9.9946	(+2) 1.9946	(+1) 9.9458	(+1) 3.9456	(+1) 1.9454	9.4496	3.4345	1.4014	(-1) 6.8050	(-1) 3.9396	(-1) 2.9388	(-1) 2.3155	(-1) 1.7814	(-1) 1.5013	(-1) 1.0707	(-2) 7.2185	2
3	(+2) 5.8700	(+2) 1.2593	(+1) 4.2622	(+1) 2.6598	(+1) 1.4124	8.6385	5.1764	2.4626	1.2322	(-1) 6.8423	(-1) 4.2966	(-1) 3.3236	(-1) 2.6874	(-1) 2.1195	(-1) 1.8119	(-1) 1.3237	(-2) 9.1208	3
4	(+2) 1.4642	(+1) 4.5766	(+1) 2.0030	(+1) 1.3929	8.5109	5.7744	3.8310	2.0827	1.1583	(-1) 6.9219	(-1) 4.5560	(-1) 3.6019	(-1) 2.9591	(-1) 2.3706	(-1) 2.0451	(-1) 1.5176	(-1) 1.0611	4
5	(+1) 6.4455	(+1) 2.5133	(+1) 1.2780	9.4665	6.2780	4.5272	3.1905	1.8802	1.1170	(-1) 7.0004	(-1) 4.7551	(-1) 3.8158	(-1) 3.1698	(-1) 2.5673	(-1) 2.2293	(-1) 1.6732	(-1) 1.1823	5
6	(+1) 3.7512	(+1) 1.6897	9.4741	7.3127	5.1172	3.8415	2.8183	1.7540	1.0907	(-1) 7.0706	(-1) 4.9138	(-1) 3.9869	(-1) 3.3393	(-1) 2.7272	(-1) 2.3799	(-1) 1.8017	(-1) 1.2838	6
7	(+1) 2.5550	(+1) 1.2732	7.6450	6.0743	4.4150	3.4105	2.5753	1.6675	1.0724	(-1) 7.1317	(-1) 5.0439	(-1) 4.1278	(-1) 3.4797	(-1) 2.8605	(-1) 2.5060	(-1) 1.9103	(-1) 1.3702	7
8	(+1) 1.9177	(+1) 1.0295	6.5029	5.2793	3.9472	3.1152	2.4041	1.6043	1.0591	(-1) 7.1849	(-1) 5.1528	(-1) 4.2461	(-1) 3.5983	(-1) 2.9736	(-1) 2.6134	(-1) 2.0035	(-1) 1.4451	8
9	(+1) 1.5349	8.7239	5.7292	4.7290	3.6142	2.9005	2.2768	1.5560	1.0489	(-1) 7.2317	(-1) 5.2458	(-1) 4.3474	(-1) 3.7000	(-1) 3.0713	(-1) 2.7064	(-1) 2.0847	(-1) 1.5106	9
10	(+1) 1.2845	7.6376	5.1732	4.3269	3.3654	2.7372	2.1784	1.5179	1.0408	(-1) 7.2727	(-1) 5.3262	(-1) 4.4352	(-1) 3.7885	(-1) 3.1565	(-1) 2.7878	(-1) 2.1561	(-1) 1.5686	10
12	9.8314	6.2488	4.4315	3.7805	3.0187	2.5055	2.0360	1.4613	1.0289	(-1) 7.3416	(-1) 5.4588	(-1) 4.5800	(-1) 3.9351	(-1) 3.2986	(-1) 2.9241	(-1) 2.2764	(-1) 1.6669	12
15	7.5168	5.1009	3.7859	3.2940	2.7006	2.2878	1.8990	1.4052	1.0172	(-1) 7.4217	(-1) 5.6082	(-1) 4.7445	(-1) 4.1027	(-1) 3.4618	(-1) 3.0811	(-1) 2.4162	(-1) 1.7821	15
20	5.7336	4.1493	3.2220	2.8594	2.4076	2.0825	1.7667	1.3494	1.0057	(-1) 7.5148	(-1) 5.7797	(-1) 4.9341	(-1) 4.2968	(-1) 3.6523	(-1) 3.2654	(-1) 2.5818	(-1) 1.9200	20
24	5.0002	3.7354	2.9667	2.6591	2.2693	1.9838	1.7019	1.3214	1.0000	(-1) 7.5677	(-1) 5.8758	(-1) 5.0408	(-1) 4.4066	(-1) 3.7607	(-1) 3.3707	(-1) 2.6771	(-1) 1.9999	24
30	4.3545	3.3572	2.7272	2.4689	2.1359	1.8874	1.6377	1.2933	(-1) 9.9438	(-1) 7.6260	(-1) 5.9805	(-1) 5.1573	(-1) 4.5269	(-1) 3.8800	(-1) 3.4869	(-1) 2.7828	(-1) 2.0891	30
40	3.7852	3.0111	2.5020	2.2880	2.0069	1.7929	1.5741	1.2649	(-1) 9.8880	(-1) 7.6899	(-1) 6.0950	(-1) 5.2854	(-1) 4.6598	(-1) 4.0124	(-1) 3.6161	(-1) 2.9012	(-1) 2.1897	40
60	3.2825	2.6938	2.2898	2.1154	1.8817	1.7001	1.5107	1.2361	(-1) 9.8328	(-1) 7.7610	(-1) 6.2216	(-1) 5.4277	(-1) 4.8079	(-1) 4.1606	(-1) 3.7615	(-1) 3.0352	(-1) 2.3043	60
120	2.8373	2.4019	2.0890	1.9500	1.7597	1.6084	1.4472	1.2068	(-1) 9.7780	(-1) 7.8407	(-1) 6.3633	(-1) 5.5875	(-1) 4.9754	(-1) 4.3292	(-1) 3.9273	(-1) 3.1890	(-1) 2.4368	120
∞	2.4422	2.1324	1.8983	1.7908	1.6402	1.5173	1.3832	1.1767	(-1) 9.7236	(-1) 7.9321	(-1) 6.5244	(-1) 5.7700	(-1) 5.1672	(-1) 4.5235	(-1) 4.1193	(-1) 3.3687	(-1) 2.5929	∞

$\nu_1 = 30$

ν_2	p = 0.0001	p = 0.001	p = 0.005	p = 0.01	p = 0.025	p = 0.05	p = 0.1	p = 0.25	p = 0.5
1	(+7) 6.2610	(+5) 6.2610	(+4) 2.5044	(+3) 6.2607	(+3) 1.0014	(+2) 2.5009	(+1) 6.2265	9.6698	2.1452
1.2	(6) 3.3154	(4) 7.1428	(3) 4.8852	(3) 1.5385	(2) 3.3378	(2) 1.0487	(1) 3.2763	6.8034	1.8555
1.5	(5) 1.7934	(3) 8.3236	(2) 9.7314	(2) 3.8592	(2) 1.1342	(1) 4.4739	(1) 1.7480	4.8237	1.6129
2	(+3) 9.9995	(+2) 9.9947	(+2) 1.9947	(+1) 9.9466	(+1) 3.9465	(+1) 1.9462	9.4579	3.4428	1.4096
3	(+2) 5.8469	(+2) 1.2545	(+1) 4.2466	(+1) 2.6505	(+1) 1.4081	8.6166	5.1681	2.4650	1.2393
4	(+2) 1.4530	(+1) 4.5429	(+1) 1.9892	(+1) 1.3838	8.4613	5.7459	3.8174	2.0825	1.1649
5	(+1) 6.3746	(+1) 2.4869	(+1) 1.2656	9.3793	6.2269	4.4957	3.1741	1.8784	1.1234
6	(+1) 3.6984	(+1) 1.6672	9.3583	7.2285	5.0652	3.8082	2.8000	1.7510	1.0969
7	(+1) 2.5118	(+1) 1.2530	7.5345	5.9921	4.3624	3.3758	2.5555	1.6635	1.0785
8	(+1) 1.8803	(+1) 1.0109	6.3961	5.1981	3.8940	3.0794	2.3830	1.5996	1.0651
9	(+1) 1.5013	8.5476	5.6248	4.6486	3.5604	2.8637	2.2547	1.5506	1.0548
10	(+1) 1.2536	7.4688	5.0705	4.2469	3.3110	2.6996	2.1554	1.5119	1.0467
12	9.5570	6.0898	4.3309	3.7008	2.9633	2.4663	2.0115	1.4544	1.0347
15	7.2707	4.9502	3.6867	3.2141	2.6437	2.2468	1.8728	1.3973	1.0229
20	5.5105	4.0050	3.1234	2.7785	2.3486	2.0391	1.7382	1.3401	1.0114
24	4.7867	3.5935	2.8679	2.5773	2.2090	1.9390	1.6721	1.3113	1.0057
30	4.1492	3.2171	2.6278	2.3860	2.0739	1.8409	1.6065	1.2823	1.0000
40	3.5868	2.8721	2.4015	2.2034	1.9429	1.7444	1.5411	1.2529	(-1) 9.9440
60	3.0894	2.5549	2.1874	2.0285	1.8152	1.6491	1.4755	1.2229	(-1) 9.8884
120	2.6480	2.2621	1.9839	1.8600	1.6899	1.5543	1.4094	1.1921	(-1) 9.8333
∞	2.2544	1.9901	1.7891	1.6964	1.5660	1.4591	1.3419	1.1600	(-1) 9.7787

p = 0.75	p = 0.9	p = 0.95	p = 0.975	p = 0.99	p = 0.995	p = 0.999	p = 0.9999	ν_2
(-1) 7.2669	(-1) 3.4714	(-1) 2.3976	(-1) 1.7961	(-1) 1.3223	(-1) 1.0894	(-2) 7.5228	(-2) 4.9771	1
(-1) 7.0790	(-1) 3.5981	(-1) 2.5476	(-1) 1.9405	(-1) 1.4502	(-1) 1.2046	(-2) 8.4290	(-2) 5.6401	1.2
(-1) 6.9461	(-1) 3.7715	(-1) 2.7444	(-1) 2.1295	(-1) 1.6190	(-1) 1.3577	(-2) 9.6487	(-2) 6.5442	1.5
(-1) 6.8852	(-1) 4.0182	(-1) 3.0159	(-1) 2.3911	(-1) 1.8551	(-1) 1.5736	(-1) 1.1398	(-2) 7.8631	2
(-1) 6.9319	(-1) 4.3935	(-1) 3.4220	(-1) 2.7860	(-1) 2.2174	(-1) 1.9088	(-1) 1.4175	(-1) 1.0006	3
(-1) 7.0205	(-1) 4.6681	(-1) 3.7180	(-1) 3.0770	(-1) 2.4889	(-1) 2.1630	(-1) 1.6328	(-1) 1.1705	4
(-1) 7.1058	(-1) 4.8800	(-1) 3.9470	(-1) 3.3041	(-1) 2.7034	(-1) 2.3654	(-1) 1.8070	(-1) 1.3102	5
(-1) 7.1824	(-1) 5.0497	(-1) 4.1314	(-1) 3.4883	(-1) 2.8789	(-1) 2.5322	(-1) 1.9523	(-1) 1.4282	6
(-1) 7.2490	(-1) 5.1897	(-1) 4.2839	(-1) 3.6417	(-1) 3.0262	(-1) 2.6727	(-1) 2.0759	(-1) 1.5296	7
(-1) 7.3073	(-1) 5.3076	(-1) 4.4127	(-1) 3.7717	(-1) 3.1520	(-1) 2.7932	(-1) 2.1827	(-1) 1.6181	8
(-1) 7.3584	(-1) 5.4083	(-1) 4.5235	(-1) 3.8841	(-1) 3.2610	(-1) 2.8981	(-1) 2.2763	(-1) 1.6961	9
(-1) 7.4036	(-1) 5.4960	(-1) 4.6198	(-1) 3.9822	(-1) 3.3567	(-1) 2.9904	(-1) 2.3592	(-1) 1.7655	10
(-1) 7.4800	(-1) 5.6411	(-1) 4.7799	(-1) 4.1459	(-1) 3.5173	(-1) 3.1459	(-1) 2.4996	(-1) 1.8841	12
(-1) 7.5683	(-1) 5.8062	(-1) 4.9633	(-1) 4.3343	(-1) 3.7034	(-1) 3.3270	(-1) 2.6647	(-1) 2.0249	15
(-1) 7.6728	(-1) 5.9977	(-1) 5.1768	(-1) 4.5554	(-1) 3.9236	(-1) 3.5423	(-1) 2.8630	(-1) 2.1959	20
(-1) 7.7322	(-1) 6.1061	(-1) 5.2983	(-1) 4.6819	(-1) 4.0504	(-1) 3.6668	(-1) 2.9787	(-1) 2.2965	24
(-1) 7.7935	(-1) 6.2247	(-1) 5.4321	(-1) 4.8218	(-1) 4.1911	(-1) 3.8055	(-1) 3.1084	(-1) 2.4101	30
(-1) 7.8722	(-1) 6.3565	(-1) 5.5810	(-1) 4.9778	(-1) 4.3493	(-1) 3.9618	(-1) 3.2556	(-1) 2.5400	40
(-1) 7.9548	(-1) 6.5036	(-1) 5.7484	(-1) 5.1546	(-1) 4.5292	(-1) 4.1406	(-1) 3.4251	(-1) 2.6908	60
(-1) 8.0489	(-1) 6.6716	(-1) 5.9400	(-1) 5.3579	(-1) 4.7378	(-1) 4.3484	(-1) 3.6238	(-1) 2.8693	120
(-1) 8.1593	(-1) 6.8662	(-1) 6.1641	(-1) 5.5969	(-1) 4.9845	(-1) 4.5956	(-1) 3.8627	(-1) 3.0860	∞

$\nu_1 = 40$

ν_2	p = 0.0001	p = 0.001	p = 0.005	p = 0.01	p = 0.025	p = 0.05	p = 0.1	p = 0.25	p = 0.5	p = 0.75	p = 0.9	p = 0.95	p = 0.975	p = 0.99	p = 0.995	p = 0.999	p = 0.9999	ν_2
1	(+7) 6.2871	(+5) 6.2871	(+4) 2.5148	(+3) 6.2868	(+3) 1.0056	(+2) 2.5114	(+1) 6.2529	9.7144	2.1584	(-1) 7.3389	(-1) 3.5268	(-1) 2.4481	(-1) 1.8437	(-1) 1.3672	(-1) 1.1328	(-2) 7.9306	(-2) 5.3566	1
1.2	(6) 3.3264	(4) 7.1666	(3) 4.9015	(3) 1.5436	(2) 3.3490	(2) 1.0522	(1) 3.2877	6.8310	1.8677	(-1) 7.1522	(-1) 3.6589	(-1) 2.6044	(-1) 1.9946	(-1) 1.5019	(-1) 1.2548	(-2) 8.9046	(-2) 6.0850	1.2
1.5	(5) 1.7971	(3) 8.3408	(2) 9.7515	(2) 3.8672	(2) 1.1366	(1) 4.4837	(1) 1.7523	4.8399	1.6225	(-1) 7.0221	(-1) 3.8399	(-1) 2.8099	(-1) 2.1929	(-1) 1.6803	(-1) 1.4176	(-1) 1.0221	(-2) 7.0830	1.5
2	(+3) 9.9995	(+2) 9.9948	(+2) 1.9947	(+1) 9.9474	(+1) 3.9473	(+1) 1.9471	9.4663	3.4511	1.4178	(-1) 6.9662	(-1) 4.0977	(-1) 3.0943	(-1) 2.4685	(-1) 1.9311	(-1) 1.6484	(-1) 1.2120	(-2) 8.5485	2
3	(+2) 5.8236	(+2) 1.2496	(+1) 4.2308	(+1) 2.6411	(+1) 1.4037	8.5944	5.1597	2.4674	1.2464	(-1) 7.0230	(-1) 4.4922	(-1) 3.5227	(-1) 2.8874	(-1) 2.3188	(-1) 2.0097	(-1) 1.5164	(-1) 1.0956	3
4	(+2) 1.4418	(+1) 4.5089	(+1) 1.9752	(+1) 1.3745	8.4111	5.7170	3.8036	2.0821	1.1716	(-1) 7.1200	(-1) 4.7826	(-1) 3.8373	(-1) 3.1989	(-1) 2.6121	(-1) 2.2863	(-1) 1.7550	(-1) 1.2888	4
5	(+1) 6.3031	(+1) 2.4602	(+1) 1.2530	9.2912	6.1751	4.4638	3.1573	1.8763	1.1297	(-1) 7.2134	(-1) 5.0080	(-1) 4.0825	(-1) 3.4439	(-1) 2.8459	(-1) 2.5088	(-1) 1.9500	(-1) 1.4495	5
6	(+1) 3.6450	(+1) 1.6445	9.2408	7.1432	5.0125	3.7743	2.7812	1.7477	1.1031	(-1) 7.2961	(-1) 5.1897	(-1) 4.2810	(-1) 3.6438	(-1) 3.0386	(-1) 2.6933	(-1) 2.1139	(-1) 1.5865	6
7	(+1) 2.4680	(+1) 1.2326	7.4225	5.9084	4.3089	3.3404	2.5351	1.6593	1.0846	(-1) 7.3687	(-1) 5.3405	(-1) 4.4464	(-1) 3.8113	(-1) 3.2012	(-1) 2.8500	(-1) 2.2545	(-1) 1.7053	7
8	(+1) 1.8425	9.9194	6.2875	5.1156	3.8398	3.0428	2.3614	1.5945	1.0711	(-1) 7.4322	(-1) 5.4678	(-1) 4.5867	(-1) 3.9543	(-1) 3.3411	(-1) 2.9853	(-1) 2.3770	(-1) 1.8097	8
9	(+1) 1.4672	8.3685	5.5186	4.5667	3.5055	2.8259	2.2320	1.5450	1.0608	(-1) 7.4884	(-1) 5.5776	(-1) 4.7081	(-1) 4.0785	(-1) 3.4631	(-1) 3.1037	(-1) 2.4849	(-1) 1.9025	9
10	(+1) 1.2222	7.2971	4.9659	4.1653	3.2554	2.6609	2.1317	1.5056	1.0526	(-1) 7.5381	(-1) 5.6731	(-1) 4.8142	(-1) 4.1873	(-1) 3.5708	(-1) 3.2085	(-1) 2.5811	(-1) 1.9856	10
12	9.2778	5.9278	4.2282	3.6192	2.9063	2.4259	1.9861	1.4471	1.0405	(-1) 7.6225	(-1) 5.8323	(-1) 4.9913	(-1) 4.3702	(-1) 3.7526	(-1) 3.3863	(-1) 2.7454	(-1) 2.1289	12
15	7.0197	4.7959	3.5850	3.1319	2.5850	2.2043	1.8454	1.3888	1.0287	(-1) 7.7208	(-1) 6.0154	(-1) 5.1962	(-1) 4.5832	(-1) 3.9657	(-1) 3.5957	(-1) 2.9409	(-1) 2.3012	15
20	5.2817	3.8564	3.0215	2.6947	2.2873	1.9938	1.7083	1.3301	1.0171	(-1) 7.8382	(-1) 6.2298	(-1) 5.4380	(-1) 4.8363	(-1) 4.2214	(-1) 3.8485	(-1) 3.1797	(-1) 2.5143	20
24	4.5669	3.4468	2.7654	2.4923	2.1460	1.8920	1.6407	1.3004	1.0113	(-1) 7.9058	(-1) 6.3528	(-1) 5.5776	(-1) 4.9828	(-1) 4.3706	(-1) 3.9968	(-1) 3.3210	(-1) 2.6418	24
30	3.9370	3.0716	2.5241	2.2992	2.0089	1.7918	1.5732	1.2703	1.0056	(-1) 7.9815	(-1) 6.4889	(-1) 5.7326	(-1) 5.1469	(-1) 4.5384	(-1) 4.1641	(-1) 3.4818	(-1) 2.7880	30
40	3.3804	2.7268	2.2958	2.1142	1.8752	1.6928	1.5056	1.2397	1.0000	(-1) 8.0665	(-1) 6.6419	(-1) 5.9074	(-1) 5.3328	(-1) 4.7299	(-1) 4.3558	(-1) 3.6673	(-1) 2.9582	40
60	2.8870	2.4086	2.0789	1.9360	1.7440	1.5943	1.4373	1.2081	(-1) 9.9441	(-1) 8.1639	(-1) 6.8157	(-1) 6.1076	(-1) 5.5469	(-1) 4.9520	(-1) 4.5792	(-1) 3.8855	(-1) 3.1604	60
120	2.4471	2.1128	1.8709	1.7628	1.6141	1.4952	1.3676	1.1752	(-1) 9.8887	(-1) 8.2781	(-1) 7.0185	(-1) 6.3428	(-1) 5.7998	(-1) 5.2159	(-1) 4.8461	(-1) 4.1489	(-1) 3.4073	120
∞	2.0516	1.8350	1.6691	1.5923	1.4835	1.3940	1.2951	1.1404	(-1) 9.8339	(-1) 8.4154	(-1) 7.2627	(-1) 6.6273	(-1) 6.1084	(-1) 5.5411	(-1) 5.1765	(-1) 4.4791	(-1) 3.7208	∞

$\nu_1 = 60$

ν_2	p = 0.0001	p = 0.001	p = 0.005	p = 0.01	p = 0.025	p = 0.05	p = 0.1	p = 0.25	p = 0.5	p = 0.75	p = 0.9	p = 0.95	p = 0.975	p = 0.99	p = 0.995	p = 0.999	p = 0.9999	ν_2
1	(+7) 6.3134	(+5) 6.3134	(+4) 2.5253	(+3) 6.3130	(+3) 1.0098	(+2) 2.5220	(+1) 6.2794	9.7591	2.1716	(-1) 7.4113	(-1) 3.5824	(-1) 2.4993	(-1) 1.8919	(-1) 1.4130	(-1) 1.1772	(-2) 8.3521	(-2) 5.7548	1
1.2	(6) 3.3375	(4) 7.1904	(3) 4.9178	(3) 1.5488	(2) 3.3602	(2) 1.0558	(1) 3.2991	6.8588	1.8779	(-1) 7.2258	(-1) 3.7204	(-1) 2.6619	(-1) 2.0497	(-1) 1.5547	(-1) 1.3064	(-2) 9.3973	(-2) 6.5529	1.2
1.5	(5) 1.8008	(3) 8.3581	(2) 9.7718	(2) 3.8753	(2) 1.1390	(1) 4.4937	(1) 1.7565	4.8562	1.6321	(-1) 7.0985	(-1) 3.9090	(-1) 2.8763	(-1) 2.2575	(-1) 1.7430	(-1) 1.4792	(-1) 1.0815	(-2) 7.6516	1.5
2	(+3) 9.9995	(+2) 9.9948	(+2) 1.9948	(+1) 9.9483	(+1) 3.9481	(+1) 1.9479	9.4746	3.4594	1.4261	(-1) 7.0432	(-1) 4.1785	(-1) 3.1742	(-1) 2.5476	(-1) 2.0091	(-1) 1.7256	(-1) 1.2874	(-2) 9.2758	2
3	(+2) 5.8002	(+2) 1.2447	(+1) 4.2149	(+1) 2.6316	(+1) 1.3992	8.5720	5.1512	2.4697	1.2536	(-1) 7.1149	(-1) 4.5926	(-1) 3.6257	(-1) 2.9918	(-1) 2.4237	(-1) 2.1146	(-1) 1.6204	(-1) 1.1972	3
4	(+2) 1.4305	(+1) 4.4746	(+1) 1.9611	(+1) 1.3652	8.3604	5.6878	3.7896	2.0817	1.1782	(-1) 7.2213	(-1) 4.8996	(-1) 3.9601	(-1) 3.3248	(-1) 2.7404	(-1) 2.4155	(-1) 1.8844	(-1) 1.4165	4
5	(+1) 6.2309	(+1) 2.4333	(+1) 1.2402	9.2020	6.1225	4.4314	3.1402	1.8742	1.1361	(-1) 7.3223	(-1) 5.1395	(-1) 4.2224	(-1) 3.5890	(-1) 2.9950	(-1) 2.6596	(-1) 2.1024	(-1) 1.6009	5
6	(+1) 3.5910	(+1) 1.6214	9.1219	7.0568	4.9589	3.7398	2.7620	1.7443	1.1093	(-1) 7.4123	(-1) 5.3342	(-1) 4.4366	(-1) 3.8060	(-1) 3.2065	(-1) 2.8639	(-1) 2.2873	(-1) 1.7596	6
7	(+1) 2.4238	(+1) 1.2119	7.3088	5.8236	4.2544	3.3043	2.5142	1.6548	1.0908	(-1) 7.4912	(-1) 5.4963	(-1) 4.6157	(-1) 3.9891	(-1) 3.3864	(-1) 3.0385	(-1) 2.4471	(-1) 1.8985	7
8	(+1) 1.8041	9.7272	6.1772	5.0316	3.7844	3.0053	2.3391	1.5892	1.0771	(-1) 7.5609	(-1) 5.6344	(-1) 4.7687	(-1) 4.1465	(-1) 3.5420	(-1) 3.1904	(-1) 2.5874	(-1) 2.0216	8
9	(+1) 1.4327	8.1865	5.4104	4.4831	3.4493	2.7872	2.2085	1.5389	1.0667	(-1) 7.6225	(-1) 5.7537	(-1) 4.9017	(-1) 4.2838	(-1) 3.6785	(-1) 3.3241	(-1) 2.7120	(-1) 2.1319	9
10	(+1) 1.1904	7.1224	4.8592	4.0819	3.1984	2.6211	2.1072	1.4990	1.0585	(-1) 7.6770	(-1) 5.8582	(-1) 5.0186	(-1) 4.4049	(-1) 3.7997	(-1) 3.4433	(-1) 2.8237	(-1) 2.2314	10
12	8.9933	5.7623	4.1229	3.5355	2.8478	2.3842	1.9597	1.4393	1.0464	(-1) 7.7700	(-1) 6.0335	(-1) 5.2154	(-1) 4.6100	(-1) 4.0062	(-1) 3.6471	(-1) 3.0163	(-1) 2.4047	12
15	6.7628	4.6377	3.4803	3.0471	2.5242	2.1601	1.8168	1.3796	1.0345	(-1) 7.8796	(-1) 6.2367	(-1) 5.4454	(-1) 4.8513	(-1) 4.2512	(-1) 3.8903	(-1) 3.2488	(-1) 2.6164	15
20	5.0463	3.7030	2.9159	2.6077	2.2234	1.9464	1.6768	1.3193	1.0228	(-1) 8.0122	(-1) 6.4788	(-1) 5.7208	(-1) 5.1427	(-1) 4.5500	(-1) 4.1890	(-1) 3.5378	(-1) 2.8834	20
24	4.3397	3.2946	2.6585	2.4035	2.0799	1.8424	1.6073	1.2885	1.0170	(-1) 8.0900	(-1) 6.6194	(-1) 5.8820	(-1) 5.3143	(-1) 4.7272	(-1) 4.3672	(-1) 3.7122	(-1) 3.0465	24
30	3.7163	2.9196	2.4151	2.2079	1.9400	1.7396	1.5376	1.2571	1.0113	(-1) 8.1773	(-1) 6.7774	(-1) 6.0639	(-1) 5.5090	(-1) 4.9298	(-1) 4.5716	(-1) 3.9140	(-1) 3.2369	30
40	3.1642	2.5737	2.1838	2.0194	1.8028	1.6373	1.4672	1.2249	1.0056	(-1) 8.2775	(-1) 6.9575	(-1) 6.2723	(-1) 5.7339	(-1) 5.1653	(-1) 4.8102	(-1) 4.1518	(-1) 3.4638	40
60	2.6723	2.2523	1.9622	1.8363	1.6668	1.5343	1.3952	1.1912	1.0000	(-1) 8.3949	(-1) 7.1674	(-1) 6.5176	(-1) 5.9995	(-1) 5.4457	(-1) 5.0963	(-1) 4.4400	(-1) 3.7420	60
120	2.2301	1.9502	1.7469	1.6557	1.5299	1.4290	1.3203	1.1555	(-1) 9.9443	(-1) 8.5361	(-1) 7.4206	(-1) 6.8152	(-1) 6.3251	(-1) 5.7927	(-1) 5.4523	(-1) 4.8028	(-1) 4.0975	120
∞	1.8250	1.6601	1.5325	1.4730	1.3883	1.3180	1.2400	1.1164	(-1) 9.8891	(-1) 8.7154	(-1) 7.7429	(-1) 7.1979	(-1) 6.7467	(-1) 6.2477	(-1) 5.9224	(-1) 5.2897	(-1) 4.5828	∞

$\nu_1 = 120$

ν_2	p = 0.0001	p = 0.001	p = 0.005	p = 0.01	p = 0.025	p = 0.05	p = 0.1	p = 0.25	p = 0.5	p = 0.75	p = 0.9	p = 0.95	p = 0.975	p = 0.99	p = 0.995	p = 0.999	p = 0.9999	ν_2
1	(+7) 6.3397	(+5) 6.3397	(+4) 2.5359	(+3) 6.3394	(+3) 1.0140	(+2) 2.5325	(+1) 6.3061	9.8041	2.1848	(-1) 7.4839	(-1) 3.6393	(-1) 2.5510	(-1) 1.9408	(-1) 1.4596	(-1) 1.2226	(-2) 8.7873	(-2) 6.1714	1
1.2	(6) 3.3486	(4) 7.2144	(3) 4.9342	(3) 1.5539	(2) 3.3714	(2) 1.0593	(1) 3.3106	6.8867	1.8892	(-1) 7.2998	(-1) 3.7824	(-1) 2.7201	(-1) 2.1056	(-1) 1.6086	(-1) 1.3592	(-2) 9.9068	(-2) 7.0438	1.2
1.5	(5) 1.8045	(3) 8.3755	(2) 9.7922	(2) 3.8834	(2) 1.1415	(1) 4.5036	(1) 1.7608	4.8725	1.6417	(-1) 7.1754	(-1) 3.9788	(-1) 2.9436	(-1) 2.3232	(-1) 1.8072	(-1) 1.5425	(-1) 1.1432	(-2) 8.2502	1.5
2	(+3) 9.9995	(+2) 9.9949	(+2) 1.9949	(+1) 9.9491	(+1) 3.9490	(+1) 1.9487	9.4829	3.4677	1.4344	(-1) 7.1306	(-1) 4.2602	(-1) 3.2554	(-1) 2.6284	(-1) 2.0892	(-1) 1.8053	(-1) 1.3659	(-1) 1.0045	2
3	(+2) 5.7766	(+2) 1.2397	(+1) 4.1989	(+1) 2.6221	(+1) 1.3947	8.5494	5.1425	2.4720	1.2608	(-1) 7.2082	(-1) 4.6948	(-1) 3.7311	(-1) 3.0989	(-1) 2.5321	(-1) 2.2236	(-1) 1.7297	(-1) 1.3058	3
4	(+2) 1.4190	(+1) 4.4400	(+1) 1.9468	(+1) 1.3558	8.3092	5.6581	3.7753	2.0812	1.1849	(-1) 7.3239	(-1) 5.0193	(-1) 4.0863	(-1) 3.4551	(-1) 2.8739	(-1) 2.5506	(-1) 2.0214	(-1) 1.5538	4
5	(+1) 6.1580	(+1) 2.4060	(+1) 1.2274	9.1118	6.0693	4.3984	3.1228	1.8719	1.1426	(-1) 7.4333	(-1) 5.2745	(-1) 4.3668	(-1) 3.7397	(-1) 3.1511	(-1) 2.8183	(-1) 2.2647	(-1) 1.7649	5
6	(+1) 3.5364	(+1) 1.5981	9.0015	6.9690	4.9045	3.7047	2.7423	1.7407	1.1156	(-1) 7.5313	(-1) 5.4831	(-1) 4.5977	(-1) 3.9755	(-1) 3.3831	(-1) 3.0442	(-1) 2.4730	(-1) 1.9484	6
7	(+1) 2.3790	(+1) 1.1909	7.1933	5.7372	4.1989	3.2674	2.4928	1.6502	1.0969	(-1) 7.6173	(-1) 5.6577	(-1) 4.7923	(-1) 4.1757	(-1) 3.5819	(-1) 3.2390	(-1) 2.6546	(-1) 2.1106	7
8	(+1) 1.7652	9.5321	6.0649	4.9460	3.7279	2.9669	2.3162	1.5836	1.0832	(-1) 7.6929	(-1) 5.8072	(-1) 4.9593	(-1) 4.3490	(-1) 3.7553	(-1) 3.4095	(-1) 2.8154	(-1) 2.2556	8
9	(+1) 1.3976	8.0014	5.3001	4.3978	3.3918	2.7475	2.1843	1.5325	1.0727	(-1) 7.7604	(-1) 5.9372	(-1) 5.1052	(-1) 4.5011	(-1) 3.9084	(-1) 3.5609	(-1) 2.9592	(-1) 2.3866	9
10	(+1) 1.1580	6.9443	4.7501	3.9965	3.1399	2.5801	2.0818	1.4919	1.0645	(-1) 7.8204	(-1) 6.0518	(-1) 5.2342	(-1) 4.6361	(-1) 4.0451	(-1) 3.6966	(-1) 3.0891	(-1) 2.5058	10
12	8.7031	5.5931	4.0149	3.4494	2.7874	2.3410	1.9323	1.4310	1.0523	(-1) 7.9233	(-1) 6.2453	(-1) 5.4535	(-1) 4.8667	(-1) 4.2803	(-1) 3.9310	(-1) 3.3155	(-1) 2.7157	12
15	6.4995	4.4750	3.3722	2.9595	2.4611	2.1141	1.7867	1.3698	1.0403	(-1) 8.0463	(-1) 6.4725	(-1) 5.7127	(-1) 5.1414	(-1) 4.5631	(-1) 4.2146	(-1) 3.5929	(-1) 2.9762	15
20	4.8031	3.5438	2.8058	2.5168	2.1562	1.8963	1.6433	1.3074	1.0285	(-1) 8.1967	(-1) 6.7472	(-1) 6.0288	(-1) 5.4798	(-1) 4.9150	(-1) 4.5702	(-1) 3.9457	(-1) 3.3135	20
24	4.1037	3.1357	2.5463	2.3099	2.0099	1.7897	1.5715	1.2754	1.0227	(-1) 8.2864	(-1) 6.9099	(-1) 6.2174	(-1) 5.6828	(-1) 5.1282	(-1) 4.7870	(-1) 4.1634	(-1) 3.5244	24
30	3.4852	2.7595	2.2997	2.1107	1.8664	1.6835	1.4989	1.2424	1.0170	(-1) 8.3886	(-1) 7.0952	(-1) 6.4338	(-1) 5.9175	(-1) 5.3763	(-1) 5.0406	(-1) 4.4206	(-1) 3.7764	30
40	2.9349	2.4103	2.0635	1.9172	1.7242	1.5766	1.4248	1.2080	1.0113	(-1) 8.5092	(-1) 7.3121	(-1) 6.6881	(-1) 6.1954	(-1) 5.6728	(-1) 5.3450	(-1) 4.7330	(-1) 4.0864	40
60	2.4405	2.0821	1.8341	1.7263	1.5810	1.4673	1.3476	1.1715	1.0056	(-1) 8.6543	(-1) 7.5740	(-1) 6.9979	(-1) 6.5364	(-1) 6.0397	(-1) 5.7244	(-1) 5.1277	(-1) 4.4842	60
120	1.9877	1.7667	1.6055	1.5330	1.4327	1.3519	1.2646	1.1314	1.0000	(-1) 8.8386	(-1) 7.9076	(-1) 7.3970	(-1) 6.9798	(-1) 6.5232	(-1) 6.2286	(-1) 5.6601	(-1) 5.0309	120
∞	1.5527	1.4468	1.3637	1.3246	1.2684	1.2214	1.1686	1.0838	(-1) 9.9445	(-1) 9.1017	(-1) 8.3850	(-1) 7.9751	(-1) 7.6313	(-1) 7.2438	(-1) 6.9876	(-1) 6.4796	(-1) 5.8940	∞

$\nu_1 = \infty$

ν_2	p = 0.0001	p = 0.001	p = 0.005	p = 0.01	p = 0.025	p = 0.05	p = 0.1	p = C.25	p = 0.5	p = 0.75	p = 0.9	p = 0.95	p = 0.975	p = 0.99	p = 0.995	p = 0.999	p = 0.9999	ν_2
1	(+7) 6.3662	(+5) 6.3662	(+4) 2.5465	(+3) 6.3660	(+3) 1.0183	(+2) 2.5432	(+1) 6.3328	9.8492	2.1981	(-1) 7.5569	(-1) 3.6962	(-1) 2.6031	(-1) 1.9905	(-1) 1.5072	(-1) 1.2691	(-2) 9.2357	(-2) 6.6065	1
1.2	(6) 3.3598	(4) 7.2384	(3) 4.9507	(3) 1.5591	(2) 3.3827	(2) 1.0629	(1) 3.3222	6.9147	1.9905	(-1) 7.3741	(-1) 3.8449	(-1) 2.7790	(-1) 2.1623	(-1) 1.6636	(-1) 1.4133	(-1) 1.0437	(-2) 7.6055	1.2
1.5	(5) 1.8083	(3) 8.3929	(2) 9.8126	(2) 3.8916	(2) 1.1439	(1) 4.5136	(1) 1.7651	4.8889	1.6514	(-1) 7.2527	(-1) 4.0494	(-1) 3.0119	(-1) 2.3901	(-1) 1.8729	(-1) 1.6076	(-1) 1.2076	(-2) 8.9520	1.5
2	(+3) 9.9995	(+2) 9.9950	(+2) 1.9951	(+1) 9.9501	(+1) 3.9498	(+1) 1.9496	9.4913	3.4761	1.4427	(-1) 7.2134	(-1) 4.3429	(-1) 3.3381	(-1) 2.7108	(-1) 2.1715	(-1) 1.8874	(-1) 1.4476	(-1) 1.0857	2
3	(+2) 5.7528	(+2) 1.2347	(+1) 4.1829	(+1) 2.6125	(+1) 1.3902	8.5265	5.1337	2.4742	1.2680	(-1) 7.3025	(-1) 4.7989	(-1) 3.8389	(-1) 3.2091	(-1) 2.6444	(-1) 2.3368	(-1) 1.8443	(-1) 1.4213	3
4	(+2) 1.4075	(+1) 4.4051	(+1) 1.9325	(+1) 1.3463	8.2573	5.6281	3.7607	2.0806	1.1916	(-1) 7.4278	(-1) 5.1417	(-1) 4.2160	(-1) 3.5896	(-1) 3.0128	(-1) 2.6917	(-1) 2.1660	(-1) 1.7012	4
5	(+1) 6.0844	(+1) 2.3785	(+1) 1.2144	9.0204	6.0153	4.3650	3.1050	1.8694	1.1490	(-1) 7.5466	(-1) 5.4133	(-1) 4.5165	(-1) 3.8964	(-1) 3.3142	(-1) 2.9852	(-1) 2.4372	(-1) 1.9421	5
6	(+1) 3.4812	(+1) 1.5745	8.8793	6.8801	4.8491	3.6688	2.7222	1.7368	1.1219	(-1) 7.6523	(-1) 5.6367	(-1) 4.7651	(-1) 4.1525	(-1) 3.5689	(-1) 3.2349	(-1) 2.6717	(-1) 2.1539	6
7	(+1) 2.3336	(+1) 1.1696	7.0760	5.6495	4.1423	3.2298	2.4708	1.6452	1.1031	(-1) 7.7459	(-1) 5.8251	(-1) 4.9761	(-1) 4.3716	(-1) 3.7889	(-1) 3.4521	(-1) 2.8781	(-1) 2.3429	7
8	(+1) 1.7257	9.3337	5.9505	4.8588	3.6702	2.9276	2.2926	1.5777	1.0893	(-1) 7.8284	(-1) 5.9873	(-1) 5.1589	(-1) 4.5625	(-1) 3.9820	(-1) 3.6438	(-1) 3.0623	(-1) 2.5135	8
9	(+1) 1.3620	7.8128	5.1875	4.3105	3.3329	2.7067	2.1592	1.5257	1.0788	(-1) 7.9026	(-1) 6.1293	(-1) 5.3194	(-1) 4.7313	(-1) 4.1540	(-1) 3.8153	(-1) 3.2284	(-1) 2.6690	9
10	(+1) 1.1250	6.7625	4.6385	3.9090	3.0798	2.5379	2.0554	1.4843	1.0705	(-1) 7.9688	(-1) 6.2551	(-1) 5.4624	(-1) 4.8821	(-1) 4.3087	(-1) 3.9701	(-1) 3.3797	(-1) 2.8118	10
12	8.4063	5.4195	3.9039	3.3608	2.7249	2.2962	1.9036	1.4221	1.0582	(-1) 8.0834	(-1) 6.4691	(-1) 5.7071	(-1) 5.1422	(-1) 4.5771	(-1) 4.2403	(-1) 3.6464	(-1) 3.0664	12
15	6.2287	4.3070	3.2602	2.8684	2.3953	2.0658	1.7551	1.3591	1.0461	(-1) 8.2217	(-1) 6.7245	(-1) 6.0010	(-1) 5.4567	(-1) 4.9056	(-1) 4.5729	(-1) 3.9791	(-1) 3.3888	15
20	4.5503	3.3778	2.6904	2.4212	2.0853	1.8432	1.6074	1.2943	1.0343	(-1) 8.3935	(-1) 7.0393	(-1) 6.3674	(-1) 5.8531	(-1) 5.3240	(-1) 5.0005	(-1) 4.4136	(-1) 3.8178	20
24	3.8566	2.9685	2.4276	2.2107	1.9353	1.7331	1.5327	1.2607	1.0284	(-1) 8.4983	(-1) 7.2296	(-1) 6.5907	(-1) 6.0968	(-1) 5.5841	(-1) 5.2679	(-1) 4.6895	(-1) 4.0947	24
30	3.2404	2.5889	2.1760	2.0062	1.7867	1.6223	1.4564	1.2256	1.0226	(-1) 8.6207	(-1) 7.4521	(-1) 6.8535	(-1) 6.3857	(-1) 5.8948	(-1) 5.5894	(-1) 5.0249	(-1) 4.4357	30
40	2.6876	2.2326	1.9318	1.8047	1.6371	1.5089	1.3769	1.1883	1.0169	(-1) 8.7689	(-1) 7.7214	(-1) 7.1736	(-1) 6.7408	(-1) 6.2802	(-1) 5.9913	(-1) 5.4494	(-1) 4.8743	40
60	2.1821	1.8905	1.6885	1.6006	1.4822	1.3893	1.2915	1.1474	1.0112	(-1) 8.9574	(-1) 8.0645	(-1) 7.5873	(-1) 7.2031	(-1) 6.7889	(-1) 6.5253	(-1) 6.0237	(-1) 5.4793	60
120	1.6966	1.5433	1.4311	1.3805	1.3104	1.2539	1.1926	1.0987	1.0056	(-1) 9.2268	(-1) 8.5572	(-1) 8.1873	(-1) 7.8839	(-1) 7.5494	(-1) 7.3330	(-1) 6.9117	(-1) 6.4403	120
∞	1.0000	1.0000	1.0000	1.0000	1.0000	1.0000	1.0000	1.0000	1.0000	1.0000	1.0000	1.0000	1.0000	1.0000	1 0000	1.0000	1.0000	∞

Table A.4 Noncentral Beta Table

Entries are $\beta_p(\nu_1, \nu_2, \phi)$, where $p = \int_0^{\beta_p(\nu_1,\nu_2,\phi)} g(\beta)\,d\beta$

$E^2_{.05}$ and the corresponding values of $P(\text{II})$ $f_1 = 1$

f_2	$E^2_{.05}$	φ = 1	1.5	2	2.5	3	4	5	6	7	8
2	.903	.862	.763	.643	.517	.395	.200	.083	.028	.008	.002
4	.658	.805	.631	.428	.247	.120	.016	.001			
6	.500	.777	.570	.343	.164	.061	.004				
7	.444	.768	.552	.319	.144	.050	.003				
8	.399	.761	.537	.302	.129	.041	.002				
9	.362	.756	.526	.288	.119	.036	.001				
10	.332	.751	.517	.278	.111	.032	.001				
11	.306	.747	.510	.269	.105	.029	.001				
12	.284	.744	.504	.262	.100	.027	.001				
13	.264	.741	.499	.256	.096	.025	.001				
14	.247	.739	.494	.251	.093	.024	.001				
15	.232	.737	.490	.247	.090	.023					
16	.219	.735	.487	.243	.087	.022					
17	.207	.734	.484	.240	.085	.021					
18	.197	.732	.481	.237	.084	.020					
19	.187	.731	.479	.235	.082	.020					
20	.179	.730	.477	.233	.081	.019					
21	.171	.729	.475	.231	.079	.019					
22	.164	.728	.473	.229	.078	.018					
23	.157	.727	.471	.227	.077	.018					
24	.151	.726	.470	.226	.076	.018					
25	.145	.725	.468	.224	.075	.017					
26	.140	.725	.467	.223	.075	.017					
27	.135	.724	.466	.222	.074	.017					
28	.130	.723	.465	.221	.073	.017					
29	.126	.723	.464	.220	.073	.017					
30	.122	.722	.463	.219	.072	.016					
60	.063	.715	.450	.205	.065	.014					
∞		.707	.437	.193	.058	.011					

ν_1 = numerator degrees of freedom; ν_2 = denominator degrees of freedom.

Table A.4 is reprinted by permission of University College, from P. C. Tang, The Power Function of the Analysis of Variance Tests with Tables and Illustrations of Their Use, Statist. Research Mem., vol. 1.

$f_1 = 2$

f_2	$E^2_{.05}$	φ 1	1.5	2	2.5	3	4	5	6	7	8
2	.950	.881	.803	.704	.595	.484	.286	.140	.064	.024	.008
4	.776	.824	.661	.460	.272	.135	.020	.001			
6	.632	.789	.579	.340	.153	.052	.002				
7	.575	.777	.551	.304	.124	.037	.001				
8	.527	.767	.530	.277	.104	.027	.001				
9	.486	.759	.513	.257	.090	.022					
10	.451	.752	.498	.241	.080	.017					
11	.420	.747	.486	.228	.072	.015					
12	.393	.742	.476	.217	.066	.013					
13	.369	.737	.468	.208	.061	.011					
14	.348	.734	.461	.201	.057	.010					
15	.329	.730	.454	.195	.054	.009					
16	.312	.727	.448	.189	.051	.008					
17	.297	.725	.443	.184	.048	.008					
18	.283	.722	.439	.180	.046	.007					
19	.270	.720	.435	.177	.044	.007					
20	.259	.718	.431	.173	.043	.006					
21	.248	.717	.428	.170	.042	.006					
22	.238	.715	.425	.168	.040	.006					
23	.229	.714	.422	.165	.039	.006					
24	.221	.712	.420	.163	.038	.005					
25	.213	.711	.417	.161	.037	.005					
26	.206	.710	.415	.159	.037	.005					
27	.199	.709	.413	.157	.036	.005					
28	.193	.708	.411	.155	.035	.005					
29	.187	.707	.410	.154	.035	.004					
30	.181	.706	.408	.153	.034	.004					
60	.095	.692	.384	.134	.027	.003					
∞		.678	.362	.117	.021	.002					

$f_1 = 3$

f_2	$E^2_{.05}$	φ 1	1.5	2	2.5	3	4	5	6	7	8
2	.966	.888	.817	.726	.624	.519	.324	.177	.084	.035	.013
4	.832	.830	.670	.468	.278	.139	.020	.001			
6	.704	.791	.574	.326	.139	.044	.002				
7	.651	.776	.540	.283	.106	.028					
8	.604	.764	.513	.251	.084	.018					
9	.563	.754	.491	.226	.068	.013					
10	.527	.745	.472	.206	.057	.010					
11	.495	.738	.457	.190	.049	.008					
12	.466	.731	.444	.178	.043	.006					
13	.440	.726	.433	.167	.038	.005					
14	.418	.721	.422	.158	.035	.004					
15	.397	.716	.414	.151	.032	.004					
16	.378	.712	.406	.144	.029	.003					
17	.361	.709	.399	.139	.027	.003					
18	.345	.705	.393	.134	.025	.002					
19	.331	.702	.388	.130	.024	.002					
20	.317	.700	.383	.126	.022	.002					
21	.305	.697	.379	.123	.021	.002					
22	.294	.695	.375	.119	.020	.002					
23	.283	.693	.371	.117	.019	.002					
24	.273	.691	.367	.114	.019	.001					
25	.264	.689	.364	.112	.018	.001					
26	.255	.687	.361	.110	.017	.001					
27	.248	.686	.359	.108	.017	.001					
28	.240	.684	.356	.106	.016	.001					
29	.233	.683	.354	.105	.016	.001					
30	.226	.682	.352	.103	.015	.001					
60	.121	.662	.320	.083	.010	.001					
∞		.642	.289	.067	.007						

$f_1 = 4$

f_2	$E^2_{.05}$	φ									
		1	1.5	2	2.5	3	4	5	6	7	8
2	.975	.892	.824	.738	.640	.537	.345	.195	.097	.043	.017
4	.865	.833	.673	.471	.279	.139	.020	.001			
6	.751	.791	.567	.314	.128	.038	.001				
7	.702	.774	.529	.265	.092	.022					
8	.657	.760	.497	.229	.069	.013					
9	.618	.748	.471	.201	.054	.008					
10	.582	.738	.449	.179	.043	.006					
11	.550	.729	.430	.161	.035	.004					
12	.521	.721	.414	.148	.030	.003					
13	.494	.714	.401	.136	.025	.002					
14	.471	.708	.389	.127	.022	.002					
15	.449	.702	.378	.119	.019	.002					
16	.429	.697	.369	.112	.017	.001					
17	.411	.693	.361	.106	.016	.001					
18	.394	.689	.354	.101	.014	.001					
19	.379	.685	.347	.097	.013	.001					
20	.364	.681	.341	.093	.012	.001					
21	.351	.678	.335	.089	.011	.001					
22	.339	.675	.331	.086	.010	.001					
23	.327	.672	.326	.083	.010						
24	.316	.670	.322	.080	.009						
25	.306	.668	.318	.078	.009						
26	.297	.665	.315	.076	.008						
27	.288	.663	.312	.074	.008						
28	.279	.661	.309	.072	.008						
29	.272	.660	.306	.071	.007						
30	.264	.658	.303	.069	.007						
60	.144	.632	.265	.049	.004						
∞		.604	.227	.036	.002						

$f_1 = 5$

f_2	$E^2_{.05}$	φ 1	1.5	2	2.5	3	4	5	6	7	8
2	.980	.894	.828	.745	.649	.549	.359	.207	.106	.048	.019
4	.887	.835	.675	.473	.280	.138	.020	.001			
6	.785	.790	.561	.304	.119	.033	.001				
7	.739	.772	.519	.251	.082	.018					
8	.697	.756	.483	.211	.059	.010					
9	.659	.743	.454	.181	.044	.006					
10	.625	.731	.429	.158	.033	.004					
11	.593	.720	.408	.140	.026	.002					
12	.564	.711	.390	.125	.021	.002					
13	.538	.703	.374	.113	.017	.001					
14	.514	.695	.360	.103	.015	.001					
15	.492	.689	.348	.095	.012	.001					
16	.471	.683	.338	.088	.011	.001					
17	.452	.678	.328	.083	.009						
18	.435	.673	.320	.078	.008						
19	.419	.668	.312	.073	.007						
20	.404	.664	.305	.069	.007						
21	.390	.660	.299	.066	.006						
22	.377	.656	.294	.063	.006						
23	.365	.653	.288	.060	.005						
24	.353	.650	.284	.058	.005						
25	.342	.647	.279	.056	.005						
26	.332	.644	.275	.054	.004						
27	.323	.642	.272	.052	.004						
28	.314	.640	.268	.050	.004						
29	.305	.637	.265	.049	.003						
30	.297	.635	.262	.048	.003						
60	.165	.604	.219	.031	.001						
∞		.567	.177	.019	.001						

$f_1 = 6$

f_2	$E^2_{.05}$	φ									
		1	1.5	2	2.5	3	4	5	6	7	8
2	.983	.895	.831	.749	.656	.557	.368	.216	.112	.052	.022
4	.902	.836	.677	.473	.280	.138	.019	.001			
6	.811	.789	.556	.296	.113	.030	.001				
7	.768	.769	.510	.239	.074	.015					
8	.729	.753	.472	.198	.051	.008					
9	.692	.738	.440	.166	.037	.005					
10	.659	.725	.412	.142	.027	.003					
11	.628	.713	.389	.123	.020	.002					
12	.600	.702	.369	.108	.016	.001					
13	.574	.693	.351	.096	.012	.001					
14	.550	.685	.336	.086	.010	.001					
15	.527	.677	.323	.078	.008						
16	.507	.669	.311	.071	.007						
17	.488	.663	.301	.065	.006						
18	.470	.657	.291	.061	.005						
19	.454	.652	.283	.056	.004						
20	.438	.648	.276	.053	.004						
21	.424	.644	.269	.050	.003						
22	.410	.639	.262	.047	.003						
23	.397	.635	.257	.045	.003						
24	.385	.632	.252	.043	.003						
25	.374	.629	.247	.041	.002						
26	.363	.625	.242	.039	.002						
27	.353	.623	.238	.037	.002						
28	.344	.620	.234	.036	.002						
29	.335	.617	.231	.034	.002						
30	.326	.615	.228	.033	.002						
60	.184	.576	.181	.019	.001						
∞		.532	.138	.010							

$f_1 = 7$

f_2	$E^2_{.05}$	φ 1	1.5	2	2.5	3	4	5	6	7	8
2	.986	.896	.833	.753	.660	.563	.374	.222	.117	.055	.023
4	.914	.837	.678	.474	.280	.138	.019	.001			
6	.831	.788	.552	.289	.108	.028	.001				
7	.791	.767	.503	.230	.068	.013					
8	.754	.749	.462	.187	.046	.007					
9	.719	.733	.427	.154	.031	.004					
10	.687	.719	.398	.129	.022	.002					
11	.657	.706	.373	.110	.016	.001					
12	.630	.695	.351	.094	.012	.001					
13	.604	.684	.332	.082	.009						
14	.580	.675	.316	.073	.007						
15	.558	.667	.301	.065	.006						
16	.538	.659	.289	.058	.005						
17	.518	.652	.277	.053	.004						
18	.501	.645	.267	.048	.003						
19	.484	.639	.258	.044	.003						
20	.468	.634	.250	.041	.002						
21	.453	.629	.243	.038	.002						
22	.439	.624	.236	.036	.002						
23	.426	.619	.230	.034	.002						
24	.414	.615	.224	.032	.001						
25	.402	.611	.219	.030	.001						
26	.391	.607	.215	.028	.001						
27	.381	.604	.210	.027	.001						
28	.371	.601	.206	.026	.001						
29	.362	.598	.202	.024	.001						
30	.353	.595	.199	.023	.001						
60	.202	.550	.150	.012							
∞		.498	.105	.005							

$f_1 = 8$

f_2	$E^2_{.05}$	φ									
		1	1.5	2	2.5	3	4	5	6	7	8
2	.987	.897	.835	.755	.664	.567	.380	.227	.121	.057	.024
4	.924	.838	.678	.474	.279	.137	.019	.001			
6	.847	.787	.548	.284	.103	.026	.001				
7	.810	.765	.497	.222	.064	.012					
8	.775	.746	.454	.178	.041	.006					
9	.742	.729	.417	.144	.028	.003					
10	.711	.714	.386	.119	.019	.001					
11	.682	.700	.359	.099	.013	.001					
12	.655	.688	.336	.084	.009						
13	.630	.677	.316	.072	.007						
14	.607	.666	.298	.062	.005						
15	.585	.657	.283	.055	.004						
16	.564	.648	.269	.048	.003						
17	.545	.641	.257	.043	.003						
18	.527	.634	.247	.039	.002						
19	.510	.627	.237	.035	.002						
20	.495	.620	.228	.032	.001						
21	.480	.615	.220	.030	.001						
22	.466	.609	.213	.027	.001						
23	.452	.604	.207	.025	.001						
24	.440	.600	.201	.024	.001						
25	.428	.595	.196	.022	.001						
26	.417	.591	.191	.021	.001						
27	.406	.588	.186	.020	.001						
28	.396	.584	.182	.019							
29	.386	.581	.178	.018							
30	.377	.578	.175	.017							
60	.219	.527	.125	.008							
∞		.466	.081	.003							

$E^2_{.01}$ and the Corresponding Values of P(II)

$f_1 = 1$

f_2	$E^2_{.01}$	φ									
		1	1.5	2	2.5	3	4	5	6	7	8
2	.980	.970	.947	.914	.874	.828	.720	.602	.484	.373	.277
4	.841	.949	.885	.784	.651	.501	.233	.077	.018	.003	
6	.696	.934	.839	.687	.498	.312	.076	.010	.001		
7	.636	.928	.822	.652	.447	.258	.049	.006			
8	.585	.924	.808	.624	.409	.221	.034	.002			
9	.540	.920	.796	.601	.379	.193	.025	.001			
10	.501	.916	.786	.582	.355	.172	.019	.001			
11	.467	.913	.777	.567	.336	.156	.015				
12	.437	.911	.770	.553	.320	.144	.012				
13	.411	.909	.763	.542	.307	.133	.010				
14	.388	.907	.758	.532	.296	.125	.009				
15	.367	.905	.753	.523	.286	.118	.008				
16	.348	.904	.749	.516	.278	.112	.007				
17	.331	.902	.745	.509	.271	.107	.006				
18	.315	.901	.741	.503	.264	.103	.006				
19	.301	.900	.738	.498	.259	.099	.005				
20	.288	.899	.735	.493	.254	.096	.005				
21	.276	.898	.732	.488	.249	.093	.004				
22	.265	.897	.730	.484	.245	.090	.004				
23	.255	.896	.728	.481	.241	.088	.004				
24	.246	.896	.726	.477	.238	.086	.004				
25	.237	.895	.724	.474	.235	.084	.004				
26	.229	.894	.722	.471	.232	.082	.003				
27	.221	.894	.720	.469	.229	.081	.003				
28	.214	.893	.718	.466	.227	.079	.003				
29	.212	.893	.717	.464	.225	.078	.003				
30	.201	.892	.716	.462	.223	.077	.003				
60	.106	.885	.696	.430	.194	.061	.002				
∞		.877	.675	.400	.169	.048	.001				

f_2	$E^2_{.01}$	φ									
		1	1.5	2	2.5	3	4	5	6	7	8
2	.990	.975	.957	.932	.901	.865	.779	.680	.577	.475	.379
4	.900	.957	.901	.810	.685	.540	.266	.095	.024	.004	.001
6	.785	.941	.850	.695	.498	.305	.068	.007			
7	.732	.934	.828	.649	.431	.235	.035	.004			
8	.684	.929	.809	.611	.379	.187	.021	.001			
9	.641	.924	.793	.579	.338	.152	.013				
10	.602	.920	.779	.552	.306	.127	.008				
11	.567	.916	.767	.528	.278	.108	.006				
12	.536	.912	.756	.508	.255	.093	.005				
13	.508	.909	.746	.491	.237	.082	.003				
14	.482	.907	.738	.476	.223	.074	.002				
15	.459	.904	.730	.463	.211	.066	.002				
16	.438	.902	.723	.452	.201	.060	.001				
17	.418	.900	.717	.442	.193	.055	.001				
18	.401	.898	.711	.433	.185	.051	.001				
19	.384	.896	.706	.424	.177	.048	.001				
20	.369	.895	.701	.417	.170	.045	.001				
21	.355	.894	.697	.410	.165	.042	.001				
22	.342	.893	.693	.404	.160	.040	.001				
23	.330	.891	.690	.399	.155	.038					
24	.319	.890	.686	.394	.151	.036					
25	.308	.889	.683	.389	.148	.035					
26	.298	.888	.680	.385	.144	.034					
27	.289	.887	.678	.381	.141	.032					
28	.280	.886	.675	.377	.138	.031					
29	.272	.886	.672	.373	.136	.030					
30	.264	.885	.670	.370	.134	.029					
60	.142	.873	.637	.324	.102	.019					
∞		.860	.601	.279	.076	.011					

f_2	$E^2_{.01}$	φ									
		1	1.5	2	2.5	3	4	5	6	7	8
2	.993	.977	.961	.939	.911	.878	.800	.709	.612	.515	.421
4	.926	.959	.907	.818	.695	.552	.276	.100	.026	.005	.001
6	.830	.943	.850	.691	.486	.290	.059	.006			
7	.784	.936	.825	.636	.408	.210	.025	.002			
8	.740	.929	.803	.590	.347	.158	.014				
9	.700	.923	.783	.550	.299	.120	.008				
10	.663	.918	.765	.517	.261	.094	.004				
11	.629	.913	.749	.487	.231	.075	.002				
12	.598	.909	.735	.463	.206	.062	.001				
13	.570	.906	.723	.441	.186	.051	.001				
14	.544	.902	.711	.422	.170	.044	.001				
15	.520	.899	.701	.406	.156	.038	.001				
16	.498	.896	.692	.391	.145	.033					
17	.478	.893	.683	.378	.135	.029					
18	.459	.891	.676	.367	.126	.026					
19	.442	.889	.669	.356	.119	.023					
20	.426	.887	.662	.347	.112	.021					
21	.410	.885	.656	.339	.107	.019					
22	.396	.883	.651	.331	.102	.017					
23	.383	.881	.646	.324	.098	.016					
24	.371	.880	.641	.318	.094	.015					
25	.359	.879	.637	.312	.090	.014					
26	.349	.877	.633	.307	.087	.013					
27	.338	.876	.629	.302	.084	.012					
28	.329	.875	.625	.297	.081	.012					
29	.319	.874	.622	.293	.079	.011					
30	.311	.872	.619	.289	.077	.011					
60	.171	.856	.571	.233	.050	.005					
∞		.836	.519	.182	.030	.002					

$f_1 = 4$

f_2	$E^2_{.01}$	φ									
		1	1.5	2	2.5	3	4	5	6	7	8
2	.995	.978	.962	.942	.915	.884	.810	.724	.631	.536	.444
4	.941	.960	.909	.822	.700	.557	.280	.102	.027	.005	.001
6	.859	.943	.849	.685	.475	.277	.053	.005			
7	.818	.936	.821	.624	.389	.191	.018				
8	.778	.928	.796	.571	.322	.136	.010				
9	.741	.922	.773	.526	.269	.098	.003				
10	.706	.916	.752	.487	.227	.073	.002				
11	.673	.911	.733	.453	.195	.055	.001				
12	.643	.906	.716	.424	.169	.042	.001				
13	.616	.901	.700	.398	.148	.034					
14	.590	.897	.687	.376	.131	.028					
15	.566	.893	.674	.357	.117	.022					
16	.544	.890	.662	.340	.106	.018					
17	.523	.886	.652	.325	.096	.015					
18	.504	.883	.642	.312	.088	.013					
19	.486	.880	.633	.301	.081	.011					
20	.470	.878	.625	.290	.075	.010					
21	.454	.876	.618	.280	.070	.009					
22	.440	.873	.611	.272	.066	.008					
23	.426	.871	.604	.264	.062	.007					
24	.413	.869	.598	.257	.059	.006					
25	.401	.867	.593	.250	.056	.006					
26	.389	.865	.588	.244	.053	.005					
27	.378	.864	.583	.239	.050	.005					
28	.368	.862	.578	.234	.048	.005					
29	.358	.861	.574	.229	.046	.004					
30	.349	.860	.570	.225	.044	.004					
60	.196	.837	.509	.165	.024	.001					
∞		.810	.443	.115	.011						

$f_1 = 5$

f_2	$E^2_{.01}$	φ 1	1.5	2	2.5	3	4	5	6	7	8
2	.996	.978	.964	.944	.918	.888	.817	.733	.642	.549	.458
4	.951	.961	.910	.824	.702	.559	.282	.103	.027	.005	.001
6	.879	.943	.848	.679	.466	.266	.048	.004			
7	.842	.935	.818	.614	.394	.177	.014				
8	.806	.928	.790	.556	.301	.121	.007				
9	.771	.920	.764	.505	.245	.083	.003				
10	.738	.914	.740	.461	.201	.058	.001				
11	.707	.908	.718	.424	.168	.042					
12	.679	.902	.699	.391	.141	.031					
13	.652	.897	.681	.363	.120	.023					
14	.626	.892	.664	.339	.104	.018					
15	.603	.888	.649	.318	.090	.014					
16	.581	.883	.636	.299	.079	.011					
17	.561	.880	.624	.283	.071	.009					
18	.541	.876	.612	.269	.063	.007					
19	.523	.873	.602	.256	.057	.006					
20	.506	.870	.592	.245	.052	.005					
21	.490	.867	.583	.234	.047	.004					
22	.475	.864	.575	.225	.044	.004					
23	.461	.861	.567	.217	.040	.003					
24	.448	.859	.560	.210	.037	.003					
25	.435	.857	.553	.203	.035	.003					
26	.423	.855	.547	.196	.033	.002					
27	.412	.853	.541	.190	.031	.002					
28	.401	.851	.536	.185	.029	.002					
29	.391	.849	.531	.180	.027	.002					
30	.381	.847	.526	.176	.026	.002					
60	.218	.819	.452	.116	.011						
∞		.784	.373	.070	.004						

$f_1 = 6$

f_2	$E^2_{.01}$	φ									
		1	1.5	2	2.5	3	4	5	6	7	8
2	.997	.978	.964	.945	.920	.891	.821	.739	.650	.558	.468
4	.958	.962	.911	.825	.704	.560	.283	.104	.027	.005	.001
6	.894	.944	.847	.675	.459	.258	.044	.003			
7	.860	.935	.815	.605	.362	.166	.011				
8	.827	.927	.784	.543	.285	.109	.006				
9	.795	.919	.756	.488	.226	.071	.003				
10	.764	.912	.730	.441	.181	.048	.001				
11	.734	.905	.706	.400	.147	.033					
12	.707	.899	.683	.365	.120	.023					
13	.681	.893	.663	.334	.100	.017					
14	.656	.888	.645	.308	.084	.013					
15	.633	.882	.628	.286	.071	.009					
16	.612	.878	.612	.266	.061	.007					
17	.591	.873	.598	.249	.053	.005					
18	.572	.869	.585	.233	.046	.004					
19	.554	.865	.573	.220	.041	.003					
20	.537	.862	.562	.208	.036	.003					
21	.521	.858	.552	.198	.033	.002					
22	.506	.855	.542	.188	.029	.002					
23	.492	.852	.533	.180	.027	.002					
24	.478	.849	.524	.172	.024	.001					
25	.465	.846	.517	.165	.022	.001					
26	.453	.844	.510	.159	.020	.001					
27	.442	.842	.503	.153	.019	.001					
28	.430	.839	.497	.147	.017	.001					
29	.420	.837	.491	.142	.016	.001					
30	.410	.835	.486	.138	.015	.001					
60	.238	.801	.401	.081	.006						
∞		.755	.311	.042	.001						

$f_1 = 7$

f_2	$E^2_{.01}$	φ									
		1	1.5	2	2.5	3	4	5	6	7	8
2	.997	.979	.965	.946	.922	.893	.824	.743	.655	.564	.475
4	.963	.962	.912	.826	.705	.561	.283	.104	.027	.005	.001
6	.906	.944	.845	.671	.452	.251	.041	.003			
7	.875	.935	.812	.598	.351	.158	.009				
8	.844	.926	.779	.532	.272	.100	.005				
9	.814	.918	.749	.474	.211	.063	.002				
10	.785	.910	.720	.423	.166	.041	.001				
11	.757	.903	.694	.379	.131	.027					
12	.730	.896	.670	.342	.105	.018					
13	.705	.889	.648	.310	.085	.013					
14	.681	.883	.627	.283	.069	.009					
15	.659	.878	.608	.259	.057	.007					
16	.638	.872	.591	.238	.048	.004					
17	.618	.868	.575	.220	.041	.003					
18	.599	.863	.561	.205	.035	.002					
19	.581	.859	.548	.191	.030	.002					
20	.564	.854	.535	.179	.026	.002					
21	.548	.851	.524	.168	.023	.001					
22	.533	.847	.513	.159	.020	.001					
23	.519	.844	.503	.150	.018	.001					
24	.505	.840	.494	.143	.016	.001					
25	.492	.837	.485	.136	.015	.001					
26	.479	.834	.477	.130	.013						
27	.468	.831	.470	.124	.012						
28	.456	.829	.463	.119	.011						
29	.446	.826	.456	.114	.010						
30	.435	.824	.450	.110	.009						
60	.256	.783	.355	.056	.003						
∞		.729	.256	.024							

$f_1 = 8$

f_2	$E^2_{.01}$	φ 1	1.5	2	2.5	3	4	5	6	7	8
2	.997	.979	.965	.946	.923	.894	.826	.746	.659	.569	.481
4	.967	.962	.912	.826	.705	.562	.284	.104	.027	.005	.001
6	.915	.944	.844	.668	.447	.246	.039	.003			
7	.887	.934	.809	.592	.343	.151	.007				
8	.858	.925	.775	.522	.261	.093	.004				
9	.829	.917	.743	.461	.199	.056					
10	.802	.908	.712	.408	.153	.035					
11	.775	.901	.684	.363	.118	.022					
12	.750	.893	.658	.324	.092	.014					
13	.726	.886	.634	.290	.073	.009					
14	.703	.880	.612	.261	.058	.006					
15	.681	.874	.591	.237	.047	.004					
16	.660	.868	.573	.216	.039	.003					
17	.641	.862	.555	.197	.032	.002					
18	.622	.857	.539	.181	.027	.002					
19	.605	.852	.525	.168	.023	.001					
20	.588	.848	.511	.156	.019	.001					
21	.572	.843	.499	.145	.017	.001					
22	.557	.839	.487	.135	.014						
23	.542	.835	.476	.127	.012						
24	.529	.832	.466	.119	.011						
25	.515	.828	.456	.113	.010						
26	.503	.825	.447	.107	.009						
27	.491	.822	.439	.101	.008						
28	.480	.819	.432	.096	.007						
29	.469	.816	.425	.092	.006						
30	.458	.813	.418	.088	.006						
60	.274	.766	.315	.039	.001						
∞		.702	.211	.014							

Index